T0290847

Gas and Steam Turbine Power Plants

Explore sustainable electric power generation technology, from first principles to cutting-edge systems, in this in-depth resource. Including energy storage, carbon capture, hydrogen and hybrid systems, the detailed coverage includes performance estimation, operability concerns, economic trade-off, and other intricate analyses, supported by implementable formulae, real-world data, and tried-and-tested quantitative and qualitative estimating techniques. Starting from basic concepts and key equipment, this book builds to a precise analysis of plant operation through data and methods gained from decades of hands-on design, testing, operation, and troubleshooting. Gain the knowledge you need to operate in conditions beyond standard settings and environment, with thorough descriptions of off-design operations. Novel technologies become accessible with stripped-back descriptions and physics-based calculations. This book is an ideal companion for engineers in the gas turbine and electric power field.

S. Can Gülen, Bechtel Fellow, is a senior principal engineer in the Engineering Technology Group of Bechtel Infrastructure and Power, Inc. He is an ASME Fellow and has authored/coauthored more than 50 papers and articles, 25 US patents, and 3 monographs.

Gas and Steam Turbine Power Plants

Applications in Sustainable Power

S. CAN GÜLEN
Bechtel Infrastructure and Power

Shaftesbury Road, Cambridge CB2 8EA, United Kingdom

One Liberty Plaza, 20th Floor, New York, NY 10006, USA

477 Williamstown Road, Port Melbourne, VIC 3207, Australia

314–321, 3rd Floor, Plot 3, Splendor Forum, Jasola District Centre, New Delhi – 110025, India

103 Penang Road, #05–06/07, Visioncrest Commercial, Singapore 238467

Cambridge University Press is part of Cambridge University Press & Assessment, a department of the University of Cambridge.

We share the University's mission to contribute to society through the pursuit of education, learning and research at the highest international levels of excellence.

www.cambridge.org
Information on this title: www.cambridge.org/9781108837910

DOI: 10.1017/9781108943475

First published 2023

Printed in the United Kingdom by CPI Group Ltd, Croydon CR0 4YY

A catalogue record for this publication is available from the British Library.

Library of Congress Cataloging-in-Publication Data
Names: Gülen, S. Can, 1962- author.
Title: Gas and steam turbine power plants : applications in sustainable power / S. Can Gülen, Bechtel Infrastructure and Power.
Description: Cambridge, United Kingdom ; New York, NY, USA : Cambridge University Press, 2023. | Includes bibliographical references and index.
Identifiers: LCCN 2022034744 (print) | LCCN 2022034745 (ebook) | ISBN 9781108837910 (hardback) | ISBN 9781108943475 (epub)
Subjects: LCSH: Electric power-plants. | Steam power plants. | Gas-turbine power-plants.
Classification: LCC TK1071 .G85 2023 (print) | LCC TK1071 (ebook) | DDC 333.793/2–dc23/eng/ 20220922
LC record available at https://lccn.loc.gov/2022034744
LC ebook record available at https://lccn.loc.gov/2022034745

ISBN 978-1-108-83791-0 Hardback

Contents

Preface

The electric power generation landscape is changing, and it is changing fast. Growing concerns about a potential (and permanent) climate change caused by global warming and its catastrophic impact on the earth's population provide the impetus for this change. The link between global warming and electric power generation is the greenhouses gases (GHG) emitted by fossil fuel–burning power plants, specifically carbon dioxide (CO_2). In terms of the "carbon footprint," i.e., intensity of CO_2 emissions as measured by kilograms of CO_2 in the stack gas per megawatt hour of electricity generated, coal is the primary culprit. Therefore, the main objective in the fight to slow or even reverse the global warming phenomenon is elimination of coal from the generation portfolio. The first step in this endeavor is the replacement of coal-fired generation by natural gas–fired generation, i.e., gas turbines in simple and combined cycle configurations.

The changing landscape of electric power generation, commonly known by the term *energy transition*, includes much more than replacement of coal with natural gas as a power plant fuel. It includes increasing the share of renewable generation resources, i.e., primarily solar and wind but also geothermal and hydro, in the generation mix. Since solar and wind are non-dispatchable or intermittent resources, they have to be supplemented by energy storage, e.g., batteries, pumped hydro, compressed air, and many other emerging technologies. Since fossil fuel–burning power plants are expected to assume a significant share of global electric power generation until completely replaced (hopefully) by CO_2-free resources (one should also include nuclear power in that group), capturing CO_2 from their stack gases becomes sine qua non. Recently, hydrogen emerged as an "energy vector" to facilitate the transition, eventually, to a world with carbon dioxide–free electricity for all.

Unfortunately, the global progress in energy transition is sporadic and haphazard without a clearly thought-out strategy agreed upon by all impacted parties, i.e., practically all nations in the world. One clear sign of the chaotic state of things is the recent energy "mini crisis" that materialized quite unexpectedly in the second half of 2021, when, since May of that year, the price of a basket of oil, coal, and gas jumped by 95%. This was especially vexing when one considers that the reduced economic activity caused by the COVID-19 pandemic in 2020 resulted in a drop in global demand to the tune of 5%. To a large extent, the big soar in energy prices can be traced back to ill-advised decisions made under a climate of panic to eliminate fossil and nuclear generating assets from the generation mix. Another clear sign of the problematic nature of energy transition is proliferation of "silver bullet" or "magic wand" (choose your favorite term) technologies, which are supposed to be the panacea to global warming caused by anthropogenic GHG emissions. They range from truly outlandish ideas to half-baked ones at very low technical readiness levels (TRLs) that would take decades to reach commercial readiness (if ever) with astronomic expenditure of financial resources (see Section 16.5 for a description of the *Commercial Readiness Index*). Even those emerging technologies that might have a decent shot at realization are peddled with outrageous marketing hype and unrealistic cost and/or performance claims that do not pass the test of rigorous technical analysis.

This book is specifically geared to address the second problem cited above. Using first principles of thermal design engineering, primarily thermodynamics (specifically, the second law), existing and emerging technologies in power generation using prime movers, carbon capture from stack gases, energy storage, different fuels such as hydrogen, and many others are investigated under a microscope from performance and operability angles. Cost considerations are addressed as much as possible with the implicit understanding that "true" CAPEX of low TRL technologies is simply impossible to predict with reasonable accuracy. One can google the horror stories from recent examples (discussed later in the book) to appreciate this simple fact. In undertaking the said investigation, the author draws upon his 25-year experience in the power generation industry in thermal design software, OEM, and EPC organizations, i.e., with real, hands-on (or foots-on-the-ground, whichever you prefer) experience in the field.

This book is not intended to be an introductory text. It is primarily geared to address questions from the practitioners in the field, and researchers in academia, government, and industrial laboratories and help them with a realistic assessment of technologies emerging in the energy transition realm. As prerequisites or, more aptly, companion texts, the author would suggest his two previous books on gas turbines and combined cycle power plants (they are mentioned in Chapter 2). A basic knowledge in power plant thermodynamics, fluid mechanics, and heat transfer obtained via a four-year degree (at least) on the part of the reader is implicitly assumed. This admittedly big onus placed on the reader is simply unavoidable to cover a broad range of material in a limited space with acceptable attention to detail.

The author has no vested interest in any of the technologies covered herein. As such, analyses and conclusions are solely based on detailed calculations and, thus, they are deemed to be as objective as possible. Any errors made in the process are fully unintentional and, in most cases, should not be more than mere typos. Having said that, the author welcomes any and all corrections from the readers and declares himself ready to address them in the Errata to be made available in the dedicated URL on the publisher's website.

Before moving on to the main body of the text, there is a simple fact that needs to be stated unequivocally. As is the case with everything I have written before and will write in future, this book is dedicated to the memory of Mustafa Kemal Atatürk (1881–1938) and his elite cadre of reformers. Without their vision, sacrifices, and groundbreaking work, there would not have been a fertile ground where my parents, teachers, mentors, family, and friends could shape me into the author of this book. Everything started with him; the rest was easy.

1 Introduction

Let us start with the subtitle of the book, *Applications in Sustainable Power*. The "power" in the subtitle is obviously electricity or electric power.[1] The term "sustainable" can be traced back to the simple fact that a gas and/or steam turbine power plant burns a fossil fuel (natural gas) to generate electric power. Combustion of fossil fuels (i.e., coal, fuel oil, natural gas) generates carbon dioxide (CO_2), a greenhouse gas, which is emitted into the atmosphere through the stack of the power plant. Greenhouse gases (e.g., ozone, methane [CH_4], CO_2, and carbon monoxide (CO) among others – they are commonly referred to by the acronym GHG) contribute to global warming via the *greenhouse effect*. This term refers to the analogy between GHG and glass in a greenhouse,[2] which allows sunlight in to keep the inside warm while simultaneously blocking the heat inside from escaping outside. A greenhouse keeps the plants inside warm; GHG in the atmosphere do the same by trapping heat radiating from Earth into space.

At the time of writing, despite overwhelming research and data accumulated over the last half century, there is still an acrimonious debate about whether anthropogenic CO_2 emissions contribute to global warming or not. As far as this book is concerned, such debate is irrelevant because what is covered herein is based on recent *concrete* developments in the electric power generation landscape, which stem from the worldwide acceptance of global warming and its dire implications, i.e., catastrophic *climate change* events, as a fact as well as a clear and imminent danger (in no way a certainty – more on this below). There are many scientific studies, reports, and papers published by reputable scientists and institutions on this subject, and these are only one click away on the internet. Unfortunately, the internet is also full of unreliable information and propaganda. Therefore, the author would like to refer the reader to arguably the most reliable resource for laypersons: David J. C. MacKay's *Sustainable Energy – Without the Hot Air* (UIT Cambridge: Cambridge, 2009). The book is also available free for download in www.withouthotair.com (last accessed on May 13, 2020). Suffice to point out a few irrefutable facts:

---

[1] Note the deliberate use of the term "power," not "energy." Since this is not a book for laypeople, I will not elaborate on the distinction between power (measured in watts, i.e., joules per second) and energy (measured in joules).

[2] A greenhouse is a glass building that is used to grow plants. Greenhouses stay warm inside, even during the winter. During the day, sunlight shines into the greenhouse and warms the plants and air inside.

- In 2018, the CO_2 concentration in the atmosphere exceeded 400 ppm.
- Until the beginning of the Industrial Revolution (late eighteenth, early nineteenth century), this number was very stable at around 280 ppm.[3]
- The dramatic increase (nearly 50%) in the concentration of CO_2 over the last two centuries, including a very rapid acceleration in the last decade or so (about 2.5 ppm per year[4]), points the finger at the ever-increasing role of fossil fuels in human activities (transportation, electric power generation, steel making, space heating, etc.).

The recorded increase in CO_2 concentration in the atmosphere coincided with an unmistakable increase in average annual temperatures recorded on earth. In particular:

- In 2019, the average temperature across global land and ocean surfaces was 1.71°F (0.95°C) above the twentieth-century average of 57.0°F (13.9°C), making it the second-warmest year on record.[5]
- The global annual temperature has increased at an average rate of 0.13°F (0.07°C) per decade since 1880 and over twice that rate (+0.32°F/+0.18°C) since 1981.[6]

Combined with the atmospheric CO_2 concentration trend since 1880, this corresponds to about a 1°C rise per 100 ppm rise in CO_2 concentration. Thus, prima facie, one could expect a 3°C rise in average temperature across global land and ocean surfaces when the CO_2 concentration reaches 700 ppm. At the current rate of carbon emissions, which is likely to increase, especially due to rapid economic development in countries like India and China, this is quite likely to happen within the next century.[7] This level of global warming can lead to significant natural disasters such as the melting of Greenland's icecap and rising sea levels. There is concern that the recent increases in severe weather events and widespread wildfires are connected to global warming.

Now, let us take a pause here. The accuracy of the statements made in the paragraph above are highly suspect! For one, weather is *not* climate. Yes, global warming is a fact. Yes, anthropogenic CO_2 emissions, which have been increasing steadily since the 1950s, play a role in global warming via the GHG mechanism. However, that mechanism is more nuanced than many published reports would make the unsuspecting public believe, and the relationship is not a direct proportionality as

---

[3] Carbon dioxide concentrations for the last millennium or so are measured using air trapped in ice cores, particularly those in Greenland and Antarctica. Since 1958, regular measurements have been made in Mauna Loa Observatory (MLO) in Hawaii. The observatory is one of the stations of the Global Monitoring Laboratory (GML), part of the National Oceanic and Atmospheric Administration (NOAA).

[4] From "State of the Climate in 2018," Special Supplement to the Bulletin of the American Meteorological Society, Vol. 100, No. 9, September 2019. Downloaded from www.ametsoc.net/sotc2018/Socin2018_lowres.pdf on May 13, 2020.

[5] "Climate Change: Global Temperature," by Rebecca Lindsey and LuAnn Dahlman, January 16, 2020. From www.climate.gov/news-features/understanding-climate/climate-change-global-temperature (last accessed (last accessed on May 13, 2020).

[6] Ibid.

[7] While this section was being written in May 2020, a screeching halt to the global economic activity was brought down by the Covid-19 pandemic. Nevertheless, this unexpected (and unintended) pause in carbon emissions is not expected to be a new trend for long.

the author (deliberately) misstated above. The devil is not only in the details, which are beyond the scope of this book, but also in the past events, which clearly show a changing climate is not a modern phenomenon. In other words, how much of the change is attributable to purely human activities is not a settled debate.

For the proverbial holes in the present "doom and gloom" predictions mixing fact with fiction, propaganda and hyperbole with science, the reader is encouraged to consult the recent book by a reputable scientist, Steven Koonin, *Unsettled* (BenBella Books, Dallas, TX, 2021). The bottom line is that accurate predictions and establishing rock-solid cause-effect relationships are notoriously difficult and subject to uncertainty. Nevertheless, governments and industrial organizations worldwide (recently, even energy giants apparently[8]) are convinced that (i) "something" is happening, and (ii) it will be even worse in the near future *unless something is done.* This is the impetus behind the so-called *energy transition* that forms the context for this book.

The *energy transition* (*die Energiewende* in German) posits that "business as usual" continuation of electric power generation via fossil fuel combustion is not a sustainable path due to (i) finite resources, and (ii) GHG, especially CO_2, emissions leading to global warming (or climate change depending on one's "political" preference).[9] Natural gas is the cleanest fossil fuel by far – for example, gas turbine combined cycle (GTCC) CO_2 emissions, measured in kg per MWh of generation, is less than half that of coal-fired boiler-turbine power plants. Thus, in the context of this book's subtitle, *Applications in Sustainable Power*, GTCC operability considerations in conjunction with the increasing share of *carbon-free* technologies in the generation portfolio (primarily, wind and solar) will be the center of attention. These considerations include the *ancillary services* such as *reserve* (spinning and non-spinning) and *regulation* to help balance the grid (i.e., matching supply and demand while maintaining the system frequency, e.g., 60 Hz in the USA). They also include new and old technologies such as:

- Energy storage (e.g., compressed air energy storage [CAES])
- Hydrogen combustion (no CO_2 emissions in the flue gas)
- Gasification (e.g., Integrated Gasification Combined Cycle [IGCC])
- Post-combustion carbon capture, sequestration, and utilization (CCSU).

Gas turbines, in simple and combined cycle configurations, constitute the best available technology, today as well as in the future, to play the supporting actor role (in some cases, the lead actor role as well) on the path to a zero-carbon emissions future. This statement is not a hyperbole. It is based on irrefutable facts:

---

[8] For example, see the climate change page on ExxonMobil's website: https://corporate.exxonmobil.com/ Sustainability/Environmental-protection/Climate-change (last accessed on November 1, 2022).
[9] In 2018, in the USA, the total electricity generation by the electric power industry of 4.17 trillion kilowatt-hours (kWh) from all energy sources resulted in the emission of 1.87 billion metric tons – 2.06 billion short tons – of CO_2. Coal-fired generation contributed 1,127 million metric tons (about 1,000 kg/MWh) whereas natural gas–fired generation's contribution was 523 million metric tons (about 420 kg/MWh). From www.eia.gov/tools/faqs/faq.php?id=74&t=11 (last accessed on May 14, 2020).

- High thermal efficiency (>40% in simple cycle, >60% net low heating value [LHV] in a combined cycle)
- High power density (a single 50 Hz gas turbine can generate 500+ MWe)
- High flexibility (fast startup and load ramps, high turndown ratio)
- Low cost (less than $1,000 per kW installed in combined cycle)
- Low emissions (no SOx, mercury, etc., low NOx, low CO_2)

All these aspects of the gas turbine technology, specifically in its combined cycle variant, will be discussed in depth in the following chapters.

It is now time to turn to the main title of the book. Gas and steam turbines are the two prime movers playing a major role in global electric power generation. There was a time when the steam turbine was the star of the show, in the USA and the rest of the world. Gas turbines, mainly burning liquid fuels, were relegated to a supporting actor load (e.g., peak shaving). In terms of fuel, coal was the king, closely followed by the number 2 fuel: oil. At the time of writing, while coal has become a four-letter word in the developed countries of the Western Hemisphere, it still reigns supreme in the rest of the world. In the USA, it has been overtaken by natural gas in the electric power generation mix, which is primarily used in GTCCs. However, steam turbines, as a major component in the GTCC power plant, still play a role as a major actor.

In addition to their conventional prime mover roles, both gas and steam turbines constitute the heart of emerging *clean* technologies in different disguises. For example, the heart of the *supercritical* CO_2 technology is essentially a closed-cycle gas turbine with CO_2 at supercritical pressure and temperatures (i.e., above 74 bar and 31°C) as the working fluid instead of air. In compressed or liquefied air energy storage, what one is dealing with, in essence, is a gas turbine with its compressor and turbine operating independently and at different times. Gas turbines burning hydrogen, with or without blending with natural gas, are the current focus of the *energy transition*. The list can be extended further. Small modular nuclear reactors with closed-cycle gas turbines or advanced steam cycles, integrated solar combined cycle, and GTCCs with post-combustion capture among the most prominent examples.

The author covered the history, design, performance, and optimization of gas and steam turbines in detail in his earlier books (see references [1–2] in Section 2.4.1). In this book, the focus will be on the application and operation of these two prime movers in their different disguises. Let us start with the dictionary definitions: *Application* is the action of putting something into *operation*. Operation is defined as the fact or condition of functioning or being active. When a *system* is in operation, *subsystems* making up the system and the *components* making up the subsystems all operate in harmony to convert system *inputs* into system *outputs*. Thus, a discourse on gas and steam turbine power plant applications is, in essence, a discourse on power plant operations. Prima facie, one would be skeptical that this subject matter would take a 500-page monograph to cover. This is so because, after all, there is one and only one power plant operation of any consequence that one can think of: burn fuel and generate electric power. Period. Come to think of it, this simplistic view is not entirely wrong either. Once the prime mover generators in a particular gas or steam turbine power plant are synchronized to the grid, no matter what spin (no pun intended!) one puts on the plant operation, this is the bottom line: Burn fuel, turn, and generate electric power. Therefore, the subject matter of this book is *not* operation per se; it is *operability*.

One final definition: Operability is the ability to keep a piece of equipment, a system, or a whole industrial installation in a safe and reliable functioning condition, according to predefined operational requirements. As far as this book is concerned:

1. The system (or industrial installation if you will) is the gas or steam turbine power plant in any way, shape, or form.
2. This system contains several pieces of major equipment and/or subsystems, i.e.,
   a. Gas turbine generator (or its equivalent)
   b. Steam turbine generator (or its equivalent)
   c. Heat recovery steam generator (HRSG) (or its equivalent)
   d. Other heat input system (e.g., a boiler or a nuclear reactor)
   e. Heat rejection subsystem
      i. Condenser (water- or air-cooled)
      ii. Cooling tower
   f. Balance of plant (BOP), e.g.,
      i. Pumps
      ii. Heat exchangers
      iii. Pipes and valves
3. Operational requirements from this system are:
   a. Startup at demand (availability)
   b. Continuous operation at varying load levels and ambient conditions (reliability)
   c. Compliance with environmental regulations
   d. Compliance with safety regulations
   e. Shutdown at demand
   f. Response to fault events in a safe manner (load rejection or trip),

Operational requirements flow down from two sources: customers or regulatory agencies (e.g., the Environmental Protection Agency in the USA). Conflicts resulting from those requirements are typically solved during design and permitting phases.

By far the largest part of operability has three constituents: *reliability*, *availability*, and *maintainability* (RAM). One can have the best power plant in the world with the most advanced components and accessories (in terms of performance, i.e., output and heat rate) but, if it can barely run a few hours at a stretch before shutting down unexpectedly and requires a lot of maintenance, labor, and parts to restart, it has essentially zero value. In other words, the system and/or its components are

- not *reliable* (they trip a lot while running)
- barely *available* (they spend a lot of time in the proverbial "shop")
- not *maintainable* (they require constant attention and immense labor and materials to upkeep)

Consider the analogy to a weapons system (e.g., a tank or an assault rifle). The main priorities in any such system on the battlefield are maximum availability (i.e., being ready to fire), reliability (i.e., firing bullets whenever the trigger is pulled without jamming), and maintainability (i.e., sturdy, quick to disassemble, clean, and reassemble in field conditions).

Safe operation of the power plant with maximum RAM within a wide envelope of site ambient and loading conditions is one of the main areas of focus of this book. Power plant operation can be classified into two major categories:

1. Steady-state operation
2. Unsteady-state (transient) operation.

Steady-state operation can be broken down into two areas:

1. Design point
2. Off-design

Strictly speaking, a power plant rarely, if ever, operates at its *design point*, which reflects site ambient and loading conditions specified for sizing of plant hardware. A typical design point definition, widely adopted for *rating performance* purposes, is ISO base load. Furthermore, especially for power plants expected to operate across a wide envelope of site ambient and loading conditions with power augmentation methods including supplementary firing in the HRSG and gas turbine inlet air cooling, major pieces of equipment (e.g., condenser or cooling tower) can be sized at different conditions than the rest of the power plant.

Off-design operation refers to operation at boundary conditions and equipment operating modes other than those at the design point. In a conventional GTCC, these boundary conditions and operating modes refer to

- Site ambient conditions (temperature and humidity mainly)
- Gas turbine and/or steam turbine load
- Gas turbine firing (base or peak)
- HRSG supplementary (duct) firing
- Gas turbine inlet conditioning (evaporative coolers or inlet chillers on or off)

For most practical purposes, unless one is looking at a rather unusual mode of operation, off-design operation can be handled by OEM-supplied[10] *correction curves*. Otherwise, one must resort to a heat and mass balance simulation model of the particular power plant. Such correction curves typically reproduce plant performance (output and efficiency or heat rate) as a function of ambient temperature and/or plant load. This, of course, is a luxury available only to established products. When one is looking at concepts in their early development phase (e.g., supercritical CO_2 turbine plants), there are no "correction" curves. Ultimately, the operability of such emerging technologies must be proven in the field. Before that, however, requisite control

---

[10] Original Equipment Manufacturer – common industry term for major manufacturers of gas and steam turbines, e.g., General Electric (GE).

systems, conceptually and in step-by-step detail, must be developed from scratch. This is a huge undertaking requiring man-years of engineering design and development with astronomic outlay of funds – with no certainty of success at the end!

In steady-state calculations, the ultimate objective is plant performance, which is quantified by net electric output and thermal efficiency. The plant performance calculation is essentially a bookkeeping exercise where one adds and subtracts individual equipment performances to arrive at the net outcome. Unsteady-state (transient) performance is not amenable to a comparably precise definition. This will be explored in more detail in the following chapters.

2   Prologue

2.1   Note to the Reader

This book is the final installation of a trilogy on gas turbines for electric power generation in simple and combined cycle configurations. Material covered in the first two books (close to 1,500 pages combined) forms the foundation on which the discussion below is based (more on those two books below). This chapter's goal is to equip the reader with the minimum knowledge required to follow the coverage in the rest of the book and also make them aware of the caveats, pitfalls, for example, so that reading the book does not turn into a hassle. Although the reader is expected to be familiar with industry jargon, the lack of a standardized technology and misuse of some terms necessitated this introductory coverage.

The author is not a fan of a large nomenclature section (glossary) in the beginning (or the end) of the book that makes the reader jump back and forth between the narrative and the glossary. All parameters used in the equations are defined where they appear first in the discussion. To the extent possible, acronyms or easy-to-guess alphabetical variables are used, e.g., TAMB for ambient temperature, and subscripts and superscripts are avoided. (For a full list of acronyms see Section 16.1.) Greek letters are used sparsely. Exceptions are those that are widely used in technical literature and textbooks, e.g., η for efficiency, ρ for density, and σ for stress. The author's hope is that this will make it reasonably comfortable to read sections with quantitative information even if the reader is not thoroughly versed in the subject matter.

Thermodynamic variables commonly used in US textbooks for pressure, temperature, specific volume, enthalpy, entropy, and *availability* (exergy) are retained here as well, i.e., P, T, v, h, s, and a, respectively. If the reader is unable to figure out that the parameter c_p designates specific heat (at constant pressure, to be precise), there is a very good chance that this book is too advanced for them.

To the extent possible, SI units will be preferred but not exclusively. The reason for that is the self-consistency of the SI system, which eliminates tedious unit conversion factors from scientific formulae – what you see is what you get. In any event, the counterpart in British units (or vice versa) will be provided in parentheses so that users more familiar with those units will not waste time with making mental conversions. No ink is wasted on unit conversions. At the time of writing (2020–2021), such information is one click away by googling on the reader's computer or smartphone.

Unless otherwise specified (mostly the case in empirical or curve-fit equations), it is assumed that the reader is aware that temperatures used in scientific formulae are in *absolute* temperature scale, i.e., degrees "Rankine" or "Kelvin."

Speaking of temperatures, a few words on the difference between *static* and *total* temperatures are in order. This is primarily a book on turbomachinery, specifically, the two prime movers, i.e., gas and steam turbines. In aero-thermo fluid theory of turbomachines, temperature is not always "temperature," pressure is not always "pressure," and enthalpy is not always "enthalpy." What is the author talking about, one might wonder? Let me explain. Pressure and temperature that one can measure with a barometer and thermometer, respectively, and plug into an *equation of state* (EOS) to calculate density are *static* values. As the designation suggests, this is the case for stagnant fluids or fluids flowing at low velocities. Inside a turbine, for example, fluid velocity can be so high that the Mach number (ratio of fluid velocity to the local sound of speed) can be close to *sonic* (i.e., 1.0) or even *supersonic* (>1.0). In that case, one must consider the *total* values of pressure, temperature, and enthalpy, which accounts for the kinetic energy of the fluid. The author covered this subject at length in his earlier book [1] – see section 3.3 therein.

2.2 Prerequisites

The leading actor in this story is the gas turbine; specifically, the "heavy duty" variant of the industrial gas turbine family. In essence, these machines are stationary jet engines "on steroids" connected to a synchronous alternating current (ac) generator for electric power generation. This book does not cover the thermo-fluid fundamentals underlying design, analysis, and optimization (performance and cost) of large stationary gas turbines. The author covered that subject in his earlier book [1]. Interested readers are advised to consult that book and references therein for an in-depth look into the gas turbine. In the remainder of the present treatise, information presented in that book, known as **GTFEPG** henceforth, will be referenced frequently. In the present book, the focus is exclusively on the gas turbine already designed, manufactured, and installed in a power plant. The goal is to ensure that it is running smoothly and safely while performing to expectations.

Heavy-duty industrial gas turbines for electric power generation rated at several hundred megawatts are typically deployed in a combined cycle configuration. In this type of power plant, the exhaust gas energy of the gas turbine is utilized in a *bottoming* Rankine cycle to generate additional electric power via a steam turbine. (Not surprisingly, the gas turbine Brayton cycle is referred to as the *topping* cycle.) Not to do that would be tantamount to a thermodynamic crime. A modern 60 Hz gas turbine rated at nearly 400 MWe has an exhaust gas flow of about 1,600 lb/s (725 kg/s) at nearly 1,200°F (650°C). As a rough estimate, this gas stream has an energy content of

$$E = 1{,}600 \times 0.3 \times (1{,}200 - 200) = 480{,}000 \text{ Btu/s} \times 1.05506 \sim 500{,}000 \text{ kWth.}$$

(The assumption in the calculation above is that the exhaust gas is cooled to about 200°F [90°C] stack temperature in a heat recovery boiler [HRB] to make steam to be used in a steam turbine.) The typical efficiency of a Rankine bottoming cycle is 40%, so the exhaust gas stream of this gas turbine is worth 200 MWe from a steam turbine generator.

For the thermo-fluid fundamentals underlying design, analysis, and optimization (performance and cost) of gas and steam turbine combined cycle power plants, interested readers should consult another recent book penned by the author and the references listed therein [2]. In the remainder of the discussion here, information presented in that book, **GTCCPP** henceforth, will be referenced frequently. In the present book, the focus will be on the smooth and safe operation of major combined cycle equipment (discussed above) in a seamlessly integrated manner. A brief description of said equipment will be presented in the next chapter to ensure the integrity of the narrative in the main body of the present book. However, the reader is cautioned that the basic premise of the coverage in this book is that he or she is thoroughly acquainted with the fundamental principles governing the thermal design and performance of gas and steam turbine power plants including the HRB (also known as the *heat recovery steam generator* [HRSG], usually pronounced as "her-sig") and important *balance of plant* (BOP) equipment.

What do we mean by the term thermo-fluid fundamentals? Three subdisciplines of mechanical engineering play an important role in the design, analysis, and optimization of fossil-fired power plant equipment:

- Thermodynamics (including chemical equilibrium)
- Fluid mechanics (including gas-dynamics)
- Heat transfer

The equipment of interest in this book are of three major types:

- Turbomachinery (gas turbine with axial compressor, steam turbine, and myriad pumps)
- Heat exchanger (including the HRSG and steam turbine condenser)
- Flow control (pipes and valves)

To this list, one could also add the combustion equipment, i.e., the gas turbine combustor and HRSG duct burner. In certain cycle variants, large centrifugal process compressors will also be included. Combustion calculations involve chemical (species) balance, which falls under the major discipline of thermodynamics.

As far as design performance calculations are concerned, the only subdiscipline involved is thermodynamics; specifically, the laws of energy and mass conservation. The first one is the well-known first law of thermodynamics. A relatively simple but highly useful perspective can also be gained by applying the less understood but famous (or *infamous*, depending on your disposition) second law. In terms of pure mathematics, one only deals with algebraic equations because the physical processes do not change with time; in other words, they are time independent. In textbooks, this is commonly described as *steady-state, steady flow* (SSSF). Furthermore, strictly speaking, little or no information regarding the actual equipment or hardware is required (e.g., size in terms of geometry, weight, material properties). In fact, one frequently hears the term *rubber* hardware, which implies that the size, cost, mechanical integrity, and similar considerations do not impose a limit on the calculated plant performance.

For *off-design* calculations, the other two subdisciplines, fluid mechanics and heat transfer, come to help. This is where the divergence between computer-based heat and mass balance simulation software and simple hand (or Excel spreadsheet) calculations becomes significant. The ultimate objective is to *size* the hardware requisite for the achievement of the calculated design performance in a feasible (economically and mechanically) manner. The off-design performance is thus limited by the *fixed* hardware with a given size, geometry, and construction materials. The most common example is the sizing of plant heat exchangers, e.g., HRSG sections or the air- or water-cooled steam condenser. Based on the thermodynamic design parameters (i.e., pressure, temperature, and flow rates of the material streams crossing the heat exchanger boundary, commonly known as the *control volume* or CV), the size and number of the tubes, the shell, the materials used for their manufacture (e.g., carbon steel, stainless steel) and their specific arrangement are determined.

Most off-design calculations, e.g., part load performance at varying boundary conditions such as site ambient temperature and humidity, are time independent or SSSF as well. Thus, calculation of the off-design performance also involves the solution of a system of algebraic equations. The problem is, unlike the energy and mass conservation equations constituting the design problem, these equations are highly *nonlinear*. (Note that, while the energy conservation equations are linear at the top level with the mass flow rate and enthalpy terms, ultimately, they are not linear either when one considers the equation of state [JANAF[1] for gases and ASME[2] steam tables for water and steam] used to calculate enthalpies from pressure and temperature.) Therefore, the solution of the system of equations constituting the off-design problem is highly iterative and time consuming. Most of the time, as mentioned earlier, this difficulty can be eschewed by using *correction curves* for quick calculations.

The thermo-fluid fundamentals briefly described above are covered in great depth in earlier books written by the author [1–2]. (Henceforth, as stated earlier, those books will be referred to as **GTFEPG** and **GTCCPP**, respectively.) They will not be repeated here at any length. Of course, to facilitate the flow of narrative in a convenient manner (convenient for the reader, that is) certain basic formulae and/ or charts will be reproduced as needed. Otherwise, jumping back and forth between two or three different books to follow the storyline would make reading the present book a chore.

---

[1] JANAF is the acronym for Joint Army Navy Air Force. The JANAF tables (originally developed for rocket scientists) are maintained by the National Institute of Standards and Technology (NIST) and provide quick access to the thermodynamic data.

[2] American Society of Mechanical Engineers.

2.3   Basic Concepts

This is a technical book and thus contains mathematical treatment of physical phenomena. However, equations, graphs, tables, etc., used in the narrative below are not piled into the mix gratuitously. They are intended for the reader to understand the physics underlying operational concepts discussed in the book in the simplest possible (but still rigorous) way. As an example, consider the painstaking process of starting a combined cycle steam turbine while paying close attention to steam flow rates and temperatures. The physical concept lying at the root of a steam turbine start is *thermal stress* induced in the metal body due to a change in temperature. If not controlled properly, thermal stresses can lead to plastic deformation and/or fracture over time.

2.3.1   Thermal Stress

There are two mechanisms that lead to thermal stress in a body: (1) restrained thermal expansion or contraction and (2) temperature gradients in thick metal parts. For example, when a metal component (say, the rotor or casing [shell] of the steam turbine) is heated or cooled in the presence of constraints preventing its expansion or contraction, respectively, a compressive or tensile stress, respectively, is produced. The magnitude of the stress resulting from a temperature change from TI (initial) to TF (final) is given by

$$\sigma = E \cdot \alpha \cdot (TI - TF), \tag{2.1}$$

where E is the elastic or Young's modulus of elasticity and α is the linear coefficient of thermal expansion. Let us do a quick calculation. Assume that the low-alloy steel rotor of a steam turbine is initially at 700°F (~370°C). Suddenly, it is subjected to steam flow at 900°F (~480°C). Ignoring the actual rotor construction details and assuming that it is basically a solid cylinder and constrained from both sides, using typical values for E and α, the resulting stress would be

$$\sigma = 200 \text{ GPa} \cdot 12 \cdot 10^{-6} \frac{1}{K} \cdot (370 - 480)K = -264 \text{ MPa}.$$

Since TF > TI (heating), σ is a negative value indicating a *compressive* stress. If TF were lower than TI, i.e., cooling, one would obtain a positive value for σ, i.e., *tensile* stress. How bad is this? A typical steam turbine rotor construction material is CrMoV (pronounced "chromolly-vee"). In the 700–900°F range, an average 0.2% yield strength of CrMoV is about 550 MPa. Thus, the magnitude of the compressive stress calculated above is close to 50% of the rotor material's strength. This, of course, is not a desirable condition.

   As a result, turbo-generators (steam or gas) have only one thrust bearing. That way, the rotor can expand or contract freely in the axial direction. Turbine casings are supported similarly, i.e., only one end is fixed in the axial direction so that the other end can slide along the guides. This can be seen in Figure 2.1, which depicts an older

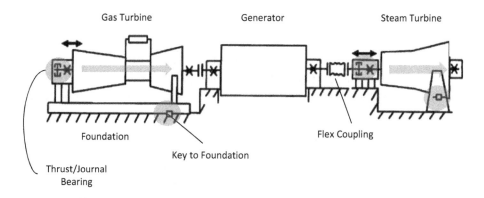

Gas Turbine   Generator   Steam Turbine

Foundation   Flex Coupling

Key to Foundation

Thrust/Journal
Bearing

Figure 2.1 Single-shaft combined cycle configuration.
Source: From "Single-Shaft Combined Cycle Power Generation System," L. O. Tomlinson
and S. McCullough, GE Power Systems, Schenectady, NY, GER-3767C.

(E Class) single-shaft gas turbine combined cycle configuration with the generator
between the prime movers. The two prime movers and the common generator have
their own shafts (rotors). The steam turbine is a simple single casing, axial flow
configuration with axial exhaust to the condenser. Note the locations of the two thrust
bearings allowing free expansion and contraction of the rotor and where the turbine
casings are "keyed" to the foundation. Also noteworthy is the flex coupling between
the steam turbine and the generator, which is more forgiving than a rigid coupling in
case of slight misalignment. (In modern single-shaft combined cycles with advanced
class gas turbines, the steam turbine is connected to the generator through a "triple S"
[SSS] *synchro-self shifting* clutch.)

A more comprehensive discussion of thermal stress can be found in chapter 13 of
GTFEPG. The point to be made herein is that the simple relationship given by
Equation (2.1) is the foundation of steam turbine stress control during plant start.
For on-line rotor stress monitoring, Equation (2.1) can be reformulated as

$$\sigma = E \frac{\alpha}{1 - \nu} K_T (TS - TB), \qquad (2.2)$$

where ν is Poisson's ratio, which converts the linear thermal expansion coefficient α
to the volume thermal expansion coefficient (for isotropic materials) and K_T is the
thermal stress concentration factor. In Equation (2.2), the temperature delta is
calculated from the difference between the surface temperature of the rotor, TS,
and the bulk or average rotor temperature, TB. For simplification, the rotor can be
idealized as a long solid cylinder whereas the actual rotor has step changes in the
diameter along its length, e.g., at the wheels, where nominal stress calculated from
solid cylinder approximation is amplified or concentrated. This effect is taken care
of by the parameter K_T. In any event, Equation (2.2) defines the *allowable* temperature difference between the steam flowing outside the rotor and the bulk rotor metal
temperature, i.e.,

$$\Delta T = \frac{1 - \nu}{E\alpha K_T} \sigma_{max}, \tag{2.3}$$

where σ_{max} is the maximum allowable stress, which is typically determined from an S-N Plot (S for stress and N for number of cycles). For a given material, e.g., CrMoV mentioned above, the S-N plot shows the applied maximum stress versus the number of load-unload cycles it took for the material to fail. The latter is also known as the fatigue life of a material. Each turbine start-shutdown event constitutes a *cycle*, during which major steam turbine components are exposed to thermal stress. The fatigue resulting from this type of cycling is commonly known as *low cycle fatigue* (LCF) because the number of cycles during operational life that can result in fatigue failure is in the range of *thousands*. In comparison, the number of mechanical load-unload cycles caused in rotating parts due to vibration can reach *millions* in a short amount of time. This type of fatigue is referred to as *high cycle fatigue* (HCF). To continue with the example, once σ_{max} is determined from S-N data, the *step change* in metal temperature given by Equation (2.3) can be converted into an allowable steam temperature ramp rate (i.e., $\partial T_S / \partial t$), via Fourier's law, which describes the heat transfer process from steam to the rotor metal. The three parameters, i.e., (1) metal temperature change (200°F, from 700°F to 900°F), (2) temperature ramp rate, i.e., how fast that amount of ΔT can happen, and (3) fatigue life of N cycles for the allowable (maximum) stress, are combined in *cyclic life expenditure* (CLE) curves, one example of which is shown in Figure 2.2.

As shown in Figure 2.2, and continuing with our example, steam at 900°F will eventually heat the rotor metal from its initial temperature of 700°F to that temperature. Naturally, this will not happen instantaneously but take time. The heat transfer rate from the steam to the metal is controlled by the convective *heat transfer*

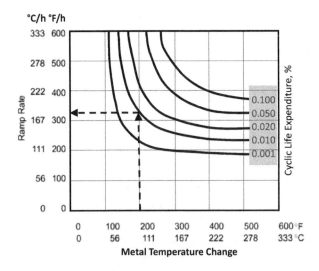

Figure 2.2 Typical turbine rotor cyclic life expenditure (CLE) curve.

coefficient (HTC), which, at a given pressure and temperature, is a function steam flow rate, MS, per

$$HTC \propto MS^{0.8}.$$

If $\partial TS/\partial t$ is controlled (via HTC, i.e., steam pressure, temperature, and flow rate) to about 330°F/h (i.e., the 200°F temperature rise in rotor metal takes about 200/330 ~ 0.6 hours or 36 minutes), CLE is 0.01%, i.e.,

$$CLE = 0.01\% = 100\%/N,$$

$$N = 10,000 \text{ cycles.}$$

In other words, the rotor life would be 10,000 cycles if all cycles (start-shutdown, load ramp up-down, etc.) were of this severity.

In conclusion, steam turbine temperature matching and loading control, which are vital components of combined cycle plant start procedure, can be understood and evaluated quite rigorously with the help of a few fundamental relationships and charts. Furthermore, the same principles can be utilized to get an understanding of operability concerns in *first-of-a-kind* (FOAK) technology components, e.g., the casing of a supercritical CO_2 turbine operating at 25–30 MPa and 700°C.

2.3.2   Load and Torque

While it is not written in the proverbial stone, load and power output are used to describe the same physical quantity, i.e., the rate of useful work production by the prime mover (gas or steam turbine in this book). The difference is that the term load is used on a *relative* basis, i.e., 100% load, 75% load, etc. whereas the term power output is used on an *absolute* basis, i.e., 125 MW, 250,000 kW, etc. In order to have a better feel for this, by no means trivial, distinction, consider how a turbomachine produces work.

Work is the product of the *force* acting on a body and the *distance* traveled by that body under the action of the said force. In other words, to produce work, there should be a distance traveled by the said body. In the context of electric power generation, the turbomachine itself, e.g., the gas turbine, does not go anywhere, i.e., no distance is traveled at all. This is where the term *torque* comes into the picture. *Torque* (τ) is the rotational analogue of the force, which acts on a body and causes it to move a distance – linearly. When a torque acts on a body, it causes the body to *rotate*. In the case of a gas or steam turbine, the body in question is the turbine shaft (rotor). The force acting on the body is the *moment* of the force, i.e., the torque. The distance traveled is the *angular* distance, e.g., 360 degrees or 2π radians in one full revolution of the shaft or $\theta = 360° = 2\pi$. Therefore, the work done by the rotating shaft is

$$W = \tau \cdot \theta. \tag{2.4}$$

Note that work and torque both have the same units, i.e., Nm (Newton-meter) in SI units, which is equal to one Joule (J), i.e., 1 J = 1 Nm. (In passing, 1 Nm is 0.7375621 lbf ft in British units; that is why it is much easier to have this discussion in SI units.) *Power* is the time rate of work production, which, for a shaft under the action of a constant torque, τ, is equal to

$$\frac{dW}{dt} = \dot{W} = \tau\frac{d\theta}{dt} = \tau\omega, \tag{2.5}$$

where the angular speed ω (in radians per second or s^{-1}) is related to the physical shaft (rotational) speed in rpm (or *frequency* in *Hertz*, Hz, with 1 Hz $= 1\ s^{-1}$) via

$$\omega = \frac{d\theta}{dt} = \frac{2\pi N}{60} = 2\pi f. \tag{2.6}$$

For heavy-duty industrial gas turbines used in electric power generation, N is either 3,000 rpm (or f $= 50$ Hz) or 3,600 rpm (or f $= 60$ Hz).

When the gas turbine is first started, it is *cranked* and/or *rolled* from a very low speed (a few rpm via the action of the *turning gear*, TG) to its full speed at a certain rate of angular acceleration, i.e., $\alpha = d\omega/dt$, by the combined action of an external driver and net shaft torque generated by fuel combustion to overcome the combined rotational inertia, I, of the *powertrain* (i.e., gas turbine, synchronous ac generator, and accessories connected to the same shaft). This is described by the following equation

$$\tau_{net} = I\frac{d\omega}{dt} = I\alpha. \tag{2.7}$$

The term *crank* is used when angular acceleration is accomplished by the action of an *external* driver. The term *roll* is used for the entire process from the TG speed to full speed. In all modern, large units, the external driver in question is the synchronous ac generator itself, which is run as a motor by the LCI (*load commutating inverter*, also known as a *static starter*). Cranking to the ignition speed (roughly 15–25% of full speed) is accomplished by the LCI. From ignition speed until the point when the unit becomes self-sustaining (about 50–60% of full speed), the gas turbine is rolled with the assistance of the LCI. Thereafter, the unit is rolled to *full speed, no load* (FSNL) under its own power. The time between ignition and FSNL is typically 8 to 10 minutes. At FSNL, when the torque generated by the turbine is just enough to compensate for torque consumed by the compressor plus all frictional losses, LCI is turned off (i.e., no external assistance) so that $\omega =$ constant, $\alpha = 0$ and the net shaft torque, $\tau_{net} = 0$. Note that the torque generated by the turbine (under the action of hot combustion gas flowing through it), τ_{turb}, is *not* zero. However, it is equal to the torque requisite to compress the air, τ_{comp} (minus losses).

Once the gas or steam turbine generator reaches FSNL, it is ready to be synchronized to the electric network, i.e., the electric grid. After the prime mover generator (the *turbogenerator*) is synchronized to the grid, i.e., the generator breakers are closed, it is ready to be loaded. The load in question is the electric power generated by the unit. The gas turbine is loaded by increasing the airflow and fuel flow through the machine in accordance with the algorithm imposed by the control system. This can be in the form of control curves or a model representing the gas turbine in the control system computer (i.e., *model-based control*). The airflow is controlled by the compressor *inlet guide vanes* (IGV). When the IGVs are at their nominal 100% open position and the fuel flow is controlled to its value corresponding to the base load turbine inlet temperature at the prevailing cycle pressure ratio, the gas turbine is said to be running at *full load* or, more precisely, at *full speed, full load* (FSFL). At any point between FSNL and FSFL, the gas turbine is running at the synchronous speed of 3,000/3,600 depending on the grid frequency, 50/60 Hz, respectively. At each such point the net shaft/mechanical torque generated by the turbogenerator is countered by the electrical torque imposed by the grid, which is of equal magnitude and in the opposite direction, so that ω = constant, α = 0. In the case of the steam turbine, power generation is controlled by the steam flow. This will be discussed in Section 3.3.

2.3.3   Simple vs. Combined Cycle

If the power plant in question has only gas turbine(s) as a prime mover, the installation and performance is commonly referred to as a *simple cycle*. To be precise, this is a misnomer because the actual gas turbine itself does not operate in a cycle per se. Air is ingested, fuel is added, and hot gas is ejected. The loop is not closed, and the working fluid is not constant. Therefore, it is also referred to as *open cycle* (yes, it *is* an *oxymoron*). Indeed there are *closed cycle* (yes, it *is* a *tautology*) gas turbines, where the term cycle is indeed apt (thermodynamically speaking), but they are not the subject of this book. (See chapter 22 of **GTFEPG** for detailed coverage.) However, closed, or semi-closed cycle systems, e.g., supercritical CO_2 technology in Chapter 10, will be covered in depth.

The term *combined* cycle refers to a power plant with a *topping* cycle and a *bottoming* cycle. The terms refer to the relative positions of the cycles on a temperature-entropy (T-s) diagram (see the next section). Heat rejected by the topping cycle is heat added to the bottoming cycle. This book's focus is primarily on a special case of combined cycle, i.e., with a gas turbine topping cycle and steam turbine bottoming cycle. On an ideal basis, on the T-s diagram, the gas turbine is described by the Brayton cycle and the steam turbine is described by the Rankine cycle. As described above, the real gas turbine does not operate in a true cycle. However, the steam turbine does. The Rankine cycle is indeed a cyclic process with a closed loop and single working fluid (H_2O).

The "official" term is *gas turbine combined cycle* (GTCC) power plant, which differentiates from *gas turbine simple cycle* power plant. For brevity, the term combined cycle (CC) is widely used in the literature, as well as in this book, and should be understood to mean GTCC.

The connection between the topping cycle and the bottoming cycle is the *heat recovery steam generator* (HRSG), pronounced as "hersig." In the author's opinion, the correct term should be *heat recovery boiler* (HRB) because the term generator also refers to the synchronous ac generator, which converts the mechanical (shaft) power of the prime movers into electric power. Another term encountered in the technical literature is *waste heat recovery boiler*, which is rarely used for electric power generation applications.

2.3.4   Performance

The sticker performance of a gas turbine is referred to as the *rating* performance. Almost without exception, it is the performance of the gas turbine at *full* load at ISO ambient conditions (59°F/15°C, 1 atm [i.e., zero altitude], and 60% relative humidity). Typically, rating performance is quoted with zero inlet and exit pressure losses and no performance fuel (100% methane, CH_4) heating (but not always). The reader is advised to check the fine print. Rating performances can be found in OEM brochures or in trade publications (e.g., *Gas Turbine World* or *Turbomachinery International*).

There is no sticker performance for a steam turbine. Its performance is dependent on myriad factors, first and foremost the gas turbine exhaust energy. Other factors include steam conditions (pressure and temperature), steam cycle (e.g., two or three pressure levels, with or without reheat), and condenser vacuum (steam turbine *backpressure*). Steam turbines are characterized by their casing configuration, last stage bucket size, and maximum steam pressure/temperature ratings. A comprehensive coverage can be found in **GTCCPP**. Key aspects are covered in depth in Section 3.3.

In a combined cycle context, the term "performance" refers to the net electric power output of the power plant, which is found by subtracting the *auxiliary power consumption* of the power plant equipment and facilities from the power output of the prime mover generators (the *gross* power output), and net efficiency. Net efficiency is the ratio of net power output to total fuel input. The latter is also referred to as *fuel* or *heat consumption*. There are two fuel consumers in a GTCC power plant: gas turbine combustors and HRSG duct burners. Duct burners increase the temperature of exhaust gas from the gas turbine by burning fuel (utilizing the approximately 11% by volume of O_2 in the exhaust gas) to increase steam production in the HRSG and thus steam turbine generator output. The technology is referred to as *supplementary firing* and is a widely used (especially in the USA) method of power augmentation on hot days (see Chapter 1). The goal is to compensate for the reduced power output of the gas turbines at hot ambient temperatures (via reduced air density and inlet airflow) by generating more power in the bottoming cycle. Supplementary firing is detrimental to GTCC efficiency via increased fuel burn (i.e., more money spent by the operator). However, increased power output at times of high demand for electric power (i.e., residential and commercial users' air conditioners going full blast) when electricity prices skyrocket more than make up for increased fuel expenditure (especially between 2010 and 2020 in the USA when natural gas prices were at historical). Combined cycle power plants equipped with duct burners are commonly referred to as *fired* power plants vis-à-vis *unfired* power plants with no supplementary firing capability.

Thermal efficiencies are expressed on a *lower heating value* (LHV) basis. As an example, natural gas, which is assumed to be 100% CH_4 (methane) has an LHV of

21,515 Btu/lb at 77°F (about 50 MJ/kg at 25°C). This is not a capricious choice. The *higher heating value* (HHV) is the *true* energy content of the fuel, which includes the latent heat of vaporization that is released by the gaseous H_2O in the combustion products when they are cooled to the room temperature. In other words, HHV is the value measured in a *calorimeter*. In a real application (e.g., a gas turbine), the combustion products, by the time they reach the exhaust gas stack (e.g., at around 180°F [~82°C] for a modern combined cycle power plant), are not cooled to a temperature to facilitate condensation, which, depending on the amount of H_2O vapor in the gas mixture, is around 110°F (~43°C). Thus, the latent heat of vaporization is *not* recovered and utilized. (It *can* be done by adding a condensing heat exchanger before the heat recovery steam generator [HRSG] stack but it would be highly uneconomical.)

Note that LHV is not measurable; it can only be calculated from the laboratory analysis of the fuel by subtracting the latent heat of vaporization. For 100% CH_4, the ratio of HHV to LHV is 1.109. Many handy formulae can be found in handbooks to estimate LHV and HHV for various fuels and fuel gas compositions.

Heat consumption is the product of fuel mass flow rate into the combustor of the gas turbine (plus fuel supplied into the duct burners of the HRSG in a combined cycle power plant, if applicable) and fuel LHV. For example, a gas turbine burning 100% CH_4 fuel at a rate of 30 lb/s has a heat consumption of $30 \times 21,515 = 645,450$ Btu/s, which is equal to about 681 MWth. If the gas turbine in question generates 275 MWe of power, its efficiency is $275/681 = 40.4\%$ (again, net or gross).

Another unit commonly used for heat consumption is MMBtu/h (million Btus per hour). For the example above, 645,450 Btu/s is 2,324 MMBtu/h. For a 40% gas turbine at different ratings between 275 and 375 MWe, heat consumption ranges between ~2,300 and 3,200 MMBtu/h.

Frequently, thermal efficiency is expressed as a *heat rate*, which is given by 3,412 Btu/kWh divided by thermal efficiency. In SI units, heat rate is measured by kJ/kWh and it is found by dividing 3,600 kJ/kWh by the thermal efficiency. The ratio $3,600/3,412 = 1.0551$ is the conversion factor for British and SI units of heat rate. Heat rate is the land-based counterpart of *specific fuel consumption*, which is a widely used metric for aircraft engines. For example, the heat rate of a gas turbine with 40% efficiency is $3,412/0.4 = 8,530$ Btu/kWh. Typically, large "frame" gas turbine efficiencies range between 36 and 42% (ISO base load rating). Thus, their heat rate range is ~8,000 to 9,500 Btu/kWh (~8,500 to 10,000 in SI units).

For a gas turbine combined cycle power plant, the difference between net and gross power output can be anywhere from 1.6% to more than 3% – mainly dictated by the steam turbine heat rejection system. This will be discussed in detail in Chapter 3.

For quoting performances of new and emerging technologies, there simply are no rules. There is no well-established rating performance criteria. On top of that, scant attention is paid to the differences (maybe out of ignorance or, maybe, deliberately for marketing hyperbole) between cycle performance and plant net and gross performance. Eye-catching numbers are liberally thrown around and they do not pass a critical examination of the underlying assumptions and details (sometimes hidden, again, whether intentionally, or unintentionally, it is hard to tell). In this book, dear reader, such hyperbole is smashed with the help of the second law of thermodynamics (and with irrepressible, scientific glee on the part of the author). Read on.

2.3.4.1   Cycle versus Plant Net Efficiency

The author would like to let the readers know that he is truly embarrassed that he felt obligated to pen this section. However, outrageous performance claims made by new technology developers and equipment manufacturers seem to be taken at their face value not only by the lay audience but also by industry practitioners and academic researchers. Unfortunately, marketing hyperbole has replaced rigorous engineering analysis and scientific truth. The author has harped on about this sad situation in his technical papers and articles [11–13]. Chapter 10 in this book contains rigorous thermodynamic analysis, busting the performance myth of two popular technologies widely touted in the trade publications and archival journals. (The same critical approach is also used in other chapters. The reader will surely recognize them when they see them.)

By far the biggest culprit in this deliberate manipulation or inadvertent (or inept) misrepresentation of the true performance of a given heat engine technology is the blurring of the line between the cycle efficiency and plant net efficiency. Cycle efficiency is a theoretical number that determines how well a given heat engine cycle approximates the ultimate *theoretical* value that is set by the *Carnot efficiency*. Plant net efficiency is a *commercial* number that measures the bottom line based on two directly and accurately *measurable* and *monetizable* quantities:

1. Amount of fuel burned in MMBtu/h or MWth (HHV)
2. Amount of net power supplied to the grid in MWh or kWh

The owner/operator pays money for the former and generates revenue from the latter. The difference between the two either sinks the ship (if negative) or floats it (if positive). This is the bottom line. Period.

Cycle efficiency is neither directly measurable nor monetizable. It can be inferred from measured parameters within an error band depending on the cycle complexity as well as precision and accuracy of available transducers. The gap between the two is primarily a function of the following:

- Cycle heat addition equipment (e.g., fired boiler, fired or unfired heat exchanger)
- Cycle heat rejection equipment (e.g., water- or air-cooled condenser and cooling tower)
- Balance of plant (auxiliary) equipment power consumption
- Power consumed by *add-on* process blocks (e.g., air separation unit to generate oxygen used in the oxy-combustor)

It is worth noting that these factors will be present and highly impactful even if the cycle performance is evaluated with the utmost care in attending to the design parameters such as key heat exchange equipment pinch points (i.e., heat exchanger *effectiveness*), component polytropic/isentropic efficiencies, and parasitic loss causes such as secondary flows for hot gas path cooling. In many cases, these cycle design intricacies are either completely ignored or optimistically set with little attention paid to size, cost, and manufacturability considerations.

Finally, one must realize that, especially in the present state of power grid mix of generating assets and operating rhythms, a gas/steam turbine power plant operates like a typical car during a normal day. It starts, stops, accelerates, runs at constant speed for a while, decelerates, stops, restarts, etc. During the normal transients encountered within this operating rhythm, the fuel consumption of the power plant is significantly different than the calculated (and published) rating performance at a prescribed site ambient and loading condition. For conventional GTCC power plants, this information is readily available (and calculable using established models, correction curves, etc.). Actual US plant data can be obtained from the statistics published by the US Energy Information Administration (EIA) – see Form EIA-923 data available online at www.eia.gov/electricity/data/eia923/. For a representative selection of US combined cycle power plants, the results for the period January–September 2018 are summarized in Table 2.1. Note that all six power plants in the table are equipped with supplementary-fired HRSGs (heavily deployed during summertime). Unfortunately, the EIA data files do not include the design performance data so back-calculating the load factor is not possible (probably somewhere around 70% or so). In any event, the disconnect between the ISO base load ratings and the reality in the field (more than 60% LHV for H and H+ Class units) is clear.

This is why pretentious efforts to make new and emerging technologies look much better than they really are eventually end up in failure. What is more, this is also why this book places great importance on RAM and operability. A doctored number at a single design point with highly optimistic assumptions and/or omissions does not a viable technology make. How to start and shut down a system, how it performs at low loads and/or extreme ambient temperatures, and whether it has a realistic RAM assessment are all vital pieces of information.

Table 2.1 Selected Form EIA-923 data (January–September 2018).

MWH_TOT: Total power generated, HC_TOT: Total fuel consumed, MEE: Mean-effective (average) efficiency.

OEM	Class	Configuration	MWH_TOT MWh	HC_TOT MMBtu HHV	Heat Rate Btu/kWh HHV	Efficiency, MEE % HHV	% LHV
A	F	2 (2×2×1)	2,315,620	16,954,109	7,322	46.60	51.68
A	F+	2×2×1	2,159,244	15,275,314	7,074	48.23	53.49
A	H	3×3×1	5,482,523	36,275,531	6,617	51.57	57.19
A	H	3×3×1	5,413,559	36,020,550	6,654	51.28	56.87
B	H+	2×2×1	2,315,067	16,682,609	7,206	47.35	52.51
B	H+	2×2×1	1,983,351	13,060,540	6,585	51.82	57.46

A very apt cautionary lesson in this sense is provided by the saga of the integrated gasification gas turbine (IGCC) technology (see Chapter 13). The underlying core technology, gasification, is more than a century old. Key subsystems on the syngas side have been successfully operational in chemical process and refinery industries for decades. The same is true for the power block, the GTCC. Nevertheless, despite a significant number of demonstration IGCC plants deployed in the last three decades of the twentieth century in the USA (Wabash, TECO Polk), Europe (Buggenum in Netherlands, Puertellano in Spain), and Japan (Negishi), the IGCC technology failed to make a commercial breakthrough mainly due to (unexpectedly) high costs, complexity of operation, and/or low reliability and availability. Even the inherent capability of pre-combustion capture of CO_2 failed to help the IGCC technology. Several such demonstration projects, e.g., the Southern Company's Kemper County project among them (an especially sobering failure), were ultimately dropped or repurposed. The interested reader can find the details of that colossal failure by simply googling the name on the internet.

2.3.5   Technology Factor

The most powerful concept in performance claim assessment is the concept of *technology factor*. The basic premise of this concept is that any practical heat engine cycle (no exceptions!) is an attempt (in most cases, quite unsuccessful) to approximate the underlying Carnot cycle. What does the qualifier "underlying" mean? It means that any heat engine cycle (no exceptions!) can be identified by two temperatures characteristic of two major heat transfer processes:

• Mean-effective temperature during cycle heat addition (METH)
• Mean-effective temperature during cycle heat rejection (METL)

This fact is a direct corollary of the Kelvin–Planck statement of the second law of thermodynamics. Note, however, that,

• METH is not necessarily the same as maximum cycle temperature; in fact, in almost all cases, METH < TMAX
• METL is not necessarily the same as minimum cycle temperature; in general, METL ≥ TMIN (see Figure 2.3 for the terminology used in the discussion of Brayton simple and combined cycle thermodynamics)

Identifying and calculating METH and METL will be described in the context of the gas turbine Brayton cycle in Section 3.1. The reader can also refer to chapter 5 of **GTFEPG**. For an in-depth look at the technology factor concept, refer to chapter 6 of **GTFEPG**. Briefly, once METH and METL are identified, the underlying or "equivalent" Carnot efficiency of the heat engine cycle in question can be found as

$$ECEFF = 1 - \frac{METL}{METH}, \qquad (2.8)$$

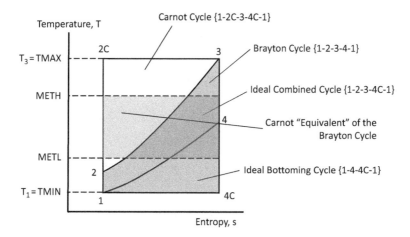

Figure 2.3 Temperature-entropy diagram of Carnot and air-standard (ideal) Brayton cycles.

which is lower than the "ultimate" Carnot efficiency given by

$$\text{UCEFF} = 1 - \frac{\text{TMIN}}{\text{TMAX}}. \tag{2.9}$$

The ratio of the two is the *cycle factor*, CF, i.e.,

$$\text{CF} = \frac{\text{ECEFF}}{\text{UCEFF}}. \tag{2.10}$$

The actual cycle efficiency can be found from rigorous heat and mass balance simulation calculations and verified in the field by conducting performance tests governed by applicable test codes. The ratio of the actual cycle efficiency to that of the equivalent Carnot efficiency is the *technology factor*, i.e.,

$$\text{TF} = \frac{\text{ACTEFF}}{\text{ECEFF}}. \tag{2.11}$$

For modern heavy-duty industrial gas turbines, the cycle factor is around 0.70. The cycle efficiency as a function of turbine inlet temperature (TIT) is plotted in Figure 2.4. For the advanced class (i.e., HA or J Class) gas turbines with inlet temperatures of 1,700°C (cycle pressure ratio [PR] of 24:1), the TF is 0.73. At the origin of the jet age, the gas turbine of the Jumo 004 engine (which powered the first operational jet Messerschmitt Me 262 in 1944–1945), TF was 0.54 at a TIT of 775°C (cycle PR of 3:1).

   A correct understanding of the technology factor concept or method by the reader is of utmost importance. It is by no means a "fudge factor." Its connection to the key cycle design parameters via fundamental thermodynamic principles and correlations has been demonstrated step by step in chapter 6 of **GTFEPG**. Furthermore, the concept has been successfully used in the past by German scientists and engineers in the design of turbomachinery, e.g., turbo-superchargers under the name of *Gütezahl* or *Gütegrad*. The technology factor quantitatively identifies the prevailing state of the

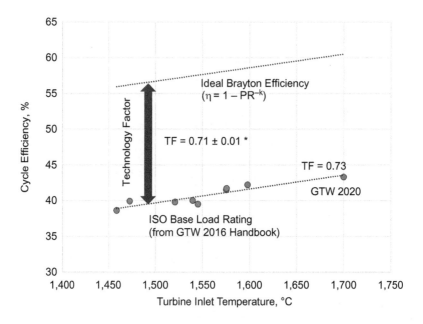

Figure 2.4 Gas turbine Brayton cycle efficiencies (ISO base load rating) as a function of turbine inlet temperature (TIT).
*: Jumo 004B (1943–1945) TF = 0.54 (TIT= 775°C)

art in heat engine design in a manner not subject to obfuscating discussion or interpretation.

Once a new heat engine technology, say, the currently (i.e., at the time of writing this book, 2020 and 2021) very much in fashion, supercritical CO_2 cycle, is identified by its temperature-entropy (T-s) diagram, its performance entitlement is easily calculable as demonstrated above (and later, in Section 3.1, for the gas turbine) for the Brayton cycle. If the implied technology factor of a published performance claim is comparable to or beyond the value established by a 75 years-old technology at the pinnacle of its development stage (i.e., a J Class gas turbine), it should be rejected out of hand. To gauge the difficulty in pushing the envelope in terms of the technology associated with translating a heat engine cycle from the T-s diagram on paper to the actual hardware in the field, simply spend a few minutes on the data plotted in Figure 2.4. It took nearly half a century to bring the TF from 0.54 (TIT of 775°C and cycle PR of 3:1) to around 0.70 (roughly representative of vintage F and introductory H Class TIT of 1,500°C with PR of around 20:1). In the following quarter of a century, a significant leap was made by the gas turbine OEMs in pushing the TIT from 1,500°C to 1,700°C but the proverbial needle barely budged. (This push was primarily driven by combined cycle efficiency, which will be discussed in Section 3.1.)

The correlation between the cycle and technology factors is illustrated in the chart in Figure 2.5. As one would expect, the higher cycle factor is an enabler of higher technology factor. This can be best explained by a financial analogy (admittedly not a perfect one but still useful). If one goes to a credit institution with say, $10K, that

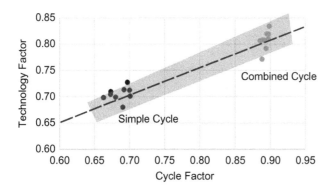

Figure 2.5 Gas turbine Brayton simple and combined cycle technology and cycle factors.

person will get a much lower interest rate than someone who brings $10 million to the same institution.

There is no question that one must acknowledge that the starting point in TF for a new technology in the first quarter of the twenty-first century does not have to be as low as 0.54. Present knowledge and experience base in materials (e.g., superalloys used in manufacturing and construction of hot gas path components), advances in aerodynamics, heat transfer and fluid mechanics, 3D computational fluid dynamics (CFD) tools, and full-scale test facilities certainly make it possible for a new technology to start its life cycle at a higher TF. Thus, if the cycle factor of the technology in question is around 0.7, a value between 0.6 and 0.7 can be taken at its face value with certain caveats in place. For a higher cycle factor, a higher technology factor can be acceptable. The evaluation should be made for any new technology on a case-by-case basis.

2.3.6   Power Augmentation

As mentioned above, on hot days, gas turbine outputs drop drastically due to reduced air density and air flow. The resulting loss in electric power generation capacity is compensated for by burning extra fuel in the HRSG to increase steam production and thus steam turbine generator output. This is the most common method of hot day power augmentation, especially in the USA with low natural gas prices.

Another commonly utilized hot day power augmentation is compressor inlet air conditioning, which refers to cooling of airflow sucked into the compressor by various means. The most common method is *evaporative cooling* (colloquially referred to as *evap cooling*), which consists of water addition in a porous media into the incoming air, which is cooled by the mechanism of evaporation of sprayed water with latent heat of evaporation supplied by incoming air. Evap coolers are especially effective in dry climates with low ambient air humidity so that the amounts of evaporation and saturation are maximized for maximum cooling effect. A reduction in air temperature leads to increased density and mass flow rate to compensate for the hot ambient temperature's detrimental effect.

A variant of evap cooling is *inlet fogging*, where high-pressure water is sprayed into the incoming air in the form of microscopic droplets (i.e., the "fog"). Cooling takes place in two steps: (1) evaporation of water droplets to the point of saturation (i.e., 100% relative humidity) and (2) evaporation of the remaining droplets carried into the compressor with airflow during the compression process. The latter effectively acts as a continuous intercooling and reduces parasitic power consumption of the compressor.

The third, less commonly deployed, inlet air conditioning method is *inlet chilling*, where cooling of inlet airflow is accomplished via sensible heat transfer. For detailed discussion of gas turbine inlet conditioning options and quantitative information, refer to section 18.1.1 in **GTFEPG**.

2.3.7 Key Parameters

A gas turbine can be fully defined by a few parameters, e.g., shaft speed, airflow, "firing" temperature and cycle (compressor) pressure ratio (PR). Airflow and firing temperature set the power output of the gas turbine (or its size). Firing temperature sets its fuel or "heat" consumption whereas cycle PR sets its exhaust temperature (and its suitability to combined cycle application). Let us look at them quickly.

Electric grids in the world are either 60 Hz, e.g., in the USA, or 50 Hz, e.g., in Europe. In some countries, e.g., Saudi Arabia and Japan, both are present. When a prime mover generator, i.e., gas or steam turbine generator, is *synchronized* to the grid, it runs either at 3,600 rpm (60 Hz grid) or at 3,000 rpm (50 Hz), where rpm denotes *revolutions per minute*. In angular terms,

- 3,600 rpm is 120π *radians per second* (rad/s)
- 3,000 rpm is 100π rad/s

Since connecting a heavy-duty industrial gas turbine rated at, say, 300 MWe and has a speed other than 3,000 or 3,600 rpm to a generator running at those speeds would require a very large, expensive, and parasitic power consuming gearbox, large *frame* machines are designed to run at 3,000 or 3,600 rpm. There are, however, some notable exceptions. In particular:

1. Smaller aero-derivative gas turbines with self-contained gas generator and power turbine, e.g., some of General Electric's (GE's) LM-2500 units, which run at 6,100 rpm.
2. Smaller industrial gas turbines such as GE's Frame 6, which runs at 5,100 rpm.
3. Alstom (now owned by GE) GT11N2 at 3,600 rpm for both 50 *and* 60 Hz versions (this is a not-so-small 115 MWe gas turbine).

Other than size, whether a gas turbine is a 50 Hz (3,000 rpm) or 60 Hz (3,600 rpm) unit is immaterial to the discussion at hand – unless, of course, the discussion is on compressor or turbine aerothermodynamics. One difference, which can be seen in combined cycle applications, is due to the larger steam flow generated in the HRSG of a larger 50 Hz gas turbine (roughly the same exhaust temperature but ~40% higher mass flow rate). Since steam turbine efficiency is a function of volumetric flow rate of steam, combined cycles with 50 Hz gas turbines are slightly more efficient than their 60 Hz counterparts (everything else being the same, of course). By the same token, at given speed, multi-gas turbine units are slightly more efficient than single-gas turbine units.

A significant source of ambiguity is associated with the definition of highest cycle temperature. In many references, terms like *firing temperature* and *turbine inlet temperature* (TIT) are used interchangeably and erroneously. This subject is covered in depth in Section 3.1.

For practical purposes, the cycle maximum temperature is the temperature of hot combustion gas at the inlet of the turbine stage 1 rotor buckets (or blades) before it starts producing useful turbine work. This temperature is commonly known as the *firing temperature* (TFIRE) and it is several hundred degrees Fahrenheit *lower* than the hot gas temperature at the combustor exit (at the inlet of turbine stage 1 stator nozzle vanes, i.e., the "true" TIT in essence). The reason for this reduction is dilution with cooling flow used for cooling the nozzle vanes and wheel spaces. Thus, another term for it is *rotor inlet temperature* (RIT) where the rotor in question is the stage 1 rotor. RIT is essentially what is defined in the ANSI Standard B 133.1 (1978),[3] which is referred to as an ISO-rated *cycle* temperature in §3.15 of API Standard 616.[4]

Firing temperature (or RIT) is about 100°C *higher* than the fictitious temperature as defined in ISO-2314 and adopted by European OEMs.[5] Based on the latter definition, all compressor airflow, including *hot gas path* cooling flows are assumed to enter reaction with the combustor fuel. Note that ISO-2314 defines *and* outlines a calculation methodology with requisite equations whereas API 616 and ANSI B 133.1 do *not*.

The thermodynamic cycle of the gas turbine (i.e., the Brayton cycle) is fully defined by TIT and cycle PR. The underlying theory is covered in great depth in **GTFEPG** and in Section 3.1. Heavy-duty industrial gas turbines are classified by their TIT, e.g., E Class (TIT = 1,300°C), F Class (TIT = 1,400°C). Fundamental thermodynamics dictate the optimal cycle PR for given a TIT as the one that maximizes specific power output of the gas turbine. Refer to Section 3.1 for details.

As mentioned earlier, the steam turbine is not amenable to a similar classification. Steam turbines are defined by their steam cycle parameters. In particular,

- Main or high pressure (HP) steam *admission* pressure and temperature (at the main stop/control valve inlet)
- Reheat steam temperature
- Intermediate (IP) steam admission pressure (at the intercept/control valve inlet)

---

[3] ANSI B113.1: Gas Turbine Terminology (1978), American Society of Mechanical Engineers, New York, USA.
[4] API Standard 616: Gas Turbines for Refinery Services, 5th ed. (2011), American Petroleum Institute, Washington, DC, USA.
[5] ISO-2314:2009: Gas Turbines – Acceptance Tests (2013), International Organization for Standardization, Geneva, Switzerland.

- Low (LP) steam admission pressure (at the LP admission valve inlet)
- Condenser vacuum (backpressure)

Modern *three-pressure reheat* (3PRH) steam cycle parameters are

- 170 bar (2,500 psi) or higher main (HP) steam pressure
- Up to 600°C (1,112°F) main and reheat steam temperature
- IP steam pressure subject to cycle optimization (usually around 25–30 bar)
- LP steam pressure subject to cycle optimization (usually around 4–5 bar)
- Condenser vacuum subject to myriad regulatory and economic factors (see Section 3.3)

HRSG design is characterized by key temperature deltas, i.e.,

- HP, IP, and LP steam evaporator pinch
- HP, IP, and LP economizer approach subcool
- HP, IP, and LP superheater approach temperature
- Condensate heater (last economizer section before the stack) terminal temperature difference

See Section 3.2 for a description of these parameters. Smaller temperature deltas result in increased steam production but at the expense of heat transfer surface area, i.e., HRSG size and cost.

2.4   Suggestions for Further Reading

2.4.1   Books by the Author

1. Gülen, S. C. 2019. *Gas Turbines for Electric Power Generation*. Cambridge: Cambridge University Press.
2. Gülen, S. C. 2019. *Gas Turbine Combined Cycle Power Plants*. Boca Raton, FL: CRC Press.

Henceforth, the first book will be referred to as **GTFEPG** and the second book will be referred to as **GTCCPP**. As mentioned in the preceding paragraphs, relevant items covered in depth in these two books will be cited frequently in the remainder of the current treatise. There are many other valuable references out there covering similar topics. An exhaustive list can be found in **GTFEPG** accompanied by a short discussion on why a particular reference book should be present in the library of someone working in this field. I will cite only three that I value highly. For the best coverage of gas turbine fundamentals, the reader is pointed to

3. Saravanamuttoo, H. I. H., Rogers, G. F. C., Cohen, H., and Straznicky, P. V. 2009. *Gas Turbine Theory*, 6th ed. Harlow, Essex, England: Pearson Education.

Older editions of this book (used, of course) can be found on-line for a low price and will serve the purpose equally well. This author is not a fan of "handbooks," which he calls "travel guides," as practically useful references. There are, however, two exceptions, one in English and one in German, which are well worth having a copy of. (They are not too expensive either.)

4. Jansohn, P. (ed.) 2013. *Modern Gas Turbine Systems: High Efficiency, Low Emission, Fuel Flexible Power Generation* (Woodhead Publishing Series in Energy) 1st ed. Cambridge, UK: Woodhead Publishing.
5. Lechner, C. and Seume, J. 2010. *Stationäre Gasturbinen, 2. neue bearbeitete Auflage*. Heidelberg: Springer.

2.4.2   Book Chapters by the Author

6. Gülen, S. C. 2017. Advanced Fossil Fuel Power Systems. In D. Y. Goswami and F. Kreith (eds.), *Energy Conversion*, 2nd ed. Boca Raton, FL: CRC Press.
7. Smith, R. W. and Gülen, S. C. 2012. Natural Gas Power. In Robert A. Meyers (ed.), *Encyclopedia of Sustainability Science and Technology*. New York: Springer.

The first of these two, with more than 160 pages, is a small book in and of itself. It is a concise guide to technologies pertinent to sustainable deployment of fossil fuel resources. Suffice it to say that the author himself uses it as a reference for a quick understanding of, say, "oxy-combustion" and "steam methane reforming." It also includes an extensive bibliography.

The second one is effectively a condensed treatise on gas turbines – also with an extensive bibliography. It was written around 2010 and updated in 2016.

2.4.3   Papers and Articles by the Author

Over the course of his career, the author has written numerous papers and articles published in prestigious archival journals such as *ASME Journal of Engineering for Gas Turbines and Power (J. Eng. Gas Turbines Power)* and trade publications such as *Gas Turbine World*. Some papers were presented at landmark industry gatherings such as *ASME IGTI Turbo Expo* and *POWERGEN International* and frequently won "Best Paper" awards. Readers interested in details of certain subjects covered in the current treatise may want to check out some of them for more in-depth information.

8. Gülen, S. C. and Lu, H. 2019. Coal-fired direct injection carbon engine (DICE) and gas turbine compound-reheat combined cycle (GT CRCC), International Pittsburgh Coal Conference, September 3–6, 2019, Pittsburgh, PA.
9. Gülen, S. C., Rogers, J., and Sprengel, M. 2019. DICE-Gas turbine compound reheat combined cycle, Presentation at: 9th International Conference on Clean Coal Technologies, June 3–7, 2019, Houston, TX (also published in the journal *Fuel*).
10. Gülen, S. C. 2019. Combined cycle technology trends, *Turbomachinery International*, 60 (May/June), 26–28.
11. Gülen, S. C. 2019. Adjusting GTW ratings for realistic expectation of plant performance, *Gas Turbine World 2019 GTW Handbook*, 34, 25–29.
12. Gülen, S. C. 2019. Disappearing thermo-economic sanity in gas turbine combined cycle ratings – a critique, ASME Paper GT2019–90883, ASME Turbo Expo 2019, June 17–21, Phoenix, AZ.
13. Gülen, S. C. 2018. Combined cycle performance ratings undergo thermodynamic reality check, *Gas Turbine World*, November/December, 23–32.
14. Gülen, S. C. 2018. HRSG duct firing revisited, ASME Paper GT2018–75768, ASME Turbo Expo 2018, June 11–15, Lillestrøm (Oslo), Norway.

15. Elliott, W. R. and Gülen, S. C. 2017. Cost-effective post-combustion carbon capture from coal-fired power plants, International Pittsburgh Coal Conference, September 5–8, Pittsburgh, PA.

16. Gülen, S. C., Adams, S. S., Haley, R. M., and Carlton, C. 2017. Compressed gas energy storage, *Power Engineering*, 121(8), 26–38.

17. Gülen, S. C. and Hall, C. 2017. Optimizing post-combustion carbon capture, *Power Engineering*, 121(6), 14–19.

18. Gülen, S. C. and Hall, C. 2017. Gas turbine combined cycle optimized for carbon capture, ASME Paper GT2017–65261, ASME Turbo Expo 2017, June 26–30, Charlotte, NC.

19. Gülen, S. C., Yarinovsky, I., and Ugolini, D. 2017. A cheaper HRSG with advanced gas turbines: when and how can it make sense, *Power Engineering*, 121(3), 35–42.

20. Gülen, S. C. 2017. Pressure gain combustion advantage in land-based electric power generation, GPPF2017–0006, First Global Power and Propulsion Forum, GPPF 2017, January 16–18, Zürich, Switzerland.

21. Gülen, S. C. 2016. Back to the future for 65%-plus combined cycle plant efficiencies, *Gas Turbine World*, May/June, 26–32.

22. Gülen, S. C. 2016. Beyond Brayton cycle: it is time to change the paradigm, ASME Paper GT2016–57979, ASME Turbo Expo 2016, June 13–17, Seoul, South Korea.

23. Gülen, S. C. and Boulden, M. S. 2015. Turbocompound reheat gas turbine combined cycle, POWER-GEN International, December 8–10, Las Vegas, NV.

24. Gülen, S. C. 2015. Turbocompound reheat gas turbine combined cycle "The Mouse That Roars," *Gas Turbine World*, November/December, 22–28.

25. Gülen, S. C. 2015. Repowering revisited, *Power Engineering*, 19(11), 34–45.

26. Gülen, S. C. 2016. Combined cycle trends: past, present and future, *Turbomachinery International Handbook*, 24–26.

27. Gülen, S. C. 2015. Powering sustainability with gas turbines, *Turbomachinery International*, 56(5), 24–26.

28. Gülen, S. C. 2015. Étude on gas turbine combined cycle power plant – next 20 years, ASME Paper GT2015–42077, ASME Turbo Expo 2015, June 15–19, Montréal, Canada. Also published in *J. Eng. Gas Turbines Power*, Vol. 138, #051701.

29. Gülen, S. C. 2014. *General Electric – Alstom merger brings visions of the Überturbine*, *Gas Turbine World*, July/August, 28–35.

30. Gülen, S. C. 2014. Second law analysis of integrated solar combined cycle power plants, ASME Paper GT2014–26156, ASME Turbo Expo 2014, June 16–20, Düsseldorf, Germany. Also published in *J. Eng. Gas Turbines Power*, Vol. 137, #051701.

31. Gülen, S. C. 2013. Constant volume combustion: the ultimate gas turbine cycle, *Gas Turbine World*, November/December, 20–27.

32. Gülen, S. C. 2013. Modern gas turbine combined cycle, *Turbomachinery International*, 54(6), 31–35.

33. Gülen, S. C. and Kihyung, K. 2013. Gas turbine combined cycle dynamic simulation: a physics based simple approach, ASME Paper GT2013–94584, ASME Turbo Expo 2013, July 3–7, San Antonio, TX. Also published in *J. Eng. Gas Turbines Power*, Vol. 136, #011601.

34. Gülen, S. C. 2013. What is the worth of 1 Btu/kWh of heat rate, *POWER*, June, 60–63.

35. Gülen, S. C. 2013. Gas turbine combined cycle fast start: the physics behind the concept, *Power Engineering*, 117(6), 40–49.

36. Gülen, S. C. 2013. Performance entitlement of supercritical steam bottoming cycle, *J. Eng. Gas Turbines Power*, 135, 124501.

37. Gülen, S. C. and Driscoll, Ann V. 2012. Simple parametric model for quick assessment of IGCC performance, ASME Paper GT2012–68301, ASME Turbo Expo 2012, June 11–15, Copenhagen, Denmark. Also published in *J. Eng. Gas Turbines Power*, Vol. 135, #010802.

38. Gülen, S. C. and Mazumder, I. 2012. An expanded cost of electricity model for highly flexible power plants, ASME Paper GT2012–68299, ASME Turbo Expo 2012, June 11–15, Copenhagen, Denmark. Also published in *J. Eng. Gas Turbines Power*, Vol. 135, #011801.

39. Gülen, S. C. and Jones, Charles J. 2011. GE's next generation CCGT plants: operational flexibility is the key, *Modern Power Systems*, June, 16–18.

40. Gülen, S. C. 2011. A more accurate way to calculate the cost of electricity, *POWER*, June, 62–65.

41. Gülen, S. C. and Joseph, J. 2011. Combined cycle off-design performance estimation: a second law perspective, ASME Paper GT2011–45940, ASME Turbo Expo 2011, June 6–10, Vancouver, Canada. Also published in *J. Eng. Gas Turbines Power*, Vol. 134, #010801.

42. Gülen, S. C. 2010. Gas turbine with constant volume heat addition, ASME Paper ESDA2010–24817, ASME 2010 10th Biennial Conference on Engineering Systems Design and Analysis, July 12–14, Istanbul, Turkey.

43. Gülen, S. C. 2010. A proposed definition of CHP efficiency, *POWER*, June, 58–63.

44. Gülen, S. C. 2010. Importance of auxiliary power consumption on combined cycle performance, ASME Paper GT2010–22161, ASME Turbo Expo 2010, June 14–18, Glasgow, Scotland, UK. Also published in *J. Eng. Gas Turbines Power*, Vol. 133, #041801.

45. Gülen, S. C. 2010. A simple parametric model for analysis of cooled gas turbines, ASME Paper GT2010–22160, ASME Turbo Expo 2010, June 14–18, Glasgow, Scotland, UK. Also published in *J. Eng. Gas Turbines Power*, Vol. 133 (2011), #011801.

46. Gülen, S. C. and Smith, R. W. 2008. Second law efficiency of the Rankine bottoming cycle of a combined cycle power plant, ASME Paper GT2008–51381, ASME Turbo Expo 2008, June 9–13, Berlin, Germany. Also published in *J. Eng. Gas Turbines Power*, Vol. 132 (2010), #011801.

47. Hofer, D. C. and Gülen, S. C. 2006. Efficiency entitlement for bottoming cycles, ASME Paper GT2006–91213, ASME Turbo Expo 2006, May 8–11, Barcelona, Spain.

48. Gülen, S. C. and Smith, R. W. 2006. A simple mathematical approach to data reconciliation in a single-shaft combined cycle system, ASME Paper GT2006–90145, ASME Turbo Expo 2006, May 8–11, Barcelona, Spain. Also published in *J. Eng. Gas Turbines Power*, Vol. 131 (2009), #021601.

49. Gülen, S. C. and Jacobs III, J. A. 2003. Optimization of gas turbine combined cycle, POWER-GEN International, Las Vegas, NV.

50. Gülen, S. C., Griffin, P. R., and Paolucci, S. 2000. Real-time on-line performance diagnostics of heavy-duty industrial gas turbines, ASME Paper 2000-GT-0312, ASME Turbo Expo 2000, May 8–11, Munich, Germany. Published in *J. Eng. Gas Turbines Power*, Vol. 124 (2002), pp. 910–921.

51. Nakhamkin, M. and Gülen, S. C. 1995. Transient analysis of the Cascaded Humidified Advanced Turbine (CHAT), ASME Paper 95-CTP-28, ASME Cogen Turbo Power Conference, August 23–25, Vienna, Austria.

3 Equipment

This chapter provides a brief introduction of the major actors in the narrative. It is not intended for the layperson (neither is this book). It is more of a "pregame warm-up" to refresh the reader's memory and provide a crib sheet for some simple calculations. The goal is to instill a sense of what is possible and what is not within reason. Trade literature is full of marketing hyperbole advertising technologies with promises of proverbial "pie in the sky," e.g., very high thermal efficiency ratings achievable under very optimistic site ambient conditions and unrealistic parasitic power consumption assumptions with no attention paid to increasingly stringent environmental regulations putting clamps on performance. For an extensive discussion of this point with detailed heat and mass balance analysis, the reader is pointed to recent publications by the author (Refs. [12–13] in Chapter 2).

3.1 Gas Turbine

Modern gas turbines are large air-breathing turbomachines with extremely large power output. For example, consider a 50 Hz (3,000 rpm) 300+ MW gas turbine. This machine ingests air at ISO conditions (15°C and 1 atmosphere at 60% relative humidity) at a rate of nearly 1,550 lb/s (705 kg/s), compresses it to a pressure that is 18 to 20 times that of the ambient, and combusts it with about 35 lb/s (16 kg/s) of natural gas (100% CH_4) generating 312 MW net electric power for a net thermal efficiency of 39.3%. At the inlet to the expander section of the gas turbine, where they produce useful shaft work, the combustion products are at nearly 1,500°C (2,732°F). This is well above the melting point of the most advanced superalloy materials that are used in the manufacture of turbine *hot gas path* (HGP) components. To ensure the survival of the turbine parts under those extreme conditions for thousands of hours of continuous operation, nickel-based superalloy components are protected by *thermal barrier coatings* (TBC) and internally cooled by using the "cold" air extracted from the turbine compressor. (Consider that without cooling and TBC, the first stage vanes of a modern gas turbine would survive barely 10 seconds before melting away.) At the time of writing, the largest 50 Hz gas turbines are rated at above 500 MWe. To appreciate the significance of this number, consider that less than two decades ago, the combined cycle output of the largest 50 Hz gas turbine was less than this (about 450 MWe).

All the pertinent details of gas turbine theory can be found in the author's earlier book, **GTFEPG**. To repeat them here would be a waste of ink at best, cheating at worst. On the other hand, the reader of the current book must be familiar with certain guiding principles to make sense of the discussion in the upcoming chapters. More importantly, the reader should be given a sense of what is marketing hyperbole and academic daydreaming and what is the stark reality in the field. At a time when overblown claims of 65% combined cycle efficiency are flying around in the media, this is indeed a pressing need. Consequently, this chapter will be a crash course in basic gas turbine thermodynamics and technology comprising simple formulae to facilitate "back of the envelope" estimates, pertinent numerical examples, and useful rules of thumb.

A modern gas turbine has three components:

1. Axial compressor
2. Dry-Low NOx (DLN) combustor
3. Axial turbine

These three components are connected through a single shaft supported by two bearings (one at each end of the shaft) to a synchronous ac generator. Practically, in all modern machines, the generator is driven from the compressor end of the shaft (commonly known as the *cold end drive*). This facilitates the smooth connection of the gas turbine exhaust to the heat recovery steam generator (HRSG) inlet duct with minimal loss and easy removal of the generator rotor for maintenance and repairs. The general layout or *architecture* of an advanced class, heavy-duty industrial gas turbine is illustrated in Figure 3.1.

A gas turbine can be described quite accurately by four parameters:

1. Power output (in kW or MW)
2. Efficiency (or heat rate in kJ/kWh or Btu/kWh)
3. Airflow (or exhaust flow), MEXH (in lb/s or kg/s)
4. Exhaust temperature, TEXH (in °F or °C).

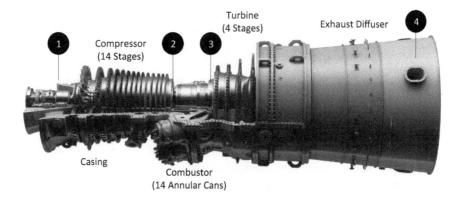

Figure 3.1 Modern heavy-duty industrial gas turbine architecture (GE's 7HA, Courtesy: General Electric).

These four parameters define the *control volume* around the gas turbine and establish the heat (energy) and mass balance thereof. In addition to these four parameters, there are two important parameters that define the gas turbine cycle: *cycle pressure ratio* (PR) and *turbine inlet temperature* (TIT).

The next important gas turbine parameter is HEXH, the exhaust gas *enthalpy*, which is a function of TEXH and exhaust gas composition. Enthalpy can be rigorously calculated using an *equation of state* (EOS) such as JANAF, which is indeed how it is done in heat balance simulation software. For quick estimates, a useful linear function is adequate, i.e.,

$$\text{HEXH} = 0.3003 \cdot \text{TEXH} - 55.576, \tag{3.1}$$

with a zero Btu/lb enthalpy reference of 59°F and H_2O in the mixture in gaseous form. With TEXH in °F, Equation (3.1) returns HEXH in Btu/lb (1 Btu/lb = 2.326 kJ/kg). Equation (3.1) can be used for the temperature range 900–1,200°F (480–650°C) with reasonable accuracy. The error resulting from using these simple equations in lieu of bona fide property calculations should not be more than ±1–2%. The information implicit in Equation (3.1) is the specific heat of the gas turbine exhaust gas, i.e., $c_p = 0.3003$ Btu/lb-R (1.2573 kJ/kg-K), which is approximately, but not exactly, constant in the range of its applicability.

This is a good place to introduce the concept of *exergy* (*availability* in US textbooks), which is a thermodynamic property and can be rigorously calculated using a suitable EOS with known pressure P, temperature T, and composition y_i (i.e., mole fraction of component i). The exergy (using the symbol *a* for availability), $a(P, T, y_i)$, of a material stream quantifies the theoretically possible maximum work production capability of that stream in a process, which brings it to an equilibrium with the surroundings (i.e., to a state defined by ambient P and T). Flow exergy of a nonreacting fluid stream (i.e., ignoring the chemical exergy) is given by

$$a = (h - h_0) - T_0(s - s_0), \tag{3.2}$$

where *a* (in Btu/lb or kJ/kg), h (in Btu/lb or kJ/kg), and s (in Btu/lb-K or kJ/kg-K) represent specific exergy, enthalpy, and entropy, respectively. The equilibrium state (sometimes referred to as the "dead state") is denoted by the subscript 0. The obvious reference / dead state, as stated earlier, is defined by the ambient conditions, e.g., 1 atm and 15°C (59°F) per ISO definition.

To a very good approximation, exhaust gas exergy of a gas turbine can be estimated from

$$\text{AEXH} = 0.1858 \cdot \text{TEXH} - 76.951, \tag{3.3}$$

with a zero Btu/lb enthalpy reference of 59°F and H_2O in the mixture in gaseous form. Zero entropy reference is defined as 59°F and 14.7 psia (since ideal gas entropy is a function of temperature *and* pressure). A very useful parameter is the ratio of exergy to energy (enthalpy). Using Equations (3.1) and (3.3), the data in Table 3.1 is generated as a function of gas turbine exhaust temperature. This is a very handy reference. It is essentially a measure of the ultimate bottoming cycle efficiency (i.e., a Carnot cycle).

Table 3.1 Gas turbine exhaust gas exergy/enthalpy ratio

Exhaust Temp., °F	Enthalpy, Btu/lb	Exergy, Btu/lb	E/Q, –
1,000	244.7	109.1	0.446
1,050	259.7	118.0	0.454
1,100	274.8	127.2	0.463
1,150	289.8	136.7	0.472
1,200	304.8	146.3	0.480

Note the improvement in maximum attainable efficiency with better exhaust gas energy "quality" (i.e., increasing exhaust gas temperature). A real cycle, e.g., a modern Rankine steam cycle would achieve only a fraction of this (0.7–0.8 depending on steam cycle design).

Let us use a vintage 50 Hz F class gas turbine, General Electric's (GE's) 9F.03 (formerly Frame 9FA), as a template for the numerical illustrations of the basic concepts. For 9F.03 with TEXH = 1,104°F (~596°C),

$$AEXH = 0.1858 \times 1,104 - 76.951 = 128.2 \, Btu\,lb = 298.1 \, kJ/kg,$$

so that the total exhaust gas exergy is 663 kg/s × 298.1 kJ/kg = 197.7 MWth. In other words, a Carnot-like bottoming cycle utilizing 9F.03 exhaust gas as a heat source would generate 197.7 MW power. This is the theoretical upper limit set by the second law of thermodynamics. In practice, only a fraction of that theoretical potential can be realized. This is given by the "exergetic efficiency" of the bottoming cycle, EEBC, which is a function of TEXH. For vintage F Class gas turbines like 9F.03, a reasonable value for the EEBC is 0.74. For advanced cycle, i.e., HA or J Class, gas turbines, the EEBC can be assumed to be 0.78 for quick estimates. Thus, continuing with the example of 9F.03, bottoming cycle output, BCMW, is

$$BCMW = 0.74 \times 197.7 = 146.3 \, MW.$$

This is the net output of the bottoming cycle, which includes the power consumed by the boiler feed pumps. For a 3PRH bottoming cycle, boiler feed pump power consumption is 1.9% of the steam turbine generator output, STGMW, which comes to

$$STGMW = 146.3/(1 - 1.9\%) = 149.1 \, MW.$$

Using Equation (3.1), 9F.03 exhaust gas enthalpy can be found as

$$HEXH = 0.3003 \times 1,104 - 55.576 = 276 \, Btu/lb = 642 \, kJ/kg,$$

so that gas turbine exhaust gas thermal power is

$$QEXH = 663 \, kg/s \times 642 \, kJ/kg = 425.6 \, MWth.$$

Thus, the first law efficiency of the bottoming cycle is

$$BCEFF = 146.3/425.6 = 34.4\%.$$

From the *air-standard* cycle analysis, the *ideal* bottoming cycle efficiency is found to be represented by

$$\text{EBCID} = 1 - \frac{\ln(X)}{X - 1},\tag{3.4}$$

$$X = \frac{T_3}{T_1}\text{PR}^{-k},\tag{3.5}$$

where PR is the gas turbine Brayton cycle pressure ratio, T_3 is the cycle maximum temperature, $k = 1 - 1/\gamma = 0.2857$ with $\gamma = 1.4$, and T_1 is the cycle minimum temperature (i.e., 15°C for ISO conditions). A good proxy for T_3 is the TIT, which for the vintage F Class is (nominally) 1,400°C. Using PR $= 16.7{:}1$ for 9F.03, from Equations (3.4) and (3.5)

$$X = \frac{1{,}673}{288}\,16.7^{-0.2857} = 2.6,$$

$$\text{EBCID} = 1 - \frac{\ln(2.6)}{2.6 - 1} = 0.403.$$

This is the efficiency of the bottoming cycle operating between the mean-effective heat rejection temperature, METL, of the gas turbine (i.e., the geometric average of TEXH, T_4, and T_1) and T_1 if it *were* a *Carnot engine*. In other words,

$$\text{METL} = T_1 \frac{X - 1}{\ln(X)},\tag{3.6}$$

$$\text{EBCID} = 1 - \frac{1}{\text{METL}}.\tag{3.7}$$

It can be readily seen that Equation (3.7) is equivalent to Equation (3.4). The actual bottoming cycle efficiency is 34.4%, or 0.344, so that the "technology factor" of this bottoming cycle is $0.344/0.403 = 0.85$. It is worth noting that although the first law efficiency of barely 35% seems to be a paltry number at first glance, it is, in fact, not too far from the theoretical entitlement.

It is also interesting to note that, from a purely computational perspective, we have not made any reference to the exact characteristics of the bottoming cycle (including the working fluid). This is also perfectly acceptable from a fundamental thermodynamic perspective. The ideal bottoming cycle is a heat engine operating in a Carnot cycle between two temperature reservoirs at METL and T_1. The parameter connecting this ideal cycle to the actual bottoming cycle is the *technology factor*. Strictly speaking, the Rankine cycle with H_2O as the working fluid is only one possibility, which has significant advantages so that it is the chosen option in practical applications. From a purely thermodynamic standpoint, however, *any* cycle with *any* working fluid is a candidate. (*Organic* Rankine cycle [ORC] is a well-known example.) With decades of design, development, and field operation under its belt, the current state of the art in the Rankine (steam) bottoming cycle is at such a stage of maturity that, as we have shown above, the technology factor, TF, in question is notably high, i.e., TF = 0.85. (Note that TF = 1.0 means a Carnot cycle, i.e., a practical impossibility.)

Let us look at the combined cycle itself with Brayton "topping" and Rankine "bottoming" cycles. This cycle can be represented by another Carnot cycle operating between two temperature reservoirs. The low temperature is the same, i.e., T_1. The high temperature is the mean effective heat addition temperature, METH, of the Brayton cycle, i.e.,

$$\text{METH} = T_2 \frac{X - 1}{\ln(X)}, \tag{3.8}$$

so that the ideal combined cycle efficiency is

$$\text{ECCID} = 1 - \text{PR}^{-k} \frac{\ln(X)}{X - 1}. \tag{3.9}$$

Continuing with the 9F.03 gas turbine example, the ideal combined cycle efficiency from Equation (3.9) is found to be

$$\text{ECCID} = 1 - 16.77^{-0.2857} \frac{\ln(2.6)}{2.6 - 1} = 0.733 \text{ or } 73.3\%.$$

In Gas Turbine World's (GTW) 2020 Handbook (**GTW2020** henceforth), 9F.03 combined cycle efficiency at ISO base load is listed as 59.1% ($1 \times 1 \times 1$ configuration, 3PRH steam cycle with 1.2 in. Hg condenser pressure). (**GTW2020** rating numbers for gas turbine, steam turbine, and net plant output are listed as 263.65, 153.4, and 412 MWe, respectively.) This translates into a TF of 59.1/73.3 = 0.81. The present state of the art of the gas turbine Brayton–Rankine combined cycle technology can be gleaned from the ISO base load rating data plotted in Figure 3.2.

The following simplified correlation can be written for the combined cycle *gross* efficiency, CCGEFF, which is the sum of gas and steam turbine generator outputs:

$$\text{CCGEFF} = \text{GTEFF} + (1 - \text{GTEFF})\text{BCEFF}. \tag{3.10}$$

Multiplying both sides of Equation (3.10) by gas turbine *heat consumption*, HC, we obtain the combined cycle gross power output, i.e.,

$$\begin{aligned} \text{CCMW} &= \text{GTMW} + \text{QEXH} \cdot \text{BCEFF}, \\ \text{CCMW} &= \text{GTMW} + \text{STMW}(1 - \text{BFP}), \end{aligned} \tag{3.11}$$

where BFP is the boiler feed pump power consumption as a fraction of steam turbine generator output. From Equation (3.11), the ratio of steam and gas turbine power outputs can be found as

$$\frac{\text{STMW}}{\text{GTMW}} = \frac{\text{QEXH}}{\text{GTMW}} \cdot \text{BCEFF} = \frac{(1 - \text{GTEFF})\text{BCEFF}}{\text{GTEFF}(1 - \text{BFP})}.$$

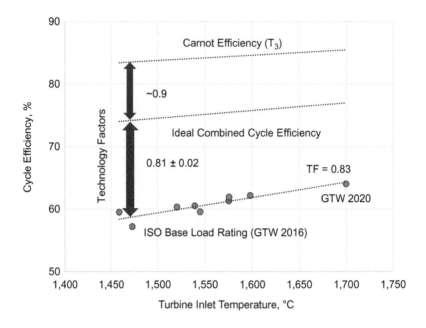

Figure 3.2 Gas turbine combined cycle efficiencies (ISO base load rating) as a function of turbine inlet temperature (TIT).

Using the values found earlier for our example with 9F.03, the right-hand side of Equation (3.11) is evaluated to be

$$(1 - 0.378)/0.378/(1 - 0.019) \times 0.344 = 0.577.$$

This simple calculation forms the basis of the well-known rule of thumb that, in a combined cycle power plant, steam turbine output is roughly half of the gas turbine output. In the 1990s and early 2000s, with vintage F Class gas turbines and steam cycles, this factor was indeed quite close to 0.5. Modern steam cycle technology with increased steam pressures and temperatures pushed this factor closer to 0.6. (From the published rating values, the ratio is 153.4/263.65 = 0.582.) For quick mental evaluations, the basic rule of thumb is still a useful tool in our proverbial toolbox.

3.1.1   Turbine Inlet Temperature

Strictly speaking, the highest temperature in the gas turbine is the flame temperature in the combustor, which is well above 3,000°F (1,650°C). In practical terms, the logical choice is the combustor exit temperature after dilution with combustor liner cooling flow and before the inlet of the turbine section, i.e., the TIT. While simple at a first glance, the definition of TIT is subject to many qualifiers and assumptions. As such, unless one is 100% sure of what is meant by a particular TIT quotation, it is impossible to have a meaningful and coherent discussion and analysis. In this document, cycle maximum temperature is the temperature of hot combustion gas at the

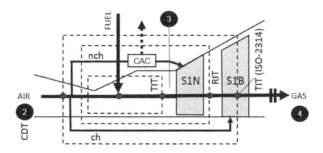

Figure 3.3 Gas turbine hot gas path (HGP) schematic (CDT: Compressor discharge temperature; CAC: Cooling air cooler; S1N: Stage 1 nozzle vane; S1B: stage 1 blade/bucket; TIT: Turbine inlet temperature; RIT: Rotor inlet temperature ["firing" temperature]; "ch" for chargeable and "nch" for nonchargeable cooling/secondary flows).

inlet of the stage 1 rotor or bucket (S1B in GE terminology) before it starts producing useful turbine work. This temperature is commonly known as the "Firing Temperature" and in E Class air-cooled gas turbines it is ~150–175°F *lower* than the hot gas temperature at the combustor exit (at the inlet of stage 1 vane or nozzle, S1N, i.e., the "true" TIT in essence). (In F Class GTs with much higher firing temperatures, the temperature delta is higher, i.e., more than 200°F). Thus, another term for it is *rotor inlet temperature* (RIT) where the rotor in question is the stage 1 rotor. Furthermore, it is about 100°C *higher* than the fictitious temperature as defined in ISO-2314 and adopted by European original equipment manufacturers (OEMs) (i.e., Siemens and Alstom). It is *imperative* to understand the distinction between different definitions to correctly interpret the severity of a particular technology upgrade with impact on firing temperature and compressor PR.

The relationship between TIT, RIT (firing temperature) and the fictitious technology indicator ISO-2314 TIT is explained graphically in Figure 3.3.[1] As shown in the graphic, TIT is at the inlet of S1N, which is cooled by airflow extracted from the compressor, which is denoted by "nch" (for "nonchargeable" flow). Thus, RIT is the hot gas temperature at the S1N exit, which is diluted by the spent cooling flow. Cooling flows downstream of RIT inlet are denoted by "ch" for "chargeable." The terms chargeable and nonchargeable should be understood in the context of work production ability of the cooling air stream. Nonchargeable cooling flows enter the gas flow path upstream of S1B so that it cannot be charged as "lost work." Chargeable flows enter the gas flow path downstream of S1B and they can be charged as lost work with the magnitude of lost work dependent on where they enter the flow path.

Due to pressure requirements, nonchargeable cooling flows are extracted from the compressor discharge. In advanced class gas turbines with high cycle PR, compressor discharge temperature (CDT) can be very high, e.g., as high as 900°F (480°C), with

---

[1] ISO 2314 "Gas Turbines – Acceptance Tests" is the international standard defining the basis and procedures for rating and testing gas turbines.

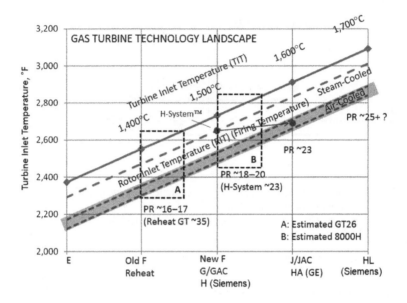

Figure 3.4 Gas turbine class hierarchy.

reduced cooling ability. One solution adopted by major OEMs (e.g., Siemens or MP^2) is to cool the extracted air in an external heat exchanger or *cooling air cooler* (CAC) by making low pressure steam to be used in the bottoming cycle.

As shown in Figure 3.3, the control volume used for calculating the fictitious TIT per ISO-2314 encompasses all HGP cooling flows (also referred to as *secondary* flows). Its advantage is that it can be easily calculated using compressor inlet airflow and fuel flow whereas the actual TIT and RIT calculation requires knowledge of secondary flows, which are highly proprietary and not disclosed by the OEMs.

Class hierarchy of heavy-duty industrial gas turbines is summarized in Figure 3.4. Simple Brayton cycle machines with single-shaft architecture are denoted by letters, e.g., E, F and H. They are also referred to as "frame" machines, e.g., GE's "Frame 6," etc. There are three major variants:

1. HGP components open loop air-cooled, which includes E, F, G, H, HA, and JAC Class machines offered by all major OEMs
2. HGP components closed loop steam-cooled, which includes GE's H-System (commercially not offered anymore), Mitsubishi Power's G and J Class gas turbines
3. Sequential (Reheat) Combustion, i.e., former Alstom's GT 24/26 (now owned by GE and Ansaldo)

---

^{2} Mitsubishi Power – formerly MHI (Mitsubishi Heavy Industries) – is the major Japanese OEM of gas turbines.

These machines come in 50 Hz (3,000 rpm) and 60 Hz (3,600 rpm) variants. They are typically rated at outputs above 100 MWe. The largest 60 Hz machines can be rated at as high as 400 MWe; the largest 50 Hz machines are rated above 500 MWe. Gas turbines with TITs of 1,600°C, i.e., H and J Class, are *advanced class* machines introduced around 2010 or later. The latest advanced class machines commercially offered at the time of writing is the Siemens HL Class. For an in-depth discussion of the best-in-class gas turbine products offered by major OEMs, refer to **GTFEPG** (chapter 21).

As a side note, gas turbine class nomenclature became somewhat confusing since 2010. The term H technology or H Class is now used for advanced F Class (air-cooled) units with TITs of 1,500°C (2,732°F) or higher. These machines are thus capable of reaching firing temperatures of 2,600°F, which were previously possible only with the steam-cooled H-System of GE (not a commercial offering anymore). In fact, Irsching power plant in Germany with its air-cooled Siemens H class gas turbine broke the 60% efficiency barrier in 2011.

The J Class (its air-cooled variant is referred to as JAC) is capable of 1,600°C (2,912°F) turbine inlet temperature (or higher). It is referred to as the HA (i.e., air-cooled H) Class by GE (for "air-cooled H Class" to differentiate from GE's steam-cooled H-System). Note that the J Class gas turbines retain steam cooling for the combustor transition piece and first stage turbine ring. Along with high firing temperatures and cycle pressure ratios (about 23:1), J Class gas turbines are characterized by high air flow rates and power ratings (almost 500 MWe for the 50 Hz units).

3.1.2   Small Industrial Gas Turbines

In addition to the simple Brayton cycle gas turbines, i.e., single-shaft, utility scale, heavy-duty industrial machines discussed above, there are smaller gas turbines (rated maximum at around 100 MWe but mostly 50–60 MWe or less), which are generally multi-shaft machines based on modified aircraft jet engines. These gas turbines are known as *aero-derivatives*, e.g., GE's LM (Land and Marine) Class or Pratt & Whitney's (now owned by Mitsubishi Power) FT Class *packaged* units. In addition to the aero-derivatives, there are also small multi-shaft industrial gas turbines, which have no aircraft jet engine pedigree but have similar architecture. This architecture typically consists of a *balanced shaft* combining an axial compressor with its driver high pressure (HP) turbine. This component is commonly referred to as a *gas generator* and in aero-derivatives it is in fact the original jet engine. In a jet engine, the useful output is *thrust*, which is created by the gas generator exhaust gas being ejected through a nozzle. In the stationary aeroderivative variant, the exhaust gas is utilized to generate shaft power via a *power turbine*. The shaft output is either used for electric power generation via a synchronous ac generator or for propulsion of a land or marine vessel (hence the LM designation).

Aeroderivative gas turbines are characterized by high cycle PR (above 30:1, 40:1 for Siemens/Rolls-Royce Trent) and high efficiency (>40%). This can be easily deduced from the ideal cycle T-s diagram in Figure 2.3. The corollary of this characteristic, as discussed earlier, is low specific power output. Since the original gas turbine is designed for aircraft propulsion applications, where a small frontal area is of prime importance to reduce drag, these machines typically have low airflow and high shaft speed (rotational) to minimize engine diameter.

Beyond the basic gas generator plus power turbine architecture with two shafts, there are multi-shaft aero-derivatives such as GE's LM6000 and Rolls-Royce Trent with HP and low pressure (LP) gas generators connected through concentric shafts and GE's LM100 with intercooled compressor. For an in-depth discussion of aeroderivative and small industrial gas turbines, the reader is referred to **GTFEPG** (chapter 23).

Due to their high cycle PR (low exhaust gas temperature) and low air/exhaust gas flow rates, aero-derivatives are not ideally suitable to combined cycle applications. (The thermodynamic signature of this characteristic is low exhaust gas exergy.) Thus, the rule of thumb introduced earlier, i.e., steam turbine output being equal to 50% (at least) of gas turbine output does not apply in this case. Let us look at this using an example, i.e., GE's 50-MWe LM6000 aeroderivative with 42.1% efficiency, 292.4 lb/s (132.6 kg/s) exhaust mass flow and 864°F exhaust temperature. From Equation (3.3), exhaust gas exergy is found as

$$AEXH = 0.1858 \times 864 - 76.951 = 83.6 \text{ Btu lb} = 194.4 \text{ kJ/kg},$$

so that the theoretical maximum of bottoming cycle power output is

$$132.6 \times 194.4/1,000 = 25.8 \text{ MW}.$$

It is not possible to design a cost-effective advanced steam cycle for such low exhaust flow and temperature resulting in low steam temperature and pressure as well steam flow (leading to low steam turbine isentropic efficiency, e.g., 78–79%). Maximum steam turbine output is unlikely to exceed 15 MW or so. Thus, the ratio is $15/50 = 0.3$ at most. In conclusion, aeroderivative gas turbines with efficiencies well above 40% in simple cycle are not players in highly efficient combined cycle realm (>60% efficiency). However, they are very well suited to small cogeneration applications.

3.1.3   Technology State of the Art

At this point, the reader should be equipped with simple but physics-based tools to assess gas turbine technology. The main driver behind the gas turbine technology advances over the last century or so for higher simple and combined cycle thermal efficiencies and power outputs is the TIT, which progressed from 1,200°C to 1,700°C over a 40 year period starting from the early 1980s (e.g., see Figure 3.5 for the evolution of the firing temperature, which is a proxy for TIT). It is thus obvious that HGP materials that can withstand such temperatures with the help of cooling air flows and *thermal barrier coatings* (TBC) constitute the foundation on which the technology is built. Equally important in component life with high resistance to creep and *low cycle fatigue* (LCF) is the casting technique used during part manufacturing.

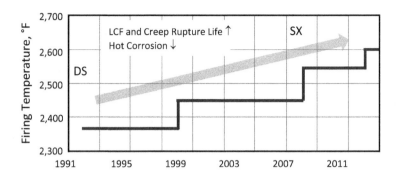

Figure 3.5 Evolution of F Class gas turbine firing temperatures.

Elimination of random grain structure (*directional solidification*, DS) and ultimately grain boundaries (*single crystal*, SX) resulted in parts with high temperature strength, increased resistance to creep fatigue, and LCF.

Turbine HGP comprises stationary and rotating components of the turbine stages. (The combustor itself and the transition piece between the combustor and the turbine can be counted as a part of the HGP.) These components are referred to as *stators/vanes/nozzles* (stationary components) and *rotors/blades/buckets* (rotating components) by different OEMs and investigators. In essence, they refer to the airfoils arranged around a circular platform (wheel) connected to the turbine shaft. Stationary airfoils convert thermal energy to kinetic energy while directing the hot gas flow into the rotating airfoils. High-speed, hot gas action on the rotating airfoils is analogous to the action of wind on the sail of a sailboat. The force resulting from the pressure difference between the "suction" and "pressure" sides of the airfoils pushes them so that the wheel on which these airfoils are attached (and the shaft on which the wheel is attached) rotates.

As mentioned above, these components operate at very high temperatures. In the direction of the hot gas flow from the combustor exit to the turbine exhaust, stage 1 experiences the hottest gas temperature and the last stage (3, 4, or 5 depending on the gas turbine model) experiences the lowest gas temperature. For the E Class machines, gas temperatures reach ~2,375°F (~1,300°C) at stage 1 stator inlet; for modern F Class machines this number is between 2,700°F (1,482°C) and nearly 3,000°F (~1,650°C) depending on the vintage of the unit in question. Survival of the HGP components under these conditions is facilitated primarily by three means:

1. Nickel-based superalloys used in manufacturing the HGP components
2. Cooling of nozzles and buckets (and their "platforms" where they are attached to the wheels and the wheels themselves)
   a. Using air extracted from the compressor discharge and intermediate stages (*open* loop)
   b. Using steam extracted from the bottoming cycle (*closed* loop)
3. Thermal barrier coating (TBC)

TBC is essentially a low thermal conductivity *ceramic* "armor" put on the "naked" component metal to increase its resistance to hot gas temperatures. In other words, TBC increases the allowable metal temperature (by about 200°F or 100°C) and, everything else being the same, it affords a concomitant rise in firing temperature, which is the key parameter associated with performance improvement of gas turbine performance as stated earlier. Depending on the gas turbine class and vintage, gas temperatures at the stage 1 rotor inlet (that is, the firing temperature) is lower by anywhere between 150°F and 250+°F (65°C to 120+°C) as a result of cooling stage 1 stator (vane).

There are two major types of turbine HGP cooling techniques:

1. Open loop air cooling
2. Closed loop steam cooling

The latter is only available in GE's now-defunct H-System™ and, to a limited extent, in Mitsubishi's G and J Class gas turbines. At the time of writing (2020–2021), most heavy-duty industrial gas turbines are cooled by open-loop systems with air extracted from the compressor discharge and intermediate stages. The label "open loop" indicates that the coolant air, once having done its job of cooling, exits the HGP component and joins the hot gas flowing through the turbine (thereby *diluting* it).

State of the art cooling techniques and the underlying fundamental physics for stationary components such as stators (vanes) are covered in depth in **GTFEPG** (chapter 9). In essence, whether a stationary vane or rotating blade or bucket, the component has a hollow core with inserts (vanes), or serpentine cooling passages lined with "rib turbulators" (blades). The coolant is extracted from the internal channel for *jet impingement* (leading edge) and *pin-fin cooling* (trailing edge). Pin-fins constitute an array of short cylinders, which increases the heat transfer area while effectively mixing the coolant air to lower the wall metal temperature of the component. After impinging on the walls of the airfoil, the coolant exits the vane and provides a protective film on the components' external surface. Similarly, the coolant traveling through the pin-fin array is ejected from the trailing edge of the airfoil. Obviously, the techniques used to cool the blades are similar to those used to cool the vanes. However, the heat transfer processes in the vanes and blades are quite different. Since the blades are rotating, the flow of the coolant in the passages is altered. Therefore, the effect of rotation on the internal heat transfer enhancement must be considered.

3.1.4   Correction Curves

Off-design performance of gas turbines can be estimated using correction curves. Typically, gas turbine OEMs provide those curves with their equipment. While there are differences between different types and classes of gas turbines and between the OEMs for gas turbines in the same product category, reasonably good estimates can be made using generic (i.e., normalized) curves. This is mainly due to the simple operating principle of a gas turbine running at a constant speed when synchronized to the grid. At constant rotational speed, the gas turbine has a constant swallowing capacity, i.e., the volumetric flow rate of air at the compressor inlet is constant. Its mass flow rate, however, changes with air density, which, in turn, changes with changing ambient conditions, i.e., temperature and relative humidity.

At constant ambient conditions, gas turbine output can be changed by (i) changing inlet airflow (via inlet guide vanes [IGVs]), (ii) by changing the fuel flow rate (i.e., turbine inlet temperature [TIT]), or (iii) a combination of the two. How this is done is mainly a question of whether the gas turbine is run in a simple or combined cycle mode. In the latter mode, the objective is to maintain high bottoming cycle efficiency, which is a strong function of gas turbine exhaust temperature (TEXH). It follows that the first control action in output reduction is to reduce the airflow by closing the IGVs progressively while maintaining the TIT. Reduced airflow, due to the constant swallowing capacity of the turbine section, will result in lower cycle pressure ratio (PR). As a consequence of constant TIT and a decreasing PR, from the basic thermodynamic relationship between turbine pressure and temperature ratios it can be easily deduced that TEXH will increase. This will ensure that the reduced combined cycle output will be generated at the best possible thermal efficiency.

The classic (or archaic, or conventional, you choose the terminology that you prefer) way of accomplishing this control is described in Figure 3.6. The gas turbine can be operated at two different levels of TIT, base or normal value for continuous operation or peak value for limited time (emergency) operation. The latter is usually an operator selected operating mode. It will differ from unit to unit but, if it is reached without operator intervention, it is a low alarm limit or close to it.

It is impossible to measure TIT directly. Consequently, inside the control system, it is an "inferred" parameter, i.e., calculated using other directly measured parameters, i.e., TEXH and PR. The performance ramification of the control philosophy described in

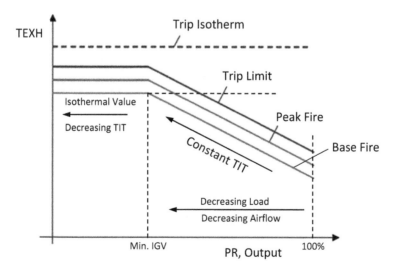

Figure 3.6 Gas turbine control curves.

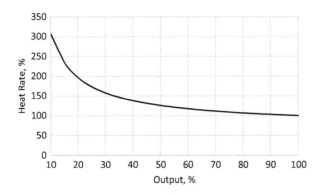

Figure 3.7 Gas turbine heat rate as a function of output.

Figure 3.6 is illustrated by the heat rate (HR) curve in Figure 3.7. (Note that HR is the inverse of thermal efficiency, i.e., HR = 3,600/η kJ/kg in SI units.) At the beginning, the efficiency deterioration is small because output and fuel flow both decrease in lock-step to maintain TIT and decreasing airflow. The main driver of efficiency reduction is component efficiency reduction at lower cycle pressure ratios. At still lower outputs, the rate of efficiency reduction (i.e., heat rate increase) increases when an increasingly larger portion of turbine output is spent on driving the compressor.

The other important off-design operation mode is at changing ambient conditions, i.e., pressure, temperature, and humidity. Pressure impact is negligible at low altitudes. It becomes significant at high altitudes via lower air density. At sea level, ambient pressure is 14.7 psia (1 atm). At an altitude of 6,000 ft, it drops to about 11.8 psia, and, at 100% load, gas turbine output, fuel consumption, and exhaust flow rate drop by about 20%. The rate of change is roughly linear, and interpolation can be used to estimate the performance loss at lower altitudes.

The impact of humidity on performance is similarly small. At ISO conditions (1 atm or 14.7 psia, 15°C or 59°F, and 60% relative humidity), specific humidity, ω, is 0.0064 lb of water vapor per lb of dry air. Increasing humidity leads to output loss via reduced air density. At ω = 0.021, output loss is 0.2%. Gas turbine efficiency decreases (i.e., HR increases) because of extra fuel burn to vaporize the added water vapor. At ω = 0.021, HR increase is about 0.55%. The trend is linear, and inter-polation (or extrapolation) can be used for different values. Specific humidity can be found from relative humidity using the formula,

$$\omega = \frac{0.622}{\frac{P_{air}}{RH \cdot P_v(T_{db})} - 1},$$ (3.12)

where P_{air} is the ambient air pressure, T_{db} is the dry bulb (ambient) temperature, and P_v is the water vapor pressure at T_{db}.

The cycle pressure ratio parameter has a strong impact on cycle thermal efficiency. Thus, pressure losses experienced at the gas turbine inlet and exhaust are critically important. In general,

- Each 4 in. H_2O (10.0 mbar) inlet pressure loss is worth −1.42% in output, +0.45% on HR, and +1.9°F (1.1°C) in exhaust temperature,
- Each 4 in. H_2O (10.0 mbar) exhaust pressure loss is worth −0.42% in output, 0.42% in HR, and +1.9°F (1.1°C) in exhaust temperature.

Generic dependences of key gas turbine performance parameters as a function of air temperature at the compressor inlet are provided in Figures 3.8 and 3.9. The trend in output and exhaust flow (sum total of airflow and fuel flow) with increasing temperature, i.e., decreasing air density, is obvious. The standard design point when normalized parameter values are unity corresponds to the ISO temperature (15°C or 59°F). At higher ambient temperatures, base load TIT is reduced in lock-step, e.g., as a rough approximation, if ISO TIT is 1,400°C, at 30°C ambient temperature, TIT becomes 1,400 − (30 − 15) = 1,385°C. At lower-than-ISO temperatures, base load TIT is held constant. The HR trend in Figure 3.8 reflects the detrimental impact of reduced cycle PR on cycle efficiency at high ambient temperatures as well as the favorable impact of higher cycle PR (due to higher gas flow at the turbine inlet with fixed swallowing

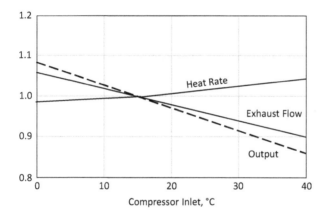

Figure 3.8 Change in output, heat rate, and exhaust flow with compressor inlet temperature (CIT).

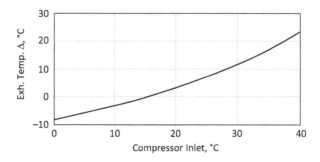

Figure 3.9 Change in exhaust temperature with compressor inlet temperature (CIT).

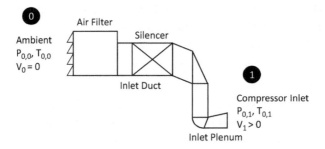

Figure 3.10 Gas turbine inlet air filter, silencer, and ducting.

capability). Similarly, the exhaust temperature trend in Figure 3.9 is directly related to the cycle PR.

Although this point is rarely elaborated upon in texts, it is rather important so that some ink is spent on it herein. Note that the x-axis in Figures 3.8 and 3.9 is labeled as "compressor inlet temperature" (CIT). In many references, these charts use the label "ambient temperature." As illustrated in Figure 3.10, ambient temperature is *not* the same as CIT.

Since the ambient air is stagnant, its temperature is the *static* temperature (i.e., air velocity, $V_0 = 0$ m/s). At the compressor inlet, on the other hand, air flows at a significantly high velocity so that its *total* temperature has two components, static and dynamic. The equation for the total (also known as *stagnation*) temperature, designated by the subscript 0, is

$$T_{0,0/1} = T_{0/1} + \frac{V_{0/1}^2}{2c_p}. \tag{3.13}$$

The first term on the right-hand side of Equation (3.13) is the *static* temperature, the second term is the *dynamic* temperature (c_p is the specific heat of air). At state-point 0, $V_0 = 0$ so that $T_{0,0} = T_0$. Using the definition of the speed of sound, $a = \sqrt{\gamma RT}$, Equation (3.13) can be rewritten as a function of Mach number, Ma, i.e., for state-point

$$T_{0,1} = T_1 \left(1 + \frac{\gamma - 1}{2} Ma_1^2 \right). \tag{3.14}$$

Assuming no loss in total temperature in the equipment and ducting upstream of the compressor inlet, $T_{0,1} = 15°C = 288.15$ K. If the Mach number at state-point 1, i.e., compressor inlet, is 0.4, we find that (with $\gamma = 1.4$, R = 287 J/kg-K)

$$288.15 = T_1 \left(1 + 0.2 \cdot 0.4^2 \right),$$

$$T_1 = 279.25 \text{ K} = 6.1°C,$$

$$V_1 = \sqrt{1.4 \cdot 287 \cdot 279.25} = 134 \text{ m/s}.$$

Clearly, when CIT and ambient temperature are used interchangeably, what is implicitly meant is the total temperature. What is noteworthy is that there is about a 9°C drop in static temperature between the gas turbine surroundings and the compressor inlet. The implication of this is that freezing can occur at the compressor inlet even if the weather conditions outside are well above freezing. This is where compressor *Inlet Bleed Heating* (IBH) system, which recirculates a fraction of the hot compressor discharge airflow into the compressor inlet, comes into play. In addition to preventing the formation of ice on the first stage stator blades, the IBH system achieves the following objectives:

- Extending emissions compliance to lower loads by allowing exhaust temperature control to occur at lower IGV angles.
- Providing sufficient compressor operating margin by reducing the compressor PR and by heating the compressor inlet air.

3.2   Heat Recovery Steam Generator

Let us start with the simple combined cycle efficiency formula derived in the preceding section, Equation (3.10), which is repeated below for convenience:

$$\text{CCGEFF} = \text{GTEFF} + (1 - \text{GTEFF})\,\text{BCEFF}. \tag{3.15}$$

This equation can be expanded as

$$\text{CCGEFF} = (\text{GTEFF} + (1 - \text{GTEFF})\text{HREFF} \cdot \text{SCEFF})(1 - \text{AUX}), \tag{3.16}$$

where the bottoming cycle efficiency, BCEFF, is the product of *heat recovery effectiveness*, HREFF, and *steam cycle efficiency*, SCEFF, i.e.,

$$\text{BCEFF} = \text{HREFF} \cdot \text{SCEFF}, \tag{3.17}$$

and AUX is the plant auxiliary load as a fraction total plant gross output, which is the sum of the prime mover generator outputs. In a combined cycle power plant, the HRSG is the connection between the gas turbine(s) and the steam turbine. It utilizes the thermal power of the gas turbine exhaust gas stream to make steam at high pressure and temperature to be utilized in a steam turbine for electric power generation. HRSG performance is quantified by the heat recovery effectiveness, HREFF, which is defined as

$$\text{HREFF} = \frac{\text{HEXH} - \text{HSTCK}}{\text{HEXH}} = 1 - \frac{\text{HSTCK}}{\text{HEXH}}, \tag{3.18}$$

where HSTCK is the gas enthalpy at the HRSG stack. Prima facie, Equation (3.18) is deceptively simple, but attention to the details is required. The most important one is the zero Btu/lb enthalpy reference temperature. The rigorous form of Equation (3.18) is

$$\text{HREFF} = \frac{\text{HEXH} - \text{HSTCK}}{\text{HEXH} - \text{HREF}}, \tag{3.19}$$

where HREF is a *reference* enthalpy. There are myriad possibilities to determine HREF, e.g.,

- Enthalpy of gas at the ambient temperature, TAMB
- Enthalpy of gas at feed water inlet temperature, TFWIN
- Zero at TREF = 59°F (15°C) or 77°F (25°C) or another value

Depending on the steam turbine condenser pressure, TFWIN is anywhere between 85°F and 110°F (~29°C and 43°C). As such, except on a very hot summer day, it is higher than TAMB. From a heat exchanger design perspective, TFWIN is the logical choice because it is physically impossible to cool the exhaust gas below that temperature. However, it is somewhat arbitrary because it assumes a specific condenser pressure. On the other hand, combined cycle ISO base load rating performances are typically quoted at 1.2 in. Hg (41 mbar) condenser pressure with an open loop water-cooled configuration (to minimize plant auxiliary load and maximize net output). Furthermore, there is also some arbitrariness in the zero enthalpy reference temperature. At the end of the day, the point that the author is trying to make here is that, when using Equation (3.19), it is of utmost importance to (i) verify the definition of HREF and (ii) ensure consistency in HEXH and HSTCK used in its evaluation with that definition. Herein, the assumptions are

- HREF = 0 Btu/lb at TREF = 59°F (15°C)
- HEXH and HSTCK are both 0 Btu/lb at TREF = 59°F (15°C)

A reliable estimation for HEXH, which is crucial to the gas turbine heat/energy balance, was provided by Equation (3.1). A similar formula for HSTCK in Btu/lb as a function of stack temperature, TSTCK, is given by

$$\text{HSTCK} = 0.2443 \cdot \text{TSTCK} - 13.571, \tag{3.20}$$

with TSTCK in °F and a zero Btu/lb enthalpy reference of 59° and H_2O in the mixture in gaseous form. Equation (3.20) is reasonably accurate for the temperature range 160–225°F (70–110°C). The information implicit in Equation (3.20) is the specific heat of the HRSG stack gas, i.e., $c_p = 0.2443$ Btu/lb-R (1.0228 kJ/kg-K), which is approximately, but not exactly, constant in the range of its applicability.

For the 9F.03 gas turbine example in the preceding section, with TEXH = 1,104°F (596°C), TSTCK in a three-pressure reheat steam cycle configuration would be around 180°F (~82°C). Thus, using Equations (3.1), (3.18) and (3.20), HREFF is calculated as

$$\text{HEXH} = 0.3003 \cdot \text{TEXH} - 55.576 = 276 \text{ Btu/lb},$$

$$\text{HSTCK} = 0.2443 \cdot 180 - 13.571 = 30.4 \text{ Btu/lb},$$

$$\text{HREFF} = 1 - \frac{30.4}{276} = 0.89 \text{ or } 89\%.$$

We had calculated the bottoming cycle efficiency as 34.4% in Section 3.1 so that, from Equation (3.17), steam cycle efficiency can be found as

$$\text{SCEFF} = \frac{0.344}{0.89} = 0.387 \text{ or } 38.7\%.$$

In-depth coverage of HRSG and bottoming cycle thermodynamics can be found in **GTCCPP** (chapter 6) and **GTFEPG** (chapter 17). In a nutshell, from Equation (3.15), it is obvious that maximizing the combined cycle efficiency is equivalent to maximizing the bottoming cycle efficiency. Maximizing bottoming cycle efficiency, as evidenced by Equation (3.17), requires striking a balance between HRSG effectiveness and steam turbine efficiency. It can be easily shown, graphically as well as numerically, that, everything else being equal, maximum combined cycle thermal efficiency is achieved at maximum heat recovery, at which point the bottoming cycle exergetic efficiency is also a maximum – as it should be (but the bottoming cycle *thermal* efficiency is not!). While this might seem counterintuitive, it reflects the dichotomy implicit in a bottoming cycle, i.e.,

- "Assistant" to the topping cycle (exhaust heat recovery)
- "Solo player" in its own right (bottoming cycle overall efficiency)

As an "assistant," the worth of the bottoming cycle is measured by the lowness of the exhaust gas stack temperature (maximum heat recovery effectiveness). For the cycle itself, however, as a "solo player," low stack temperature means low mean-effective cycle heat addition temperature, which, of course, hurts the cycle efficiency. Since the goal is the best possible *combined* cycle efficiency, exhaust gas heat recovery is more important than the bottoming cycle efficiency. This is quantitatively demonstrated in references cited above. Even with a high conversion (thermal) efficiency, however, a low heat recovery bottoming cycle with high TSTCK, can only achieve a low fraction of the theoretically possible maximum quantified by the lower-right triangular area {1-4-4C-1} on the T-s diagram in Figure 2.3 (i.e., low bottoming exergetic efficiency).[3] Only a high heat recovery bottoming cycle with the lowest possible TSTCK can approach the theoretical maximum (i.e., highest bottoming cycle exergetic efficiency) even with a more modest thermal efficiency.

In a coal-fired power plant with a large utility boiler, steam is generated at only one pressure and superheated to a high temperature (usually twice in cycles with *reheat*). The highest possible efficiency is achieved by feed water heating to maximize mean-effective heat addition temperature, i.e., ignoring reheat

$$\overline{\text{T}} = \frac{\text{TSTM} - \text{TFWIN}}{\ln\left(\frac{\text{TSTM}}{\text{TFWIN}}\right)}. \tag{3.21}$$

Obviously, this goal requires highest possible TFWIN. This, however, is very detrimental to combined cycle efficiency because high TFWIN translates into high TSTCK and low heat recovery. If the HRSG is also limited to steam generation at only one

---

[3] The theoretical maximum combined cycle efficiency is when the gas turbine exhaust exergy (lower-right triangular area below the topping cycle) is completely converted into useful work – a practical impossibility.

pressure, this requires that the steam pressure should be as low as possible to ensure that TSTCK is as low as possible, which, unfortunately is detrimental to steam cycle efficiency, SCEFF. If, on the other hand, steam pressure is kept as high as possible to maximize SCEFF, TSTCK becomes very high and hurts the heat recovery effectiveness and overall bottoming cycle efficiency. The solution to this dilemma is obvious: steam generation at multiple pressures, i.e., at least two, one high for high SCEFF, the other low for high HREFF. This is the reason that bottoming cycles with heavy-duty industrial gas turbines have at least two steam generation pressures: high pressure (HP) and low pressure (LP). The other steam cycle enhancer, via increasing mean-effective heat addition temperature, is reheat, where HP turbine exhaust steam is superheated to the same level as the HP steam temperature before admission into the LP turbine.

To have enough "oomph" to accomplish steam generation in meaningful quantities accompanied by reheat superheating, the gas turbine must be large enough (i.e., high exhaust gas mass flow rate, MEXH) with high enough TIT (i.e., high exhaust gas temperature, TEXH). This fact determines the HRSG and steam cycle design hierarchy from E Class to the most advanced H Class gas turbines:

- One-pressure with reheat (1PRH) or no reheat (1PNRH) for small industrial or aeroderivative gas turbines with low MEXH and TEXH
- Two-pressure with reheat (2PRH) or no reheat (2PNRH) for E Class gas turbines
- Two-pressure with reheat (2PRH) or three-pressure no reheat (3PNRH) for vintage F Class gas turbines
- Three-pressure with reheat (3PRH) for advanced class gas turbines

The main components of a 3PRH HRSG are identified in Figure 3.11. The HRSG shown in the figure is the most common design with horizontal arrangement of heat exchange sections in the direction of gas turbine exhaust gas flow from left to right. Steam boiling sections (evaporators) are of the *drum* type. There are other variants as well, e.g., the *vertical* design with vertical arrangement of heat exchange sections in the direction of gas turbine exhaust gas flow from bottom to top. There is also the *Benson* type HRSG with the HP evaporator of *once-through* type with the steam drum replaced by a much smaller *separation bottle*. In the drum-type evaporator of a horizontal, *natural* circulation of boiling water is accomplished by gravity (downward) and buoyancy (upward). In the older variants of vertical designs, *forced* circulation via a *circ pump* was necessary. Separation of saturated steam and water is accomplished in the steam drum. In the once-through evaporator design, such separation is not necessary, and circulation does not rely on gravity in the vertical variants. As such, once-through evaporator designs such as the Benson can accommodate supercritical steam pressures. Modern HRSGs of either design can accommodate subcritical pressures up to about 185 bar.

The basic theory outlined above can be graphically seen in the "heat release" diagram of a 3PRH HRSG in Figure 3.12. There are three steam production regions in the HRSG: at high, intermediate (or medium), and low pressure, HPEVAP, IPEVAP, and LPEVAP, respectively. Note the proximity of the LP section to the

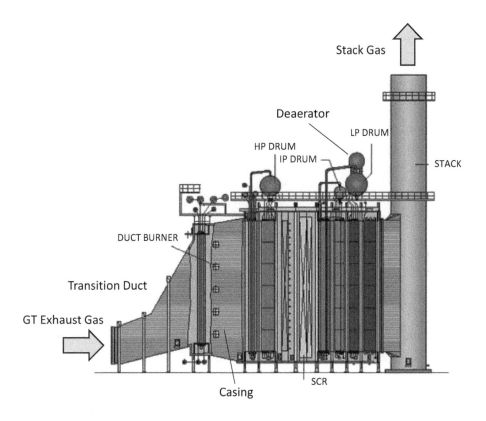

Figure 3.11 Three-pressure reheat (3PRH) HRSG.

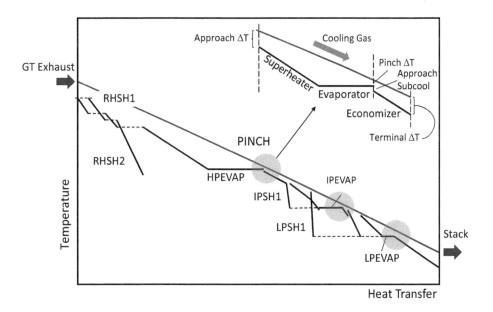

Figure 3.12 Typical heat release diagram of a 3PRH HRSG.

stack to ensure lowest possible TSTCK. The location of the HP section is determined by the highest possible steam pressure, i.e., about 190 bar (2,756 psi), and corresponding saturation temperature, which is around 690°F (365°C) vis-à-vis 630°C or higher for modern gas turbine exhaust gas temperatures. Each steam production process comprises three types of heat exchanger: economizer (feed water heating), evaporator (boiling at constant pressure and temperature), and superheater. There are two types of superheaters: HP steam superheater (heating HP steam) and reheat steam superheater (heating HP turbine exhaust steam mixed with IP steam produced in the HRSG).

The three evaporators are easy to identify by the constant temperature boiling lines and the "pinch" points or, simply, pinches. The pinch refers to the minimum temperature difference (or Δ, *delta*) between the *hot* fluid (i.e., gas turbine exhaust gas) and *cold* fluid (i.e., boiling water/steam). It has two contributors: (1) temperature delta between gas temperature at the evaporator exit (saturation temperature at the boiling pressure) and boiling water, which is the "pinch delta," (2) temperature delta between saturated steam and the feed water temperature at the exit of the economizer, which is the "approach subcool."

There are three pinches in the heat release diagram corresponding to the three steam evaporators. Typical values for pinch delta for advanced designs is 12–15°F (6.7–8.3°C); for approach subcool, it is 5–10°F (2.8–5.6°C). Strictly speaking, the economizer can be designed for 0°F approach subcool but the downside is the possibility of water boiling in the economizer tubes or "steaming." For a once-through boiler, approach subcool is 0°F by definition.

Superheaters are specified by their "approach" delta, which is the temperature difference between the exhaust gas at the superheater inlet and steam temperature at the exit. Typically, 30°F (16.7°C) is the minimum for the front-end steam superheaters (HP and reheat steam) and 20°F (11.1°C) for the internal ones (IP and LP superheaters). Designs with smaller temperature deltas (pinch, approach, etc.) result in more steam production with higher temperatures but at the expense of heat transfer surface area, i.e., size and cost. A fine balance also exists between the HRSG size and the gas pressure drop, which has a negative impact on gas turbine via higher backpressure. Typically, modern designs with large gas turbines are designed to have gas pressure losses between 12 and 15 in. H_2O (about 30 and 37 mbar).

To achieve maximum performance at the lowest possible cost, heat exchange sections, which are vertically oriented tube banks with gas crossflow, are interleaved as shown in Figure 3.12. The heat transfer process is enhanced by *finned tubes* to minimize the tube number per row and/or tube rows per heat exchanger. Sometimes, certain sections are configured to be parallel rather than in series. Ultimately, the actual design is determined by the HRSG OEM to deliver specified steam flows at specified pressures and temperatures with lowest possible gas path pressure drop. Consequently, the numbers provided here should be considered as rough guidelines for thermodynamic design.

Since the gas turbine exhaust temperature is known, the parameter that defines the HRSG effectiveness is the stack temperature. For typical HRSG designs with two-pressure or three-pressure reheat, HRSG effectiveness (as a percentage) as a function of TEXH in °F can be found using the following linear correlations:

$$3\text{PRH: HREFF} = 73.728 + 0.0142 \cdot \text{TEXH} \quad \text{for } (1{,}000 - 1{,}200°\text{F}), \tag{3.22}$$

$$2\text{PRH: HREFF} = 58.585 + 0.0240 \cdot \text{TEXH} \quad \text{for } (1{,}000 - 1{,}125°\text{F}). \tag{3.23}$$

The correlations are obtained from detailed heat and mass balance simulations with the following assumptions:

- 1.2 in. Hg (about 41 mbar) condenser pressure (i.e., the condensate reaches the HRSG at less than 90°F).
- Gas enthalpies are evaluated with 59°F (15°C) zero enthalpy reference.
- 100% methane-fired gas turbine.

Once the HRSG effectiveness is obtained from Equation (3.22) or (3.23), one can calculate HSTCK from Equation (3.18) and TSTCK from Equation (3.20).

The key steam cycle parameters are main or HP steam pressure and temperature, which depend on the gas turbine exhaust temperature. Front-end superheater approach delta requirements set the steam temperature with limits imposed by HRSG tube, steam pipe and steam turbine casing, and rotor materials. For vintage E Class gas turbines from the 1980s and 1990s, the steam temperature limit was a maximum of 1,000°F (538°C) or less; for F Class gas turbines in the 1990s and early 2000s, this limit was 1,050°F (565°C). Advanced F, H, and J Class gas turbines with exhaust temperatures well above 1,100°F and new materials made steam temperatures as high as 1,112°F (600°C) feasible. At the same time, higher gas turbine exhaust gas temperatures paced by increasing TIT for improved simple and combined cycle efficiency led to increasing HP steam pressures as illustrated in Figure 3.13.

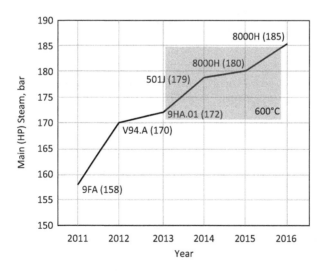

Figure 3.13 Main steam pressure progress.

Recently, super-heavy advanced class gas turbines with large exhaust gas mass flow rates (>2,000 lb/s, i.e., ~900 kg/s) and exhaust temperatures as high as 1,200°F (~650°C) or more, especially in multi-shaft configurations with two or three gas turbines, brought supercritical HP steam pressures into the realm of feasibility. This was not the case earlier and it is still not clear whether it can ever be cost-effective. For an in-depth discussion of supercritical steam generation in HRSGs, the paper written by the author is a good source ([36] in Section 2.4.3).

There are several important auxiliary systems inside an HRSG. For power augmentation on hot days when gas turbines lose power output via reduced airflow, duct burners installed in the front of the HRSG (usually between reheat superheater tube banks) increase exhaust gas temperature for increased steam production and steam turbine output. This technique, commonly known as supplementary firing, is widely deployed in the USA. For a detailed coverage of supplementary firing, the reader should consult chapter 6 in **GTCCPP** and [14] in Section 2.4.3.

Increasingly stringent regulations limit NOx emissions to levels as low as 2 ppmvd. Although modern DLN combustors are capable of single-digit NOx emissions at a broad range of loads and ambient temperatures, ultimately, *selective catalytic reduction* (SCR) is necessary to achieve the permitted levels. The SCR system (including the CO catalyst) is installed in the HRSG. The SCR catalyst design and location in the HRSG are based on the required NOx removal and expected flue gas temperature at the SCR location (300–350°C, thus, usually between the HP evaporator tube bundles). For a detailed description of the SCR technology and hardware, the reader should consult chapter 12 in **GTCCPP**.

Deaeration is the process to remove dissolved oxygen and carbon dioxide from the condensate/make-up water sent to the HRSG. In most modern combined cycles, deaeration is taken care of in the condenser. In some designs, the drum of the LP evaporator is equipped with an *integral deaerator*. The deaerator is a direct-contact feedwater heater that heats the boiler feedwater and removes the dissolved O_2 and CO_2. Heating and scrubbing functions are accomplished by utilizing steam from the LP evaporator, a steam turbine extraction point, or from the HRSG. If LP steam is not sufficient, additional *pegging* steam from a higher-pressure system (typically the IP evaporator/drum) is used to supplement it. For detailed information on deaeration, refer to chapter 10 of **GTCCPP**.

3.3   Steam Turbine

3.3.1   Combined Cycle Power Plant

Combined cycle steam turbines differ from the steam turbines used in coal-fired boiler plants in several aspects. For one, steam flow increases from the inlet of the HP turbine to the inlet of the LP turbine via two steam admissions at two different pressure levels, i.e., IP and LP. Except in seal packing leakages, there is no steam extraction for feed water heating. (In cogeneration power plants, steam can be extracted for other uses

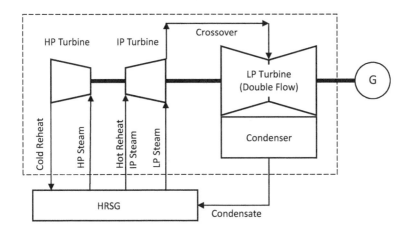

Figure 3.14 Steam turbine control volume for thermodynamic heat and mass balance.

such as district heating.) There are many excellent references that the reader can use for a deep dive into steam turbine technology going back to the monumental work of Aurel Stodola from the early twentieth century. For a reasonably short coverage with a focus on basic aerodynamics, performance, and architecture, including a list of important references, the reader is advised to consult **GTCCPP** (chapter 5).

A typical steam turbine *control volume* for thermodynamic analysis is shown in Figure 3.14. The important characteristics are:

• Opposed flow HP-IP turbine section (separate casings as shown)
• Double-flow LP turbine (to maximize exhaust annulus area)
• Generator connected at the LP end
• LP steam admission at the IP turbine exit
• Crossover pipe from the IP turbine exit to the LP turbine inlet
• IP steam mixed with hot reheat steam (IP turbine admission)

The actual hardware corresponding to the control volume in Figure 3.14 is shown in Figure 3.15. Interesting features are:

• Two casings
• Combined HP-IP turbine (casing 1)
• Double-flow LP turbine with down exhaust (casing 2)
• Three bearings (bearing #1 is combined journal-thrust bearing)

Depending on the turbine rating (steam flow and power output), which is a direct function of gas turbine size and number, there are different configurations, which are summarized in Figure 3.16. For smaller ratings (typically, 1×1×1 combined cycle with vintage E or F Class gas turbines), a single-flow LP turbine with axial exhaust is a feasible configuration (ST-3000). For the highest ratings (typically, 2×2×1 combined cycle with large J or H Class gas turbines), two double-flow LP turbines (i.e., a four-flow LP turbine) are required to provide the annulus area for such large steam flows

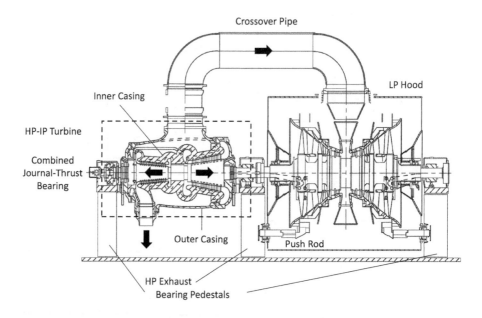

Figure 3.15 Steam turbine with combined HP-IP and double-flow LP casings.

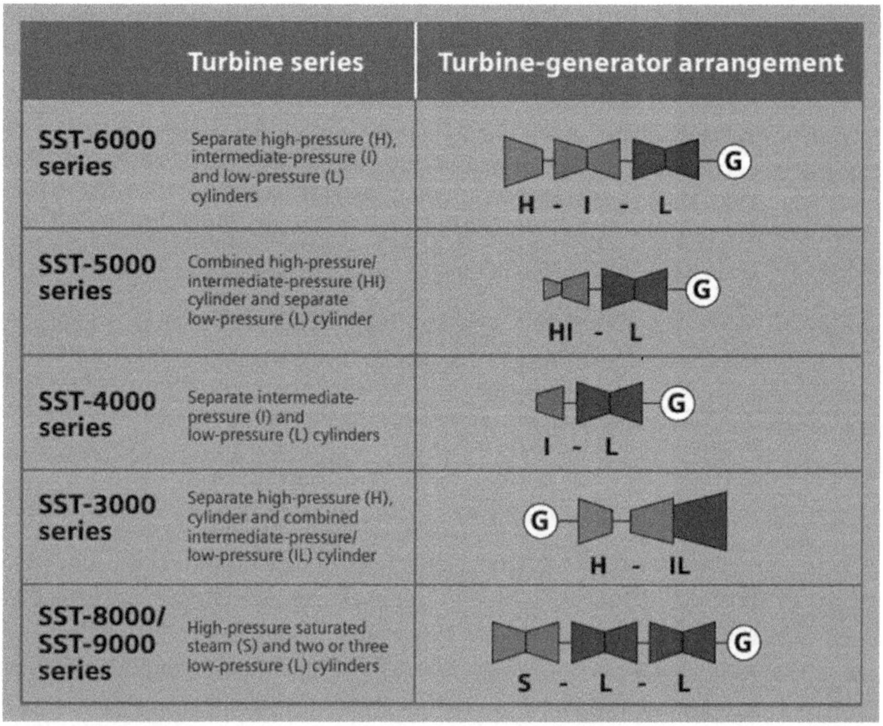

Turbine series		Turbine-generator arrangement
SST-6000 series	Separate high-pressure (H), intermediate-pressure (I) and low-pressure (L) cylinders	H - I - L
SST-5000 series	Combined high-pressure/intermediate-pressure (HI) cylinder and separate low-pressure (L) cylinder	HI - L
SST-4000 series	Separate intermediate-pressure (I) and low-pressure (L) cylinders	I - L
SST-3000 series	Separate high-pressure (H), cylinder and combined intermediate-pressure/low-pressure (IL) cylinder	H - IL
SST-8000/ SST-9000 series	High-pressure saturated steam (S) and two or three low-pressure (L) cylinders	S - L - L

Figure 3.16 Steam turbine generator arrangements (Courtesy: Siemens Energy).

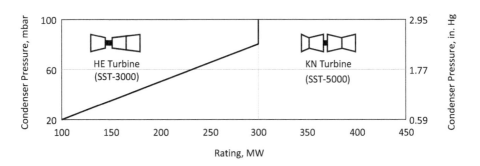

Figure 3.17 Applicability range of most common steam turbine configurations.

(SST-8000/9000). The applicability range of the two most common configurations is shown in Figure 3.17, which clearly illustrates the impact of condenser pressure on the size of exhaust annulus to limit exhaust losses at a given rating (i.e., steam flow).

Steam produced in the HRSG is a function of the thermal power transmitted to it by the gas turbine exhaust, which is a function of exhaust gas mass flow rate and temperature. Higher steam flows lead to higher steam turbine generator outputs and efficiencies (via higher volume flow) but require larger LP turbines with long last stage buckets. The driver behind this effect is the decreasing steam density and increasing volumetric flow rate as steam expands through the steam turbine to vacuum conditions at the LP turbine exit (i.e., less-than-atmospheric pressure). LP turbine exhaust pressure is determined by the steam condenser at the LP turbine exhaust, wherein steam is condensed using water or air as a coolant. Basic thermodynamics dictates the lowest possible condensing steam pressure (and temperature) to maximize thermal efficiency. Low exhaust pressures translate into large volume flows, which translate into high steam velocities at the LP turbine exit. For given steam mass flow rate, keeping the steam velocity at subsonic levels to minimize losses requires a large flow annulus area. This requirement drives the last stage bucket length beyond material strength (large centrifugal forces acting on the bucket). The obvious remedy is to have multiple LP steam flow paths so that the requisite large flow area is achieved with manageable bucket lengths. The dynamic described above is quantified by the chart in Figure 3.18. The calculation of annulus velocity, exhaust loss, and exhaust volume flow is described below.

The three steam turbine sections shown in Figure 3.14, i.e., HP, IP, and LP, are characterized by total section efficiencies including associated piping and valves. For example, HP and IP sections have their own steam stop-control valves at the inlet. Although these valves are combined valves with different duties, they are sometimes colloquially referred to as a "throttle valve." In contrast, LP steam joins the steam flow at the exit of the IP turbine. The inlet of the steam flow path of each section is referred to as "bowl." The isentropic expansion efficiency of each turbine section is referred to as "bowl-to-exhaust" steam path or "cylinder" efficiency. Section efficiency, on the other hand, is measured from the throttle valve inlet to the casing exit (including valve, inlet, and exit losses). Typical section efficiencies are listed in Table 3.2.

Table 3.2 Steam turbine section efficiencies

	HP, %	IP, %	LP, %
BASE	89.0	90.0	91.0
Unfired (see Ref. [1])	89.5	92.2	94.2
Fired (see Ref. [1])	89.4	92.3	92.7
2×1 50-Hz F (±1.5%)	90.1	91.1	91.4
Min–Max	86.9–91.9	87.8–92.9	88.1–93.2
2×1 60-Hz F (±2.5%)	87.8	88.8	89.1
Min–Max	84.0–92.0	85.0–93.1	85.2–93.4
1×1 50-Hz F (±1.6%)	88.7	89.7	90.0
Min–Max	85.9–90.7	86.8–91.8	87.1–92.1
1×1 60-Hz F (±2.6%)	86.5	87.5	87.7
Min–Max	82.6–90.3	83.6–91.3	83.8–91.6

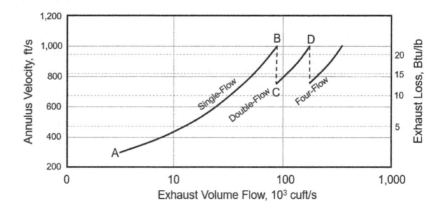

Figure 3.18 Accommodation of large steam flows with multiple LP flow paths.

For the last steam turbine section (the LP section) exhausting to the condenser, the efficiency is bowl to *used energy end point* (UEEP) and implicitly contains the exhaust losses. It can be modeled as a base value with a correction for the exhaust loss, i.e.,

$$\eta_{LP} = \eta_{base} - \lambda \cdot \text{TEL}, \tag{3.24}$$

where λ denotes the decrease in the base bowl-to-UEEP LP efficiency per unit increase in the total (dry) exhaust loss (TEL) (see [46] in Section 2.4.3). A good value is $\lambda = 0.25\%$, i.e., the η_{LP} decreases by 0.25 points for each Btu/lb (2.326 kJ/kg) increase in TEL, which is obtained from the exhaust loss curve (see Figure 3.19). The UEEP can be calculated from the *expansion line end point* (ELEP) as follows:

$$\text{UEEP} = \text{ELEP} + \text{TEL} \cdot (1 - 0.01 \cdot y) \cdot 0.87 \cdot (1 - 0.0065 \cdot y). \tag{3.25}$$

ELEP is the enthalpy found from the definition of the isentropic efficiency for the LP stage with known LP bowl (i.e., LP turbine inlet) conditions and exhaust (condenser) pressure.

In general, the optimum design point for the steam turbine exhaust end is not at the minimum TEL point (i.e., the *thermodynamic optimum*) but slightly to the right for the

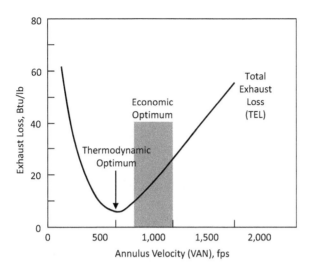

Figure 3.19 Typical exhaust loss curve.

steam velocity at the LP section exhaust (VAN for *annulus velocity*) of 600–700 ft/s (183–213 m/s, TEL of about 10 Btu/lb or 23 kJ/kg), which is the *economic optimum*. For generic calculations, this is sufficient. For a given steam turbine with known hardware, VAN can be calculated using the following equation

$$\text{VAN} = \frac{\text{MSTM} \cdot \text{VG} \cdot (1 - y)}{3{,}600 \cdot \text{AAN}}, \tag{3.26}$$

where VG is the specific volume of saturated vapor at the condenser pressure from the ASME steam tables in cuft/lb or m³/kg, and AAN is the exhaust annulus area in sqft or m². If not known exactly, exhaust moisture, y, can be taken as 8%. The annulus area in sqft can be estimated from a curve-fit to exhaust end curves from [44] in Section 2.4.3 as a function of the last stage bucket (LSB) length L in inches and the number of LP flow paths NLP as follows (from [46] in Section 2.4.3):

$$\text{AAN} = \text{NLP} \cdot 0.129 \cdot L^{1.7722}. \tag{3.27}$$

The total steam mass flow rate, MSTM, is proportional to the gas turbine exhaust flow (for all gas turbines if there is more than one), i.e.,

$$\text{MSTM} = \mu \cdot \text{MEXH}. \tag{3.28}$$

For advanced class gas turbines, the proportionality factor μ can be taken as 1.7. For example, for some large 60 Hz units, the exhaust flow rate can be 1,600 lb/s or even higher. Thus, for a 2×2×1 configuration, from Equation (3.25),

$$\text{MSTM} = 0.17 \times 2 \times 1{,}600 = 544 \text{ lb/s}.$$

For a condenser pressure of 1.2 in. Hg (~0.6 psia), VG is 540 ft³/lb. For an optimum VAN value of 700 ft/s, from Equation (3.26),

$$\text{AAN} = 544 \times 540 \times (1 - 0.08)/700 = 386 \text{ ft}^2.$$

The largest 60 Hz bucket available is probably 42 in. so that, from Equation (3.27),

$$\text{NLP} = 386/\left(0.129 \times 42^{1.77122}\right) \sim 4.$$

In other words, a four-flow LP turbine, i.e., similar to the SST-8000/9000 with two double-flow LP turbines in Figure 3.16, is required. If a larger last stage bucket is available, say, 48 in., can one get by with a double-flow LP turbine? Let us check:

$$\text{NLP} = 386/\left(0.129 \times 48^{1.77122}\right) \sim 3.$$

In theory, a three-flow LP turbine is possible but not seen in practice. In any event, the goal is to save installed equipment cost and reduce size by eliminating a second double-flow LP turbine. This does not seem possible with the specified condenser pressure. In fact, although a standard in calculating performance ratings for marketing "chest thumping" purposes, presently, it is practically impossible to permit a combined cycle power plant with an open loop water-cooled condenser that can achieve such a low pressure (1.2 in. Hg or 41 mbar). Especially, in the USA, stringent environmental regulations stemming from water scarcity and ecological preservation considerations make this patently impossible. For details, the reader is referred to chapter 12 in **GTCCPP**.

Modern combined cycle steam turbines designed for use with the advanced F, G, H, and J Class gas turbines have the following capabilities:

- Main (HP) steam pressures up to 190 bar (2,756 psi)
- Main (HP) and reheat (IP) steam temperatures up to 600°C (1,112°F)
- Up to 600 MWe generator output
- Last stage bucket (LSB) sizes up to 56 in. length (usually from titanium, not all OEMs) in 50 Hz
- LSB sizes up to 48 in. length (usually from titanium, not all OEMs) in 60 Hz

Construction characteristics (refer to Figure 3.15) are

- Large crossover pipe to minimize pressure loss
- Down or side exhaust options (to minimize pedestal and building height)
- Large selection of LSB blades to ensure optimal fit with the plant heat sink
- Single bearing between the HP-IP and LP shafts (better alignment)
- 3D aero design of HP and IP turbine blades with variable reaction for high aerothermal efficiency and optimal vibration damping (reaction designs characterized by large number of stages and shorter blades with larger rotor diameter)
- 3D aero design of long LSB with coatings to prevent erosion (via impact of condensing steam droplets)
- Double shell design (inner and outer casings) with spring-loaded labyrinth seals between the inner casing and rotating shaft

3.3.2   Boiler-Turbine Rankine Cycle Power Plant

These power plants used to be the workhorses of electric utilities until recently. They mostly utilized coal as the fuel in a giant industrial boiler to convert water to steam. That steam was used in a large steam turbine to generate electric power. Their capacities ranged from a few hundred megawatts to gigawatts. In fact, in the beginning of the twenty-first century, more than 50% of electric power generated in the USA was from coal-fired power plants with a capacity factor of more than 70%. At the time of writing (early 2021 for this chapter), that share has dropped to the low 20s as a percentage with a capacity factor of about 40%. Worldwide, however, coal remains the primary energy source for electricity generation, with a share of 36% due to its widespread use in Asian economies.

Due to its high CO_2 emissions (not to mention other quite nasty pollutants such as mercury), fossil-fired power plant technology is pretty much an *anachronism* in the developed world. (If it is to survive deeper into the twenty-first century, one would most likely have to add a carbon capture block to the mix.) Nevertheless, it still provides a useful guide to the understanding of the actual power plant performance as distinct from the performance of the underlying thermodynamic cycle. The key principles highlighted in this section should provide a useful guide to evaluating highly exaggerated claims made by the marketers of new "wonder cycles."

When one discusses the operation of a gas turbine, one does not make a distinction between the "heat engine" and the (ideal) thermodynamic cycle it is operating in, i.e., the *Brayton* cycle. They are one and the same. Admittedly, this is a problematic statement. For starters, the term *cycle* implies that the working fluid completes a *closed loop* during the process. This is indeed the case for the ideal, *air-standard* Brayton cycle. The actual gas turbine, however, (i) does not operate in a closed loop[4] and (ii) the working fluid changes during the process (i.e., from air to combustion products). Nevertheless, the Brayton cycle provides a quite reliable albeit idealized framework to analyze and quantify the operation of the gas turbine, i.e., the heat engine itself.

When it comes to the analysis of a steam turbine, the situation becomes somewhat less clear-cut. The ideal thermodynamic cycle describing the operation of the steam turbine is the *Rankine* cycle. As is the case with the ideal Brayton cycle, the working fluid, H_2O, completes a closed loop during the process. In contrast to the gas turbine Brayton cycle, however, this is indeed the case for the *real* system as well. The working fluid (i) stays the same (but it undergoes a phase change from liquid to vapor and then back to liquid) and (ii) starts and ends at the same thermodynamic state. The big difference between gas and steam turbine cycles is that the cycle heat addition process is an integral part of the heat engine in the former. Therefore, the

---

[4] Closed-loop or closed-cycle gas turbines were actively investigated in the past, and even made their way from the drawing board to the field installation. Ultimately, however, they ended up as commercial and (arguably) technical flops. For an in-depth discussion, refer to chapter 22 of **GTFEPG** and the monograph by Hans Frutschi [11].

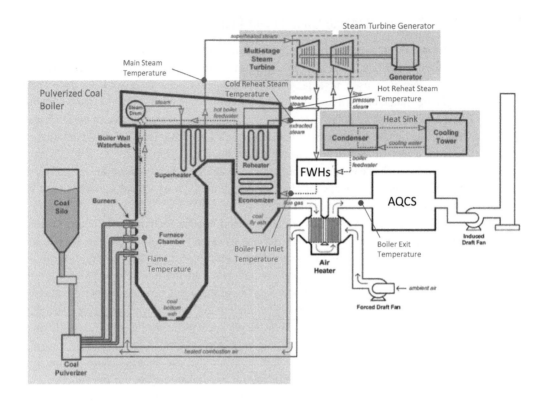

Figure 3.20 Pulverized coal-fired conventional steam power plant.

gas turbine is classified as an *internal combustion engine*. In the Rankine cycle, however, the cycle heat addition process forms the basis of a major piece of equipment, i.e., the *boiler*, which must be studied on its own and this can be done *outside* the cycle itself. Hence, the conventional steam plant operating in a Rankine cycle is an *external combustion engine*, wherein the steam turbine is only one of the *three* major pieces of equipment (the other two are the boiler and the condenser/cooling tower, i.e., the *heat sink*).

A typical Rankine steam cycle power plant with a coal-fired boiler is shown schematically in Figure 3.20. The corresponding cycle diagram on a temperature-entropy (T-s) surface is depicted in Figure 3.21. Key cycle state-points are identified on both. Note that there is in fact a *fourth* and very important subsystem in Figure 3.20, i.e., the AQCS,[5] which is critical to sustainable power generation from coal. However, the AQCS is strictly an *auxiliary* plant system and does not play a direct role in steam cycle thermodynamics. It is also worth noting that, from a purely thermodynamic cycle and heat engine analysis perspective, the nature of the fuel or heat source (e.g., solar radiation, nuclear reaction) is immaterial. All one needs to

---

[5] Air Quality Control System – to scrub the boiler flue gas from the pollutants to minimize the environmental impact of the power plant.

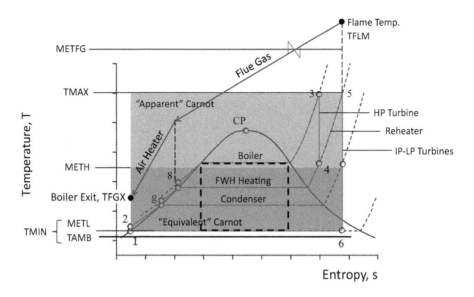

Figure 3.21 Temperature-entropy diagram of Rankine steam cycle with reheat and feed water heating.

know is (i) the boiler (or heat transfer) efficiency and (ii) the impact on plant auxiliary power consumption.

IMPORTANT: In summary, ideal Brayton cycle analysis provides direct information on the heat engine itself, i.e., the gas turbine. In contrast, ideal Rankine cycle analysis does not provide direct information on the steam turbine. This distinction must be kept in mind in thermodynamic analysis of steam turbines.

Before proceeding, a few words on the cycle diagram in Figure 3.21 are in order. First of all, it is a qualitative *sketch*, i.e., it is not based on precise thermodynamic property data. The intention is to illustrate the relative positions of key cycle temperatures and their (logarithmic) averages. What is depicted is a very basic cycle with only one *feed water heater* (FWH) and *subcritical* main steam pressure. In contrast, modern steam cycles with *supercritical* main steam pressure have eight or nine FWHs and showing them all on the T-s diagram would make it unintelligible without adding anything to a solid understanding of the fundamental concepts. Also shown in the diagram is a dashed rectangle inside the H_2O phase dome. This is a commonly used "Carnot cycle" *proxy* in explaining the Rankine cycle to laymen. It is hoped that the discussion below will make clear that this is a totally useless construct. (For example, the interested reader can verify readily that, for a subcritical cycle, METH is *less than* the saturation temperature at main steam pressure.)

Let us state the obvious first. Everything that was discussed within the context of gas turbine Brayton cycle, i.e., in Sections 2.3.5 and 3.1, applies to the discussion of the steam Rankine cycle here. Thus, even if one can design a perfect cycle with zero

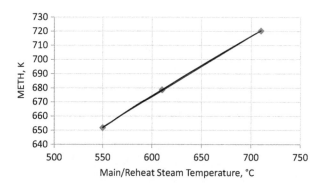

Figure 3.22 Rankine steam cycle mean-effective heat addition temperature (for 250 bara steam). For each 50 bara increase in steam pressure, add 6 K to METH.

losses and isentropic pumps and turbines, the resulting efficiency is much less than the *ultimate* Carnot efficiency given by Equation (2.9). In fact, the efficiency of such a *perfect* cycle is equal to the efficiency of the *equivalent* Carnot cycle defined by Equation (2.8).

By virtue of the constant pressure-temperature heat rejection process via steam condensation, the Rankine cycle METL is indeed equal to the real temperature TMIN. However, METH is a hypothetical temperature, which, for a hypothetical isothermal heat addition process between states 8 and 5 in Figure 3.21, results in the same amount of heat addition, which is the sum of non-isothermal main and reheat heat addition processes. Thus, starting from the general expression for an arbitrary number of constant pressure heat addition processes (i = 1, 2, ..., N),

$$\text{METH} = \frac{\sum_{i=1}^{N} \Delta h_i}{\Delta s_{\text{overall}}}, \tag{3.29}$$

where h is enthalpy and s is entropy (from a suitable EOS using pressure, P, and temperature, T, e.g., *the ASME Steam Tables*), for the specific reheat boiler in Figure 3.21, one arrives at

$$\text{METH} = \frac{(h_3 - h_8) + (h_5 - h_4)}{(s_5 - s_8)}.$$

For typical subcritical, supercritical (SC), and ultra-supercritical (USC) cycles, the relationship between METH and the cycle's main and reheat steam temperatures (use the average if they are not the same[6]) is shown in Figure 3.22.

---

[6] In high-efficiency designs, the reheat steam temperature is higher than the main steam temperature. The reason for that is to enable low condenser pressures without excessive moisture in the LP turbine last stages (i.e., higher efficiency with reduced erosion). (This can be visually deduced easily from the simple T-s diagram in Figure 3.21.) It also increases the load range with minimal moisture erosion of the LP last stage blades.

Let us consider an advanced steam turbine cycle with 595°C main and reheat steam temperatures and a condenser pressure of 50 mbar (the saturation temperature is 33°C). The *ultimate* Carnot cycle efficiency is

$$UCEFF = 1 - (33 + 273)/(595 + 273) = 0.647 \text{ or } 64.7\%.$$

With 675 K for METH from Figure 3.22, the *equivalent* Carnot cycle efficiency is

$$ECEFF = 1 - (33 + 273)/675 = 0.547 \text{ or } 54.7\%.$$

The cycle factor is thus CF = 0.547/0.647 = 0.845. For generality, it is preferable to use a standard temperature for TMIN and the logical choice is the ISO ambient temperature of 15°C. This convention eliminates one design specification from the discussion/analysis, i.e., condenser pressure (steam turbine backpressure), which is highly dependent on site conditions, project economics, and prevailing regulations. In that case, UCEFF = 66.8% and ECEFF = 57.3% so that CF = 0.858. In conclusion, assuming a CF of 0.85 is sufficiently accurate for most practical purposes.

Let us open a parenthesis here. The analysis above was exclusively based on the steam cycle and steam temperatures. Also shown in Figure 3.21 are three boiler temperatures:

- burner flame temperature, TFLM,
- boiler flue gas exit temperature, TFGX, and
- mean-effective average of the two, METFG.

The typical flame temperature is around 2,000°C (3,600°F) and the flue gas temperature upstream of the AQCS is around 130°C (266°F). The logarithmic mean of these two temperatures is 808°C. Rewriting Equations (2.8) and (2.9) with TFLM, METFG, and TAMB, the Carnot efficiencies and CF are recalculated as

$$UCEFF = 1 - \frac{TAMB}{TFLM} = 1 - \frac{288}{2{,}000 + 273} = 87.3\%,$$

$$ECEFF = 1 - \frac{TAMB}{METFG} = 1 - \frac{288}{808 + 273} = 73.4\%,$$

$$CF = 73.4/87.3 = 0.84.$$

One would argue that this is the true measure of the steam cycle's thermodynamic potential and the "true" TMAX is TFLM. This argument is fundamentally correct. However, for the high-level thermodynamic analysis described herein, burner flame and gas exit temperatures are not readily available numbers. Obtaining these numbers requires in-depth boiler thermodynamic and heat transfer analysis. This involves not only the heat transfer between hot combustion gas from the furnace and the heater coils (i.e., evaporator, superheaters, and reheaters) but also heat recovery via combustion air heating (to increase the boiler efficiency), flue gas recirculation, etc. (See Chapters 4 and 22 in Ref. [3] for detailed boiler combustion calculations.) For the second law analysis, it is easier to leave this tedious exercise out and take care of it later using a boiler efficiency (typically, 0.90–0.95) in the first law roll-up of efficiencies. We can now close the parenthesis.

Today's state-of-the-art in SC and USC design can achieve a *technology factor* (TF) of 0.80–0.85 at METH range of 670–675 K (steam cycle only, i.e., steam turbine generator (STG) output minus boiler condensate/feed pump motor input, excluding boiler efficiency and other plant auxiliary loads). For quick estimates, one can assume that each 30 K in METH is worth 0.01 (equivalent to 1% point) in TF. Modern utility boiler LHV efficiencies are in the low to mid-90s (percent). The plant auxiliary load is largely a function of the heat rejection system and AQCS. Both systems are highly dependent on existing environmental regulations and permits. Optimistic values are 5–6% of STG output but can be higher. Electric motor-drive boiler condensate and feed pumps consume about 3–4% of STG output. In some designs, the large *boiler feed pump* (BFP) is driven by a mechanical drive steam turbine (in a steam power plant rated at 600 MWe, the BFP consumes more than 20 MWe). This reduces the electric motor power, but steam extracted to run the BFP steam turbine reduces STG output, i.e., the bottom-line plant thermal performance is not affected much. Going with the example above, if TF = 0.82 the steam Rankine cycle efficiency is found as

$$RCEFF = 0.82 \times 0.573 = 0.47 \text{ or } 47.0\%.$$

This number, however, does not tell the *whole* story and requires a deeper look into it.

When it comes to defining a cycle efficiency for a real machine, precise definitions become of the utmost importance. Alas, articles and papers published in trade and/or archival journals rarely provide a diligent list of assumptions used in arriving at a stated efficiency. In some cases, this is a deliberate attempt to hide true technology features or a marketing tactic. In other cases, it is just inattention to detail and/or insufficient knowledge of a real electric power plant operating in the field. In any event, as was demonstrated by the examples above, there is a hierarchy of efficiencies. There is no ambiguity about the definition of the *denominator* of the efficiency formula, which is a simple ratio;[7] it is the rate of fuel burn in the boiler, either in LHV or in HHV.[8] Since fuel is purchased on an HHV basis, industry convention in the USA has always been to quote the plant efficiencies in HHV (in Europe and Japan, LHV is preferred). Herein, the LHV benchmark is adopted as well (which results in a higher efficiency number – HHV can be higher than LHV by up to 10%) because there is practically no value to the additional heat content from a power generation perspective.

The Rankine cycle efficiency of 47% calculated above was (implicitly) based on the following criteria:

- The denominator of the efficiency formula is the heat picked up by feed water and steam in the boiler.
- The numerator is the STG output minus power consumed by all cycle pumps (typically, on the 4 kV bus in Figure 3.23), which can be identified on the Rankine cycle T-s or the *heat and mass balance* (HMB) diagram.

---

[7] Thermal Efficiency = Power Output / Fuel Input.
[8] LHV is also known as "net calorific value" and HHV as "gross calorific value." The difference is the latent heat of condensation of the water vapor, which is a constituent of the combustion products.

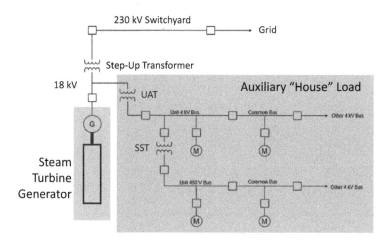

Figure 3.23 Simple power plant one-line diagram.
UAT: Unit Auxiliary Transformer, SST: Station Service Transformer.

First make a correction to the denominator so that it reflects the fuel energy supplied to the boiler. Assuming a boiler efficiency of 92%, the plant efficiency becomes

$$\text{PLNTEFF} = 47\% \times 0.92 = 43.3\%.$$

Let us then make a correction to the numerator for the "house load" (including heat rejection system pumps, fans, etc.) and transformer losses (see Figure 3.23). This adds up to roughly 2% of the net cycle output so that

$$\text{PLNTEFF} = 43.3\% \times (1 - 2\%) = 42.4\%.$$

From a practical perspective, the key performance metric is the last one – after all, this is the *raison d'être* of the whole enterprise, i.e., net electric power supply to the electricity network (the grid). That number is highly dependent on (i) site ambient conditions (i.e., geographical location), (ii) project financing criteria and prevailing economic conditions, and (iii) environmental and other regulatory requirements in force (for permitting). The efficiency hierarchy is summarized in Table 3.3 for quick reference.

3.3.3 Geothermal Power

Geothermal power plants are steam turbine power plants. There are three types of geothermal power plants:

- *Dry steam plants* use steam directly from a geothermal reservoir to drive the steam turbine and generate electric power. The first geothermal power plant was built in 1904 in Tuscany, Italy, where natural steam erupted from the earth.

Table 3.3 Boiler-turbine (Rankine cycle) steam power plant efficiency hierarchy

	°C	K	Carnot Efficiency, %	Factor	Cycle/Plant Efficiency, %	Remarks
TMIN	15	288				See Figure 3.21
TFGX	130	403				See Figure 3.21
TFLM	2,000	2,273	87.3			See Figure 3.21
METFG	808	1,081	73.4	0.840		See Figure 3.21
TMAX	595	868	66.8	0.911		See Figure 3.21
METH	402	675	57.3	0.858		Cycle factor
				0.82	47.0	Technology factor
				0.92	43.3	Boiler efficiency
				0.98	42.4	2% plant auxiliary load

- *Flash steam plants* take high-pressure hot water from deep inside the Earth and convert it to steam to drive the STG. When the steam cools, it condenses to water and is injected back into the ground to be used again. Most geothermal power plants are flash steam plants.
- *Binary cycle power plants* transfer the heat from geothermal hot water to another liquid. The heat causes the second liquid to turn to vapor, which is used to drive an expander (i.e., "gas" turbine) generator.

Geothermal energy is a renewable resource. Through proper reservoir management, the rate of energy extraction can be balanced with a reservoir's natural heat recharge rate. Modern closed-loop geothermal power plants emit no GHGs. According to the US DOE, life cycle GHG emissions (about 50 kg/MWh in CO_2 equivalent) are 4 times less than solar PV, and 6–20 times lower than natural gas.[9] Geothermal power plants consume less water on average over the lifetime energy output than the most conventional generation technologies.

As of 2020, worldwide geothermal power capacity is about 14 GW, whereas electricity generation is about 92,000 GWh (as of 2019). This corresponds to a capacity factor of 75%. In addition to electricity generation, geothermal energy is also amenable to use for heating. In Iceland, for example, more than 90% of the heating demand is met by geothermal energy. In August 2021, Icelandic power generator HS Orka announced plans for a green hydrogen project using geothermal energy with an initial phase of 30 MW. Green hydrogen thus produced will be used in a plant for producing green methanol to fuel the marine sector, as well as private and commercial vehicles.

One novel geothermal power concept is utilizing supercritical CO_2 in a quasi-closed cycle to make use of the heat deep in the ground. The idea is elegantly simple. Carbon dioxide captured at a power or industrial process plant is stored underground

---

[9] Argonne National Lab, Life Cycle Analysis Results of Geothermal Systems in Comparison to Other Power Systems; figure 16, page 43. August 2010.

where it forms a plume and heats up geothermally in the reservoir. Thereafter, heated CO_2 rises via a thermosiphon effect through a production well and is expanded in a turbine to generate electric power. Exhaust CO_2 is cooled and reinjected with the main CO_2 stream through an injection well [4].

Assuming a geothermal gradient of 35°C/km, at 2.5 km depth, temperatures around 100°C at about 250 bar are achievable. Typical design parameters for about 5×10^{-13} m^2 reservoir permeability are as follows:

- Injection to 276 bar, 50°C (density around 800 kg/m^3) at the well bottom
- 245 bar, 102°C at the production well bottom (density around 500 kg/m^3)
- 119 bar, 60°C at the turbine inlet (85% isentropic efficiency)
- 61 bar, 22.7°C at the turbine exhaust (quality, x = 0.79)

Pressure loss in the production well between the reservoir and the turbine inlet is quantified by Darcy's law. At a flow rate of 500 kg/s, about 6.5 MWe is produced by the turbine. Heat rejection in the cooler (CO_2 condenser) is about 54 MWth with a temperature delta of 7°C (ISO ambient). Net output is 3.24 MWe after subtracting the power consumption of the injection pump (1.67 MWe) and the heat rejection system (1.59 MWe). Performance is strongly dependent on the reservoir conditions. If the reservoir depth is increased to 3.5 km, the net output becomes 8.45 MWe. If the permeability is doubled, 10.18 MWe net output is calculated [5].

3.3.4 Operability

In the past, conventional steam turbine power plants, coal-fired and nuclear, were the base load workhorses of the power grid. They were rarely shut down and operated at very high capacity factors. The same could be said of the early CC power plants. In fact, in the 1990s, F Class gas turbine CC power plants were designed for base load duty and operated as such. With increasing penetration of renewable technologies, primarily solar and wind, into the generation portfolio in the 2000s, however, all fossil fuel–fired power plants were required to cycle more than before. The reason is the inability of solar or wind resources to be dispatched on demand. They operate at low capacity factors because they are dependent on the prevailing weather conditions. If there is demand for power but there is not enough sunshine or wind power supply to meet it, dispatchable fossil fuel resources are required to take up the slack. Some of those instances can be planned, e.g., at night when solar power generation is inactive. In those cases, operators can plan their plant startup beforehand to prevent wear and tear of equipment due to thermal stresses and/or shocks. In other instances, the need can arise unexpectedly, e.g., when there is a sudden drop in wind speed or cloud cover over the solar field at a time when there is high demand for power. In that case, the sudden drop in power supply should be made up by rapidly responding resources, either starting from "cold iron" or ramping to full load from a low-load state (i.e., drawing upon a "spinning reserve"). For the latter type of response, the fossil fuel resource must be able to operate reliably and efficiently at a low load in compliance with emissions requirements.

Running the steam turbine (normally operating in a "valves wide open" or VWO mode) with throttled main steam *stop-control valve* (SCV) is an option for fast frequency response as required by the applicable grid code. At a grid underfrequency event, SCV is rapidly opened so that energy bottled in the HRSG is released for an extra power "shot" from the STG as the *primary* response (say, within 10–15 seconds). Due to its much faster dynamic characteristics, the primary response is easier to obtain from the gas turbine via *over-* or *peak-fire* (with some sacrifice in hot gas path component life – as specified by the OEM), opening of the inlet guide vanes (IGVs) or even activation of compressor on-line wash sprays (extra mass flow through the machine). This would contribute to a *secondary* response from the steam turbine as well, but it is slower in kicking in due to the large thermal inertia of the HRSG. In order to not exceed the maximum allowable main steam pressure specified by the OEM, the steam turbine can be equipped with an *overload* valve, which would open and redirect some of the main (HP) steam into a downstream stage in the HP turbine.

Typically, gas turbines can load ramp up or down at a rate of 8% per minute. In a vintage F Class GTCC with two 200 MWe units, this is equivalent to 32 MWe from the gas turbines only. The steam turbine contribution follows with the characteristic time lag of the HRSG and the turbine itself. In recently introduced advanced class gas turbines, this has been increased so that OEMs quote combined cycle ramp rates as high as 50–60 MWe per minute for plants rated at 500 MWe.

The turndown capability of GTCC power plants is limited by the lowest load at which the gas turbines can maintain their base/full load NOx and CO emissions. This load level is referred to as *minimum emissions compliant load* (MECL). In vintage F Class units, MECL used to be as high as 50–60%, which limited the lowest allowable GTCC load to about 40–50%. This means that a $2\times2\times1$ GTCC rated at 600 MWe could not run at loads below 300 MWe in the spinning reserve to deliver 300–350 MW in about 8–10 minutes (typical ISO base load basis – exact numbers are dependent on-site ambient conditions). In contrast, advanced class gas turbines with MECLs quoted at as low as 20% (gas turbine) load can bring the GTCC spinning reserve load and fuel consumption down significantly. The exact lowest allowable load is also a function of operability limits imposed by the steam turbine. Combined cycle steam turbines are limited by the minimum allowable steam flow through the LP turbine. Typically, that lower limit is about 15–20% of the steam flow at the maximum rated load. Low steam flow through the last LP turbine stages with long buckets, accompanied by high exhaust pressure, can lead to flow reversal and separation at the bucket root (i.e., the last turbine stage acting almost as a "compressor") with resulting overheating and/or flutter.

The steam turbine itself and the bottoming steam (Rankine) cycle comprising the HRSG, large steam pipes, and valves (i.e., thick-walled metal components with large thermal inertia) are crucial in limiting the GTCC plant's startup time in much the same way as their counterparts in the conventional steam power plants. The critical item in steam cycle startup is steam–metal temperature matching. In other words, the longer the layoff time between the last shutdown and pushing the start button, the colder the steam turbine casing, rotor, valves, and steam pipe metal (thick-walled components) so the startup time is longer to allow for slow and uniform warming of the components. Typically, startup regimes are divided into three categories based on the length of layoff time, i.e., (i) *hot* start (8–12 hours or less since shutdown), (ii) *warm* start (up to 48–72 hours since shutdown), and (iii) *cold* start (more than 48–72 hours since shutdown). The critical component is the turbine rotor, whose temperature cannot be measured directly and is instead inferred by proxies (e.g., HP and IP inner bowl). Metal temperature, T_m, is a direct function of unit downtime and ambient temperature, as shown in Figure 3.24 (for a combined cycle steam turbine). (Note the difference in cooling characteristics of the HP and IP turbines.) Depending on the OEM and turbine type, the combined cycle steam turbine is rolled by steam admission into the IP or HP turbine. In older fossil boiler plants, the unit is typically started by HP steam admission (if equipped with simple steam bypass from boiler – upstream of the HP valves – to the condenser). More modern units are equipped with a "cascaded" steam bypass system, in which steam is routed through the cold reheat line, the reheater, and the hot reheat line before being dumped into the condenser. In this case, the steam turbine is rolled by IP steam admission [6].

One classification of a hot start for SC/USC units is metal temperature at 400°C (752°F) or higher. For warm start classification, the minimum metal temperature is 200°C (392°F) or higher. The *natural* cooling time depicted in Figure 3.24 is represented by the exponential decay law

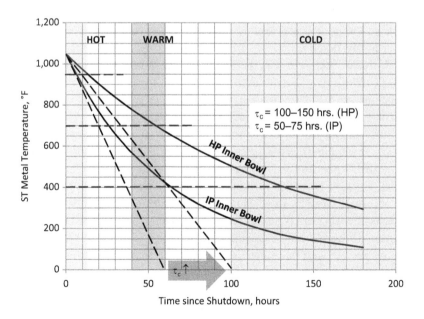

Figure 3.24 Typical steam turbine cool-down profiles (as measured at HP and IP inner bowls). Shaded regions indicate typical time windows for "hot," "warm," and "cold" start classifications. Horizontal dashed lines indicate average metal HP bowl temperature corresponding to the same.

$$\frac{T_m - T_{amb}}{T_{m,0} - T_{amb}} = \exp\left(-\frac{t}{\tau_c}\right), \tag{3.30}$$

with a characteristic *cooling time constant*, τ_c, as a function of the ambient temperature, T_{amb}, and the starting value (denoted by subscript 0). This temperature is the main plant startup classification gauge instead of widely used but fuzzy terms such as "hot" or "warm," whose definitions vary from one source to another.

Steam turbine startup has two steps: (i) roll up from *turning gear* (TG) speed to *full speed no load* (FSNL), i.e., 3,000 or 3,600 rpm depending on grid frequency (50 or 60 Hz, respectively), followed by synchronization; (ii) load ramp from FSNL to *full speed full load* (FSFL) or dispatch load. It is difficult to generalize but, for conventional steam cycles, as a rule of thumb, a hot start from TG to FSFL takes about 1 hour, a warm start takes about 4 hours, and a cold start takes about 12 hours. A one-hour hot start time enables some units to operate in a "two-shift" cycle, i.e., startup in the morning, operation for 12–16 hours, and shutdown for the night. This type of operation should be within the guidelines of turbine OEM regarding minimum downtime and minimum runtime. For a detailed description of large steam turbine pre-warming, steam bypass, turbine roll, and loading processes, refer to [6, Part II].

In GTCC power plants, there are two control mechanisms during hot, warm, or cold startup: (i) gas turbine exhaust energy (flow rate and temperature) and (ii) steam bypass. There are two control methods: (i) the old (conventional) method where steam production in the HRSG is controlled by the gas turbine load (i.e., exhaust gas energy) and (ii) the *fast* or *rapid* new method where gas turbine loading is decoupled from steam turbine roll and loading. Details of each method and underlying physics are succinctly described in an article by Gülen [7]. More extensive information and additional references can be found in the relevant chapters of **GTFEPG** and **GTCCPP**. From a steam turbine perspective, the problem is well defined: minimize thermal stresses resulting in the rotor (i.e., "cold metal") via excessively fast heating with hot steam. An integral part of this problem, especially when the gas turbine is started and loaded as fast as possible (e.g., well under 30 minutes in modern advanced class gas turbines), is the management of flow and temperature of steam generated in the HRSG via *cascaded bypass* and *terminal attemperators*. The former redirects HP and IP steam via reheat superheaters into the condenser to bypass the steam turbine until steam conditions are ready to start the steam turbine roll. (This is also done in conventional steam plants as described earlier.) The latter ensures precise control of steam temperature downstream of the HRSG superheaters, which would not be possible with *interstage* attemperators. Controlled heating of the steam turbine rotor during roll to FSNL is accomplished by precise control of the steam admission flow into the IP turbine (ST roll can also be done via HP turbine steam admission). After synchronization, during load ramp, controlled heating to minimize thermal stress is accomplished by combined steam flow and temperature control in conjunction with steam pressure control.

In any event, steam turbine stress control as described above and concomitant control actions (e.g., prevention of HP turbine heating during turbine roll to FSNL via windage heating, transition to HP turbine steam admission, prevention of LP turbine windage heating) are well known and widely implemented by major OEMs in the field. Further refinements to shave off extra minutes from the startup time can certainly be achieved by exercising sophisticated dynamic simulation models in conjunction with 3D CFD analysis (a steam turbine *digital twin* maybe) to account for thermal stresses across the rather complex rotor geometry.[10] Welded rotors made from materials suitable to each turbine section's operating steam conditions (up to 40% lower thermal stresses during transients vis-à-vis monobloc rotors at the same steam conditions), HP turbine inner casing shrink-ring design for radial symmetry, optimization of shaft seals (e.g., *advanced clearance control* with spring-loaded seals), bearing configuration (e.g., how many thrust bearings, where to put them) and other mechanical refinements (e.g., HP turbine exhaust end *evacuation line*) are also investigated and implemented by major OEMs (e.g., see Saito et al. [8]).

As far as future trends are concerned, one can point to the increasingly larger size of *super heavy-duty* H/J Class gas turbines with more than 500+ MWe output (50 Hz) in simple cycle. A $2\times2\times1$ GTCC with one of these machines would be rated at 1.5 GWe with a 500+ MWe steam turbine. This brings the size of combined cycle STGs into the same league as coal-fired boiler-turbine, i.e., conventional, steam plants. The ramification of this trend is increasing bucket sizes, not only in the LP turbine, which can be somewhat controlled by increasing LP exhaust ends – at extra installed cost of course – but also in the IP turbine. Protecting windage heating in IP and LP turbines creates a tug of war between the steam flow required to roll the unit (e.g., via IP turbine steam admission) and the steam flow required to keep the long last stage buckets cool. This is so because controlled acceleration from turning gear (a few rpm) to FSNL (3,000 or 3,600 rpm) requires precise control of steam flow, which may not be enough for overheating protection. This is especially true if the steam turbine is rolled to FSNL via HP steam admission. Accomplishment of this balance puts a big onus on the admission as well as bypass valves and precise control schemes.

Thermal stress and temperature ramp rate considerations become increasingly limiting as steam conditions rise to SC (for GTCC), USC, and ultimately A-USC (for conventional fossil fuel–fired steam power plants). At those demanding conditions with temperatures above 600°C, turbine, steam pipe, and steam valve construction materials transition from ferritic to austenitic steels. The allowable steam temperature ramp rate is given by

$$\frac{dT_{stm}}{dt} \propto \left(\frac{k}{\rho\, c\, \alpha}\right)\left(\frac{\sigma_{max}}{E}\right), \tag{3.31}$$

where k is thermal conductivity [W/m-K], c is specific heat [kJ/kg-K], and α is the linear coefficient of thermal expansion [1/K] with σ_{max} designating the maximum

---

[10] From the 1950s to the 1980s, steam turbine startup schemes were devised with thermal stress analysis with the assumption of cylindrical rotor. Stress concentration factors were used to account for deviations from that simplified geometry.

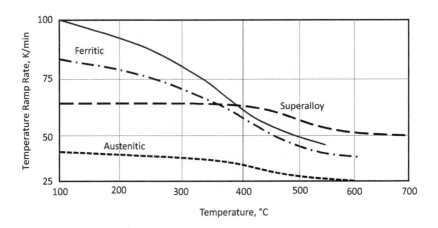

Figure 3.25 Turbine rotor thermal ramp rates for different materials.

allowable stress (E is the Young's modulus with the same units as stress). For the derivation of this relationship, property data on construction materials and mathematical description of the thermal behavior of steam turbines, refer to the highly informative VGB book on the subject [9]. Ferritic and austenitic steels have similar specific heat, but thermal conductivity of ferritic steels is much higher. Consequently, components made of ferritic steels can heat up two to three times faster. This is illustrated in Figure 3.25, which compares the ramp rates of the IP rotor of a large steam turbine as a function of starting metal temperature [10]. The difference between ferritic and austenitic steels (and superalloys) is significant for cold starts but not so much for warm or hot starts. Interestingly, for the latter type of starts, superalloys have higher ramp rates than steels.

3.4   Heat Sink

In a fossil fuel–fired power plant, cycle heat rejection takes place through the steam turbine condenser. From the ideal cycle theory and simplified analysis presented in Section 3.1, the importance of the cycle heat rejection temperature for cycle thermal efficiency is unmistakable. To have the lowest possible cycle minimum temperature, one should achieve the lowest possible pressure (i.e., vacuum) in the steam condenser. This requirement can only be met in the presence of a suitably cold heat sink. In practice, there are two such heat sinks: water and air (ambient). The piece of equipment where heat rejection from the Rankine cycle working fluid (steam) to the heat sink (water or air) is the steam condenser, which is a vital component of gas turbine combined cycle.

There are two types of combined cycle steam turbine condensers:

• Water as coolant
• Air as coolant

Water-cooled condensers can be of two types: open loop and closed loop. In the former, cooling water is drawn from a natural reservoir (e.g., river, lake, or ocean) and returned to the same location after picking up the heat rejected by condensing steam (i.e., at a higher temperature). In the latter type, the cooling water rejects the heat picked up from condensing steam to the ambient in a cooling tower. In other words, the cooling water circulates in a closed loop (with the aid of a water circulation or *circ* pump) between the cooling tower and the condenser.

Air-cooled condensers (ACC) can be thought of as condenser-*cum*-cooling tower systems. Heat transfer from condensing steam to the cooling medium, i.e., air in an ACC, takes place via convection instead of evaporation (mostly) in the wet cooling towers. Due to the lower specific heat of air vis-à-vis water (about one-fourth of it), achieving low condenser pressures require a large heat transfer surface area with large airflows, i.e., large fan power. The heat sink temperature is the ambient *dry bulb temperature* (TDB), which is higher than the *wet bulb temperature* (TWB) by up to 10°F (the heat sink temperature in a wet cooling tower).

Although detrimental to plant net power out and expensive to procure, install, and maintain (and with a large footprint), the ACC systems are advantageous in dry climates and other locations where water scarcity and environmental regulations such as water pollution restrictions prohibit use of the water-cooled systems. The elimination of make-up water supply, blowdown disposal, and water vapor plumes makes ACC an environmentally friendly option at the expense of lower steam turbine performance. In many cases, especially in the USA, the choice is not between an open-loop and closed-loop water-cooled system; it is between a closed-loop system with a ZLD (zero liquid discharge) component and ACC.

For detailed information on condensers and cooling towers, design and off-design calculations, and optimization, the reader is advised to consult chapter 7 in **GTCCPP,** and references cited therein. For more extensive coverage of the heat rejection systems and components for fossil power plants, refer to the book by Black & Veatch [11]. Another useful source for system optimization and performance considerations is the paper by the author ([44] in Section 2.4.3). The typical impact of the heat sink system on combined cycle performance is summarized in Table 3.4. Due to the strong dependence of the cooling tower and ACC performance on ambient conditions, these

Table 3.4 Impact of heat sink on combined cycle performance

CL-CT: closed loop with wet cooling tower, OL-OT: open loop, once-through

Coolant	Water	Water	Air
Type	CL-CT	OL-OT	A-Frame
Aux. Load, % of Gross	2.8%	2.0%	2.5%
Output	−0.30%	Base	−1.25%
Heat Rate	+0.30%	Base	+1.25%
PCOND, in. Hg (mbar)	1.90	1.20	3.00
	(65)	(41)	(103)

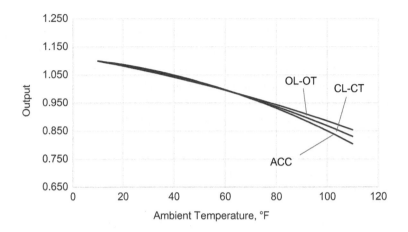

Figure 3.26 Combined cycle ambient output lapse for different heat sinks.

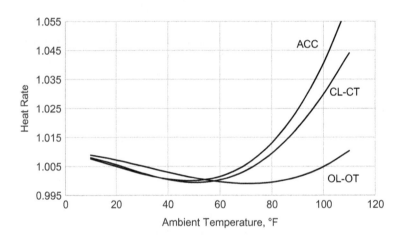

Figure 3.27 Combined cycle ambient heat rate lapse for different heat sinks.

systems have stronger impact on combined cycle performance lapse (reduced output/ efficiency and increased heat rate) with varying ambient temperature. This is illustrated in Figures 3.26 and 3.27,[11] which are typical of vintage F Class combined cycles.

3.4.1   Operability

Controls for the water-cooled condensers and wet cooling towers are pretty straightforward. For the once-through systems, pretty much the only control action is to turn

---

[11] From "Siemens Gas Turbine SGT6–5000F – Application Overview," E50001-W210-A104-V2–4A00 (2008).

one of the circ water pumps off at low loads. The reader should consult chapter 11 of [18] for a detailed discussion of circ water types, design criteria, and operability aspects. In general, these pumps are low head and high flow pumps designed for continuous operation. The most common type employed in once-through as well as closed-loop (i.e., with a wet cooling tower) systems is the vertical wet pit pump. This is a single-stage, single-suction pump with an impeller of the mixed flow design. Typically, two 50% flow pumps are operated to provide the cooling water flow to the condenser. For the closed-loop systems with a wet cooling tower, the main control action is turning off cooling tower cells at low loads and/or cold ambient conditions. This can be combined with parallel circ water pump on/off action as well.

A key criterion for circ pump selection is the *suction specific speed*, NSS. According to the Hydraulic Institute,[12] NSS should not exceed 8,500 in US customary units per the formula below:

$$NSS = N\sqrt{Q}NPSHR^{-0.75},$$

where N is the pump speed (rpm), Q is the flow rate (gpm), and NPSHR is the required net positive suction head in ft.

During the system operation, the main concern is cooling performance degradation due to factors such as tube fouling and leaks. The degradation in cooling performance manifests itself in increased steam turbine backpressure. As a rough rule of thumb, each 10% increase in backpressure reduces the steam turbine generator (STG) output by 0.5%. This may seem a trifle but for a system designed for, say, 2 in. Hg during normal operation when new and clean, fouling can easily increase it to, say, 2.5 in. Hg, an increase of 25%, i.e., 0.5 in. Hg, resulting in STG output loss of 1.25%. In a combined cycle power plant, this translates into a plant heat rate increase of about 0.45% (equivalent to more than 0.25 percentage points of thermal efficiency for a modern GTCC). For a modern H/J Class GTCC, this is equivalent to 25 Btu/kWh.

Fouling can take place inside the tubes (coolant side) or, less likely, outside them (steam side). Its causes can be inorganic or organic (especially with cooling water drawn from a river or the ocean in the open-loop systems). Inorganic fouling agents are typically calcium compounds, i.e., carbonates, sulfates, phosphates, and magnesium silicate, but they can also include silt (open-loop systems) or cooling tower debris (closed-loop systems). Organic slime, marine life (e.g., barnacles, sea anemones), and microorganisms are typically controlled by injecting a biocide into the circ water (continuously or in a "shock" treatment). The preferred biocide is chlorine. It should be noted that open-loop systems are quite rare in GTCC applications and are highly unlikely to be an option in the future due to stringent environmental regulations.

In the closed-loop systems, in addition to the biological control, methods such as pH control and inhibitor feed are utilized to prevent scaling. The exact nature of the

---

[12] ANSI/HI 9.6.1-2017 – Rotodynamic Pumps Guideline for NPSH Margin.

treatment system is dependent on the cooling water characteristics in the particular site. Makeup is required to replace coolant loss to the atmosphere in the cooling tower via evaporation, blowdown, and drift. (See chapter 7 in **GTCCPP** for estimation of such losses and cycles of concentration to determine the amount of blowdown.) In general, scaling is minimized via injection of an alkaline inhibitor, alkalinity is controlled by feeding sulfuric acid, and calcium concentration is controlled via blowdown. Corrosion monitoring is implemented (using weight-loss coupons or other inline systems, e.g., electrical resistance or galvanic current) to protect the tubes, pipes, and other parts coming in contact with the circ water.

The list of fouling agents can be expanded by including *total dissolved solids* (TDS), *total suspended solids* (TSS), chlorides, sulfates, and manganese. On a case-by-case basis, the best control methods are prevention and using appropriate materials in construction. This is done during the FEED based on a detailed makeup water analysis.

There are two types of leaks that can hurt the condenser performance severely. Air leakage from the atmosphere into the condenser operating in a vacuum is closely monitored to maintain the proper vacuum. Air is removed from the system using a *steam jet air ejector* (SJAE) or vacuum pump. Typical culprits in air leak problems are incomplete welds (usually detected during startup and fixed), leaking valves or valve packings, weld cracks, and broken gauge glasses among others. One detection method for small leaks involves the installation of a freon detector at the vacuum pump or SJAE discharge. One then blows a stream of freon into the suspected leak spot. If there is a leak, freon entering the condenser via that route shows up itself in the freon detector.

3.4.2   Dry Air-Cooled Systems

As mentioned earlier, open-loop water-cooled condensers are essentially things of the past. Increasingly strict environmental regulations, driven by concerns of global water scarcity, practically preclude water-cooled heat rejection systems, open or closed loop, in thermal power plant applications. This leaves the designers with two options, i.e., ZLD systems and air-cooled condensers. Unfortunately, ZLD systems add significant operational complexity in addition to higher CAPEX and OPEX (see Section 3.5). For all practical purposes, the preferred heat rejection system in GTCC power plants nowadays is the *air-cooled condenser* (ACC), a once-through system that uses the ambient air as the heat sink. The ACC systems are advantageous in dry climates and other locations where water scarcity and environmental regulations such as water pollution restrictions prohibit the water-cooled systems. The elimination of make-up water supply, blowdown disposal, and water vapor plumes makes ACC an environmentally friendly option at the expense of lower steam turbine performance.

A schematic description of an *A-Frame* ACC is shown in Figure 3.28. As shown in the figure, finned-tube bundles are sloped downward to reduce the plot area. Forced-draft fans located under the A-frame push the ambient air across the finned-tube bundles, causing the steam ducted to the apex of the frame to condense as it flows down the inclined tubes.

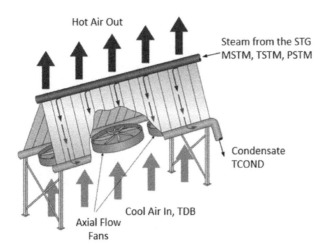

Figure 3.28 A-Frame air-cooled (dry) steam condenser.

Steam exhausting the LP turbine is piped to the condenser by a large-diameter duct (approximately 10% pressure drop). The A-frame structure is mounted on a steel structural support. Non-condensable gases are drawn off by the air ejection equipment. The condensate drains into a collection tank and is sent back to the HRSG via the condensate pump. A typical $2\times2\times1$ GTCC with two F Class gas turbines (about 750 MW ISO base load rating) can have an ACC with 25–30 such cells in a 5×5 or 6×5 configuration. The size of an ACC is determined as a result of the cost–performance trade-off between heat transfer surface area and fan power consumption.

Due to the size/cost restrictions and large parasitic power consumption, ACC systems are not feasible for condenser pressures below 68 mbar (2 in. Hg). A reasonable estimate of the ACC fan power consumption is given by

$$\dot{W}_{FAN} = \frac{0.12}{\eta_{fan}\eta_{mot}} \frac{MSTM\,(x\cdot HFG)}{(TSTM - TDB)}\Delta p_{air},\qquad(3.32)$$

with steam flow rate, MSTM, in kg/s, steam and air (dry bulb) temperatures, TSTM and TDB, respectively, in °C, steam quality, x, as a fraction, latent heat of condensation, HFG, in kJ/kg, and air pressure drop Δp_{air} in mbar. Typical values for fan and motor efficiencies are 80% and 90%, respectively. If not readily available, for front-end estimates, air pressure drop can be assumed to be 1.5 mbar and steam quality can be set to 0.95. With consistent units, Equation (3.32) returns the fan power in kW.

Normal ACC design is based on a 40°F to 50°F difference between TSTM and TDB. For a plant with an ambient design condition of 90°F and 45% relative humidity (73°F wet bulb temperature), a feasible back pressure with an ACC is about 4.5 in. Hg (150 mbar, cf. 2.1 in. Hg or 70 mbar with a mechanical wet cooling tower at the same ambient conditions).

For ISO ambient conditions and PCOND values between 68 to 170 mbar, the following power law is found to represent the ACC power consumption reasonably well:

Table 3.5 ACC performance data

9.5°C TDB, 252 kg/s steam flow, 105.7 mbar BP

Steam Flow, %	Ambient Temperature, °C						
	5	9.5	15	20	25	30	35
80	−41	−30	−6	18	50	89	138
90	−25	−14	11	38	73	116	170
100	−13	0	29	61	100	149	211
110	0	17	50	86	133	193	253
120	16	34	72	113	167	237	308

$$\frac{\dot{W}_{FAN}}{\dot{W}_{STG}} = \frac{1.58\%}{(0.03\ \text{PCOND} - 1.036)^{0.365}} \tag{3.33}$$

For different site conditions, Equation (3.32) should be used. A reasonable approximate estimate is 1.7% of the steam turbine generator output. Equation (3.33) clearly shows that below PCOND of about 68 mbar, the ACC fan power consumption increases exponentially, and the ACC becomes an unfeasible alternative to the wet cooling systems with or without a cooling tower in terms of net electric power maximization.

A typical off-design performance curve for an ACC is shown in Table 3.5. The design conditions are 251.748 kg/s steam flow (y = 6.8%), 105.7 mbar condenser pressure at TDB = 9.5°C. The data is adopted from Table B-1.2.3 in ASME PTC 30.1 [12]. Power consumption of the ACC fans is found as 3,000 kW from Equation (3.32). The data in Table 3.5 can be used as a guide to estimate off-design performance for different applications. Typical degradation in ACC performance is about 4% drop in condenser pressure after 10 years of operation (with regular cleaning of the finned tubes as prescribed by the vendor).

As it turns out, as well as its size/footprint, added cost, and negative performance impact, ACC is also an extremely finicky piece of equipment to operate. Thus, it requires a closer look from the operability perspective.

To begin with, from a fluid flow perspective, ACC is a very unstable and complex system. In other words, the deck is stacked against the designer, EPC contractor, and plant operator from the get-go. A typical ACC for a large GTCC comprises 25–30 individual cells in a rectangular arrangement with N rows and M "streets." A typical 5×5 arrangement for a 750 MW (fired) GTCC is shown in Figure 3.29.

The ideal ACC fan model assumptions used in design calculations are based on the arrangement depicted in Figure 3.30. In other words, the ideal thermodynamic design model, encapsulated by the simple formula in Equation (3.32), is based on a perfect distribution of airflow (i.e., the "coolant") among N×M cells with identical inlet pressure, temperature, and density. The reality in the field is much different, as shown in Figure 3.31.

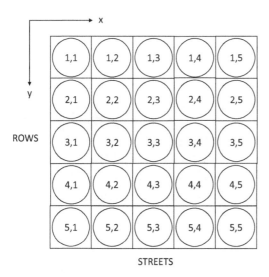

Figure 3.29 Typical ACC arrangement (25 cells with 5 rows and 5 streets).

IDEAL ACC MODEL

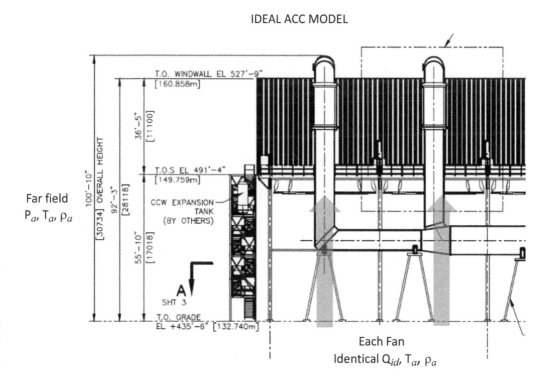

Figure 3.30 Ideal ACC fan model; P, T, ρ, Q denote pressure, temperature, density, and volume flow rate, respectively, with subscripts *a* for air (far-field properties) and *id* for ideal.

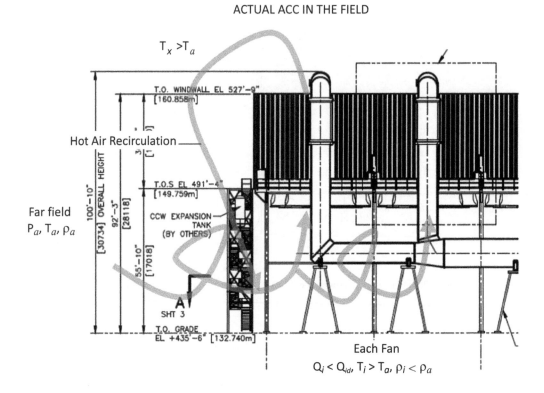

ACTUAL ACC IN THE FIELD

$T_x > T_a$

T.O. WINDWALL EL 527'–9"
[160.858m]

Hot Air Recirculation

T.O.S EL 491'–4"
[149.759m]

CCW EXPANSION
TANK
(BY OTHERS)

Far field
P_a, T_a, ρ_a

100'–10" [30734] OVERALL HEIGHT

92'–3" [28118]

55'–10" [17018]

A
SHT 3

T.O. GRADE
EL +435'–6" [132.740m]

Each Fan
$Q_i < Q_{id}, T_i > T_a, \rho_i < \rho_a$

Figure 3.31 Actual ACC fan model in the field; refer to Figure 3.30 for variable symbols; subscripts *i* and *x* refer to cell inlet and exit, respectively.

In the field, each individual fan has different flow rate and inlet conditions due to the interactions between individual neighboring fans as well as between recirculating hot air and the far-field air flow. Even without wind, there is inlet flow distortion caused by the crossflow of far-field ("cold") air entering the ACC superstructure from below in a parallel-to-ground (i.e., horizontal) manner and then turning upward to a vertical entry into the axial fans. This situation is turned into a mild chaos by wind (especially wind gusts) and nearby structures (e.g., the turbine building), which impede free flow of air and, in some cases, bounce back the hot air plume exiting the ACC at the top back into the ACC.

ACC performance (i.e., heat transfer from condensing steam and low steam temperature-pressure) is hurt by two key mechanisms:

- reduction in airflow (i.e., less "coolant" flow)
- increase in inlet temperature (i.e., low LMTD[13]).

The ACC vendor is fully aware of this situation and has two *knobs* that provide a margin to cover the "heat transfer gap" between ideal design and field performances.

---

[13] Log-mean temperature difference; the "driving force" behind the heat transfer.

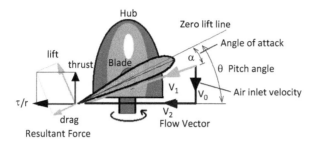

Figure 3.32 Fan blade angles and flow vectors (τ = torque, r = blade radius).

Both increase airflow through the fans through the fan blade pitch angle and the fan speed. The pitch angle is adjusted in the field to increase the airflow. Understanding this requires a quick look into fan blade aerodynamics. An axial fan, once designed, manufactured, and implemented in the field, can modulate its airflow via two means:

- Increasing the rotational speed of the fan (rpm)
- Changing the pitch angle of the fan blades (see Figure 3.32)

The first one is obvious: the fan is a constant volume flow machine. Its volumetric flow rate, Q, is directly proportional to its rotational speed, N, as dictated by one of the so-called fan laws, i.e.

$$Q \propto N.$$

The effect of the pitch angle can be best understood by a visual analogy to a venetian blind. Fan blades are analogous to blind slats. If the angle of the slats with respect to the horizontal plane, when looking at a venetian blind directly, is zero, one gets a full view. When the angle is 90°, the blind is fully closed, i.e., there is no view. Similarly, when the pitch angle is 0°, the fan blades cannot *scoop* air in and the volume flow rate is very low; as the pitch angle increases, the blades start scooping air and volume flow rate increases. (This point will be returned to later.)

There are physical limitations to increasing airflow via either method:

- Clearance between fan blades and fan casing
- Stable operation of the fan (indicated by the fan curves)
- Rated power of the fan motor

In pretty much all cases, an ACC fan driver in a thermal power plant, e.g., a GTCC power plant, is an AC induction motor controlled by a *variable frequency/speed drive* (VFD or VSD). A simplified VFD working principle diagram is shown in Figure 3.33. The three major sections of the controller are as follows:

- **Converter**, which rectifies the incoming three-phase AC power and converts it to DC.
- **DC filter** (also known as the DC link or DC bus), which provides a smooth, rectified DC voltage; and

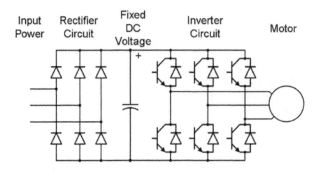

Figure 3.33 Simplified VFD diagram.

- **Inverter**, which switches the DC on and off so rapidly that the motor receives a pulsating voltage that is similar to AC. The switching rate is controlled to vary the frequency of the simulated AC that is applied to the motor.

The voltage applied to the AC motor is proportionally adjusted by the VFD whenever the frequency is changed. For example, if a given motor is designed to operate at 460 volts at 60 Hz, the applied voltage must be reduced to 230 volts when the frequency is reduced to 30 Hz. In other words, the ratio V/F = 460/60 = 7.67 is maintained constant. Consequently, the motor does not draw *inrush current* when starting. This is probably the biggest advantage of the VFD, especially for large motors. The VFD also changes the speed of the AC motor by changing the frequency of the voltage powering the motor.

There are three important power measurement points in the system:

- VFD input
- VFD output
- Motor input terminals

It is thus *extremely important* to ensure that one understands precisely where exactly power is measured and/or quoted. This is so because,

- Steam turbine backpressure (i.e., ACC pressure) is dictated by heat rejection.
- Heat rejection in the ACC is controlled by airflow (Q).
- Airflow is controlled via fan blade pitch angle and/or rotational speed (N).
- Fan motor power consumption (HP) is directly proportional to airflow via well-known fan laws,[14] i.e.,

$$Q \propto N,$$

$$HP \propto N^3.$$

---

[14] In the field, the exponent is slightly less than 3, e.g., 2.79.

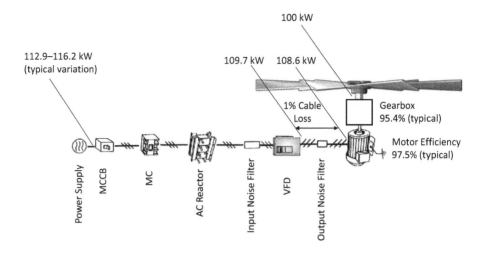

Figure 3.34 Generic VFD-motor wiring diagram to illustrate the power input/output relationship.

- Since there is no measurement of airflow (very difficult/imprecise), the only indication of airflow through the ACC is motor power consumption.

Thus, the number one task in assessing ACC performance is to get a good handle on fan power measurement. This is easier said than done. The power input/output summary for a typical VFD-motor-fan shaft system is provided in Figure 3.34. Field measurements indicate that VFD input power is, on average, about 3% higher than VFD output power. (Indeed, as a rule of thumb, the efficiency of most VFDs is in the mid-90s range.)

In summary, one needs a solid understanding of the power system and proper documentation, i.e., motor and VFD data sheets and performance curves. The other critical piece of information comes from the fan curves and data sheet along with the knowledge of the blade pitch angle (not the angle noted on the design curves, but the actual angle set during the commissioning).

Once the fan motor power consumption is thus determined, as accurately as possible, the next important step is to gauge the operating conditions under prevailing wind conditions on site. There are two basic mechanisms causing the effect of wind on ACC performance: (i) hot air circulation and (ii) degraded fan performance.

Recirculation occurs when the plume of heated air exiting the ACC, which normally rises vertically above the unit under quiescent conditions, is bent over by the wind to the point where a portion of it is entrained by the inlet air stream entering around the sides of the ACC below the fan deck (see Figure 3.35). This results in the cooling air entering the condensing tube bundles being at a higher temperature than the far-field ambient air temperature. The recirculation effect is larger on the downwind cells.

Degraded fan performance occurs when wind passing underneath the fan deck distorts the inlet velocity field to the fans (see Figure 3.36). ACC fans are typically

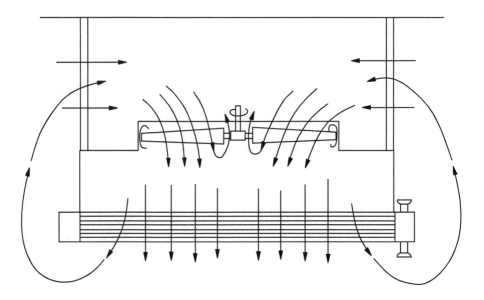

Figure 3.35 Schematic depiction of ACC hot air recirculation.

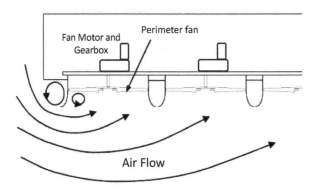

Figure 3.36 Sketch showing the distorted air inlet conditions to a fan at the perimeter of the ACC cell block.

large, low speed fans with relatively low static pressure rise and are particularly susceptible to inlet losses. (Typical fans are almost 10 m in diameter and rotate at slightly higher than 60 rpm.) The effect, usually occurring the *upwind* fans, can range from a modest decrease in inlet air flow to causing a stall condition on all or a portion of the fan blades with a very large reduction in air flow.

In a detailed study, it was found that the combination of the two effects could cause the steam turbine backpressure to increase by *about 2 to 2.5 in. Hg* above the expected value obtained from the performance curves for the same steam flow and ambient temperature (see Figure 1-1 in Ref. [13]). Condenser vendors are cognizant of this fact and provide wind effect correction curves for their equipment. However, experience shows that these corrections can be rather optimistic. The reason for that is that the wind effect on performance can show wide variations as a function of site topography, presence of nearby obstructions, ACC orientation relative to prevailing winds, and any other factor that can influence wind speed, direction, turbulence, and gustiness at the ACC inlet.

In a recent paper, van Rooyen and Kröger looked at the problem in depth [14]. They considered a 30-cell ACC with 6 rows and 5 streets. They looked at ACC performance at different wind speeds in the x and x-y direction (45° with respect to each direction) with the aid of commercial CFD code FLUENT. The key performance parameter considered by the authors is the *net volumetric effectiveness* of each fan and ACC as a total system. The volumetric effectiveness (VE) is defined as the ratio of actual air flow rate through the fan divided by the ideal volume flow rate (i.e., when inlet conditions to the fan are undisturbed by wind or other external factors):

$$VE = \frac{\dot{Q}_{act}}{\dot{Q}_{ideal}}.$$

Van Rooyen and Kröger found that:

1. The wind has a pronounced and detrimental effect on fan VE with increasing speed.
2. Some fans are affected more by wind than the others and, furthermore, some fans are actually *positively* influenced, i.e., their VE increases with wind (e.g., see Figure 7 in [18] when the wind is in the x direction).
3. Overall, the VE of the whole ACC suffers from wind (see Figure 8 in [18]). On average, at about 10 m/s wind speed, VE drops by about 20%.

These findings are corroborated qualitatively by fan motor power data obtained in the field.

It is worth noting that hot air recirculation does not require wind as an "instigator." Under certain conditions, "self-recirculation is possible." This is demonstrated clearly in a 1971 paper by Gunter and Shipes [15]. Referring to the diagram in Figure 3.37, self-recirculation is determined by the criterion

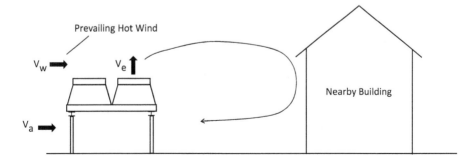

Figure 3.37 Force diagram and nomenclature for induced or forced-draft fan. Also shown is the hot air recirculation due to nearby building obstruction.

$$\frac{V_e^2 \rho_e}{2g} - \frac{V_a^2 \rho_a}{2g} = H.$$

If H > 0, there is no recirculation, otherwise, i.e., if H < 0, there is. In other words, if the hot plume exit velocity, corrected by density, is smaller than the air velocity, even if the wind velocity is zero, recirculation will take place.

With *steady* wind, the criterion becomes

$$\frac{V_e^2 \rho_e}{2g} - \left(\frac{V_a^2 \rho_a}{2g} + \phi \right) = H,$$

where ϕ is the negative pressure on the downwind side of air cooler given by

$$\phi = C \frac{V_w^2 \rho_w}{2g}$$

and C has a value between 0.2 and 0.6. As noted by Gunter and Shipes, the type of wind is important. With a steady wind, stable conditions usually prevail around the air coolers. Gusty, rolling turbulent winds cause unstable conditions and are not amenable to accurate prediction.

Equally important is the plot plan layout of the system, especially in prevailing summer wind conditions. A bad example is shown in Figure 3.37. When the wind direction is across coolers toward the nearby building, the hot air recirculates back from the building to the air coolers. The situation would be worse for *forced-draft* units (i.e., like those in the ACC of a power plant) with a lower hot exit gas trajectory.

For commercial performance tests, ACC performance assessment is governed by the ASME Performance Test Code (PTC) PTC 30.1. The code prescribes four adjustments or corrections to be made to measured steam flow rate:

- Steam quality, x
- Barometric pressure
- Fan power
- ACC pressure and inlet temperature

Based on these corrections, the ASME code defines a "capability," which is used as the metric for a pass/no pass decision. The capability is the ratio of the corrected measured steam flow rate to the design steam flow rate. This, however, is a very misleading (not incorrect or erroneous) metric for determining whether or how ACC performance impacts STG power output via backpressure, i.e., condensing steam pressure in the ACC. Furthermore, while the code prescribes a method for wind measurement and an upper limit of 5 m/s (11.1 mph), there is ample data in the literature (from field studies, wind tunnel tests, and CFD models) that wind can adversely affect ACC performance even at speeds significantly less than 5 m/s. The code does not include guidelines to account for any wind effect.

A proper investigation of the field performance of an ACC with all the ambient factors including prevailing wind conditions (speed, direction, gustiness) and site characteristics (i.e., neighboring man-made and/or natural structures and obstructions) requires a *rigorous CFD model* of the equipment and its surroundings. In the absence of requisite resources for an in-depth CFD simulation, the next best option is a *heat and mass balance model* of the ACC that can help answer "what if" questions in a *qualitative* manner. Such a model of the ACC should have the following capabilities:

- Thermodynamic model with design data
- Translation of thermodynamic model into hardware design
- Translation of hardware design into an "off-design" model

The as-built (i.e., off-design) ACC model enables the following analysis steps:

- ACC performance (i.e., steam pressure and fan power) with measured data, e.g.,
  - Site ambient conditions
  - Steam flow rate
- Impact of operating conditions, e.g.,
  - Number of operating cells
  - Volumetric effectiveness
  - Air recirculation
  - Component degradation

The THERMOFLEX software (see Section 16.2) model library includes an ACC icon with the aforementioned capabilities. Using the best possible measurement for steam flow rate (from the condensate flow), fan power, and steam quality (ASME PTC 30.1 appendix E can be used for that), the model can be used to investigate the field performance. Performance derivatives calculated with such a model are as follows:

- each 1% increase in air recirculation is worth 0.04 psi increase in condenser steam pressure;
- each 1% decrease in volumetric efficiency is worth 0.03 psi increase in condenser steam pressure;
- each 1% decrease in volumetric efficiency leads to 72 kW reduction in total fan power consumption;
- each 25 kW increase in fan power is worth 0.01 psi in ACC steam pressure.

3.5 Zero Liquid Discharge

Water is life. A cliché that is as true for fossil fuel–fired power plants as for living organisms. The steam cycle heat rejection temperature is as important for power plant thermal efficiency as gas turbine TIT. Lowering the former is as powerful a "remedy" as increasing the latter. Achieving the lowest possible cycle heat rejection temperature depends on two things:

- availability of naturally available coolant source and
- availability of hardware (and money) to take advantage of it.

There are two naturally available coolants: water and air. Due to its higher specific heat, water is a much better coolant than air. Thus, for maximum thermal efficiency, the ideal heat rejection system is a water-cooled condenser utilizing cold water drawn from, e.g., a lake, river, or ocean, to condense the steam turbine exhaust steam and return the warm water back to where it came from. This is why the *once-through* or *open-loop* water-cooled condenser is the default heat rejection system for rating performances (typically at 1.2 in. Hg or 41 mbar condenser pressure). Not surprisingly, it is also the heat rejection system of the *world record* setting GTCC power plants (e.g., see Refs. [11–13,28] in Section 2.4.3).

The benefit of open-loop water cooling is twofold:

- achieving low condensing pressure (i.e., steam turbine backpressure) for high steam turbine generator power output and
- doing this without incurring excessive parasitic power consumption via cooling tower fans.

The problem is also twofold:

- availability of a natural water source nearby (e.g., lake or river) and
- ability to obtain the requisite environmental permits to make use of that water source.

Five basic configurations represent the available options. Selecting the most optimal system (i.e., the one with the best performance versus overall life-cycle cost trade-off) requires a careful evaluation of boundary conditions (raw water quality, site availability for solids disposal and evaporation ponds, water recovery requirements [i.e., desired product water quality], etc.) Unfortunately, because of disruptions to local ecosystems due to the large withdrawal amount (which can be as high as 20,000 gallons per MWh), permitting the once-through cooling systems is increasingly difficult. Alternatives are closed-loop systems such as with a mechanical-draft cooling tower (water consumption of about 200 gallons per MWh) or dry (air-cooled) systems (water consumption of only a few gallons per MWh) with extra parasitic power consumption by air circulation fans and high steam pressures.

Increasingly, though, even the much smaller water usage of closed-loop systems (mainly cooling tower blowdown) becomes a burden on scarce water resources. This is where ZLD systems come into play. ZLD systems are designed to process plant wastewater and reclaim it for reuse after discarding the solid waste. In a GTCC with a mechanical-draft (wet) cooling tower, by far the largest source of wastewater is the cooling tower blowdown.

3.5.1   Technology Options

The available ZLD technologies can be classified broadly in two groups (evaporation ponds and subsurface injection are not considered technologies per se) (from a WateReuse Foundation [henceforth, WF] study [16]), i.e.,

1. thermal evaporation (brine concentrator and crystallizer)
2. membrane technology (reverse osmosis, RO)

a. conventional RO
b. high-recovery RO (i.e., two-stage RO used in seawater desalination)
c. high-efficiency reverse osmosis or HERO™ or RO/EDI (Electrodeionization)

These two technologies can be deployed alone or combined resulting in *five* basic configurations:

1. thermal evaporation
   a. brine concentrator only (**Type 1A**)
   b. brine concentrator plus crystallizer (**Type 1B**)
2. combined
   a. softener and reverse osmosis plus brine concentrator (**Type 2A**)
   b. softener and reverse osmosis plus brine concentrator and crystallizer (**Type 2B**)
3. softener and reverse osmosis only (**Type 3**)

The five configurations listed above are described schematically in the figures below (from the WF study). Note that the pretreatment or volume reduction in the combined configurations, 2A and 2B, and the RO system in configuration 3 are based on a high-recovery approach combining lime softening (to alleviate the RO recovery limitation due to less soluble salts and silica) with a standard (conventional) RO step. (That's why the latter is referred to as the "2nd stage RO.") The study claims that the recovery performance is similar to that of HERO™ technology.

While not considered in the study, type 2A and 2B pretreatment systems can be softener-only (i.e., no RO and no volume reduction).

Another item worth pointing out is the absence of storage tanks (brine and concentrate) in the system schematics in Figures 3.38–3.42. This, of course, is a significant omission, especially for the single-train design employed for each configuration.

One option (namely, a *spray dryer*) to remove the remaining moisture from the concentrate in the absence of a crystallizer or evaporation pond (or to supplement them) is not considered. Due to its consumption of fuel (natural gas or fuel oil), it is

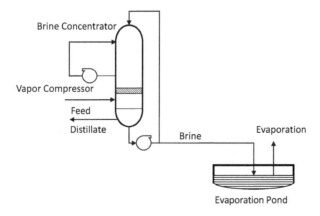

Figure 3.38 Brine Concentrator (BC) to Evaporation Pond (EP) – Type 1A [14].

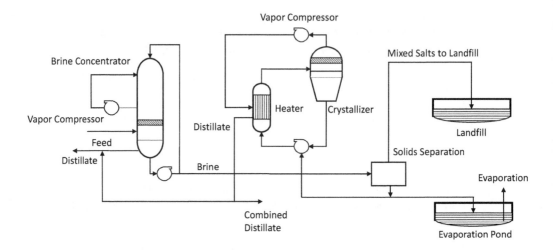

Figure 3.39 BC plus Crystallizer (XT) to EP and Landfill (LF) – Type 1B [14].

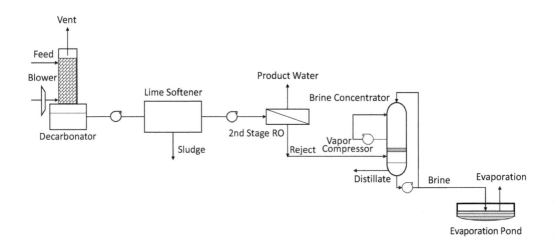

Figure 3.40 Lime Softener (LS) plus Reverse Osmosis (RO) plus Brine Concentrator (BC) to Evaporation Pond (EP) – Type 2A [14].

dubious that it would be feasible from a cost and environmental (air permit) perspective.

One variant of the thermal evaporation system is the *CoLD®* process (Crystallization of high-solubility salts at Low temperature and Deep vacuum) by HPD LLC, a Veolia Water Solutions & Technologies company. It comprises a crystallizer designed to operate at low temperature and pressure. The heat required to boil the solution and the cooling necessary to condense the water vapor are supplied by a closed-cycle heat pump (see below). Not enough data or experience is available to make a direct comparison with existing technologies.

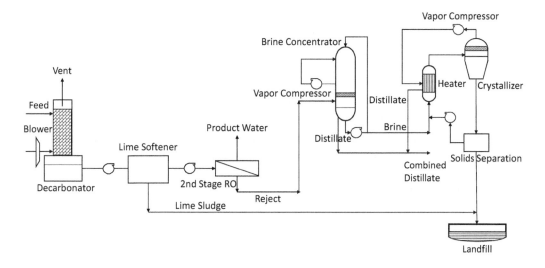

Figure 3.41 LS plus RO plus BC plus XT to LF – Type 2B [14].

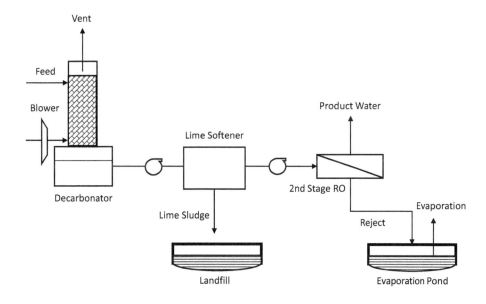

Figure 3.42 LS and RO to EP and LF – Type 3 [14].

3.5.2   Cost-Performance Trade-Off

The selection of a particular ZLD system configuration requires a careful performance-cost trade-off study on a case-by-case basis. Performance herein is the rate of recovery, i.e., water recovered for reuse in the power plant as a percentage of the ZLD plant feed water. The main drivers are quantity of feed water and water quality (i.e., *total dissolved solids* [TDS], composition of cations and anions, *total hardness* [TH], silica content, etc.)

The WF study evaluated the 5 basic configurations using 12 different cases with different flows, salinity (i.e., TDS), and composition (to represent several US sites). For the assumptions and approaches pertaining to equipment sizing, costing, and performance evaluation, refer to the study. Herein, several key findings are going to be highlighted for guidance in undertaking similar studies.

Unit annualized cost is the metric combining system performance and cost (capital plus operating costs). It is the annualized cost divided by the gallons per day (GPD) of feed water. Note that operating costs are *inclusive* of the cost of electricity to run the ZLD pumps and compressors. As such, they account for the differences in power consumption of thermal and membrane technologies (the latter consumes much less power). Selected salinity and composition parameters for cases 1 to 12 of the WF study, along with the unit annualized costs for the five ZLD plant configurations, are shown in Table 3.6.

One key finding of the cited WF study was the absence of economies of scale (feed flow range was 1–20 million gallons per day [MGD]). Essentially, multiple trains were required for the larger sizes.

Recovery rates for ZLD plant configurations are shown in Table 3.7. Not surprisingly, higher rates are for Type 1B and 2B plants with crystallizers (100% means "true" ZLD). Lowest recovery is for Type 3 with RO as the final recovery step.

The effect of salinity (quantified by TDS) on unit cost was evident and stronger for ZLD plant configurations with membrane technologies (see Figure 3.43).

The costs generally increase with hardness for ZLD plants comprising lime softener as shown in Figure 3.44. (Thermal evaporation plants are relatively insensitive to TH.) Except for the highest TH case, plant type 2A has the lowest unit annualized cost with recovery rates comparable to 1A and 1B (thermal evaporation plants). Plants with crystallizers were more expensive than others; not surprisingly, their better performance (i.e., higher recovery) came at a price.

Caveat 1. The crystallizers in ZLD plant types 1B and 2B in the WF study include a vapor compressor, which adds to the energy costs.

Caveat 2. The WF study capital costs include evaporation pond (brine disposal) and dedicated landfill (solids disposal) costs, which are the largest individual capital cost items in most cases.

Since types 1B and 2B with crystallizer include dedicated landfills ($250,000 per acre capital cost and 1% of CAPEX plus $10/ton for hauling), the results in the WF study might be misleading for specific projects with, say, steam-driven crystallizers and no dedicated landfill (i.e., only hauling).

To illustrate this point, type 1B and 2B unit annualized costs are recalculated by excluding the dedicated landfill capital cost. The results are shown in Figure 3.45 (ZLD plant type 3 is excluded due to its much lower recovery rate). The dramatic improvement in ZLD plant type 2B (thermal evaporation including crystallizers with pretreatment [volume reduction] via lime softening and reverse osmosis, see Figure 3.41) makes it by far the most cost-effective option.

Table 3.6 Selected parameters for cases 1 through 12 along with ZLD plant unit costs (annualized) [16]

CASE		1	2	3	4	5	6	7	8	9	10	11	12
TDS	mg/L	4,000	4,000	8,000	12,000	12,000	8,000	8,000	8,000	8,000	8,000	8,000	8,000
Ca^{++}	mg/L	365	365	731	1,096	1,096	912	967	574	941	75	944	488
Mg^{+++}	mg/L	178	178	355	533	533	677	212	147	310	36	338	206
Hardness	mg/L (as $CaCO_3$)			3,283			5,056	3,287	2,038	3,624	335	3,746	2,065
HCO_3^-	mg/L			928			336	2,176	161	2,212	5,882	920	184
Silica (as SiO_3)	mg/L			22			130	134	180	194	0	64	29
Cl^-	mg/L			1,111			798	1,158	4,141	1,640	95	1,323	1,376
1A	$/yr/GPD	4.19	4.07	4.53	4.74	4.92	4.51	4.62	4.99	4.95	4.27	4.84	4.72
1B	$/yr/GPD	4.38	4.18	4.74	5.21	5.21	4.70	4.78	5.44	5.12	4.96	5.04	4.84
2A	$/yr/GPD	2.43	2.22	3.53	5.12	4.83	4.84	3.07	3.12	3.27	2.35	3.67	2.97
2B	$/yr/GPD	3.11	2.71	4.53	6.59	6.30	6.03	3.96	3.37	4.21	2.64	4.54	3.71
3	$/yr/GPD	2.38	2.37	3.85	5.44	5.39	5.21	3.26	3.35	3.47	2.45	4.09	3.47

Table 3.7 Recovery rates (100% is "true" ZLD) [16]

Process Scheme	% Recovery for Case No.											
	1	2	3	4	5	6	7	8	9	10	11	12
1A	98.4	98.4	96.9	95.4	95.4	96.9	96.5	97.8	96.5	95.2	96.9	96.0
1B	99.9	99.9	99.8	99.6	99.6	100	99.9	98.8	99.6	100	99.8	100
2A	97.5	97.5	94.9	92.3	92.5	94.2	95.9	97.7	95.9	96.8	95.9	95.8
2B	100	100	100	100	100	100	100	100	100	100	100	100
3	93.9	94.0	87.9	81.0	81.3	85.5	90.0	87.0	90.8	87.4	87.6	86.5

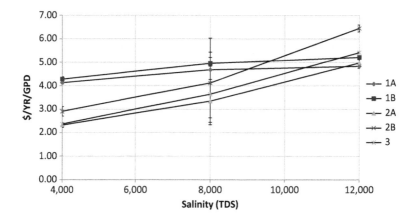

Figure 3.43 Unit annualized cost as a function of salinity (TDS). Error bars for 8,000 TDS indicate the variation due to composition (i.e., cations, anions, and silica). Refer to Table 3.7 for the labels.

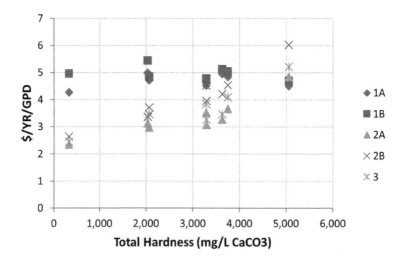

Figure 3.44 Unit annualized cost as a function of total hardness (8,000 TDS cases). Refer to Table 3.7 for the labels.

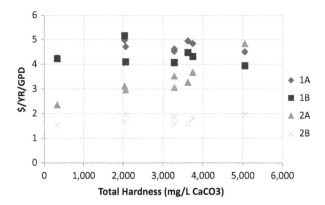

Figure 3.45 Unit annualized cost as a function of total hardness (8,000 TDS cases). Dedicated landfill capital cost is excluded from types 1B and 2B.

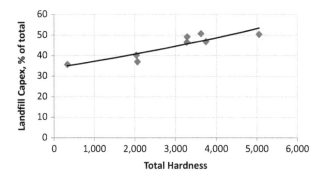

Figure 3.46 ZLD plant type 2B – dedicated landfill capital cost.

For type 2B, the dedicated landfill CAPEX comprises a large portion of the total plant CAPEX in direct proportion to the feed water TH (see Figure 3.46). This is because both the lime softener and the crystallizer produce solids for disposal.

For cases where the feed water has low salinity (TDS), one can consider a membrane system without thermal evaporation (Type 3) – see Figures 3.47 and 3.48. The numbers herein pertain to a high-recovery system comprising a lime softener (pretreatment) followed by RO (Figure 3.42). The cost numbers include a dedicated landfill; thus, in the case where simply hauling the sludge from the softener off-site is an option, they will be even lower.

High efficiency options such as HERO™ and RO/EDI can also be considered. A detailed cost and performance analysis is required to decide.

3.5.3   System Reliability

Based on the (admittedly simplistic) cost-performance trade-off picture summarized in the preceding section, one can conclude that

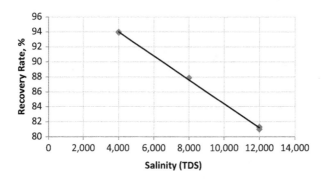

Figure 3.47 Performance of membrane system (Type 3).

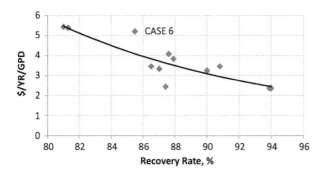

Figure 3.48 Unit cost of membrane system (Type 3).

1. For low TDS feed water, the best option is a membrane-only system (Type 3) with an evaporation pond.
2. For all others, a combined system comprising pretreatment with a membrane system followed by thermal evaporation (Type 2B) is the best option.

However, in single-train systems, the higher the number of components or subsystems, the lower the overall system reliability (product of individual component reliabilities). When one component breaks down, the entire ZLD system breaks down. (The term "breaking down" is used in the sense of a forced outage or decreased processing capacity.) Since the GTCC power plant cannot run when the ZLD system is down, this becomes a huge problem for the plant owner.

The simplest way to address the reliability problem is storage. This is best explained in a paper using an actual system (actually a Type 2B ZLD plant) and numerical examples [17].

Factoring in the capital cost of sufficient storage (brine and concentrate holding tanks and/or storage ponds – the latter will essentially disconnect the ZLD system from the power plant to enable continued power generation while the ZLD system is in forced or planned outage) and including design margins to account for system

Table 3.8 ZLD manpower analysis in terms of time spent per day [18]

	BC	XT	HERO
Supervisory	BASE	1.0x	1.5x
Troubleshooting	BASE	0.4x	3.0x
Chemical Feeds and Online Cleaning	N/A	BASE	1.5x
Solids Handling	N/A	BASE	1.2x
Control Board	BASE	1.0x	1.0x
Lab Work	BASE	1.1x	3.0x

degradation (a key observation in the aforementioned paper) can change the cost picture in Figure 3.44 or Figure 3.45 significantly.

Consequently, taking into consideration the probability of a forced outage and capacity degradation when calculating annualized unit cost, one may very well end up with a simple Type 1A (brine concentrator with evaporation pond) or Type 3 system as the most optimal solution. It should be pointed out that the latter may not be an option for sites with very high TDS feed water.

Another (guaranteed to be more expensive in terms of upfront capital expenditure) solution is multi-train design with parallel components, i.e., 2 × 50% or 3 × 50%, or whatever. Depending on the resulting overall system reliability, without resorting to large storage capacity this may also be a feasible path.

The ultimate solution can only be determined by a detailed analysis on a case-by-case basis.

As a final note, two industry papers on the subject went out of their way to emphasize the absolute necessity for a *qualified operating team* to run and maintain the ZLD plant (see Table 3.8 for a breakdown of typical ZLD operational duties). Reliability without requisite manpower is impossible.

3.5.4   Recommendations

Generic, high-level recommendations are rather obvious:

1. Go for the *simplest* solution based on source/raw and product water quality (*Occam's Razor* principle).
2. The *simplest single-train* system should be pursued first, i.e.,
   a. Brine concentrator with evaporation pond;
   b. Membrane system with evaporation pond;
      i. High-recovery system (two-stage RO or RO following treatment);
      ii. High-efficiency system (HERO™ or RO/EDI).
3. If the simplest, single-train system turns out not to be feasible, the remaining variants should be considered with *reliability enhancing features* such as those listed below (individually or in combination):
   a. Maximum possible storage capacity (e.g., tanks);

     b. Possibility of a pond (reservoir) to separate the ZLD and power plants;

     c. Temporary volume reduction systems such as portable RO and tanks to treat and/or wastewater when the ZLD is in outage;

     d. Multiple, parallel trains for selected subsystems;

     e. Sufficient design margin to preclude reduced capacity (due to fouling, aging, etc.).

4. Ensure to make allowance for a well-rounded operating team – in sufficient numbers *and* qualifications.

General industry experience (i.e., ZLD systems are notoriously difficult, costly and time-consuming processes to operate and maintain) made it abundantly clear: Consider ZLD the "last resort" in water recovery (i.e., air-cooled condenser is probably the better option). The items listed below should be considered first.

1. Conventional options such as surface discharge are most likely not available (e.g., environmental regulations – otherwise, why would one consider ZLD in the first place?).
2. Does it have to be a "true" ZLD? That is, all ZLD feed water (essentially cooling tower blowdown for a GTCC) is reduced to dry solids and sent to a landfill and/or recycled.
3. Investigate the possibility of other options:
     a. Subsurface (deep well) injection,
     b. Evaporation ponds.

If (and only if) ZLD ultimately emerges as the only available option, the system selection should be made after a diligent analysis of raw water chemistry (average, best, and worst cases) and environmental regulations and/or permits pertaining to the final disposal of concentrate (liquid and/or solid). Based on lessons learned in the field, the following should be considered when choosing and/or designing a ZLD system:

1. System capacity/sizing with ample margin (i.e., consider the worst-case scenario under *fouled* conditions – especially brine concentrator and crystallizer heat exchangers)
2. Redundancy of certain components (i.e., 2 × 50 or 3 × 50 trains, or whatever)
3. Material selection (do not skimp on materials to save a few bucks up front – e.g., alloy steel for crystallizer feed tank fabrication)
4. Availability of ample spares (e.g., smaller pumps, pump motors, vapor compressor [VC] impeller)
5. Availability of storage tanks (to ensure continued operation when the system or a subsystem is down)
6. Availability of substitute steam source (in case of brine concentrator VC outage)

3.6   Heat Exchangers

Pretty much all power generation systems discussed in this book contain one or more heat exchangers (in addition to the HRSG and steam turbine condensers discussed earlier in the chapter). By far the most common type of heat exchanger used in process and power applications is the shell-and-tube type heat exchanger. The literature on this type of heat exchanger is so extensive that the reader can easily find such information by a simple search of the internet. As it turns out, however, recently proposed advanced technologies such as supercritical CO_2 (sCO2) power cycle (Chapter 10), nuclear power cycles with working fluids such as CO_2 and helium (Chapter 9), and thermal/cryogenic energy storage (Sections 5.1 and 5.3) impose onerous requirements on the design of heat exchange equipment, i.e., low pressure drop and high rate of heat transfer with tight temperature approaches (a few degrees Celsius at each end). These requirements result from cycle characteristics such as low cycle pressure ratio (e.g., about 3:1 in a supercritical CO_2 cycle) and maximum recuperation for the highest possible efficiency. Shell-and-tube exchangers are not able to meet such demanding design requirements at a reasonable size and cost, if at all.

Such applications typically require *plate* (or *plate-frame*) type heat exchangers or *printed circuit heat exchangers* (PCHE) with compact construction, high heat transfer coefficients, the capability to withstand high pressures, and a wide range of operating temperatures. The latter, also known as a *diffusion-bonded micro-channel* heat exchanger, is a compact "platelet" heat exchanger. It can thus be considered a more advanced variant of a basic plate heat exchanger. Its invention goes back to early 1980s at the University of Sydney and was commercialized by *Heatric* in Australia in 1985 [19]. Nowadays, other OEMs such as *Alfa-Laval* provide their own designs of PCHEs among other compact heat exchangers. The reader is referred to the paper by Chai and Tassou [19] for a comprehensive review of the technology.

To get an idea about the difficulty of the heat transfer problem, let us consider the high-temperature recuperator (HTR) for the sample sCO2 cycle in Chapter 10 (Figure 10.7). The ideal heat transfer diagram of the HTR, i.e., effectiveness, $\varepsilon = 100\%$, is shown in Figure 10.4. The HTR in the sCO2 cycle has an effectiveness of $\varepsilon = 98.5\%$, i.e., it is quite close to the ideal HTR in Figure 10.4. Its design performance from the heat and mass balance shown in Figure 10.7 is as follows:

- Heat transfer, Q = 702,924 kWth
- Overall thermal conductance, UA = 27,158 kW/°C (U is the overall heat transfer coefficient [HTC], A is the heat transfer surface area)
- Log-mean temperature difference, LMTD = 25.9°C
- Number of transfer units, NTU = 10.9 ~ 11 (NTU is the ratio of UA to the smaller heat capacity rate)
- Pressure drop on the cold side is 1.8 bar
- Pressure drop on the hot side is 1 bar

This performance is practically impossible to achieve in a single shell-and-tube heat exchanger. It would most likely require two or even three of them in series with unacceptable pressure drop, size, and cost. Specifically, consider that the overall HTC, U, is of the order of probably 1,000 W/m^2-°C. Thus the requisite heat transfer area, A, is about 30,000 m^2!

The HTR operates at high pressure and temperature, i.e., maximum values of 250 bar and 450°C. The pressure drop allowance on the hot and cold sides are less than 2 bar. The design heat duty is of the order of 700 MWth. The most likely choice would be a PCHE. However, clearly this would not be an "off-the-shelf" product that one can order from the catalog of an OEM like Heatric. The onerous design criteria cited above would necessitate a unique design for a FOAK application. (Existing PCHE designs are probably for heat duties at least one order of magnitude smaller than this example.) The reader can consult Ref. [19] for PCHE design process details including material selection (for this example, 316L/H stainless steel would be appropriate), manufacturing and assembly, types of flow passages, and thermohydraulic performance.

Another emerging heat exchange technology, enabled by the advances made in additive manufacturing (commonly known as 3D printing), is based on micro-trifurcating core structures and manifold designs. One example of the technology is the *Ultra Performance Heat Exchanger Enabled by Additive Technology* (UPHEAT) developed by GE's R&D organization with support from the US DOE's *Advanced Research Projects Agency-Energy* (ARPA-E). The UPHEAT heat exchanger is made from a novel, high-temperature capable, crack-resistant nickel superalloy (GE's AM303), designed specifically for additive manufacturing. The equipment is capable of operation at 900°C (1,652°F) and 250 bar (3,626 psi) with a pressure loss of only 0.5% and power density of 4 kWth/kg. A detailed description of the furcating flow heat exchangers can be found in the US Patent Application Publication US 2016/0202003 A1.[15] The key underlying principle is to split and recombine one fluid flow continuously (into a *trifurcating* network of ducts, each less than 2.5 millimeters in diameter) while the other fluid moves through a similar structure in the opposite direction. These intertwined flows help to achieve superior heat transfer performance. Fluid streams are kept physically separate, but their proximity allows for efficient heat exchange. The UPHEAT design is said to be inspired by the operation of human lungs.

In any event, the real challenge in the design of any type of superefficient heat exchange equipment is to ensure that it does not fail due to the thermal stresses imposed on welded joints and headers (or on thin walls separating micro channels) during transients such as startup, shutdown, and load changes. The resolution of such problems requires extensive CAD/FEA modeling and analysis followed by long-term testing and operation in the field with equipment sized for typical commercial applications. Even then, scaling to larger sizes is not guaranteed to result in acceptable life and performance with high RAM. It is almost a certainty that technologies such as the sCO2 cycle (to replace the Rankine steam cycle for utility-scale power generation applications) are decades away from a CRI of 6 (see Section 16.5), if ever. Caveat emptor.

---

[15] Heat Exchanger Including Furcating Unit Cells, by Gerstler, W. D., Erno, D. J., Kenworthy, M. T., Rambo, J. D., and Sabo, N. K. (July 14, 2016).

3.7   Pumps, Compressors, and Expanders

Pumps, compressors, and expanders (also referred to as *turboexpanders*) are the types of turbomachinery that have been widely used in chemical process and refinery applications for decades. In particular, the latter two are also the key building blocks of new technologies coming to the fore in the era of *Energy Transition*, e.g., supercritical CO_2 (sCO2) power cycle (Chapter 10), nuclear power cycles with working fluids such as CO_2 and helium (Chapter 9), and thermal/cryogenic energy storage (Sections 5.1 and 5.3). We have already come across them within the framework of a gas turbine (the axial compressor and the hot gas path) and steam turbine. In more recently proposed technologies, they appear as individual components with specific duties within the framework of the specified process. Similar to the shell-and-tube heat exchangers discussed in the preceding section, there is a large body of literature available on design, operation, and accessories (e.g., valves, bearings, lubrication systems, and controls) of process compressors and expanders. For a quick and handy reference, the reader is pointed to Chapters 12 (Pumps and Hydraulic Turbines) and 13 (Compressors and Expanders) of *Gas Processors Suppliers Association* (GPSA) Engineering Data Book (FPS Volumes I and II).[16] API Standard 617, *Axial and Centrifugal Compressors and Expander-Compressors for Petroleum, Chemical and Gas Industry Services*, contains the data most relevant to practitioners in the field. A good source of information is the handbook, *Compression Machinery for Oil and Gas* (edited by K. Brun and R. Kurz, Gulf Professional Publishing, 2019).

Thermodynamic design calculations of pumps, compressors (with inter- and aftercoolers), and expanders are straightforward. With the help of an introductory thermodynamics textbook (see below for a recommendation) and/or the sources cited above, they can be easily done in an Excel spreadsheet. Aeromechanical design calculations are more involved but basic concepts can be gleaned from standard turbomachinery books. Several of them are highly recommended by the author. In particular:

1. Shepherd, D. G., *Principles of Turbomachinery* (New York: Macmillan, 1968)[17]
2. Moran, M. J., and Shapiro, H. N., *Fundamentals of Engineering Thermodynamics* (New York: John Wiley & Sons, 1988)
3. Logan Jr., Earl, *Turbomachinery*, 2nd ed. (New York: Marcel Dekker, 1993)
4. Lewis, R. I., *Turbomachinery Performance Analysis* (London: Arnold, 1996)
5. Dixon, S. L., *Fluid Mechanics, Thermodynamics of Turbomachinery*, 5th ed. (Burlington: Elsevier Butterworth–Heinemann, 2005)

---

[16] The fourteenth edition of the GPSA data book is available for purchase in www.gpsamidstreamsuppliers .org/databook/order-register (last accessed on December 26, 2021).
[17] By far the best book for self-learning along with Dixon's book.

6. Sultanian, B., *Logan's Turbomachinery: Flowpath Design and Performance Fundamentals*, 3rd ed. (Boca Raton, FL: CRC Press, 2019)

7. Van den Braembussche, R., *Design and Analysis of Centrifugal Compressors* (Hoboken, NJ: ASME Press and John Wiley & Sons, 2019)

For worked-out examples of compressor and expander stage-by-stage preliminary design examples (for nuclear closed-cycle gas turbine applications – see Chapter 9), the reader can consult the papers by McDonald (Refs. [17,20] in Chapter 9). For operability considerations, consult chapter 13 in the GPSA data book cited above. For pumps consult chapter 12 in the GPSA data book and chapter 10 (Section 10.3) in **GTCCPP**.

Thermodynamic design and preliminary design of compressors and expanders are rather straightforward. Caveats like those outlined for heat exchangers in the preceding section apply here as well. Especially in closed cycles with the need for hermetic sealing, leakage problems can be tricky to solve. As an example, consider the dry gas seal problems encountered in preliminary designs in sCO2 pilot projects (Section 10.2). Since the components typically operate at higher-than-synchronous speeds, vibration and HCF issues can regularly bug the field operation. In order not to be repetitive, the reader is directed to the last two sentences in Section 3.6.

3.8   References

1. Narula, R., Zachary, J., and Olson, J. 2004. Matching steam turbines with the new generation of gas turbines, *PowerGen Europe*, Barcelona, Spain.

2. Frutschi, H. 2005. *Closed-Cycle Gas Turbines: Operating Experience and Future Potential*. New York: ASME Press.

3. The Babcock & Wilcox Company. 2015. *Steam – Its Generation and Use*, 42nd ed. Akron, OH USA: The Babcock & Wilcox Company.

4. Randolph, J. B. and Saar, M. O. 2011. Combining geothermal energy capture with geologic carbon dioxide sequestration, *Geophysical Research Letters*, 38, L10401.

5. Glos, S. 2019. Assessment of performance and costs of CO2 based next gen geothermal power (NGP) systems, 3rd European sCO2 Conference, September 2019, Paris.

6. Leyzerovich, A. S. 2008. *Steam Turbines for Modern Fossil Fuel Power Plants*. Lilburn, GA: The Fairmont Press.

7. Gülen, S. C. 2013. Gas turbine combined cycle fast start: the physics behind the concept, *Power Engineering*, www.power-eng.com, pp. 40–49.

8. Saito, E., Nishimoto, S., Endo, H. et al. 2017. Development of 700°C class steam turbine technology, *Mitsubishi Heavy Industries Technical Review*, 54(3), 10–15.

9. VGB PowerTech Guideline. 1990. *Thermal Behaviour of Steam Turbines*, Revised 2nd ed., VGB-R105e. Essen, Germany: VGB PowerTech Service GmbH.

10. Zörner, W. 1994. Steam Turbines for Power Plants Employing Advanced Steam Conditions, 10th CEPSI, September 19–23, Christchurch, New Zealand.

11. Black & Veatch. 1998. *Power Plant Engineering*. New Delhi, India: CBS Publishers & Distributors.

12. ASME. 2018. Air-cooled steam condensers, PTC 30.1-2007. Issued June 24.

13. Maulbetsch, J. S. and DiFilippo, M. N. 2010. Effect of Wind on the Performance of Air-Cooled Condensers, CEC-500-2013-065, Report Prepared for California Energy Commission.

14. van Ruyen, J. A. and Kröger, D. G. 2008. Performance trends of an air-cooled steam condenser under windy conditions, *Journal of Engineering for Gas Turbines and Power*, 130, 023006-1.

15. Gunter, A. Y. and Shipes, K. V. 1971. Hot air recirculation by air coolers, Twelfth National Heat Transfer Conference AIChE – ASME, August 15–18, Tulsa, OK.

16. Mickley, M. 2008. *Survey of High-Recovery and Zero Liquid Discharge Technologies for Water Utilities*. Alexandria, VA: WateReuse Foundation.

17. Sampson, D. 2012. No easy answers: ZLD improvement options for a 720-MW power generation facility, International Water Conference, November 4–8, San Antonio, TX.

18. Como, V. A. 2013. Operating a ZLD, what does it take?, International Water Conference, November 17–21, Orlando, FL.

19. Chai, L. and Tassou, S. A. 2020. A review of printed circuit heat exchangers for helium and supercritical CO2 Brayton cycles, *Thermal Science and Engineering Progress*, 18, 100543.

4   Operation

The first thing to do before operating any machine is to start it from standstill. A gas turbine is an internal combustion engine. In some ways it is much simpler than a typical piston-cylinder engine similar to the one in a car. In others, especially due to its size and large metal mass, it is more complicated. Nevertheless, like any internal combustion engine, it must be *cranked* to get it going. What does that mean? Well, think about the engine in your car. During its normal operation, combustion air is sucked into the cylinders by the action of pistons, which are driven up and down by the rotation of the crankshaft to which they are connected by their connecting rods. When the engine must start from a standstill with no crankshaft rotation and no up and down movement of the pistons to suck air into the cylinders, that rotation must be started by external means. That external means is the starter motor, which is a small electrical motor that turns the crankshaft via gears connecting the motor shaft to the flywheel. This is essentially what happens when the driver turns the ignition. Once the crankshaft reaches a certain rotational speed, say, a few hundred revolutions per minute (rpm), the motor disengages automatically. Air thus sucked into the cylinder (and pressure inside the chamber in diesel engines) is enough to sustain combustion with injected fuel and power generation to turn the crankshaft so that the engine is self-sustaining. It should be emphasized that what is described above is true for all types of large turbomachinery, e.g., large process compressors and turboexpanders.

In modern gas turbines rated at several hundred megawatts, the process is almost the same. Instead of a separate starter motor, the synchronous ac generator of the gas turbine is run as a synchronous ac motor by the *load commutating inverter* (LCI), which is also known as a *static starter*. (In the past, when gas turbines were relatively small compared to the state of the art behemoths rated anywhere from 300 MWe to almost 500 MWe, electric motors, small diesel engines, torque converters, and expansion turbines were used for this purpose.) For a detailed description of the gas turbine starting system, the reader is referred to chapter 19 (section 19.3) in **GTFEPG**.

4.1   Simple Cycle Start

During gas turbine startup, there is a preset schedule of acceleration rates and speed set points, which are programmed into the turbine controller (e.g., see Figure 4.1). Once the operator pushes the START button, the turbine controller relays this information to

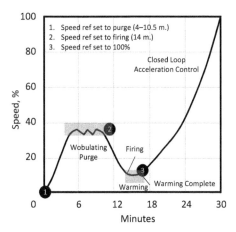

Figure 4.1 Typical gas turbine startup sequence.

the LCI controller. By controlling the generator field voltage and stator current, the LCI controller adjusts the torque produced by the generator (which now runs as a *motor*) and thus controls the acceleration and speed of the turbine-generator set.

Prior to the start command, the gas turbine-generator set is spinning at turning gear (TG) speed to prevent a bow in the gas turbine rotor. For General Electric (GE) gas turbines, TG speed is typically less than 10 rpm. In contrast, TG speed for Siemens gas turbines is more than 100 rpm. The reason for the high TG speed (as claimed by the OEM) is to eliminate the turbine rotor blade *chatter* that is (claimed) to be common to other gas turbines when running on TGs with a much lower rpm. In other words, by eliminating the chatter via the centrifugal force generated by the higher rpm, i.e., extending the blades radially from the center of the rotor and into a position much closer to their normal operating position, any wear that would result from the blades chattering and rubbing would be negated.

At startup, the LCI connects to the generator stator and assumes control of the exciter field voltage reference. The LCI then accelerates the turbine to the purge speed set point (typically 25% of synchronous speed). The turbine is held at purge speed for approximately six minutes, during which time combustible gases are expelled from the non-combustion portions of the system. Upon completion of the purge cycle, the LCI is turned off and the turbine coasts down to the ignition speed of 15%. Once at the firing speed, the LCI is turned on again and the ignition commences. The turbine is briefly held at constant speed to allow for warming. Thereafter, the LCI accelerates the turbine to its self-sustaining speed, at which point the LCI is turned off and disconnected from the generator. From that point on, the controller accelerates the machine to FSNL (Full Speed No Load) and synchronization to the grid.

The power of the LCI is primarily determined by the required startup time of the turbogenerator from standstill until synchronization at 3,000 or 3,600 rpm. As the LCI must only be connected until ~60% speed is reached (i.e., 1,800–2,160 rpm depending on grid frequency, which is referred to as the *self-sustaining* speed), the

LCI active time is somewhat shorter. For state of the art "fast start" or "rapid responses" gas turbines, the LCI rating should be about 5% of the gas turbine base load rating.

Once the gas turbine has been synchronized to the grid, it can be loaded to any desired load up to FSFL (Full Speed Full Load). What is "Full Load," (also known as 100% load)? Full load refers to the power generated by the gas turbine generator with its inlet guide vanes (IGV) at their *fully open* position and fired as dictated by its control curve at measured cycle pressure ratio (PR) and exhaust temperature.

This is admittedly an archaic explanation – at least for certain advanced class gas turbines with model-based adaptive control algorithms. Nevertheless, it is still useful for understanding the concept of the otherwise ambiguous term "full load." Let us consider the ISO base load rating performance. This is the performance of a gas turbine when it is running at 100% load at site ambient conditions of 1 atm, 59°F (15°C) and 60% relative humidity. Let us say that this operation corresponds to IGVs open at X degrees and the controller-set fuel flow is such that at the cycle PR and with turbine inlet temperature (TIT) resulting from the IGV-specified airflow and controller-set fuel flow results in the turbine exhaust temperature (TEXH) listed in the rating data. In other words, *base load* is equivalent to *full load* at ISO conditions.

The industrial gas turbine synchronized to the grid at 50 Hz (3,000 rpm) or 60 Hz (3,600 rpm) is a fixed volume flow machine. Consequently, when the inlet air density changes with changing site ambient conditions (primarily, the ambient temperature), its mass flow rate will change, i.e., it will increase when the air outside becomes cooler and decrease when the air outside becomes hotter. Typically, the controller counters this effect by opening or closing the IGVs (effectively, changing the IGV angle from the ISO base load value). Thus, at each different ambient air temperature, there is a specified airflow. Based on the design of the gas turbine in question, i.e., compressor map, turbine "swallowing capacity" and lifetime expectations of the parts therein, the controller will set a fuel flow to achieve a specified TIT. In the field, this TIT will be determined from the measured cycle PR (via measurements of inlet pressure and compressor discharge temperature, CDT) and exhaust temperature. To a good approximation,

- at ambient temperatures (TAMB) colder than 59°F (15°C), TIT is kept at its ISO base load value,
- at ambient temperatures hotter than 59°F (15°C), TIT is set to TIT at ISO minus the temperature deviation from TAMB.

Consequently, at any given ambient temperature TAMB, full load refers to the power generated with IGVs at an angle per the built-in IGV opening schedule and fuel flow set to the TIT from the built-in PR-TEXH schedule. In general, full load output at colder-than-ISO ambient temperatures will be higher than ISO base load (limited by shaft torque capability and the generator capacity) whereas at hotter-than-ISO ambient temperatures it will be lower than ISO base load. This is illustrated for a vintage F Class unit in Figure 4.2.

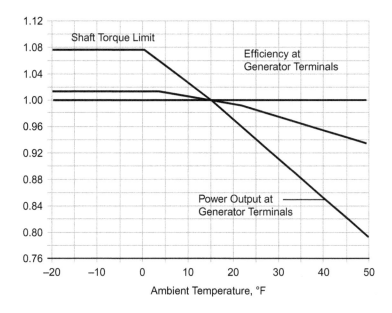

Figure 4.2 Gas turbine output and efficiency as a function of ambient temperature.

The TIT (or firing temperature) schedule mentioned above actually refers to a *collection* of schedules, which are commonly known as *temperature control curves* (TCC). The collection includes different TCCs for

- Gas and liquid (usually number 2 fuel oil as backup) fuels
- Simple or combined cycle operation
- Base, part, or peak loads
- Water/steam injection for NOx control or power augmentation (not applicable with modern Dry-Low-NOx (DLN) combustors)

In practice, the TCCs are composites of line segments or line *pieces*. There are two, three, or more pieces/segments depending on the type of TCC. The base load TCC is a *three-piece* curve as shown in Figure 4.3.

Base load TCC is constructed based on the philosophy of maintaining the nominal TIT. As the ambient temperature and, consequently, *compressor inlet temperature* (CIT) decrease, airflow increases, leading to a rise in *cycle/compressor PR* (CPR). The controller adjusts the fuel flow to arrive at the TEXH indicated by the first segment of the TCC, which implies constant TIT (which cannot be measured directly). Similarly, as the ambient temperature and CIT increase, airflow decreases leading to a decrease in CPR. The controller adjusts the fuel flow to arrive at the TEXH indicated by the second segment of the TCC. At a sufficiently high ambient temperature, the exhaust temperature limit is reached, and the control adjusts the fuel flow to maintain it.

Also known as the "isotherm," the exhaust temperature limit is dictated by the materials used in the exhaust end of the gas turbine, i.e., the exhaust frame, the exhaust diffusers, the struts, etc. Its value is set to protect the turbine hardware from damage

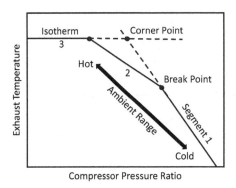

Figure 4.3 Base load temperature control curve.

caused by excessive temperatures. Its typical value for the F Class gas turbines is around 1,200°F (650°C). In modern H/J Class gas turbines with TITs exceeding 2,912°F (1,600°C), it is higher. In some cases, though, especially in combined cycle power plants built in the late 1990s and early 2000s, heat recovery steam generator (HRSG) concerns around the lifetime of parts, mainly due to extended low load operation and frequent cycling, can require a reduction in the isotherm value. This requires construction of new TCCs to be installed in the controller.

4.2   Combined Cycle Start

In a combined cycle startup, there are many considerations in addition to the basic task of bringing the gas turbine to FSNL, synchronizing it to the grid, and loading it to FSFL at OEM-prescribed rates. Most important of all, there is a *second* turbogenerator that needs to be rolled to FSNL, synchronized, and loaded to FSFL: the steam turbine generator. In addition, correct steam chemistry, establishment of steam seals, and maintaining proper vibration, overspeed, and thrust controls are all vital for acceptable component life and reliability, availability, and maintainability (RAM). When all said and done, however, the single most important issue from a fast start perspective is steam turbine *thermal stress management*. Furthermore, if the HRSG is a drum-type design, high pressure (HP) drum thermal stress management becomes an integral part of the problem.

Combined cycle start can be described as the process to reach the dispatch power (e.g., full load or a specific part load) without breaking anything in the process – literally. The failure mode to avoid is crack initiation and propagation. Failure to control thermal stresses results in cracks via *low/high cycle fatigue* (LCF and HCF) and brittle fracture. In fact, LCF is found to account for roughly two-thirds of the steam turbine rotor life with the remainder attributable mainly to creep. In particular, thick-walled components such as the HP drum, steam turbine valves, casings, and rotor are exposed to LCF due to thermal cycling (start-stop sequence or load up-down ramps) and the associated thermal stress-strain loop.

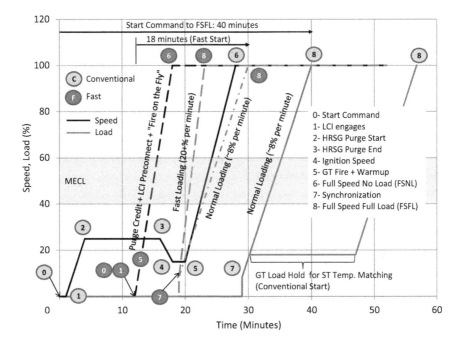

Figure 4.4 Typical gas turbine startup diagram (conventional and fast versions).

There are two basic combined cycle startup methods (see Figure 4.4):

1. Conventional or standard, i.e., *slow*
2. Accelerated, i.e., *fast*

These methods are described below in detail. The fast/rapid method gained acceptance in the last 10–15 years (these lines are written in 2020) for two reasons:

1. Increasingly stringent air permits that put strict limits on NOx and CO emissions during plant start;
2. The ability to act as a backup to wind and solar resources with intermittent nature gaining an increasing share in generation portfolio in North America and Europe (in particular).

It is safe to assume that new combined cycle power plants with advanced class gas turbines will not be designed for conventional (slow) start technology. Even many vintage E and F Class technology power plants out there are (or will be) equipped with new DLN technology. Nevertheless, it benefits us to describe it below for a better understanding of the improvement in start technology over the last two decades in the interest of sustainable power generation.

4.2.1   Conventional Gas Turbine Start

In a conventional start, the following sequence of events happen when the START button is pushed (0):

1. The LCI engages.
2. The gas turbine is rolled to 25% speed (e.g., 900 rpm for a 60 Hz machine) and purging of the HRSG starts.
3. HRSG purging ends after 12 minutes and the LCI is turned off.
4. The gas turbine rolls down to 15% speed, at which point the ignition starts.
5. After a warm-up period when all combustor cans are cross-fired, the gas turbine starts rolling again.
6. With the help of LCI, FSNL is reached at the 28-minute mark.
7. Synchronization is complete and the breakers are closed one minute later; the gas turbine load is about 10%.

Once the generator breakers are closed, the gas turbine is rolled to a very low load, less than 20%, and held there by the DCS (distributed control system) for a prescribed amount of time, which is a function of the time elapsed since the last shutdown. The longer that time, the colder the bottoming cycle metal, i.e., the HRSG (heat exchanger tubes, headers, steam drums, etc.), steam pipes and valves between the HRSG and the steam turbine, steam turbine casing and rotor. The colder the metal, the longer the hold time at low load so that it warms up gradually to prevent excessive thermal stresses, especially in thick-walled components (e.g., the HP steam drum of the HRSG or the steam turbine casing).

Before moving on, we have to open a parenthesis and introduce the concept of *minimum emissions compliant load* (MECL). Modern DLN combustors (i.e., lean, premixed combustion technology) with multiple fuel nozzles can hold the full load NOx and CO emissions down to a certain level of load. The reason for that lies in the basic design philosophy of modern DLN combustors with fuel-air premixing, which are designed to run near the lean limit for low emissions. This is accomplished by piloted, multi-nozzle fuel injectors via sequential activation of fuel flow through individual nozzles (known as *staging*) to prevent lean blow-out and combustion dynamics while staying within the narrow equivalence ratio band to control NOx and CO emissions. In older E and F Class gas turbines with early generation DLN combustors, MECL was quite high, i.e., 60% load. For later variants with next-generation DLN technology, MECL was pulled back to around 40–50% (maybe 35% for the most advanced systems – see Figure 4.5). The exception to the rule is *sequential combustion* (reheat) gas turbines (i.e., GE/Alstom GT24 and GT26), which can turn off their second combustors (low-pressure or SEV combustor in OEM parlance) to operate at 20% or lower load while emissions compliant. For an in-depth discussion of DLN combustor technology and operation, the reader is referred to **GTFEPG** (chapter 12). Most modern DLN combustors with *axial fuel staging* can go as low as 20% in MECL.

Let us now resume the startup narrative from where we left off. The aforementioned hold time (i.e., the gas turbine running at a very low, constant load) is required for steam temperature matching. If the gas turbine is loaded to its MECL, its exhaust temperature will be either very high (e.g., at the *isotherm* for GE F Class gas turbines, i.e., 1,200°F) or at its base value (e.g., for Siemens gas turbines). In either case, steam

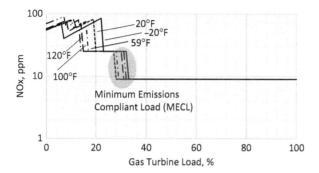

Figure 4.5 Modern F Class gas turbine NOx emissions as a function of load and ambient temperature.

generated in the HRSG will be too high for admission into the steam turbine. Interstage steam *attemperators* (also referred to as *desuperheaters*) between superheater rows are not suitable to accurately control the steam temperature. Thus, HRSG steam temperature is controlled at the requisite level (as dictated by the steam turbine controller) by controlling the gas turbine exhaust temperature. Steam admitted at that low temperature and OEM-prescribed low flow (and controlled by the DCS via *steam bypass* to the condenser) into the steam turbine gradually warms up the rotor and casing metal to allowable levels. Obviously, this warm-up process takes a longer time for colder startups, e.g., up to 3 hours. We can now continue with the gas turbine startup event list.

8. Once the controller deems that the steam turbine is warm enough, the gas turbine is loaded to FSFL at an OEM-prescribed rate, which is about 8% per minute.

This concludes the gas turbine start from TG to FSFL. As described, both prime movers are thermally coupled during the process. In other words, how the gas turbine is loaded from the FSNL to FSFL is dictated by thermal stress management requirements of the steam turbine.

4.2.2   Fast/Rapid Start

To reduce startup emissions, it is critical to shorten the time to MECL. Two steps are instrumental in reducing gas turbine and combined cycle start time:

1. elimination of HRSG purge sequence (by performing it right after shutdown in compliance with NFPA® 85) and
2. elimination of hold time at low load with reduced exhaust energy (flow and temperature) to control HRSG steam production rate and steam temperatures (at the HP drum and HP superheater exit).

Elimination of direct HRSG steam temperature control via gas turbine load and exhaust energy is the *thermal decoupling*, which is the key enabler of fast start.

It can be accomplished via a *bypass stack* and modulated damper controlling the exhaust flow to the HRSG. A recently proposed technique is *air attemperation* of the gas turbine exhaust gas flow via air injection into the transition duct. Ignoring the obvious but wasteful practice of *sky venting*, the currently accepted method is a *cascaded* steam bypass system with *terminal (or "final stage") attemperators* (TA). Steam generation and temperature-pressure ramp rates in the HP drum are dictated by gas turbine exhaust energy whereas final steam temperature control is accomplished by the TAs, which are located at the exit of the HRSG (in addition to the conventional attemperators located between superheater sections). Until steam temperatures reach acceptable levels for admission into the steam turbine, steam is bypassed via a route including the reheat superheater so that the latter is pressurized and "wet" (i.e., cooled by steam flow obviating the need for expensive alloys). This is commonly known as a cascaded bypass, which will be discussed in more detail later in this chapter.

The fast start in Figure 4.4 has the following steps after the START button is pushed (step 0):

1. The LCI is pre-connected and engages immediately (no time lag).
2. Eliminated
3. Eliminated
4. Eliminated
5. "Fire-on-the-fly" ignition[1]
6. With the help of LCI, FSNL is reached in 6 minutes.
7. Synchronization is complete and the breakers are closed one minute later; the gas turbine load is about 10%.
8. Loading to FSFL
   a. At normal rate (~8% per minute)
   b. At faster rate (~20% per minute)

In comparison to the conventional start, which took 40 minutes from START to FSFL, the fast mode takes only 18 minutes, which can be even shorter (12 minutes) if the load ramp rate is increased to 20% per minute. Fast start benefits vis-à-vis the conventional start are:

1. Faster time to full load with about 15% less fuel consumption
2. Reduced startup emissions
3. Improved dispatch ranking

There is a third startup option, which is inbetween the conventional and fast startups. Note that the fast startup capability requires additional hardware to facilitate the elimination of the steam temperature matching hold: terminal attemperators and parallel bypass pipes.

In a conventional cascaded bypass, excess HP steam is first routed to the reheater and from there to the condenser. In a combined cycle capable of fast start, extra steam

---

[1] The warm-up hold period immediately after ignition shown in Figure 4.1 is eliminated. Ignition occurs at the standard rotor speed and airflow. Acceleration is paused until flame is detected and resumes thereafter.

bypass capacity is needed, which is accommodated by a *parallel* bypass line directly from the HP superheater exit to the condenser. This is known as a *hybrid* bypass system. It requires significantly longer and expensive piping with an additional desuperheating station.

By loading the gas turbine to MECL and holding it there for steam temperature matching (albeit for a much shorter duration vis-à-vis the conventional method), parallel bypass piping can be eliminated. This is accomplished by reduced steam production in the HRSG, which can be handled by the standard cascaded bypass piping.

Furthermore, even a less expensive variant is possible for highly cycled power plants, with daily shutdown and hot starts. Due to the limited cooling of the steam turbine during an overnight shutdown (e.g., see), the terminal attemperation system can be eliminated as well for even more reduction in capital cost. This, of course, requires an advanced gas turbine with low MECL capability, i.e., 40% or lower. A combined cycle hot start of this type is shown in Figure 4.6.

The description of the combined cycle startup is provided in Figure 4.6.

1. Steam and gas turbines are on turning gear (TG); the plant is ready to start; operator pushes START button.
2. The LCI accelerates the gas turbine to the purge speed (the steam turbine is still on TG; in a single-shaft configuration, it is separated from the rest of the powertrain via the synchronous, self-shifting [SSS] clutch).
3. Purge completed; the gas turbine slows down to ignition speed; light off and acceleration to FSNL.
4. The gas turbine synchronizes to the grid and is ramped to MECL (45% load as shown) and held there while the HRSG is warmed up.
5. HRSG warm-up is complete; the steam turbine bypass valves are open; terminal attemperators are in temperature matching mode.
6. Steam turbine roll-off to FSNL starts by admitting steam into the IP section (via the ICVs); the SSS clutch engages.

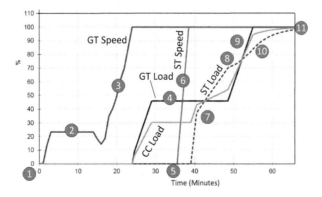

Figure 4.6 Combined cycle hot start with gas turbine hold at MECL for temperature matching. See main text for description of stage numbers.

7. The steam turbine is loaded to the *Forward Flow Transfer* point (where HP steam is admitted into the HP turbine via the MCVs)
8. Steam turbine loading continues to the *Inlet Pressure Control* mode, at which point the bypass valves are closed and gas turbine loading to full load starts.
9. The plant (i.e., the steam turbine – the gas turbine is already at full load) is ramped to full load.
10. Terminal attemperators are turned off, the steam turbine is in full steam admission mode.
11. The plant at 100% (full) load.

Figure 4.6 illustrates another startup definition issue. The first one discussed earlier pertained to the definition of time t = 0, i.e., when to *start* the chronometer. This one pertains to when to *stop* it. Due to the inverse exponential decay nature of the bottoming cycle *heat soak* process, the last part of reaching the "true" 100% combined cycle load takes a long time. A commonly used definition for quantifying the combined cycle start time is to stop the chronometer when

1. bypass valves are closed,
2. terminal attemperators are off, and
3. the steam turbine is in full admission mode.

This is a quite logical choice because, beyond that point, there is no control knob left available to the DCS to impact the combined cycle output. The gas turbine is at full load with IGVs fully open and the temperature control active. All bypass valves are fully closed, and all admission valves are fully open. In other words, there is nothing left except to wait until all components reach their fully heat-soaked, steady-state operating level. In any event, based on the definition used for beginning and ending points, the start time can change by 10–20 minutes for a hot start.

4.2.3   Example

Let us look at some numbers. A 50 Hz (3,000 rpm) F Class gas turbine rated at 275 MWe weighs about ~390,000 lb (about 175 metric tons) including the generator and load coupling. Rotational inertia of the powertrain is about 1,400,000 lbf-sqft (about 60,000 kg-m^2). The torque required to crank this powertrain in, say, 3 minutes from 10 rpm (TG speed) to 15% speed (roughly the ignition speed) can be calculated as follows:

$$\omega = 2 \times \pi \times 50 = 314 \text{ rad/s,}$$
$$\alpha = 0.15 \times 314/(3 \times 60) = 0.2618 \text{ rad/s}^2,$$
$$\tau = 60,000 \times 0.262 \sim 15,700 \text{ Nm.}$$

The torque generated by this gas turbine at its rated output is found from Equation (2.5) as

$$\tau = 275,000 \times 1,000/314 = 875,350 \text{ Nm.}$$

What we calculated above for the full load torque, i.e., 875,350 Nm, is the running torque generated by the turbogenerator at constant 3,000 rpm when connected to the grid. Thus, the acceleration torque is about 1.8% of the rated torque output of the gas turbine generator.

If this gas turbine (175 mt powertrain with 60,000 kg-m^2 rotational inertia) were completely at standstill, i.e., the shaft resting on the bearings at 0 rpm, one would require a very large amount of breakaway torque.

4.2.4 Cascaded Bypass

The steam turbine is not an internal combustion engine. As such, it cannot be *cranked* to a certain speed where it becomes self-sustaining via ignition. It is *rolled* from TG to synchronization and FSNL by having steam flow through it in the requisite quantity (flow rate) and quality (i.e., pressure and temperature) to overcome the rotational inertia of the turbine itself and its generator. This requires the passage of a certain amount of time so that the requisite steam conditions are achieved in the HRSG. In conjunction with this requirement are three independent parameter controls:

1. **Steam flow control**. It is impossible to control the amount of steam generation precisely to that requisite for admission into the steam turbine. Thus, excess steam generated by the HRSG bypasses the steam turbine and is sent to the condenser. This is achieved by the cascaded bypass arrangement of steam pipes and valves between the HP superheater, reheat superheater, and the steam turbine.
2. **Steam pressure control**. Steam pressure in the steam system is created by the flow passing ability of steam admission and bypass valves and the pressure "information" propagates upstream (as an acoustic wave) through the steam pipes and HRSG superheaters to the HRSG evaporator steam drums. The critical piece of equipment is the HP steam drum with thick walls. What is more important than the actual steam pressure in the drum is the rate of increase, i.e., $\partial P/\partial t$, which is crucial for maintaining thermal stress in the drum walls at an acceptable level.
3. **Steam temperature control**. As described earlier, in the now-outdated conventional startup technology, the steam temperature was controlled by the gas turbine exhaust energy (i.e., flow and temperature) pumped into the HRSG. In modern combined cycle power plants, precise steam temperature (HP and hot reheat) control is achieved by the use of TAs. In other words, the superheated steam temperature from the *exit* of the HRSG superheaters is reduced by condensate spray into the steam.

A modern cascaded steam bypass system is schematically illustrated in Figure 4.7. This system is also known as a "wet bypass" system (also referred to as *European* bypass system). The term "wet" alludes to the fact that the HP steam bypassing the steam turbine is first passed through the reheat superheaters before being directed to the condenser. The advantage of this method is the cooling effect of steam flowing through the reheat superheater tubes subjected to hot exhaust gas. In the older bypass systems with HP steam directly sent to the condenser, reheat superheater tubes were

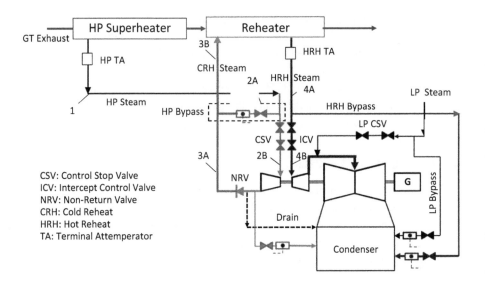

Figure 4.7 Cascaded steam bypass system.

"dry," i.e., there was no steam flow through the tubes. The disadvantages of the older dry or parallel bypass system are twofold:

1. Lack of steam cooling necessitating expensive alloy materials for reheat superheater tubes, which adds cost to the HRSG.
2. Long lengths of expensive alloy HP steam piping from the HRSG to the condenser.

To understand the second drawback, refer to Figure 4.7, where each bypass system is characterized by a pressure letdown (throttle) valve and attemperating water spray station, which are located close to the condenser. The water spray system that injects a controlled quantity of water into the steam flow to prevent the throttled steam's temperature/enthalpy from exceeding the condenser design limit (about 1,200 Btu/lb or per condenser OEM's specs). Consequently, the direct HP bypass pipe between the superheater exit and the bypass conditioning system inlet would be quite long and expensive because of the high steam temperature.

As shown in Figure 4.7, until steam admission into the steam turbine, HP steam goes through the pipe segment 1, HP bypass line 2A, and then segment 3B of the cold reheat (CRH) pipe into the reheater (reheat superheater). Thereafter, reheated *hot reheat* (HRH) steam goes through segment 4A of the HRH pipe and then through the HRH bypass pipe to the condenser. Both HP and HRH steam temperatures at the exit of respective superheaters are reduced in the TA.

4.2.5　Steam Turbine Roll

The steam turbine generator is *rolled* to FSNL under the action of steam flowing through the turbine. This can be accomplished via HP or IP (hot reheat to be precise)

Table 4.1 Steam turbine startup categories

	Down Time, h	Metal Temperature, °C
Cold	140	150
Warm	48	275
Hot	8	400

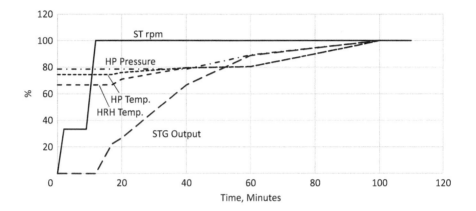

Figure 4.8 Steam turbine cold start.

steam admission. Until that point, in multi-shaft configurations, the steam turbine is idle, i.e., rotating at the TG speed (a few rpm).

Steam turbine startup schedules are also categorized in terms of hot, warm, and cold. The deciding parameter is the temperature of the HP-IP rotor. A typical classification is shown in Table 4.1. A sample cold steam turbine startup schedule is shown in Figure 4.8.

As shown in Figure 4.8, the steam turbine roll to FSNL is accomplished in two steps, with a break for allowing steam temperature matching. (Another key point in this process is that the critical speed ranges are passed through quickly.) Once the generator is synchronized to the grid (assuming that the steam turbine has its own generator), the unit is loaded to FSFL by controlled increase of steam admission and steam temperatures. This process is governed by the *turbine stress controller* to ensure that the thermal stresses are kept in check. The underlying principles are described in detail in the article by the author ([35] in Section 2.4.3).

In a single-shaft powertrain, an SSS clutch is installed between the steam turbine and the generator. During the combined cycle start, the gas turbine is started first and the generator is synchronized to the grid. During this period, the steam turbine is disconnected from the rest of the powertrain by the SSS clutch. The steam turbine is rolled, as described above, to FSNL via HP or IP steam admission. Once it reaches the synchronous speed (3,600 or 3,000 rpm), the SSS clutch engages and connects the steam turbine to the generator.

4.2.6   How Long?

In several recent combined cycle projects ($2\times2\times1$ multi-shaft configuration) with conventional drum-type HRSG, start time commitments from two different OEMs were

- Cold Start: 226/220 minutes
- Warm Start: 128/115 minutes
- Hot Start: 38/58 minutes

Listed times are commercial values selected to match the requirements of the plant air permit. For example, the first set is defined from gas turbine ignition to HRSG stack emissions compliance (including the selective catalytic reduction [SCR]) with the gas turbines at MECL. In another project ($3\times3\times1$ multi-shaft configuration), OEM-estimated (*not* guaranteed) start times (from first GT ignition to steam turbine at base load) were

- Cold Start: 197 minutes
- Warm Start: 136 minutes (72 hours down time)
- Hot Start: 85 minutes (16 hours down time)

The sample start times listed above are significantly longer (especially for warm and cold starts) than the fast start capability advertised by OEMs in the trade literature, e.g.,

- Cold Start: 60 minutes
- Warm Start: 45 minutes
- Hot Start: 30 minutes

In fact, they are more in line with the conventional combined cycle start times, which can be found in papers and articles from 10 to 20 years ago, i.e.,

- Cold Start: 3 hours
- Warm Start: 2 hours
- Hot Start: 1 hour

The reason for the discrepancy is not that the new technologies are not capable of achieving what they promise. There is simply very little commercial need for fast start after an extended shutdown (e.g., for maintenance), which are events planned in advance and take place as scheduled. Therefore, there is little or no financial incentive to put the plant equipment through unnecessary thermal stresses. Consequently, after an extended shutdown (several days or longer), operators choose to start their equipment at a slow pace with ample time for gradual warming *as long as they are compliant with their air permits.* A typical permit (for a $3\times3\times1$ advanced F Class gas turbine combined cycle power plant) reads something like this:

Pursuant to the best available technology requirements of [applicable code sections], the total emissions from the combined cycle gas turbines combined (including HRSG duct burners) shall not exceed the following totals in any 12 consecutive month period. (These emissions limits also include those during startup and shutdown events.)

- Nitrogen Oxides – 230.16 tons
- Carbon Monoxide – 388.9 tons
- Volatile Organic Compounds – 71.6 tons
- Sulfur Oxides – 54.3 tons
- Total Particulate Matter (PM) (including PM10 and PM2.5) – 197.0 tons
- Sulfuric Acid Mist – 26.6 tons
- Ammonia – 213.7 tons

For a gas turbine with a fired HRSG (including an SCR), typical NOx emission guarantee is 2 ppmvd @ 15% O_2. This corresponds to about 75 tons per year of NOx as NO_2 emissions (8,760 hours at full load with duct-firing) for an F Class gas turbine with ~1,200 lb/s (~545 kg/s) exhaust flow. For three gas turbines, one is looking at 225 (short) tons/year. Prima facie, this leaves 5 tons/year for emissions during startup and shutdowns. Obviously, the actual plant is going to be run much less severely than the non-stop, round-the-clock duct–fired assumption above. Thus, operators will have a good idea about their emissions budget when deciding how to start the plant at a given time in a given year.

4.3   Other Features

In addition to the startup and stable operation of the prime movers and plant systems, other important functions of the control system are monitoring of operating parameters, issuing warnings and alarms when protective limits are approached or breeched, and safe shutdown. The control system features accomplishing these tasks are (i) sequencing, (ii) load-speed control, and (iii) monitoring and protection.

As described in the preceding sections, speed-load control via modulating IGVs and fuel flow is critical during startup phases, i.e., acceleration from the TG to synchronous speed, synchronization, and loading from FSNL to FSFL. This is accomplished by the controller commonly referred to as the "governor." The governor actively controls the turbine speed even during normal operation at synchronous speed when the unit is connected to the grid.

The governor controls the turbine speed by adjusting the fuel flow. Any changes in the load torque will momentarily result in an imbalance between the shaft and load torques. This imbalance causes a speed change that results in the speed governor making the requisite adjustment to the fuel flow. The imbalance exists until the shaft torque matches the load torque again. The details of this process, e.g., *isochronous* or *droop* modes, are described in detail in chapter 18 in **GTFEPG**.

4.3.1   Dual-Fuel Combustion

In some cases, industrial gas turbines have a dual-fuel combustion system, which allows the operator to switch from the primary natural gas fuel to the secondary or backup fuel, i.e., typically #2 distillate. While owners are eager to forgo a dual-fuel system for CAPEX savings, certain grid operators such as PJM in the USA require the backup fuel capability for system reliability. Furthermore, the gas fuel system must be able to handle variations in fuel gas composition and heating value. This capability is quantified by the modified *Wobbe* Index, which is defined as

$$\text{MWI} = \frac{\text{LHV}}{\sqrt{\text{SG} \cdot \text{T}_{\text{fuel}}}} = \frac{\text{WI}}{\sqrt{\text{T}_{\text{fuel}}}}$$

where WI is the acronym for the *Wobbe Index*, SG is the *specific gravity* of the fuel relative to air, LHV is in Btu/scf (i.e., it is the volumetric heating value) and fuel temperature is in degrees Rankine. (In SI units, use kJ/m^3 and degrees Kelvin, respectively.) The Wobbe Index is a relative measure of the energy injected to the combustor at a fixed fuel nozzle PR. Note that WI is a dimensional number and care must be given to the unit system used in calculating it to prevent gross errors. Allowable MWI variation used to be ±5% in the early F Class gas turbines. Modern H and J Class gas turbines have an extended operability range commensurate with the fuel flexibility requirements (e.g., blending natural gas with ethane or propane, limited natural gas reserves in some countries) up to ±15% or even higher.

In principle, the gas turbine can be started on either fuel. However, due to stringent emissions requirements, liquid fuel use is avoided unless absolutely necessary. Transfers from one fuel to the other fuel may be initiated by the operator at any time after the completion of the start sequence. In general, the control system automatically transfers from gas to distillate if a low gas supply pressure is detected to ensure stable, continuous operation. The transfer back to gas is usually only operator initiated. Operator initiated transfers prevent unstable operation if the gas supply pressure is marginal at the transfer initiation pressure.

During the transfer from one type of fuel to the other, the key is to maintain the total energy flow (in Btu/s or kWth) as a function of the individual fuel flows. This ensures equal energy release in the combustors from the two fuels. The transfer sequence is divided into two parts: a line fill period and the actual transfer. The incoming fuel command signal is increased to a level that will allow the system to fill with fuel. At the same time, the outgoing fuel command signal is decreased by an equal energy amount. A slight decrease in power output may be observed until the incoming fuel system is filled. When the line fill period has been completed, the incoming fuel command signal is increased to equal the total fuel command. Simultaneously, the outgoing fuel command signal is decreased to zero. Thus, the total energy to the gas turbine is held roughly constant to minimize the load variation during the fuel transfer. The next step in the fuel transfer process is to purge the outgoing fuel system. Because a purge of the outgoing fuel system results in additional fuel injection into the gas turbine, a potential for a load disturbance is possible if the purge is initiated too abruptly. Once the outgoing fuel system is cleared of fuel, the potential for load variations disappears. The purge sequence is designed to minimize the effects of load variations.

4.3.2 Turbine Protection

Gas turbine protection systems are designed to automatically react to conditions that could cause damage to the machine. Protection systems receive input signals from the transducers measuring critical operating parameters. They include

- Rotational speed
- Vibration
- Flame presence
- Compressor inlet pressure drop
- Turbine wheel space temperatures
- Exhaust temperature (multiple measurements)
- Exhaust temperature spread
- Bearing temperatures
- Lube oil temperature

There are typically two preset limits for critical parameters:

- *Low* (when the parameter value is decreasing) or *high* (when the parameter value is increasing) resulting in a warning or alarm (it can be visual or audible) when exceeded
- *Low-low* or *high-high* resulting in an automatic trip when exceeded

Lube oil and hydraulic control/trip oil pressures are the two parameters that can initiate a gas turbine trip. The lube oil pressure detector senses the lube oil pressure at the extreme end of the lube oil header that supplies the gas turbine bearings. Typical low and low-low values are 12 and 8 psig, respectively. They can change from unit to unit. If the hydraulic oil pressure decreases to about 1,050 psig, a low alarm is issued and, if one is available, the auxiliary hydraulic oil pump will kick in. The low-low level is typically 980 psig, leading to a trip.

The overspeed control system is independent of the turbine speed control. Typically, a minimum of one electric or one mechanical trip device per shaft is provided. In a fully electronic system is a triple redundant system with three shaft probes and two-out-of-three voting logic. In multi-shaft gas turbines such as small industrial turbines, each shaft has its own overspeed trip protection. On-line testing is required without over-speeding the turbine.

While the description above is for conventional heavy-duty gas turbines, the reader should be aware that the same considerations are equally valid for any new or innovative system incorporating rotating components, e.g., expanders.

4.3.3 Plant Normal Shutdown

Upon receiving the shutdown command, the gas turbine starts unloading with a normal ramp rate. In the case of duct-fired operations, the duct burner is switched off with an OEM-prescribed ramp rate, e.g., 10%. Once the exhaust gas temperature is below the HP steam temperature, the gas turbine is kept at the load for an

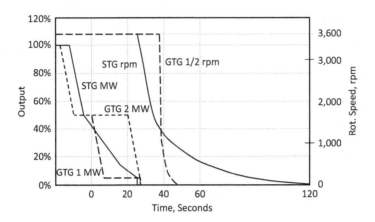

Figure 4.9 Normal shutdown of a $2\times2\times1$ GTCC power plant.

OEM-prescribed duration, e.g., 5 minutes, during which the HP bypass control valve takes over the HP steam pressure control from the steam turbine. All the HP steam is directed through the cascaded bypass system to the condenser. The gas turbine is further unloaded to FSNL with the exhaust gas temperature at around 750–800°F (~400–425°C). The gas turbine is held at this state for a prescribed period (e.g., 10 minutes) to soak the HRSG. Thereafter, the gas turbine is decelerated until the unit is completely shut down and placed on the turning gear.

During the steam bypass, the set point of the HP bypass control valve is tracking the process value, but not lower than the HP floor pressure (typically, 40% of the normal operation value). The set points of the IP and LP steam admission valves are fixed to their respective floor pressures.

The process described above is for a $1\times1\times1$ GTCC. For a multi-gas turbine GTCC, it is similar with the exception of the staged shutdown of the gas turbines. A typical normal shutdown sequence for a $2\times2\times1$ GTCC is shown in Figure 4.9.

4.4   Power–Speed–Frequency Control

A rotating machine (a prime mover) in electricity generation duty, via its ac generator, is connected to an interconnected network, which is commonly referred to as an electrical or power grid or, simply, "the grid." The ultimate duty of the grid is to transmit electric power generated by a multiplicity of generators (fossil fuel–fired, nuclear, renewable) to an order-of-magnitude larger number of end users (industrial and residential) via a complex network of transmission lines at the highest possible reliability and the lowest possible cost. Since electric power must be consumed when generated, this requires a continuous balancing act of matching demand from the end users (which fluctuate from minute to minute, hour to hour, and day to day) with the supply from generators with very different performance and operability characteristics.

Consequently, operation and maintenance of the grid is subject to rigorous rules and regulations, which constitute the "grid code." A grid code specifies the technical criteria pertaining to the design and operation of electric power generation plants such as

- Quality of supply (e.g., voltage and frequency)
- Protection (e.g., backup in case of system failure)
- Generating unit specifications (e.g., power factor, frequency response)
- Metering and monitoring requirements

Different grid codes are developed and enforced by different countries/groups of countries (e.g., the European Union). In the USA, for example, there are three main power grids (the Eastern, Western, and Texas Interconnects), and ten regional reliability councils of the *North American Energy Reliability Council* (NERC). Members of those councils come from all segments of the electric industry, e.g., investor-owned utilities, rural electric cooperatives, independent power producers.

Frequency stability and control are the most important grid code requirements. Grid frequency is the single most important measure of the balance between electric power supply and demand. If power generation and consumption (the "load") in the grid are exactly matched, the system frequency is exactly equal to the rated frequency (i.e., 50 or 60 Hz depending on the region). Unexpected malfunctions in any given part of the grid such as an emergency trip of a large generation station creates an imbalance between supply and demand, which is reflected by a change in system frequency.

In fact, the grid frequency is a continuously changing variable. If demand or load is greater than generation, the system frequency falls, i.e., generators connected to the grid slow down. If generation is greater than demand, the system frequency goes up, i.e., generators connected to the grid speed up.

In general, three types of events cause a change in grid frequency:

- Loss of generation (supply)
- Loss of load (demand)
- Normal variations in load and generator output

For stable operation, grid frequency should be held within narrow limits defined by the applicable grid code. Minor deviations such as 200 mHz (0.2 Hz) or the absence thereof indicate that there is a balance between generation and load. Unexpected, sudden events such as plant trips and load losses result in system faults and lead to larger deviations, which, unless corrected, can destabilize and collapse the grid.

There are two types of faults:

- Faults *within* a controllable range
- Faults *outside* the controllable range

Faults in the former category lead to fluctuations within a small band (e.g., ±200 mHz) and can be ridden out by adjustments on the generation side. In principle, it must be possible to ride out the loss of the largest generator in the system without a frequency excursion outside the controllable band.

Faults in the latter category cannot be alleviated by only increasing or decreasing power generation. In severe cases of underfrequency (i.e., loss of significant amount of generation), the system survives by "load shedding," i.e., throwing certain segments of users into "darkness." Even if the underfrequency event takes place very fast, i.e., within a few seconds, as long as it remains above a grid code-specified limit, generators must continue operating stably as required by the grid code.

The control of power generation by the prime movers, e.g., steam, gas and other types of rotating equipment as well as piston-cylinder engines, and frequency is referred to as *load-frequency control* (LFC). A grid or network comprises many generators of those types. Recently, the mix increasingly included renewable resources, solar and wind, with different characteristics. This added to the complexity of an already complex balancing act. In this section, we will take a brief look at the governing principles of LFC. Readers who want to study the subject further are pointed to two superb books, one by Kundur [1], which dives deep into the power system stability and control, and the other by Klimstra [2], which is more oriented to the lay audience. The coverage below heavily draws upon those two sources.

Let us start with the basics, i.e., a prime mover generator connected to an isolated load as shown in Figure 4.10.

To determine the fundamental relationships between the parameters in Figure 4.10, we have to start with the basic equations of motion. When a torque, T, is applied to a body with moment of inertia, J, the angular acceleration, α, is given by

$$\alpha = T/J,$$
$$\alpha = d\omega/dt,$$
$$\omega = d\theta/dt,$$

where θ is the angular displacement (in radians, e.g., 180 degrees are equal to π radians). For a prime mover rotating at N rpm, the equations of motion are

$$\omega = 2\pi \frac{N}{60} = 2\pi f, \tag{4.1}$$

$$\dot{W} = \omega T, \tag{4.2}$$

where f is the frequency in Hz, ω is the angular rotor speed, $\dot{W}$ is power, and T is torque. The subscripts in Figure 4.10 are as follows: m is for *mechanical* (i.e., shaft), e is for *electrical*, and L is for *load*. During normal operation, $T_m = T_e$ or $\dot{W}_m = \dot{W}_e$.

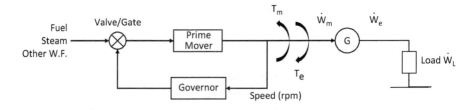

Figure 4.10 Simple prime mover generator-load system.

If there is a small deviation or disturbance in one of the parameters (*perturbation* in mathematical lingo), from the differentiation of Equation (4.1) and neglecting higher order terms we find that

$$\Delta \dot{W}_m - \Delta \dot{W}_e = \Delta T_m - \Delta T_e. \tag{4.3}$$

Now, if there is an imbalance between the two torques acting on the rotor of the prime mover, the resulting net torque causes an acceleration or deceleration, i.e., from the equation of motion,

$$J\frac{d\omega}{dt} = T_m - T_e = T_a, \tag{4.4}$$

where J is the combined (total) moment of inertia of the prime mover and its generator in kg-m^2, and T_a is the *accelerating torque* in Nm. The kinetic energy of a rotating system is given by

$$E_k = \frac{1}{2}J\omega^2. \tag{4.5}$$

The ratio of the kinetic energy to the rated power, $\dot{W}_0$, is the *inertia constant*, H, in units of seconds, i.e.

$$H = \frac{1}{2}\frac{J\omega_0^2}{\dot{W}_0}, \tag{4.6}$$

where ω_0 is the rated angular speed, so that

$$J = \frac{2H\dot{W}_0}{\omega_0^2}, \tag{4.7}$$

and with substitution into Equation (4.4), results in

$$\frac{2H\dot{W}_0}{\omega_0^2}\frac{d\omega}{dt} = T_m - T_e = T_a, \tag{4.8}$$

$$2H\frac{d}{dt}\left(\frac{\omega}{\omega_0}\right) = \frac{T_a}{\dot{W}_0/\omega_0} = \frac{T_a}{T_0}. \tag{4.9}$$

Denoting the nondimensionalized parameters by a bar,

$$2H\frac{d}{dt}(\bar{\omega}) = \bar{T}, \tag{4.10}$$

which, by integration from 0 to the rated values, i.e., unity, results in the *mechanical starting time*, M = 2H. Equation (4.4) defines the transfer function relating speed and torque or power, which is depicted in Figure 4.11.

The inertia constant, H, is a highly useful yardstick that helps one to assess the capability of a prime mover generator to change its speed in response to an imbalance between power supply and demand manifesting itself in a change in frequency. Quantitatively, H is a measure of the energy stored in the rotating mass of the

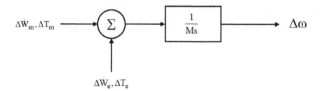

Figure 4.11 Transfer function representation of the relationship between speed and power/torque.

equipment, i.e., just enough to supply the rated load of the generator for H seconds. For large F or H/J Class gas turbines, H is of the order of 10 seconds. For smaller, industrial gas turbines or gas-fired recip engines, H is about 5 seconds.

Let us take the time derivative of both sides of Equation (4.5), i.e.,

$$\frac{dE_k}{dt} = J\omega \frac{d\omega}{dt}, \tag{4.11a}$$

$$\frac{dE_k}{dt} = 4\pi^2 Jf \frac{df}{dt}. \tag{4.11b}$$

The rate of change of kinetic energy of a rotating system is the net power acting on the system, thus

$$\frac{dE_k}{dt} = \Delta\dot{W} = \dot{W}_m - \dot{W}_e = 4\pi^2 Jf \frac{df}{dt}. \tag{4.12}$$

Consequently,

- if $\dot{W}_m > \dot{W}_e$, df/dt > 0 and the generator accelerates,
- if $\dot{W}_m < \dot{W}_e$, df/dt > 0 and the generator decelerates.

Substituting Equation (4.7) into Equation (4.12), gives

$$\frac{\Delta\dot{W}}{\dot{W}_0} = \left(\frac{H}{f_0^2}\right)\frac{df^2}{dt}. \tag{4.13}$$

Noting that at time, t = 0, f = f_0, when the power imbalance just occurs, one can write that

$$\left.\frac{df}{dt}\right|_{t=0} = \left(\frac{f_0}{2H}\right)\frac{\Delta\dot{W}}{\dot{W}_0}. \tag{4.14}$$

As an example, consider a 60 Hz large gas turbine generator experiencing a sudden load drop of 10%. If H = 10 s,

$$\left.\frac{df}{dt}\right|_{t=0} = \left(\frac{60}{20}\right)0.1 = 0.3 \ \frac{Hz}{sec}.$$

In other words, the initial tendency of the unit is to accelerate at a rate of 0.3 Hz (18 rpm) per second.

For a constant power imbalance, $\Delta \dot{W}$, between power supply and demand, Equation (4.13) can be integrated to find f(t), i.e.,

$$f(t) = f_0 \sqrt{1 + \frac{\Delta \dot{W}}{\dot{W}_0} \frac{t}{H}}. \tag{4.15}$$

Power system loads comprise different types of devices. Some of them, e.g., lighting and heating, are *resistive* loads, which are independent of frequency. Others, e.g., motors driving fans, pumps, and compressors, are dependent on frequency, which impacts the motor speed. Combining the two types of loads, we have

$$\Delta \dot{W}_e = \Delta \dot{W}_L + \frac{D}{100} \dot{W}_L \left(\frac{\Delta f}{f_0}\right), \tag{4.16}$$

where D is the *load damping constant* and $\dot{W}_L$ is the initial value of the load. A value of $D = 2$ means that a 1% change in frequency would cause a 2% change in load. The load damping constant is a measure of the grid's *self-regulation*. The primary cause of self-regulation is the synchronous electric motors in the grid. When the grid frequency decreases, motor speeds decrease along with their power consumption. This can be as high as 6% per unit Hz of frequency decrease in motor driven pumps. As stated above, the load mix of a large grid is an eclectic mix so that a typical value for D is around 1.

Combining Equation (4.16) with Equation (4.13) gives

$$\frac{df^2}{dt} = \left(\frac{f_0^2}{\dot{W}_0 H}\right)\left(\Delta \dot{W} + \frac{D}{100} \dot{W}_L \left(\frac{\Delta f}{f_0}\right)\right). \tag{4.17}$$

Once the imbalance event takes place, the *primary control* reserve, $\dot{W}_{PC}$, will kick in to rectify the disturbance and must be accounted for in the frequency change equation, Thus, Equation (4.17) becomes

$$\frac{df^2}{dt} = \left(\frac{f_0^2}{\dot{W}_0 H}\right)\left(\Delta \dot{W} + \frac{D}{100} \dot{W}_L \left(\frac{\Delta f}{f_0}\right) + \dot{W}_{PC}\right). \tag{4.18}$$

In the general case of $\Delta W \neq$ const., Equation (4.18) must be integrated numerically with $\dot{W}_{PC} = f(t)$. This can be easily done using, say, fourth-order Runge–Kutta integration.

Typically, primary reserves react with a constant ramp rate, e.g., starting a few seconds after the event, to reach 100% in, say, 30 seconds, i.e.,

$$\bar{\dot{W}}_{PC} = 0, t \leq t_{lag}, \tag{4.19a}$$

$$\bar{\dot{W}}_{PC} = \frac{1}{30}(t - t_{lag}), t > t_{lag}, \tag{4.19b}$$

where $\bar{\dot{W}}_{PC}$ is the normalized value of the primary control reserve. During those first few seconds, ignoring the contribution of self-regulation, i.e., $D = 0$, the system will behave as described by Equation (4.15).

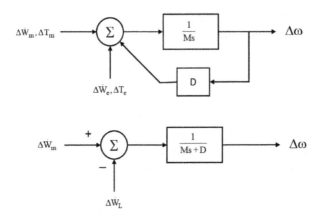

Figure 4.12 Transfer function with inertia and damping (self-regulation) constants.

In the absence of a speed governor, the prime mover generator will respond to a load change as determined by the inertia and the damping constants, i.e., $M = 2H$ and D, respectively, as shown in Figure 4.12.

Consider a 300 MW generating unit operating at $f = 50$ Hz (3,000 rpm) with $H = 10$ s. System self-regulation is represented by $D = 1.5$. There is a sudden drop in load by 5 MW, i.e., $\dot{W}_L = -5$ MW or $5/300 = 0.0167$. Two new parameters are defined, i.e.,

$$K = 1/D = 1/1.5 = 0.667,$$
$$T = M/D = 2 \times 10/1.5 = 13.33 \text{ s (time constant)},$$

so that on a *per unit* (p.u.) basis,

$$\bar{\Delta f} = \frac{\Delta \dot{W}}{\dot{W}_0} K \left(1 - e^{-t/T} \right), \qquad (4.20a)$$

$$\bar{\Delta f} = 0.011139 \left(1 - e^{-t/13.33} \right). \qquad (4.20b)$$

Equation (4.20b) is plotted in Figure 4.13. For comparison, Equation (4.15), without damping (self-regulation), is also plotted. As shown in the figure, without a speed governor in action, the unit accelerates due to the sudden drop in load. Grid self-regulation arrests the frequency/speed increase at $\Delta f = 0.011139 \times 50 = 0.56$ Hz. Without the damping action of the self-regulation, the acceleration would continue until the unit trip at the maximum allowable speed is initiated by the DCS.

In the case of a sudden load jump of 5 MW, the behavior of the system would be a mirror image of what is shown in Figure 4.13 around the x-axis. Grid self-regulation would arrest the frequency/speed decrease at $\Delta f = -0.011139 \times 50 = -0.56$ Hz.

Let us now look at the situation with a speed governor in action. There are two types of speed governor:

1. droop or "straight proportional" speed control
2. isochronous or "integrated" (proportional *plus* reset) speed control

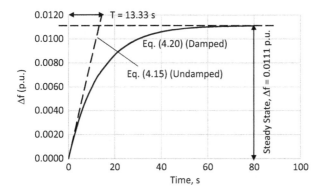

Figure 4.13 Generating unit (prime mover plus ac generator) response to a load loss without speed governor.

In general, droop governor is applicable to operations when connected to a grid (i.e., fixed system frequency and shaft rotational speed). Isochronous governor is applicable when the gas turbine is *not* connected to the grid and operates in an isolated or *islanded* mode. It is essentially a *zero droop* governor.

The droop is defined as the ratio of the relative change in system frequency to the relative change in generator power output, i.e.,

$$d = \frac{\frac{\Delta f}{f_0}}{\frac{\Delta \dot{W}}{\dot{W}}} \times 100, \tag{4.21}$$

where

$$\Delta f = \text{Change in system frequency in Hz}$$
$$f_0 = \text{Rated system frequency in Hz}$$
$$\Delta \dot{W} = \text{Change in generator output in kW or MW}$$
$$\dot{W} = \text{Rated generator output in kW or MW}$$
$$d = \text{Droop in percent}$$

For example, if a 1% change in frequency (i.e., 0.5 Hz for a 50 Hz grid) causes a 25% change in output, the droop is

$$d = 0.01/0.25 \times 100 = 4, \text{i.e.}, 4\%.$$

The droop governor operates as follows:

1. It compares the actual speed to a reference (set point) value, which is also known as the "speed changer."
2. The difference between the two is referred to as the "speed error."
3. To maintain the actual speed at its set point value, the droop governor changes the fuel flow (i.e., the generator output) in proportion to the speed error.
4. The proportionality factor is d.

Thus, for a 4% droop governor, to run a prime mover, e.g., a gas turbine, at 100% output/load, the operator sets the speed changer to 4% because, as per Equation (4.21),

$$\frac{\Delta \dot{W}}{\dot{W}} = \frac{\frac{\Delta f}{f_0}}{d} \times 100 = \frac{0.04}{4} \times 100 = 1 \text{ or } 100\%.$$

Similarly, for a 4% droop governor, to run the gas turbine at 50% output/load, the operator sets the speed changer to 2% because, as per Equation (4.21),

$$\frac{\Delta \dot{W}}{\dot{W}} = \frac{\frac{\Delta f}{f_0}}{d} \times 100 = \frac{0.02}{4} \times 100 = 0.5 \text{ or } 50\%.$$

The droop governor curve for 50% output at 100% speed is shown in Figure 4.14. From the preceding discussion, it should be obvious that 4% droop operating lines for 75% and 100% load are constructed by shifting the operating line in Figure 4.14 to the right (i.e., 3% and 4% speed error, respectively).

The concept of a droop governor is not intuitive; in fact, it is *counterintuitive*. In contrast, the isochronous governor concept is perfectly intuitive. In the example above, when the system/grid frequency goes up, the isochronous governor would simply decrease the fuel flow and generator output to maintain the rated frequency. Why is it not chosen? Here is why:

Without some form of droop, turbine speed regulation would always be unstable in a networked system, i.e., the electric grid supplied by many generators.

This can be best understood by an analogy to a simple physical system shown in Figure 4.15.

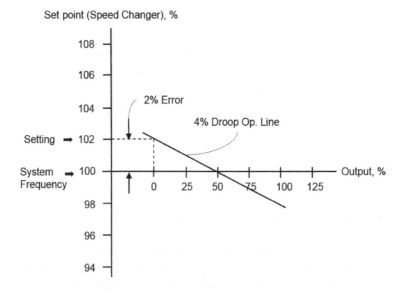

Figure 4.14 Droop governor – 50% load at 100% speed/frequency.

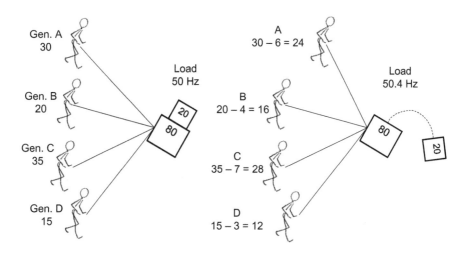

Figure 4.15 Physical analogy to generators with droop governor responding to a loss in load.

In Figure 4.15, four men pull a 100 unit load, which is analogous to a 50 Hz electric grid. Consequently, the four men are analogous to four gas turbine generators. Each man (generator) provides a certain fraction of the load. All of a sudden, a 20 unit load block falls off. The mismatch between the force exerted by the four men, 100 units, and the load, 80 units, would increase the speed of the system, which, in grid analogy, corresponds to a frequency increase of 0.4 Hz. Here is the catch:

Without droop, generators would not know how to divide the new load (80 units) among themselves!

If the generators/men were all operating as an isochronous governor, they would all try to correct the frequency/speed on their own, which would lead to utter chaos and system breakdown.

With 4% droop governor, however, each generator knows exactly how much output correction it needs to make. For generator A, for example, using Equation (4.21),

$$\Delta \dot{W} = \frac{\frac{\Delta f}{f_0}}{d} \times 100 \times \dot{W} = \frac{\frac{0.4}{50}}{4} \times 100 \times 30 = 6 \text{ units.}$$

A similar correction is made by generators B, C, and D *independently* and their total output matches the new system load, 80 units, and the system goes back to its stable running speed/frequency.

If, on the other hand, a particular gas turbine generator is the only unit on the grid, an isochronous governor is indeed the logical choice because the generator can set its speed and, thus, the grid frequency. But for a generator connected to a large electrical grid of tens of gigawatts, it is the grid that determines the frequency and speed of the particular electric generator. For almost all cases, the grid disturbance and frequency

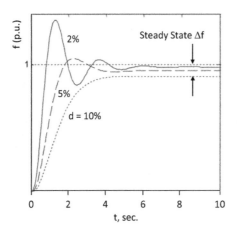

Figure 4.16 Gas turbine speed governor response with different droop settings.

variation are too big to be influenced by a single generator. The droop governor, however, does *not* try to change the speed/frequency of the system. It simply decreases or increases the governor speed reference as the load increases or decreases. This allows the governor to vary the gas turbine load since the speed, which is determined by the grid frequency, cannot change.

The droop setting, d, can be adjusted in the field between 2 and 10 (i.e., a 2% or 10% droop). Most national grids in the world operate at 4% or 5% droop. A low value of d means a large load change for a small change in system frequency. In the same vein, a high value of d means a small load change for a large change in the system frequency. In the former case, load control stability is not very good (i.e., very large load swings in response to a small disturbance in grid frequency). In the latter case, the grid frequency restoring or regulating contribution from the particular generator is marginal. Simulations of the speed governor response to a unit load change (i.e., from 0% to 100%) for different droop settings is conceptually shown in Figure 4.16. The situation shown in the figure is typical of a smaller gas turbine (e.g., GE's Frame 5 or 6); for larger units (e.g., Frame 9) with much higher inertia, the oscillations are damped quickly. The difference between the "stiff" (d = 2%) and stable cases (d = 10%) is striking.[2]

The interesting takeaway from Figure 4.16 is that the droop governor does *not* completely restore the frequency. As the droop setting increases the steady-state gap increases but the response is stable. As it decreases, the steady-state decreases but the response becomes increasingly unstable. In other words, *primary control* reserves operating with a set droop cannot restore the system frequency to its nominal value. The remaining gap or steady-state deviation has a defined band of minimum and

---

[2] In mathematics, a stiff equation is a differential equation for which the numerical solution is unstable (i.e., it oscillates), unless the step size is taken to be extremely small. In analogy, with d = 10%, the full load range is divided into ten steps vis-à-vis only two steps with d = 2%.

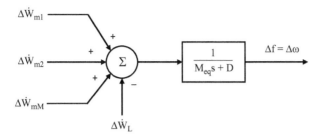

Figure 4.17 Grid system transfer function (simplified).

maximum values set by the transmission system (grid) operators. Restoring the grid frequency to its nominal value, e.g., 50 Hz or 60 Hz, within a narrow band (e.g., ±20 mHz in Europe) requires the engagement of *secondary control* reserves. Once the secondary control reserves come on-line, the primary control reserves return to their initial load setting so that they are ready for the next disturbance.

In a typical grid, many generators participate in meeting the load demand. The simplest way to analyze this highly complicated system for illustrating the basic principles of load-speed or LFC is the equivalent generator approach shown in Figure 4.17. There are M generators in the system, and they respond collectively to changes in the system load. They are represented by an equivalent inertia constant, M_{eq}, which is equal to the sum of the inertia constants of M units. Self-regulation of the entire system is represented by a single damping constant, D.

For this composite system, the steady-state D is represented by an equivalent droop, i.e.,

$$\Delta f_{SS} = \frac{-\Delta \dot{W}_L}{\left(\frac{1}{d_1} + \frac{1}{d_1} + \cdots + \frac{1}{d_M}\right) + D} = \frac{-\Delta \dot{W}_L}{\frac{1}{d_{eq}} + D}. \tag{4.22}$$

Let us consider a system running at a load of 1,260 MW at 50 Hz with a damping constant of 1.5%. The spinning reserve of the system is 240 MW. Suddenly, a 60 MW load is lost. The equivalent droop setting of this system is 4% and only 80% of the governors respond to load-frequency events. When 60 MW is lost, the system load becomes 1,200 MW so that

$$D = 1.5 \times 1{,}200/50 = 36 \text{ MW/Hz.}$$

If there were no governor response, the frequency increase would be

$$\Delta f_{SS} = \frac{-\Delta \dot{W}_L}{D} = \frac{-(-60)}{36} = 1.67 \text{ Hz.}$$

The total system capacity with spinning reserve is $1{,}240 + 260 = 1{,}500$ MW. The generation contributing to the frequency regulation is 80% of this, i.e., $80\% \times 1{,}500 = 1{,}200$ MW. The equivalent droop of the system is 4%, thus, using Equation (4.21), we find that

$$\frac{1}{d_{eq}} = \frac{1,200}{4\% \times 50} = 600 \text{ MW/Hz},$$

so that the denominator of Equation (4.22) becomes

$$\frac{1}{d_{eq}} + D = 600 + 36 = 636 \text{ MW/Hz}.$$

Consequently, with speed governor control, the steady-state frequency increase is found as

$$\Delta f_{SS} = \frac{-\Delta \dot{W}_L}{D} = \frac{-(-60)}{636} = 0.094 \text{ Hz}.$$

Finally, Equation (4.18) is numerically integrated to illustrate how the frequency is restored with 4% droop speed governor (see Figure 4.18). Droop control engages at $t = 5$ seconds.

The increasing penetration of renewable resources into the generation mix reduces the system inertia. This can significantly reduce the resilience of the grid to frequency disturbances. This can be illustrated quite dramatically by changing the inertia constant of the last example from 10 seconds to 3 seconds as shown in Figure 4.19.

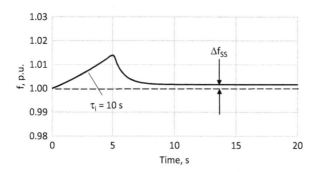

Figure 4.18 Frequency restoration with 4% droop.

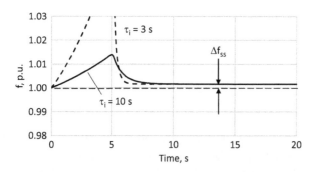

Figure 4.19 Impact of inertia constant on system response.

The steep rise in frequency could destabilize the grid before the primary control (PC) reserves could rectify the situation. (In the case of a generating unit loss, the steep dip in grid frequency would be the mirror image around the x-axis.) The remedy in this case would be to include more and/or much faster reacting PC resources in the generation mix.

Primary control reserves constitute the first line of defense in grid disturbances. If there are no *secondary* or *tertiary* control reserves, once PC reserves are fully on-line, the system would be vulnerable to another load loss or generator trip. Such cascading events can happen and lead to a complete collapse of the grid and severe blackouts. There is also the matter of Δf_{ss} remaining due to the nature of droop governor described above. This gap will be taken care of by the secondary reserves going into action. Once this happens, PC reserves will automatically go back to their nominal set point, and they will be ready to go into action when the next frequency event happens. The exact nature of the response of a generating unit, e.g., a gas turbine or gas engine, in PC duty will be explored in Chapter 7. See Section 16.3 for more on power system reserves.

4.5   Fuel Flexibility

Gas turbines have the inherent capability to burn a wide variety of fuels, both gaseous and liquid. (In fact, closed cycle gas turbines can even utilize solid fuels like coal as the feedstock.) The problem is emissions, specifically, NOx emissions when burning typical liquid fuels such as number 2 fuel oil (distillate) or diesel fuel. The standard method to reduce NOx emissions when burning distillate is injection of a diluent, e.g., water or steam. From a sustainability perspective, this method is frowned upon due to significant expenditure of water, which is an increasingly scarce resource in many places in the world. Due to this reason, such fuels are strictly considered as backup fuels to be used only when absolutely necessary.

The ideal gaseous fuel is hydrogen because it is carbon-free, and its combustion generates no CO_2. Burning hydrogen is covered in its own chapter later in the book (Chapter 8). Burning hydrogen in modern DLN combustors also has the same problem as burning distillates, i.e., high NOx generation. This problem can be easily alleviated by using the diffusion flame combustor technology with steam or water injection. As noted above, due to strict environmental regulations in place to save scarce water resources, this option is off the table.

For more details on the combustion of distillates and hydrogen as well as other fuels such as ethane, the reader is referred to **GTFEPG**. Herein, the focus will be on biofuels, which are considered to be carbon neutral. There are three types of biofuel:

- Solid, i.e., wood and other biomass fuels
- Biogas
- Liquid, e.g., biodiesel, ethanol, and methanol

Table 4.2 Gas turbine runs with standard fossil fuels and biofuels

	Natural Gas (NG)	Biogas (BG)	NG + BG	#2 Distillate	Biodiesel
Output, kW	238,635	237,530	240,176	240,880	240,889
Efficiency, %	39.2	39.31	39.25	36.35	36.13
Exhaust CO_2, kg/s	34.3	52.58	37.28	49.64	49.62
Exhaust CO_2, kg/MWh	517	797	559	742	742
Lower Heating Value, kJ/kg	46,280	18,891	38,063	42,557	38,090
Modified Wobbe Index, $kJ/m^3\text{-}K^{0.5}$	2,525	1,256	1,973	NA	NA
Turbine Inlet Temperature, °C	1360.6	1314.8	1359.1	1277	1257.1
Turbine Inlet Pressure, bara	17.2	17.48	17.28	17.66	17.74

The prefix "bio" emphasizes the organic nature of these fuels derived from plant or animal matter. Strictly speaking, biofuels are also carbon based and, in that sense, they are not different from conventional fossil fuels. The difference is in the length of the carbon life cycle, which is hundreds of millions of years for the latter. Thus, biofuels are considered to provide a convenient path to carbon neutrality. In other words, carbon dioxide generated by burning, say, biomass is matched and implicitly offset by the carbon dioxide absorbed by the plants growing the biomass.

Nevertheless, there are detractors of biofuels, who claim that they are not "green" at all. One argument against the biofuels is that they actually generate more CO_2 than conventional gas or liquid fossil fuels. This can be easily checked using a typical advanced F Class gas turbine as an example using the following fuels:

- Natural gas (87% methane by volume)
- Number 2 distillate
- Digester gas[3] (62% methane, 36% CO_2 by volume)
- Biodiesel (77% C, 12% H, and 11% O by mass)

Calculations are done in THERMOFLEX using built-in engine library model #683 (GE's 7FA.05). The results are summarized in Table 4.2. Water injection at a water-to-fuel ratio of 1.5:1 is assumed to control NOx with liquid fuels. Clearly, biofuels emit similar or higher CO_2 vis-à-vis their fossil counterparts. In the case of 100% biogas, due to the low heating value of the gas (about 550 Btu/scf), firing temperature is dropped by the model to prevent excessive rise in turbine inlet pressure (due to fixed S1N inlet area, i.e., "swallowing capacity" of the turbine). In all likelihood, in an application like this, the gas turbine would be modified by staggering S1N vanes to open up the inlet are similar to syngas applications. Another option is to blend the low-Btu biodiesel with natural gas.

---

[3] Digester gas is produced by anaerobic digestion at wastewater treatment plants. The composition of wastewater digester gas varies, though the primary constituents are methane (CH_4) and carbon dioxide (CO_2).

The counter-argument is, of course, that the CO_2 generated by burning biofuels, as already explained above, is CO_2 absorbed from the atmosphere and, thus, should be discounted. In the case of digester gas and biodiesel, the source organic matter is waste material to begin with. Thus, to argue against them is difficult. In the case of ethanol, produced from plants such as sugar cane or corn, the situation is somewhat different. In that case as well as in the case of solid biomass, e.g., wood, the argument of the detractors has several points, i.e.,

- Those crops, plants, trees, etc. were going to grow anyway so that turning them into biomass and burning them is not a zero-sum approach.
- When natural forests are felled to make biomass fuel, greenhouse gas emissions go up.
- Repurposing the farmlands used for crop growth to biofuel production reduces food stock.

In 2010, GE converted the combustors of two LM6000 aero-derivative gas turbines in a plant in Brazil to burn ethanol. The conversion proved to be successful and the gas turbines operated without a problem. However, rated at 43.5 MW, each gas turbine burns 18,000 liters (about 4,700 gallons) per hour, corresponding to about 6.7 million gallons of ethanol every 30 days for both machines.

To check the feasibility of burning ethanol in a gas turbine, an example case is run using GE's LM 6000 aero-derivative operating 2,000 hours per year as a peaker as the basis. The engine is modeled in THERMOFLEX using the built-in engine library model #565. For NOx control in the ethanol case, water injection at a diluent-to-fuel ratio of 1.27:1 is assumed. A performance comparison is summarized in Table 4.3. Carbon dioxide emissions with ethanol are about 40% higher than with natural gas. At a cost of about $1.40 per gallon, the annual fuel cost for the ethanol case is nearly $14 million. For natural gas at $3 and $6 per MMBtu (HHV), the annual fuel cost is $2.6 and $5.2 million, respectively. If the gas turbine is equipped with a carbon capture system at $100/ton, at $6 natural gas, the total cost is $9.6 million – still almost $4.5 million cheaper than ethanol.

Table 4.3 GE's LM 6000 aero-derivative with natural gas and ethanol

	Natural Gas	**Ethanol**
Output, MW	43.5	43.7
Fuel Mass Flow, kg/s	2.348	4.241
CO_2 Exhaust Flow, kg/s	6.138	8.398
Fuel HHV, kJ/kg	51,237	31,360
Fuel Volume Flow, gph		5,096
CO_2 Emissions, kg/MWh	508	692
Total Generation, MWh (2,000 hours/year)	87,000	87,346
Total CO_2 Emissions, mt	44,194	60,466
Total Fuel Consumption	866,196 MMBtu	10,192,755 gallons

4.5.1   Biomass

Biomass is promoted as a carbon-neutral fuel for use in gas turbines and thermal power plant (steam turbine) boilers. As mentioned earlier, solid biomass, e.g., wood pellets, can be used as a fuel in closed cycle gas turbines, e.g., a supercritical CO_2 (sCO2) cycle. As an example, consider a nominal 300 MW sCO2 split-flow recompression cycle burning wood pellets in a furnace to heat sCO2 working fluid to 715°C. The cycle heat and mass balance (modeled in THERMOFLEX) are shown in Figure 4.20.

Fuel analysis is shown in Table 4.4. Furnace efficiency is assumed to be 93% with 2% heat loss. Fuel transport, preparation, and storage are not included in the model. For an overview of solid biomass handling equipment, the reader is referred to the paper by Dafnomilis et al. [3]. No flue gas cleaning equipment is specified or accounted for in plant auxiliary loads. (Typically, biomass fuels exhibit high NOx and PM emissions.) Key performance data is as follows:

Table 4.4 Heating value and ultimate analysis (by weight) of wood pellets

LHV (moisture and ash included), kJ/kg	16,784
HHV (moisture and ash included), kJ/kg	18,197
Moisture, %	8.7
Ash, %	0.5
Carbon, %	45.8
Hydrogen, %	5.5
Nitrogen, %	0.08
Chlorine, %	0.01
Sulfur, %	0.01
Oxygen, %	39.4

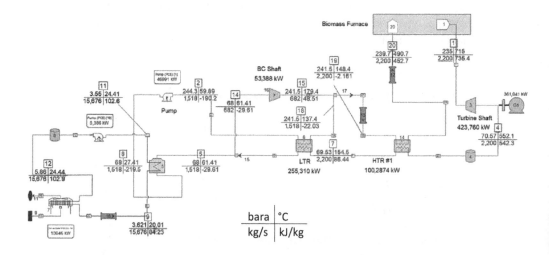

Figure 4.20 Split-flow recompression sCO2 cycle with wood pellets – heat and mass balance (THERMOFLOW).

- Furnace fuel consumption 643.4 MWth
- Net output 294,254 kWe
- Net efficiency 45.74%
- CO_2 in exhaust flow 64.44 kg/s (15.91%(v))
- CO_2 footprint 788.4 kg/MWh

Supercritical CO_2 (closed cycle gas turbine) technology is apparently an ideal fit to utilizing solid biomass fuels for efficient electric power generation and low carbon footprint. Nevertheless, the technology is still in development and its commercial viability is uncertain (see Chapter 10 for more on this subject). However, burning biomass in thermal power plants originally designed for coal-fired boilers is a present reality. A prime example in this respect is the *Drax Power Station* in North Yorkshire, England. Four coal-fired units (total capacity of 2,600 MW) in the station started co-firing biomass in 2010; by 2018 all four units were converted to burning 100% biomass. The fuel, in the form of wood pellets, is imported from the USA.

In a thermal power plant utilizing biomass fuel, the critical design aspect is the selection of the boiler. In utility-scale power plants (i.e., those with a capacity of several hundred megawatts), there are two alternatives: suspension fired (or *entrained bed*, i.e., pulverized coal–fired boiler) or *fluidized bed* (e.g., circulating fluidized bed). Unlike the Drax biomass units rated at 600 MW each, most biomass power systems in commercial operation are typically below 50 MW in size. In those plants, stoker grate or underfire stoker boilers can be used as well.

Two other biomass utilization options are *co-firing* (e.g., with coal or petcoke) and *gasification*. Co-firing makes the most economic sense in near-term because much of the existing power plant equipment can be used without major modifications. On the plus side, by replacing part of the coal feedstock, firing biomass reduces SOx, NOx, and certain other air emissions.

Gasification of biomass can be advantageous vis-à-vis direct burning because the resulting syngas (also referred to as wood gas, producer gas, or biogas) can be cleaned and filtered to remove problem chemical compounds before it is burned. The syngas can be burned efficiently in a gas turbine in simple or combined cycle configuration. Gasification technologies using biomass byproducts such as hogged wood, bark, and spent black liquor are popular in the pulp and paper industry for process steam and electric power generation.

Gasification can also be accomplished using chemicals or biologic action (e.g., anaerobic digestion to produce digester gas); however, thermal gasification is currently the only option close to commercial reality. Due to the wide variability in biomass fuel characteristics, gasification is a versatile technology that can handle a wide variety of feedstocks without major changes in the basic process. It can also be applied to waste fuels, providing safe removal of biohazards and entrainment of heavy metals in non-reactive slag. Syngas is a more versatile fuel than solid biomass feedstock because it can be used in boilers, process heaters, turbines, engines and fuel cells, distributed in pipelines, and blended with natural gas or other gaseous fuels.

Table 4.5 Biomass gasifiers [4]

	Fixed Bed	Fluidized Bed
Fuel size, in. (mm)	0.4–4 (10–100)	up to 0.8 (up to 20)
Fuel ash, %(w)	<6	<25
Operating temperature, °F (°C)	1,450–2,250 (790–1,230)	1,350–1,750 (730–950)
Controls/Operability Complexity	Simple	Average
Turndown ratio	4:1	3:1
Construction material	Mild steel + refractory	Heat-resistant steel
Capacity, MWth (mt/day)	<5 (<30)	>5 (>30)
Startup time	Minutes	Hours
Tar content, lb/MMBtu	<1.2	<2
HHV, Btu/scf	130	150

Consequently, while coal gasification is most likely a "dead" technology, biomass gasification is still a viable option.

Biomass gasifiers can be one of two types, i.e., fixed bed and fluidized bed. *Fixed bed* gasifiers are typically simpler, less expensive, and produce a lower heat content syngas. *Fluidized bed* gasifiers are more complicated, more expensive, and produce a syngas with a higher heating value. Typical gasifier characteristics are summarized in Table 4.5. For a detailed discussion of biomass gasifier operational issues and concerns, the reader is referred to Ref. [4].

For more information on biomass fuel characteristics, combustion, and boiler technologies for power generation, the reader is referred to Refs. [5–7]. For a historical example of biomass (wood) gasifier as part of a prime mover, i.e., car engine, see Section 13.2.1.

4.6   Reliability, Availability, and Maintainability

RAM aspects of gas and steam turbine power plants have been covered in detail in the earlier books by the author, **GTFEPG** and **GTCCPP** (Refs. [1] and [2], respectively, in chapter 2). The reader is advised to consult those books and references therein. Brief definitions of the terms are as follows:

- *Reliability* is the probability that a particular piece of equipment or system *will* operate for its designed interval under specific conditions (the ability to maximize *uptime*).
- *Availability* is the probability that a particular piece of equipment or system *can* operate at any given moment (e.g., the ability of a power plant to operate and generate electricity when dispatched).
- *Maintainability* is the probability that a particular piece of equipment or system *can* be restored to operating condition in a specified amount of time (the ability to minimize *downtime*).

To maximize power plant availability, one must (i) maximize plant *uptime* and (ii) minimize plant *downtime*. Consequently, high availability requires (i) high reliability and (ii) high maintainability. The ultimate goal of a power plant operator is to maximize (control) availability, which can only be done by maximizing (controlling) reliability and maintainability. No matter how efficient and clean (i.e., low greenhouse gas and criteria pollutant emissions) a particular power plant is, if it is not able to start and run when dispatched, it is of no use to anybody. Unfortunately, marketing hype surrounding most emerging technologies completely ignore the RAM aspect of system design and development (not to mention operability).

As mentioned above, reliability is the probability of operation without failure as a function of time, t. It is typically expressed by the exponential distribution function[4]

$$R(t) = \frac{1}{e^{\frac{t}{MTBF}}} = e^{-\lambda t}, \tag{4.23}$$

where MTBF is the *mean time between failures* and $\lambda = 1/MTBF$ is the *failure rate*. Another term that can be used in the definition of the failure rate is the *mean time to failure* (MTTF). For components that are repaired when they fail and put back into service, one uses the term MTBF. For components that are discarded and replaced when they fail, the term to use is MTTF.

Consider a new product recently introduced by an OEM. To gauge the reliability of that machine, we have to wait for a period of time and count the times when the units installed in the field (the fleet) fail (r) over that period of time (t) to calculate $\lambda = r/t$. Let us assume that over a period of 15,000 *collective* hours of operation, five (5) failures are recorded in a fleet of three (3) machines. Thus,

$$\lambda = 5/15,000 = 0.000333 \text{ h}^{-1}$$

and, on average, one machine failure can be expected every year, i.e., $0.000333/3 \times 8,760 \sim 1$. Thus, over a period of, say, one year,

$$R(1) = 2.7183^{-(1 \times 1)} = 0.368 \text{ or about } 37\%.$$

In other words, after one year in the field, about 63% of the population of *identical machines* operating in the field *under similar conditions* can be expected to fail. If this sounds a high number, consider that the highest rate of failure occurs either early in the lifetime of a product (*infant mortality*) or near the end of it (*wear-out mode*). When the proverbial *bugs* are discovered and fixed, the equipment will (or, at least, is expected to) operate with higher reliability.

For a system comprising multiple subsystems, system reliability is the product of the reliabilities of the subsystems. For example, for a gas turbine

---

[4] The exponential distribution is the probability distribution of the time between events in a Poisson process, i.e., a process in which events occur continuously and independently at a constant average rate. It is used in reliability calculations to model data with a constant failure rate.

$$R_{GT}(t) = R_{Comp}(t) \times R_{Comb}(t) \times R_{Turbine}(t). \tag{4.24}$$

This is also the formula for a system comprising components *in series*. For a system comprising, say, three components, A, B, and C, *in parallel*, the system reliability becomes

$$R_{Sys}(t) = 1 - (1 - R_A(t)) \times (1 - R_B(t)) \times (1 - R_C(t)). \tag{4.25}$$

The latter is the approach used in plant design when specifying the number of critical pieces of equipment, e.g., feed water pumps in a GTCC power plant. For example, either $3 \times 50\%$ pumps in operation or $2 \times 100\%$ pumps with one in operation and the other one in stand-by. If single pump reliability is, say, $R = 0.95$, the reliability of the feed water pump system in the latter case is

$$R = 1 - (1 - 0.95) \times (1 - 0.95) = 1 - 0.05 \times 0.05 = 0.9975.$$

Let us say that *unreliability* of a given system is $Q = 1 - R$, i.e., $Q + R = 1$. Let us consider a system comprising N identical units (installed). At least M units must be operating to meet the system's reliability requirement. (An example of this is a power plant with N prime movers or powertrains.) Thus, for N units we have

$$(R + Q)^N = 1. \tag{4.26}$$

Expanding the left-hand side, we obtain

$$\begin{aligned}(R + Q)^N &= R^N + NR^{N-1}Q + \frac{N(N-1)}{2!}R^{N-2}Q^2 \\ &+ \frac{N(N-1)(N-2)}{3!}R^{N-3}Q^3 \dots + Q^N.\end{aligned} \tag{4.27}$$

Hence, the probability of at least $N - 1$ units operating (no failure) is given by

$$P(N - 1) = R^N + NR^{N-1}Q. \tag{4.28}$$

Similarly, the probability of at least $N - 2$ units operating (no failure) is given by

$$P(N - 2) = R^N + NR^{N-1}Q + \frac{N(N-1)}{2!}R^{N-2}Q^2, \tag{4.29}$$

and so on and so forth.

Returning to the redundant pumps example above, for the option of $3 \times 50\%$ pumps, at any given time, two $(M = 2)$ out of three $(N = 3)$ pumps must be working. In other words, using Equation (4.28), the reliability of the feed water pump system becomes

$$R = R^3 + 3R^2Q = 0.95^3 + 3 \times 0.95^2Q = 0.99275.$$

Thus, the $3 \times 50\%$ option is slightly less reliable than the $2 \times 100\%$ option. Which one to go with is a question of cost-performance trade-off with cost of failure thrown in.

System *availability* is a function of both the *frequency* and *duration* of outages, i.e.,

$$A = \frac{1}{1 + \lambda t} = \frac{1}{1 - \ln R}. \tag{4.30}$$

Three factors play a role in achieving high availability:

• High MTBF;
• Low downtime due to repairs or scheduled maintenance; and
• accomplishing the two in a cost-effective manner.

High availability is of prime importance to the plant owner/operator because it translates into more opportunities for making money (the equipment is in service for longer periods of time). While there are several different expressions proposed for availability, the most widely used one is the inherent availability given by

$$A = \frac{MTBF}{MTBF + MTTR}, \tag{4.31}$$

where MTTR represents the *mean time to repair*. The denominator of the fraction on the right-hand side of Equation (4.31) is known as *mean time between repairs*, i.e.

$$MTBR = MTBF + MTTR. \tag{4.32}$$

If it takes a long time to repair the system in case of failure, system availability decreases. This is where *maintainability* enters the picture. Maintainability must be built into the design as well as construction (installation) of the equipment and/or the system. The best example of this in everyday life is that, especially in some exotic cars with a complex engine and cramped space under the hood, it takes your neighborhood mechanic the whole day to replace, say, a broken alternator. In comparison, passenger cars built for large number of sales and low cost of ownership are designed such that most critical parts can be replaced in a matter of hours (not to mention that the replacement parts are much cheaper).

Quantitatively, maintainability is the total down time for maintenance including all the time required for: diagnosis, trouble-shooting, dismantling, removal/replacement, active repair time, verification testing that the repair is adequate, logistic movement delays, and administrative tardiness. It is expressed as

$$M(t) = 1 - e^{-\mu t}, \tag{4.33}$$

where $\mu = 1/MTTR$ is the *maintenance rate*.

Another definition of availability, specifically for power plant applications, is

$$A = \frac{AH}{PH} = \frac{PH - FOH - POH}{PH}, \tag{4.34}$$

where

PH = Period Hours (number of hours a unit was in the *active* state)
AH = Available Hours (number of hours a unit was in the *available* state)

Table 4.6 Availability calculation example

YEAR	1	2	3	4	5
AH	8,760	8,760	8,760	8,760	8,760
PH	7,500	1,500	22,500	30,000	37,500
POH	168	168	336	240	504
PO (Days)	7	7	14	10	21
POH (%)	1.92	1.92	3.84	2.74	5.75
FOH	240	240	240	240	240
FO (Days)	10	10	10	10	10
FOH (%)	2.74	2.74	2.74	2.74	2.74
Availability (%)	95.34	95.34	93.42	94.52	91.51

FOH = Forced Outage Hours (number of hours a unit was in the *unplanned* outage state)

POH = Planned Outage Hours (number of hours a unit was in the *planned* outage state)

A sample availability calculation for a typical prime mover is illustrated in Table 4.6. Planned operating hours are based on the maintenance schedule set by the OEM. Unplanned operating hours are based on the reliability statistics of the particular equipment. Average unit availability is calculated as 94%.

At the time of writing, one of the key initiatives in many industrial sectors, e.g., chemical process industry, is to increase the *resilience* of major facilities such as refineries, aluminum smelters, or LNG stations to power outages caused by major weather events such as hurricanes or floods. In the period covering 2012 through 2016, *Hurricane Sandy* was responsible for about one-third of major electricity disturbances in the USA (in terms of the share of total customer-hours disrupted). The rest was mostly due to other severe weather events.

Resilience is different from reliability, which is defined by NERC as a combination of sufficient resources to meet demand (adequacy) and the ability to withstand disturbances (security). The measure of *adequacy* is the percentage of capacity in excess of projected or historical peak demand for that system (adequacy standards differ between reliability regions, subject to NERC approval). Adequacy also includes essential reliability services like frequency and voltage support and, increasingly, flexibility as wind and solar resources increase their share in the generation portfolio.

Security means preparedness to endure uncertain external forces. It is difficult to measure since impacts of low-probability, high-impact events remain difficult to predict. NERC promotes cybersecurity, emergency preparedness and operations, and physical security standards to ensure that grid operators and utilities are prepared for attacks or blackouts.

FERC definition of resilience is the ability to withstand and reduce the magnitude and/or duration of disruptive events, which includes the capability to anticipate, absorb, adapt to, and rapidly recover from such an event. Resilience accounts for consequences to the electricity system and other critical infrastructure from increasingly likely high-impact external events by using metrics like cumulative customer-hours of outages.

One obvious solution to increase the resilience of large industrial facilities is to have a dedicated power plant to serve the power demand of the facility and, if necessary, export power to external customers, i.e., the grid. Total independence from the grid in terms of power import/export requires an "islanded" power system. Such a system should comprise mature, proven technologies and redundancy to ensure the highest possible probability that the facility continues uninterrupted operation. While efficiency is critical in terms of reduced operating costs (i.e., fuel expenditure) and low emissions, it cannot be the sole determinant at the expense of reduced reliability and availability. A custom-designed power island can comprise multiple technologies, e.g., solar PV or wind, fossil fired generators, and battery energy storage systems to ensure that the carbon footprint is minimized.

Let us illustrate the redundancy requirement for such an islanded power plant. The facility in question in this example requires 180 MWe power. Nominally, this can be supplied by 10 gas-fired reciprocating internal combustion engines rated at 18 MWe each, i.e., $N = 10$. (For simplicity, equipment and plant auxiliary power is ignored.) In operation, however, 12 engines are running at 83.3% load with a $12 \times 18 \times (1 - 0.833) = 36$ MWe spinning reserve. In other words, even if *two* engines fail unexpectedly at the same time (a rare occurrence indeed; perhaps, by the failure of a common bus in the switchyard), the remaining engines can be ramped up to 100% load and continue delivering 180 MWe to the host facility. Furthermore, one engine is in a warm stand-by mode and, at any given time, one engine is down for planned maintenance. Thus, there are a total of $N = 12 + 1 + 1 = 14$ engines with a rated capacity of $14 \times 18 = 252$ MWe installed in the power island. At any given time, at least ten units must be running at 100% load to meet the host facility's power demand, i.e., $M = 14 - 10 = 4$.

The availability of a system with N components where the system is considered to be available when at least $N - M$ components are available is given by the following formula:

$$A = \sum_{i=1}^{M} \frac{N!}{i!(N-i)!} A^{N-i}(1 - A)^i. \tag{4.35}$$

Hence, for the example with $N = 14$ and $M = 4$, A is calculated as 0.999 or 99.9%.

In conventional, fossil fired thermal power systems, high RAM is a result of decades of design, development, and field operation (more than a *century* in the case of boiler-turbine power plants). Some newer technologies have been unable to reach that level even after decades of tinkering. The most glaring example is the integrated gasification combined cycle power plant. This is a direct result of the sheer number of exceedingly complex subsystems with a high level of interaction with each other.

Even systems comprising proven components, which have been used in other applications successfully for many decades, should be considered First of a Kind. A striking example of this is the considerable difficulty experienced by the operators to make post-combustion capture systems work reliably in the field (see Section 11.2 for details). Unfortunately, the RAM aspect of "miracle" technologies widely touted in academic and trade publications is completely ignored by developers, researchers and, most importantly, by the public. One useful tool to address this facet of technology development is the concept of *Advancement Degree of Difficulty* (AD^2) – see Section 16.6.

4.7   References

1. Kundur, P. 1994. *Power System Stability and Control*, New York, NY: McGraw-Hill, Inc.
2. Klimstra, J., 2014. *Power Supply Challenges*, Arkmedia, Vaasa, Wärtsila Finland Oy.
3. Dafnomilis, I., Lanphen, L., Schott, D. L., and Lodewijks, G., 2015. Biomass handling equipment overview, Proceedings of the XXI International Conference MHCL'15.
4. U. S. Environmental Protection Agency Combined Heat and Power Partnership, 2007 Biomass Combined Heat and Power Catalog of Technologies, Version 1.1.
5. Van den Broek, R., Faau, A., and van Wijk, A., 1996. Biomass Combustion for Power Generation, *Biomass and Bioenergy*, 11(4), 271–281.
6. Obernberger, I., 1998. Decentralized biomass combustion: State of the art and future development, *Biomass and Bioenergy*, 14(1), 33–56.
7. Nussbaumer, T., 2003. Combustion and co-combustion of biomass: Fundamentals, technologies, and primary measures for emission reduction, *Energy & Fuels*, 17, 1510–1521.

5   Energy Storage

The Achilles' heel of electric power generation from wind and solar is the intermittency and low predictability of their ability to be online on demand. They generate power when the wind blows and the sun shines. If a wind or solar farm is oversized to provide a certain generation capacity, curtailing excess power will be unavoidable. In other words, power generation from the particular facility will be reduced below its generation capacity to meet the existing demand. In some cases, this reduction can be avoided by sending the surplus to a neighboring grid depending on the availability of the transmission infrastructure and its congestion status. The logical solution to the elimination of curtailment for the full utilization of solar and wind resources is energy storage. At any given time, surplus generation is used to "charge" the storage unit. Stored energy is "discharged" at those times when the renewable generation facility is unable to meet the demand. In that way, the particular renewable generation facility is utilized at its full capacity. This is commonly referred to as "time shifting" and it is also the basis of capacity or grid "firming." The energy stored can be used when there is a dip in renewable power output due its inherent intermittency so that the plant owner can quote a "firm" generation capacity.

A well-known graphic widely used to illustrate the need and the benefit of energy storage is the so-called duck curve, which is depicted conceptually in Figure 5.1. The generic curve, originally conceived to illustrate the impact of solar PV on the California Independent System Operator (CAISO) net load, shows what happens to the load of the other generation assets on a grid when solar or other non-fossil (renewable) generation meets a large portion of the demand. As shown in the figure, as the day progresses and the solar contribution increases, other assets on the grid ramp down, reaching a minimum midday when solar irradiation is at its apex. The balance can be so skewed that even a "back feed" condition might arise when the nonsolar assets must be shut down.

In general, nonrenewable generation assets such as fossil fuel–fired (coal and gas) and nuclear are not designed for daily cycles involving shut down and restart. This is especially true for coal-fired and nuclear power plants. Even in the present "energy transition" modus operandi where coal is being rapidly phased out and nuclear is the perennial stepchild in the public eye, when natural gas fired gas turbine combined cycle (GTCC) is the preferred "bridge" technology, this is a highly undesirable scenario. As far as the GTCC power plants are concerned, the ideal scenario is to

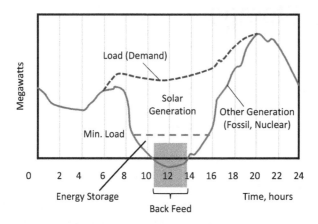

Figure 5.1 Generic "duck curve" illustrating the trade-off between solar (non-fossil, renewable) and other generation assets (mostly fossil).

run at a *minimum emissions-compliant load* (MECL) when NOx and CO emissions are under control. New Dry-Low NOx combustors allow MECL turndown to go as low as 20% load. Thus, a situation may arise when excess solar generation must be curtailed. This is when energy storage presents itself as a useful component of the generation portfolio to absorb the curtailed energy, which otherwise would be wasted, and use it later in the day for clean power generation when the solar generation asset is down. This is commonly known as "time shifting." In operational dynamics, time shifting is analogous to "energy arbitrage." In the latter, charging is done when energy is available for free or cheap (or even at negative price, i.e., the operator is paid to use the energy to charge the system) at off-peak hours and discharging is done when energy is expensive at peak hours.

There are a large variety of energy storage technologies available. A graphical illustration of their comparative ranking in terms of capacity (kW or MW) and duration (hours) is provided in Figure 5.2 [1]. Other properties of selected energy storage technologies are summarized in Table 5.1 [2]. They include mechanical (pumped hydro and compressed air), electrochemical (batteries), chemical (fuel cell), electrical (superconducting magnetic coil), and thermal (molten salt). (A similar, more detailed comparison matrix in the form of a table can be found in Ref. [3] that provides a comprehensive review of the existing technologies.) Until recently, from maturity, installed capacity in megawatt-hours, and utility scale (i.e., several hundred megawatts) generation perspectives, only two could be considered as feasible technologies: *Pumped Hydro Energy Storage* (PHES) and *Compressed Air Energy Storage* (CAES). At the time of writing (2020–2021), one could cautiously add Li-Ion battery technology to this group with certain, but by no means trivial, caveats.

As reported by the US Department of Energy (DOE), grid-scale energy storage in the USA was dominated by pumped hydro at 23.4 GW, corresponding to 95% of the total. The remaining 5% (1.2 GW) was divided roughly equally among compressed

Table 5.1 Key characteristics of selected energy storage technologies

	Technology Maturity	Environmental Impact	Response Time	Lifetime (years)	Energy Density (Wh/l)	Roundtrip Efficiency (%)
PHES	Mature	Large	sec – mins	40–60	0.2–2	75–85
CAES	Mature	Large	mins	20–40	2–6	40–70
Molten Salt	Mature	Moderate	mins	30	70–210	80–90
Li-Ion Battery	Comm.	Moderate	ms – secs	5–15	200–400	85–95
NaS Battery	Comm.	Moderate	ms	10–15	50–100	80–90
Flow Battery	Early Comm.	Moderate	ms	5–10	20–70	60–85
Flywheel	Early Comm.	Almost None	ms – secs	15+	20–80	93–95

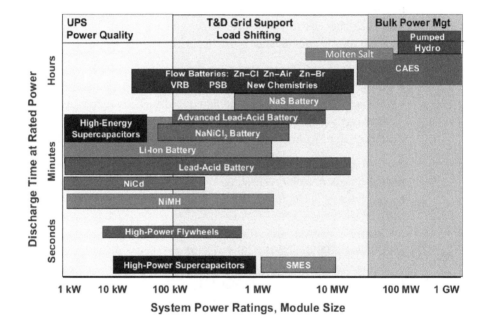

Figure 5.2 Capacity and duration ranking of energy storage technologies. (SMES: Superconducting Magnetic Energy Storage, VRB: Vanadium Redox Battery, UPS: Uninterrupted Power Supply, T&D: Transmission and Distribution.)

air, thermal, and battery storage. Thermal energy storage (TES) is typically used to make ice using excess electricity (e.g., at night in a nuclear or fossil power plant), which is then used for space cooling or for gas turbine inlet air cooling.

There are also "emerging" storage technologies worth mentioning. Several of them are listed below:

Flywheels store energy in the angular momentum of a spinning mass. During charge, the flywheel is spun up by a motor with the input of electrical energy; during discharge, the same motor acts as a generator, producing electricity from the rotational energy of the flywheel.

Flow Batteries are electrochemical conversion devices whose operating principle is based on the energy difference between the oxidation states of certain metals. There are several types of flow batteries, of which the most common is the redox flow battery, e.g., the *vanadium redox flow battery* (VRFB). What is flowing in a flow battery is the liquid electrolyte between the positive and negative electrolyte tanks. The main disadvantage of the VRFB, i.e., its low electrolyte energy density (about 70 kJ/kg) vis-à-vis lithium ion (as high as 720 kJ/kg), is more than countered by its scalability to very large sizes (in MWh) simply by increasing the size and/or number of electrolyte storage tanks. In contrast, lithium-ion batteries store the electrolyte in the cell, which also contains the electrodes and the separator. Thus, unlike a flow battery, its capacity (MWh) can only be increased by stacking the cells, i.e., increasing the power output simultaneously. However, it is not clear whether the trade-off between increasing the electrolyte storage tanks and increasing the battery packs is cost effective for long duration applications, i.e., beyond 4–6 hours.

- *Chemical* energy storage systems come in several varieties:
- utilization of excess electricity to produce *hydrogen* via water electrolysis
- gasification of coal or other solid/liquid feedstock to produce hydrogen or *synthetic natural gas* (SNG)
- production of hydrogen or ammonia (NH_3) via steam methane reforming (SMR)

The end products are universal energy carriers and can be used in different sectors, such as transport, mobility, heating, and the chemical industry. Hydrogen's role as an energy "vector," including as an energy storage medium, will be discussed in detail in Chapter 8.

Gravity energy storage (GES) is extremely simple in principle. Any college student with Physics 101 under his or her belt can easily grasp the concept. Charging consists of lifting a large weight (e.g., a big lump of rock or concrete) from the ground to a substantial height, i.e., conversion of electro-mechanical energy to potential energy. Discharging is simply letting the raised weight fall back to the ground, i.e., conversion of potential energy to electro-mechanical energy. Easier said than done, of course.

Several companies are promoting GES technology using different methods. One option is pumping water into a cylinder and raising a piston (rock, concrete, metal, etc.). When power is needed, the piston drops and pushes the water in the cylinder through a hydraulic turbine to generate electricity. In another variant, the piston is replaced by a multiplicity of weights, which are raised mechanically, e.g., by an electric motor-driven winch, and placed at the top of a structure. During generation, the weights are lowered sequentially while turning electric generators (like a dropping weight turning a pulley via a cable tied to it). In principle, GES is analogous to PHES. It has the advantage of being modular and scalable, without the need for the availability of conveniently located natural and/or man-made reservoirs. Whether the technology can become commercially viable remains to be seen.

5.1   Thermal Energy Storage

Liquid Air or Cryogenic Energy Storage (LAES or CES) can be loosely described as a version of TES. The proposed technology uses electricity to cool the air until it liquefies and stores the liquid air in a tank. This is the storage or charge phase. During the discharge phase, liquid air is brought back to a gaseous state (by exposure to the ambient air or with waste heat from an industrial process) and expanded in a turboexpander to generate electricity. This technology will be examined more closely in Section 5.3.

Other TES systems can be classified into two major categories: storage of (i) sensible heat and (ii) latent heat. The former is based on successive heating (charge) and cooling (discharge) of a storage medium, which can be liquid or solid. The latter involves a phase change at constant temperature (and pressure), e.g., liquid to vapor (charge) and vapor to liquid (discharge). A comprehensive review of TES technologies can be found in Sarbu and Sebarchievici [4]. From a practical perspective, there are two types of TES technologies that have been developed to a commercial readiness level (see Section 16.5 for *Commercial Readiness Index*):

1. Gas turbine (GT) inlet air cooling with a chilled and stored medium (e.g., chilled water)
2. Storage of excess solar energy in molten salt

In the first technology, chilled water (sensible heat) or ice (latent heat) is produced by electricity supplied from the grid at night-time hours (at low, off-peak pricing) and stored for use the next day during the peak demand for GT compressor inlet air cooling. The economic justification for this application is strongly dependent on the price arbitrage between off-peak and peak times. For an in-depth discussion of how the inlet chiller plus TES system works, the reader is referred to the paper by Ebeling et al. [5]. A comprehensive review of the variants of this type of TES is provided in the paper by Gkoutzamanis et al. [6]. Briefly,

1. In chilled water storage, conventional chillers are used to cool water to temperatures of 4°C–5.5°C, which is stored in an insulated tank with cooler denser chilled water stratified below warmer less dense chilled water. Commercial examples can be found on the website of *Turbine Air Systems* (TAS).[1]
2. In the ice storage, chillers that can provide low-temperature brine (−5.6°C minimum) are used to freeze the water in the ice tank, which is then melted to provide inlet air cooling. For an example of the field application of this technology, refer to Ebeling et al. [5].

The ice storage option requires a fraction of the space required for chilled water storage (about 3 cubic feet or 85 liters per ton-hour) and takes advantage of the latent heat of ice (144 Btu/lb or 335 kJ/kg). It is also claimed that the chillers and heat

---

[1] http://turbineairsystems.com/thermal-storage/ (last accessed on March 22, 2021).

rejection equipment required for ice storage technology comprise about 60% of the chilled water storage variant. As a result, it also has a smaller footprint. Cold water supply temperatures for GT inlet air cooling range from 1°C to 6.7°C.

Instead of using excess energy to *cool* a medium, one could also use it to *heat* a medium. The most prominent example of the latter, which also found its way to field application, is utilizing molten salt to store solar energy as heat for power generation in a Rankine steam cycle. This is demonstrated in Figure 5.3, which shows the schematic diagram of the *Solar Two* demonstration project. Solar Two was a 10 MW concentrated-thermal solar facility, which operated from 1996 to 1999, and successfully demonstrated solar molten salt technology [7].

Typical molten salt used in storage applications is a 60%/40% (by weight) combination of *sodium nitrate* ($NaNO_3$) and *potassium nitrate* (KNO_3). Nitrate salts are widely used in chemical and fertilizer industry and there are many suppliers of these salts. (Typically, industrial grade $NaNO_3$ and KNO_3 is purchased separately and mixed at the site to assure good quality control.) Prima facie, this particular molten salt composition can be used over a temperature range of 260°C to approximately 621°C. However, the stability of the salt above 565°C is not firmly established. As the salt temperature decreases, it starts to crystallize at 238°C and solidifies at 220°C. The heat of fusion (based on the average of heat of fusion of each component) is 161 kJ/kg. The change in volume density upon melting is $\Delta V/V_{solid} = 4.6\%$.

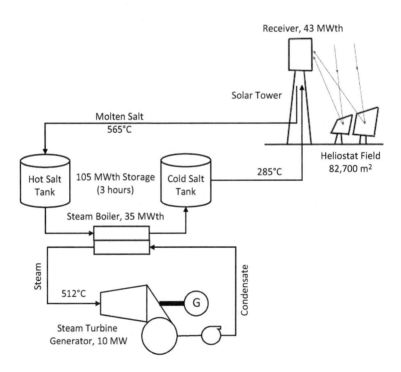

Figure 5.3 Schematic diagram of the molten salt *Power Tower* project (Solar Two).

Average thermal and fluid properties at 400°C are as follows: density of 1,834 kg/m^3, specific heat of 1.516 kJ/kg-K, dynamic viscosity of 0.0018 kg/m-s and thermal conductivity of 0.5198 W/m-K. The specific heat, CP, as a function of temperature is given by:

$$CP \text{ [in J/kg-K]} = 1,449.8 + 0.1688 \times T \text{ [in °C]}.$$

One of the difficulties of operating the molten salt-based systems is the high freeze temperature of the typical salts used for energy storage. Provisions must be made to ensure that the salt does not freeze inside the components of the storage tanks, particularly during extended plant shutdowns. Typically, immersion heaters are included to compensate for the heat loss from the storage tanks during plant outages. (Immersion heaters are electrical heaters that are inserted into the storage tanks as heating elements. In addition, electrical heat tracing is requisite for molten salt pipes, valves, and other BOP equipment.) Alternatively, natural gas or LPG heaters can be used for long-term temperature control of the molten salt storage inventory. However, fired gas heaters will require an environmental permit. Furthermore, such fired heaters are not suitable for frequent start and stop cycles. Due to the purging requirement before every start, this introduces significant inefficiencies into the operation of the system. Initial capital costs can be higher due to the custom design required for this particular application. If the fuel is not available on site, it will have to be trucked in or have a pipeline supply natural gas to remote areas where most solar plants are located.

The storage tanks are field constructed such that the tank size can be project specific. However, to utilize standard sheet metal cuts, tank sizes are typically determined by assuming that the metal sheets of which the tanks are constructed can be readily manufactured. To find the optimal tank size, a range of heights (up to the maximum shaft length of the molten salt pumps available) and diameters are calculated based on the required inventory volume, and then checked for an appropriate wall thickness and maximum diameter. If a single pair of tanks is unable to provide sufficient tank volume without exceeding the maximum wall thickness and diameter, then multiple pairs of tanks are designed for compliance with engineering codes. The smallest number of tank pairs capable of satisfying the storage requirements is normally selected to keep the final installed cost to a minimum.

The generic layout of a typical molten salt tank is presented in Figure 5.4. Molten salt flow into the tank is facilitated by a ring header with eductors located at the bottom of the tank. This arrangement effectively eliminates thermal stratification in the tank. Tank walls are insulated with mineral wool on the outside. The immersion heaters maintain a minimum temperature at least 20°C above the freezing point of the salt. During the charging phase, a cold salt pump located in the cold salt tank delivers nitrate salt to the solar receiver. During the discharge phase, a separate hot salt pump, located in the hot salt tank, delivers nitrate salt to the steam generator (boiler). Long-shafted pumps are typically used for molten salt storage systems, with a maximum shaft length of 12–14 m. The height of the tank will depend on the salt pump configuration so that, in this case, it will be limited to 14–15 m. Parasitic power

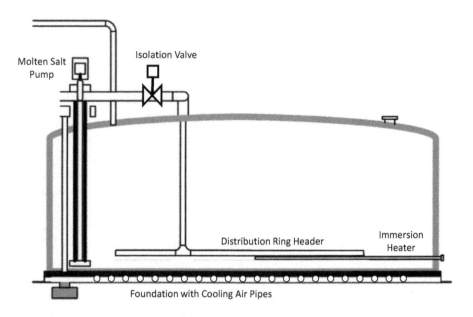

Figure 5.4 Internal arrangement of a typical molten salt tank.

consumption of immersion heaters and salt pumps can be significant in molten salt storge systems. In a 50 MW parabolic trough CSP plant, it was found to be as high as 12% in summer and 16–24% of the total generated power in winter [8].

The tanks are designed per *American Petroleum Institute* (API) 650 standard that establishes minimum requirements for material, design, fabrication, erection, and testing for vertical, cylindrical, aboveground, closed top, welded carbon, or stainless-steel storage tanks in various sizes and capacities. (The API 650 is applicable for tank diameters of 60 m or larger.)

Proper design of the tank foundation is critical. Concrete will begin to fail if its temperature reaches over 100°C and if the water of hydration is removed. One solution is a passive cooling system comprising rows of parallel carbon steel pipes, which are installed on top of the concrete mat. The foundation mat insulation is installed on top of the passive cooling system and consists of a sandwiched insulation system that is constructed of "foam glass" and refractory brick.

Solar Two employed 2,000 *heliostats*, rectangular glass mirrors that track the sun, all pointed at different angles toward the receiver atop the *solar tower*. Once the receiver collected enough heat, molten salt – kept at a constant 285°C – was heated to 565°C and sent through an insulated hot salt storage tank to a steam generator (boiler). The solar tower consisted of a 300 ft (about 100 m) lattice structure supporting a newly designed nitrate salt central receiver. About 36% of the solar energy absorbed by the central receiver was converted to electricity.

The power block comprised two insulated storage tanks – one used for cold salt storage, one for hot salt storage – which held 1.6 million kg of salt each. Each 30 × 40 (in ft) tank was well insulated to minimize the energy loss. Electric heaters offset any heat loss in the storage tanks and prevented the salt from solidifying during unanticipated outages. The molten salt heated the water in three shell-and-tube heat exchangers to produce steam, which in turn drove a steam turbine generator to produce electricity. The turbine generator consisted of a preheater, evaporator, and superheater.

Molten salt heat storage concept is very simple and can be applied in different configurations. One of them is the integration of molten salt TES with a GTCC, as depicted schematically in Figure 5.5. In the diagram, the heating source for the molten salt storage system is concentrated solar power. In general, this source can be anything, e.g., an electric heater utilizing surplus power during off-peak hours available for free or very cheaply.

As shown in Figure 5.5, during discharge, hot molten salt is used to generate steam in a boiler parallel to the heat recovery steam generator (HRSG) evaporator. This steam is added to the bottoming cycle at the evaporator drum to increase the overall steam flow to the steam turbine for additional power generation. It is also possible to replace the HRSG evaporator completely with the molten salt steam generator. One can use water/steam or a heat transfer fluid instead of molten salt as the storage medium. This technology is covered in Chapter 12 (*Integrated Solar Combined Cycle*, ISCC).

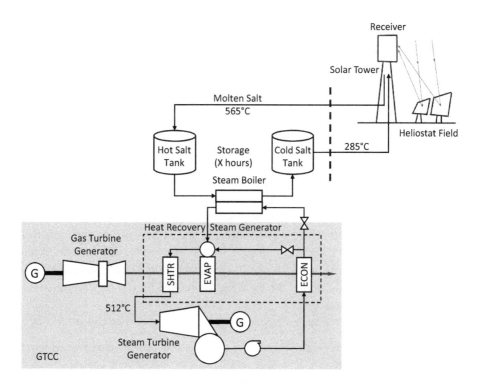

Figure 5.5 Gas turbine combined cycle with molten salt energy storage.

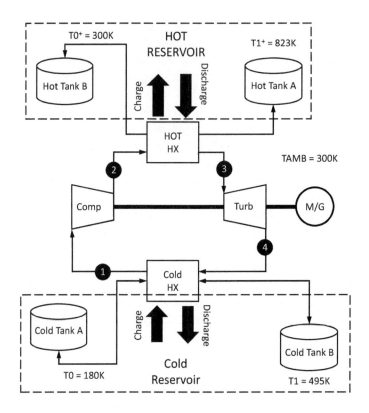

Figure 5.6 Conceptual schematic of a closed cycle gas turbine for pumped thermal storage.

Another variant of molten salt heat storage is the *Electro-Thermal Energy Storage System* promoted by a startup company, Malta, Inc. (a spin-off of Alphabet's (parent company of Google) Moonshot Factory, X). The idea is a variant of *"pumped thermal" storage* and is described theoretically in a paper by Nobel laureate Robert Laughlin [9]. Conceptually, it can be described as a closed cycle GT as shown in Figure 5.6. During charging, the synchronous ac machine acts as a motor and helps the turbine to drive the compressor. Compression heat is transferred to a heat storage medium (e.g., molten salt). During the discharge phase, the heat stored in the molten salt is transferred to compressed working fluid. In this mode, the synchronous ac machine acts as a generator because the turbine power output is higher than the compressor power consumption. This is ensured by the "hot temperature reservoir" comprising two tanks, A and B. During the charge mode, molten salt flows from tank B to tank A. During the discharge mode, the flow is in the reverse direction, from tank A to tank B.

In order for this energy storage scheme to work, two things must happen:

1. During the *charge* phase, the compressor discharge temperature should be as *high* as possible.
2. During the *discharge* phase, the compressor discharge temperature should be as *low* as possible.

This is ensured by the flow of the storage medium (e.g., a suitable *heat transfer fluid*, HTF) in the "cold reservoir" also comprising two tanks, A and B. In particular,

1. During the *charge* phase, working fluid from the turbine exit is *heated* by the HTF flowing from tank B to tank A.
2. During the *discharge* phase, working fluid from the turbine exit is *cooled* by the HTF flowing from tank A to tank B.

For an ideal system, using the nomenclature in Figure 5.6, it can be shown that the round trip efficiency (RTE) should satisfy the following condition [9]:

$$RTE < 1 - 2\left(\frac{TAMB}{T1 - T0}\right)\left(\frac{1}{EFFC} - EFFT\right)\left(\frac{\ln(E)}{E - 1}\right), \qquad (5.1)$$

$$E = \frac{T0^+}{T0} = \frac{T1^+}{T1}. \qquad (5.2)$$

Using the temperature values in Figure 5.6, with compressor and turbine polytropic efficiencies of EFFC = 0.91 and EFFT = 0.93, respectively, it can be shown that E = 1.667 and the right-hand side of Equation (5.1) is 0.753. Prima facie, this calculation suggests the possibility of achieving an RTE of 70% in the *actual* system. This number is patently unrealistic. To show where this assertion comes from, the full charge-discharge cycle calculation is reproduced in THERMOFLEX flowsheet simulation software (see Section 16.2) with *nitrogen* (N_2) as the working fluid (one of the two working fluids used in the Laughlin paper; the other was *argon*). Fluid properties are calculated with the REFPROP package (see Section 16.2). Compressor and turbine inlet pressures are set to 10 and 65 bars, respectively (from the original paper). With the specified pressure ratio of 6.5:1, the *isentropic* component efficiencies corresponding to the *polytropic* efficiencies used in the paper are 0.884 and 0.945, respectively. With these component efficiencies, 1 bar pressure loss in the high- and low-pressure legs of the cycle, and a heat exchanger design with 5 K pinch, charge, and discharge loops are calculated in THERMOFLEX. The results are summarized in Table 5.2.

In the basic conceptual scheme shown in Figure 5.6, the turbine exhaust temperature of 219.1 K in the charge cycle cannot support a cold reservoir T_0 of 180 K. This requires an additional component, i.e., a cryogenic chiller. The power consumption of this chiller, with a coefficient of performance of 3, is subtracted from the powertrain generator output to arrive at the net power generation in the discharge mode. The roundtrip efficiency is found as RTE = 148.3/269.5 = 55%. This suggests a technology factor of 0.55/0.753 = 0.73 and is in good alignment with past experience in similar technologies. The thermal efficiency of the system during discharge/generation is 148.3/521.5 = 28.4%.

5.1.1   Other TES Options

The two-tank molten salt TES system is currently the most advanced storage technology, and several solar thermal power plants are in operation, under construction, or in

Table 5.2 Pumped thermal energy storage performance

	Charge		Discharge	
	P, bara	**T, K**	**P, bara**	**T, K**
State Point 1	10	470	10	185
State Point 2	66.3	832.4	66.3	337.8
State Point 3	65	355	65	811
State Point 4	11	219.1	11	516.6
Compressor, kJ/kg	399.0		154.7	
Turbine, kJ/kg	136.8		323.7	
Motor, kJ/kg	269.5		NA	
Chiller, kJ/kg	NA		16.3	
Generator, kJ/kg	NA		164.6	
Net, kJ/kg	269.5		148.3	
Hot Reservoir Duty, kJ/kg	521.3		521.5	

the planning stages. (The *Andasol-1* power plant that was completed in 2010 in Spain is a notable example. It uses a two-tank indirect storage system, with synthetic oil as the HTF and molten salt as the thermal storage medium.) There are other options for TES systems and these are briefly mentioned below.

5.1.1.1   Solid Media Sensible Heat Storage

Solid media are typically incorporated into the TES system as a packed bed with a fluid used to exchange heat between the TES system and the rest of the plant. As such, they are a prime example of passive energy storage. The advantage of solid media is the option to use inexpensive solids such as rock, sand, or concrete as energy storage materials in conjunction with more expensive heat transfer fluids like thermal oil. However, the pressure drop and, thus, parasitic energy consumption may be high in a dual system and must be addressed in the storage system design. For one particular example of this technology in solar power applications, see Section 12.5.1.

5.1.1.2   Liquid Media

Both organic oils and molten salts are feasible liquid media for sensible heat storage. The salts generally have a higher melting point than oil, and parasitic heating is required to keep them liquid during long periods without insolation, or during extended plant shutdowns. Silicone oil is quite expensive, though it does have environmental benefits because it is a nonhazardous material, whereas synthetic oils may be classified as hazardous materials. Nitrites in salts present potential corrosion problems, though these are probably acceptable with use of the right material of construction at the temperatures required for TES.

5.1.1.3   Single Tank Thermocline System

Liquids maintain natural *thermal stratification* because of density differences between hot and cold fluids. A single tank system can capitalize on this characteristic. In such a system, the hot fluid is supplied to the upper part of a storage system during charging and the cold fluid is extracted from the bottom part during discharging. Another possible mechanism is to ensure that the fluid enters the storage at the appropriate level in accordance with its temperature (density) to avoid mixing. This can be done by some stratification devices (floating entry, mantle heat exchange, etc.). The other option is to simply separate the hot and cold liquids into two different tanks. Using two tanks is simpler and more efficient, but more expensive than a single tank system.

5.1.1.4 Latent Heat Storage

Thermal energy can be stored nearly isothermally in some substances as the latent heat of phase change, i.e., as *heat of fusion* (solid-liquid transition), *heat of vaporization* (liquid-vapor), or *heat of solid-solid crystalline phase transformation*. All substances with these characteristics are called *phase change materials* (PCMs). Since the latent heat of fusion between the liquid and solid states of materials is rather high compared to the sensible heat, storage systems utilizing PCMs can be reduced in size compared to single-phase sensible heating systems. However, heat transfer design and media selection are limited and more difficult, and experience with low-temperature salts has shown that the performance of the materials can degrade after a moderate number of freeze-melt cycles. Other design considerations for the PCM require that due to the constant melting temperature, PCM requires cascading configuration to manage the gap between the hot and cold temperature reservoirs. Furthermore, the heat transfer rate degrades during course of discharge due to solids formation and the resulting structure fatigue cycle issues have not been fully investigated.

5.1.1.5 Miscibility Gap Alloys

Thermal storage materials discussed above, i.e., solid, liquid, and phase change, have one or more disadvantages, e.g., low thermal conductivity, low energy density, or temperature limitations. For example, graphite has excellent characteristics in terms of high thermal conductivity and high maximum temperature but suffers from low energy density. PCMs have high energy density but suffer from low thermal conductivity. These disadvantages manifest themselves in complex designs to facilitate heat transfer with penalties incurred in parasitic power consumption (i.e., pumping large amounts of fluid) and size and cost (CAPEX). Generic characteristics of selected thermal storage materials are summarized in Table 5.3.

Miscibility gap alloys (MGAs) present themselves as a solution to all of the aforementioned shortcomings of existing thermal storage materials simultaneously. MGAs contains discrete, fully encapsulated microstructures of a metal, say, copper (Cu), trapped in a dense matrix of another metal, e.g., iron (Fe) [10]. The melting point of the former is lower than that of the latter. A popular analogy to an MGA is a muffin with chocolate chips. The bulk of the muffin represents the metal matrix with chocolate chips corresponding to the microstructure metal. When heated in an oven, chocolate chips melt but stay trapped inside the spongy bulk of the muffin. The advantages offered by the MGA materials are very high thermal conductivity,

Table 5.3 Key characteristics of selected thermal storage materials (energy density is evaluated at 100°C sensible heat capacity) [10]

	Material Type	Maximum Temperature, °C	Thermal Conductivity, W/m-K	Energy Density, MJ/l
Solid Media	Concrete, rocks, sand	400–450	1–1.5	0.18–0.22
	graphite	2,000	20–100	0.15
Liquid Media	Thermal oils	300–345	0.10	0.18–0.22
	molten salt	290–565	0.57	0.27
PCM	Waxes	60–150	0.30	0.50
	alkali metal salts	400–1,000	0.50–2	0.8–3

Table 5.4 Properties of selected MGA materials [10]

Material Type (Microstructure – Matrix)	Maximum Temperature (Active Phase), °C	Thermal Conductivity, W/m-K	Energy Density, MJ/l
Sn 50% – Al	230	120	0.43
Zn 50% – C	420	70	0.65
Mg 50% – Fe	650	100	0.58
Cu 50% – Fe	1,085	200	1.2
Si 50% – SiC	1,410	75	2.5

moderate to high energy density, and large temperature range. Properties of selected MGA materials are listed in Table 5.4.

In applications, MGA materials can be deployed as solid blocks. There are no moving parts and no change in the physical structure of the block. HTF tubing can be integrated into the MGA blocks to complete the system. A recently proposed energy storage system technology is based on a module comprising MGA blocks made of graphite and aluminum (to store heat), radiating electric heaters (to heat the blocks), and steam generator tubes in an enclosure. During charging, surplus electric power is used to raise the temperature of the MGA blocks (melting the aluminum) via radiating heaters. During the discharge mode, heat stored in the MGA blocks is transferred to the feed water flowing through the tubes, first boiling then superheating it (aluminum solidifies during this process). Generated steam is used in a steam turbine to generate electric power.

5.1.1.6   Thermochemical Storage

Yet another mechanism involves fully reversible chemical reactions. The heat produced by the solar receiver is used to induce an endothermic chemical reaction in the storage medium. The energy in storage can then be recovered by reversing the chemical reaction. Usually, catalysts are necessary to release the heat, which offers the advantage of using the catalyst to control the reaction speed and timing. Other advantages inherent in using a *reversible thermochemical reaction* for storage are high storage energy densities, indefinitely long storage duration at near-ambient temperature, and heat-pumping capability. Drawbacks may include complexity, uncertainties in the thermodynamic properties of the reaction components and of the reaction's kinetics under the wide range of operating conditions, high cost, toxicity, and flammability.

5.1.2   Liquid Salt Combined Cycle

This technology, with the acronym LSCC, is a straightforward combination of a conventional GTCC with TES as described in Figure 5.7 (Conlon, 2018, *Dispatchable Combined Cycle Power Plant*, US Patent 10,113,535). In essence, the evaporator is taken out of the HRSG and replaced by the TES system comprising a HTF loop. During charging, surplus, or inexpensive (or even free) energy in the form of, say, electric power, is utilized to heat the HTF (e.g., molten salt) and store it in a hot tank. During discharge, stored energy, i.e., hot HTF is utilized to boil feed water in the evaporator. The HRSG is used for feed water heating (economizer) and steam superheating.

For a detailed description of the system, the reader is referred to a recent article by Conlon [11]. As stated in the article, moving steam production outside the HRSG increases the available steam flowrate by 2.5–3 times vis-à-vis a conventional

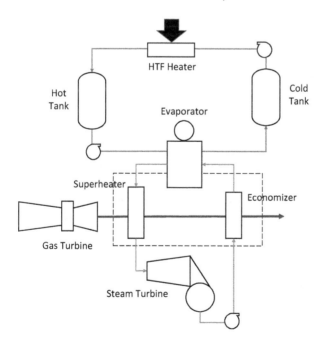

Figure 5.7 Simplified process diagram of the liquid salt combined cycle (LSCC), HTF: Heat Transfer Fluid (e.g., molten salt).

Table 5.5 LSCC performance with different gas turbines

	GT Output	GT Heat Rate	Optimal GTCC Output	Optimal GTCC Net Efficiency	LSCC Net Output	LSCC Net Fuel Efficiency
	MW	Btu/kWh	MW	%	MW	%
Siemens Trent WLE	66	8,150	86.7	55.0	113.5	72.0
Siemens SGT 800	57	8,502	83.1	58.5	105	73.9
GE LM6000 SPRINT	50	8,392	67.4	54.8	91.6	74.5
MP H-25	41	9,432	63.2	55.8	82.3	72.6
BHGE Nova LT16	16.2	9,456	24.0	53.5	33	73.5
Solar Taurus 65	6.5	10,375	10.2	51.7	13.3	67.3

Table 5.6 LSCC roundtrip efficiency

	Electrical Rate	Electric Heater Power	GT Fuel Consumption	LSCC Net Output	Roundtrip Efficiency
	–	MWe	MWth	MW	%
Siemens Trent WLE	0.93	105.6	157.6	113.5	59.0
Siemens SGT 800	0.76	79.8	142.0	105	64.4
GE LM6000 SPRINT	0.95	87.0	123.0	91.6	59.3
MP H-25	0.82	67.5	113.3	82.3	63.0
BHGE Nova LT16	0.91	30.0	44.9	33	61.1
Solar Taurus 65	0.96	12.8	19.8	13.3	57.9

bottoming cycle. Consequently, the bottoming cycle gross power output (i.e., the steam turbine generator output) should increase by the same ratio. This would translate into about a 30–35% increase in the net combined cycle output. Sample calculations from Ref. [11] are summarized in Table 5.5. The optimal GTCC output in the table is calculated using the exergy method described in Gülen (Ref. [45] in Section 2.4.3). Based on that, bottoming cycle output improvement is by a factor of 1.8–2.4. Corresponding net CC output improvement is by a factor of 1.25 to about 1.4. Roundtrip efficiency (see the definition of RTE1 in Section 6.2.1) is provided in Table 5.6, where the parameter "Electrical Rate" is the ratio of the electric power used to heat the molten salt to the LSCC net output.

As the HTF, LSCC is envisioned to use a low freezing point 288°F (142°C) eutectic salt, which is a mixture of water-soluble, inorganic salts of potassium nitrate, sodium nitrite, and sodium nitrate. It is safe, non-flammable, non-explosive, and non-toxic. During charging, the HTF is heated to 800°F (427°C) as it circulates through parallel trains of medium-voltage electric heaters, which offer high turndown for demand response, single-cycle response for frequency control, and provide accurate

Table 5.7 LSCC with Siemens Trent WLE with H_2 fuel

H_2, %(v)	H_2, kg/h	Electrolyzer, MW	CO_2, kg/MWh
0	0	0.0	278
10	150	8.3	270
20	330	18.1	260
30	539	29.6	248

temperature regulation. High-temperature salt is kept in an insulated carbon steel hot tank, where thermal losses are estimated to be less than 1°C per day [11].

During discharge, the salt is circulated through a molten salt steam generator (boiler) with the molten salt on the shell side. The system is elevated to facilitate gravity drainage of salt when not in use. It is claimed that the boiler can be kept hot to provide instant startup or can ramp from a cold condition to full power in less than 30 minutes [11]. The salt exits the steam generator at roughly 100°C above its freezing point.

The LSCC concept is very similar to the ISCC covered in Chapter 12. Instead of heating the HTF, e.g., molten salt, directly with solar energy, the LSCC uses electric heaters powered by a solar field (or a wind farm, or the grid), to store low-cost or even free energy (otherwise, it would have to be curtailed). At low loss rates (estimated to be less than 1% per day), this energy can remain stored in the LSCC tank(s) for a long time (even several days) until needed. Unlike the case in ISCC (see Section 12.6), stored energy can be used for fast startup of the system. During charging, variations in wind and solar generation can be accommodated rapidly. The steam turbine generator could provide voltage regulation as a synchronous condenser by inserting a synchronous, self-shifting (SSS) clutch between the steam turbine and its synchronous ac generator.

The LSCC concept can be expanded to include green hydrogen with electrolysis (see Chapter 8). This is illustrated by using Siemens Trent WLE as an example. The impact of increasing the H_2 content of fuel gas on CO_2 emissions and the requisite electrolyzer power to generate it (at 55 kWh/kg) are summarized in Table 5.7. It is interesting to note that, even with zero hydrogen content in the fuel gas, CO_2 emissions of the LSCC is lower than that of a conventional GTCC with an advanced class gas turbine (about 325 kg/MWh).

5.2 Pumped Hydro Energy Storage

PHES refers to a hydroelectric power plant comprising a pump and a turbine. During off-peak hours, water is pumped (by means of reversible pump-turbines or dedicated pumps) from a lower reservoir to an upper reservoir; energy is thus stored for later production during peak hours.

According to a 2012 NHA report [12], globally, there are approximately 270 pumped storage plants either operating or under construction, representing a combined generating capacity of over 127,000 MW. Of these total installations,

36 units consist of adjustable-speed machines, 17 of which are currently in operation (totaling 3,569 MW) and 19 of which are under construction (totaling 4,558 MW). Adjustable-speed pump-turbines have been used since the early 1990s in Japan and the late 1990s in Europe. The main reason that adjustable-speed pumped storage was developed in Japan in the early 1990s was the realization that significant quantities of oil burned in combustion turbines could be reduced by shifting the responsibility for regulation to pumped storage plants.

The operational efficiency of a storage plant is measured in terms of *roundtrip efficiency*. Another name for the roundtrip efficiency is cycle efficiency, which, as defined in the ASCE hydropower planning and design guide, is the ratio of the generating output of the pumped storage plant to the pumping input, including pump turbine, generator motor, and hydraulic losses. Dividing the recorded energy output (from generation) by the recorded energy input (for pumping) would provide cycle (roundtrip) efficiency. A review of the historical data for seventeen different large PHES projects in operation in the USA showed that the overall cycle efficiencies ranged from a low of 60% to a high of 80% [13]. For planning purposes, an overall cycle efficiency value of 75%, including hydraulic losses, is probably adequate for state of the art in PHES. It may be possible to achieve 80% or more in a controlled test of evacuating and refilling the reservoir within a short time, with the facility operated at optimized output. Nowadays, however, plants must often be operated at off-design conditions for overall system economy and market reasons, and the best theoretical cycle efficiency cannot always be attained.

Strictly speaking, PHES is outside the scope of the current book with a focus on prime movers. Nevertheless, as an energy storage system, the "load smoothing" or "load leveling" capability of PHES, illustrated in Figure 5.8, is the same as of other

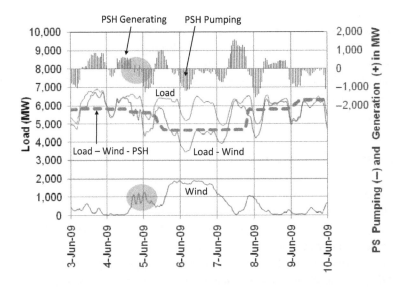

Figure 5.8 Pumped hydro energy storage integrated with wind to smooth the load curve [13].

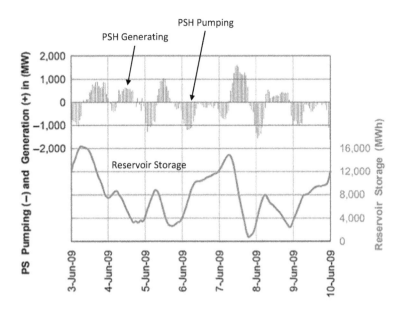

Figure 5.9 Pumping, generation, and reservoir storage change patterns [13].

storage technologies. It will thus be used as an example herein. Especially interesting is the highlighted period when the hourly variation in wind generation is "absorbed" and "smoothed out" by PHES [13]. (The corresponding reservoir storage variation is shown in Figure 5.9.) In that time window, wind generation is highly variable. Since the load demand does not show a similar trend, without power supply from the PHES the remaining generating assets, i.e., fossil-fired power plants, would also show the same fluctuating trend. By utilizing the PHES's ability to absorb (pumping) or deliver (discharging) power with a rapid transition from one mode to the other, the generation profile of those assets has been smoothed out. This is favorable from both thermal efficiency and RAM (power plant components life) perspectives.

5.3   Cryogenic Energy Storage

CES (or LAES) comprises three processes: charging (air liquefaction), storage, and discharging (regasification and expansion). It can be used for firming the renewable power by utilizing the excess energy, which otherwise would be curtailed and wasted, to liquefy ambient air and store it in a tank at cryogenic temperature (around *minus* 150°C). Later, when the demand exceeds the renewable resource's ability to meet it, stored liquid air is reheated and expanded in a turboexpander to generate electric power. Alternatively, the technology can also be used for arbitrage, i.e., using cheap energy for charging (air liquefaction) and discharging (power generation) when demand and electricity prices are high. A simplified schematic diagram of the CES process is shown in Figure 5.10 [14]. The process in the figure includes the two

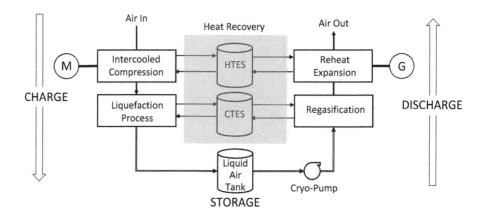

Figure 5.10 A high-level, conceptual schematic diagram of CES.

enhancements to the basic CES, *hot thermal energy storage* (HTES) and *cold thermal energy storage* (CTES). HTES recovers the heat rejected from the intercoolers of the air compressor. CTES is used as a heat sink to reduce the energy demand of the total charging process by cooling about 70% of the compressed air. The remaining 30% is cooled down by the internal recycle flows.

There are two major processes that can be used to liquefy air: The *Linde–Hampson* (LH) and *Claude* cycles. The LH cycle uses a throttle valve and the *Joule–Thompson* effect (isenthalpic) to liquefy air. This process is characterized by a large loss of exergy. A recent variant, the *Solvay* cycle, which replaces the throttle valve by a cryogenic expander (isentropic), is thus more efficient. The Claude cycle combines the throttle valve with an expander and is more efficient than the LH cycle. It is also less complicated than the Solvay cycle, which results in two-phase fluid flow in the cryogenic expander and is quite difficult to design and control. For a comprehensive review of the liquefaction technologies, refer to the paper by Damak et al. [15]. A detailed thermo-economic evaluation of the LH and Claude processes (including their variants) utilizing the exergy concept can be found in Hamdy et al. [16].

A generic CES process diagram adopted from the paper by Hamdy et al. [16] to evaluate the different liquefaction processes is depicted in Figure 5.11. It includes all the major processes, including heat recovery, with the liquefaction process treated as a "black box." The authors evaluated six different liquefaction technologies, three based on the LH cycle and three based on the Claude processes. The six cases are evaluated on the basis of four key parameters:

- Exergetic efficiency
- Liquid yield
- Specific charging power
- Roundtrip efficiency

Before delving into the details of these parameters and the findings, let us look at the basic LH and Claude processes for liquefaction of air first.

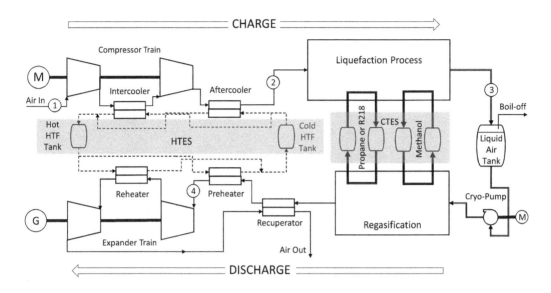

Figure 5.11 Generic CES with intercooled compression, reheat expansion, HTES, and CTES.

The basic LH process is quite simple and straightforward as shown in Figure 5.12. Air is compressed in a multi-stage, intercooled compressor train to a very high pressure (e.g., up to 200 bar). The main cooling takes place in two steps: (1) heat transfer to the recirculated gaseous air at cryogenic temperature in a heat exchanger (the *regenerator*) and (2) pressure reduction across a throttle valve in an isenthalpic Joule–Thompson process, which results in partial condensation of air. Liquid and vapor phases are separated in a flash tank. The vapor is sent to the regenerator to cool the high-pressure air. The liquid air is sent to the storage tank. The efficiency (liquid yield) of the LH process is a function of the pressurized air at the inlet of the regenerator (state 4 in Figure 5.12), which reflects itself in the air temperature at the throttle valve inlet. Without CTES, at 200 bar, air temperature at the throttle valve inlet is around $-100°C$ and the liquid yield, γ, is barely 10%. In other words, for each 100 kg/h of ambient air supply, only about 10 kg/h of liquid air is produced. With CTES, the air temperature at the throttle valve inlet can be reduced to about $-125°C$ and the liquid yield increases to about 30%.

The exergetic efficiency of the LH process is not stellar either. The key to understanding air liquefaction is recognizing that, while the internal energy and enthalpy of the air is *reduced* in this process, its *exergy*, i.e., its ability to produce useful work, is *increased*. This assertion may be met with incredulity by a reader, who is unfamiliar with the concept of exergy. Thus, it behooves us to spend a few words on this vital aspect of the air liquefaction. Let us start with the definition of exergy, i.e.,

$$e = (h - h_0) - T_0(s - s_0), \tag{5.3}$$

where the subscript 0 denotes the "dead state," i.e., the state of the fluid in question when it is in perfect equilibrium with its surroundings and, as such, does not have any

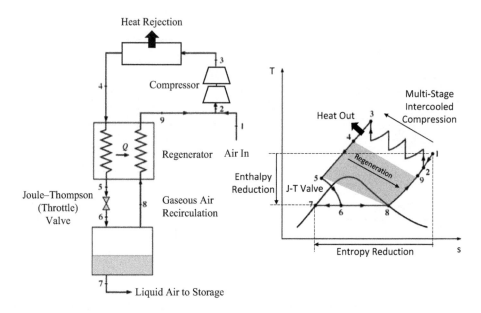

Figure 5.12 Schematic description of the basic Linde–Hampson air liquefaction process and its conceptual T-s diagram.

ability to produce useful work. If one writes the change in exergy of air between states 1 and 7 in the T-s diagram in Figure 5.12, we obtain the following relationship:

$$e_7 - e_1 = (h_7 - h_1) - T_0(s_7 - s_1), \tag{5.4}$$

$$\Delta e = \Delta h - T_0 \Delta s. \tag{5.5}$$

Furthermore, also from the T-s diagram, one can *qualitatively* infer that

$$\Delta h < 0, \Delta s < 0 \text{ and } |\Delta s| > |\Delta h|. \tag{5.6}$$

Thus, combining Equations (5.5) and (5.6), one can easily realize that

$$\Delta e = -|\Delta h| + T_0 |\Delta s| > 0. \tag{5.7}$$

The exergetic efficiency of the CES process is defined by Hamdy et al. [16] as

$$\varepsilon = \frac{\dot{E}_{LA} + \dot{E}_{HTES}}{\dot{W}_C - \dot{W}_E + \dot{E}_{CTES}}, \tag{5.8}$$

where (using the T-s diagram in Figure 5.12 as a reference) $\dot{E}_{LA}$ is the rate of exergy transfer associated with the liquefied air (state point 7); $\dot{W}_C$ is the compressor power consumption; $\dot{W}_E$ is the expander power production; $\dot{E}_{HTES}$ is the exergy transfer associated with the heat transfer, $\dot{Q}_{HTES}$, from the compressor inter- and aftercoolers to the HTES; $\dot{E}_{CTES}$ is the exergy transfer associated with the heat transfer, $\dot{Q}_{CTES}$, from the CTES to the pressurized air to cool it to cryogenic temperatures. The requisite formulae are provided below.

$$\dot{E}_{LA} = \dot{m}_7 e_7, \tag{5.9}$$

$$\dot{W}_C = \dot{m}_2 \omega_c, \tag{5.10}$$

$$\dot{W}_E = \dot{m}_7 \omega_e = \gamma \dot{m}_2 \omega_e, \tag{5.11}$$

with ω_c denoting the characteristic specific power consumption of an intercooled, multi-stage compressor train comprising state of the art centrifugal process compressors and ω_e denoting the characteristic specific power output of a multi-stage expander train comprising state of the art cryogenic and/or gas expanders with interstage reheat. The liquid yield, γ, is the ratio of the mass flow rates liquefied air (produced in the particular liquefaction process) and the compressed air fed to the liquefier (essentially, ambient air after "drying" and intercooled compression). (Clearly, γ is going to be less than one for the LH process shown in Figure 5.12.)

$$\gamma = \frac{\dot{m}_7}{\dot{m}_2}. \tag{5.12}$$

The exergy transfers associated with heat transfers via HTES and CTES can be calculated as shown below:

$$\dot{E}_{HTES} = \left(1 - \frac{T_0}{\overline{T}_H}\right) \dot{Q}_{HTES}, \tag{5.13}$$

$$\dot{E}_{CTES} = \left|\left(1 - \frac{T_0}{\overline{T}_C}\right) \dot{Q}_{CTES}\right|. \tag{5.14}$$

The terms in parentheses on the right-hand sides of Equations (5.13) and (5.14) are the Carnot factors. The temperatures, $\overline{T}_H$ and $\overline{T}_C$, are the "mean-effective" temperatures at which the heat is taken from the air (for HTES) and supplied to it (low-temperature energy from the CTES), respectively. They can be calculated as logarithmic averages of supply and discharge temperatures, T_S and T_D, respectively, i.e.,

$$\overline{T} = \frac{T_D - T_S}{\ln\left(\frac{T_D}{T_S}\right)}. \tag{5.15}$$

For example, let us assume that the heat rejected in the intercoolers of the air compressor to a particular HTF is stored in the hot storage tank at 168°C. Let us also assume that the HTF supply temperature from the cold storage tank is 20°C. Thus, from Equation (5.15), it is found that

$$\overline{T}_H = \frac{168 - 20}{\ln\left(\frac{168 + 273}{20 + 273}\right)} = 362 \text{ K} = 89°\text{C}.$$

Hence, with the assumption of $T_0 = 15°$C, the Carnot factor, CF_H, on the right-hand side of Equation (5.13) is calculated as

$$CF_H = 1 - \frac{T_0}{T_H} = 1 - \frac{288}{362} = 0.20.$$

The interpretation of this factor is as follows: If one could build a hypothetical Carnot engine that receives heat, $\dot{Q}_{HTES}$, from a high-temperature reservoir at, $\overline{T}_H$, and rejects it to a low-temperature reservoir at T_0, the efficiency of that (perfect) heat engine would be 20%. In other words, the heat rejected from the intercoolers of the multi-stage air compressor has little value from a work production perspective. (Note that, if one really tries to make use of the heat using, say, an *organic Rankine cycle* device, the actual efficiency will barely amount to 10%.) Nevertheless, from an exergetic bookkeeping point of view, every kilowatt counts.

From a purely pedagogical perspective, $\dot{E}_{CTES}$, as defined by Equation (5.14), is even more problematic because $\overline{T}_L < T_0$. (There is also the practical difficulty of "guessing" the supply and discharge temperatures, T_S and T_D, respectively, without a bona fide heat and mass balance simulation of the process.) Conceptually, the CTES, as described by Hüttermann et al. [14], is a multi-loop, multi-tank process as shown in Figure 5.13. The temperature range dictates the usage of two different HTFs for energy storage. One of them is *methanol* (CH_3OH), which freezes at 176 K. The other one is *propane* (C_3H_8), which freezes at 85 K and vaporizes at 231 K. Thus, propane is used between 126 K and 216 K and methanol is used for higher temperatures. Hamdy et al. uses methanol and *octafluoropropane* (R-218, with the formula C_3F_8), instead of propane, for the same purpose in their study [16]. The reason for the intermediate heat transfer loop utilizing supercritical N_2 (an inert fluid) is safety because O_2 and liquid hydrocarbons have to be isolated from each other.

In the LH process shown in Figure 5.12, the charge mode of the CTES (on the left in the diagram in Figure 5.13) would be placed between the compressor and the regenerator. Compressed air is cooled by N_2 at cryogenic temperatures, which is heated to near-ambient temperatures. Heated N_2 is then cooled down back to a cryogenic temperature in a second, parallel heat exchanger system heating first

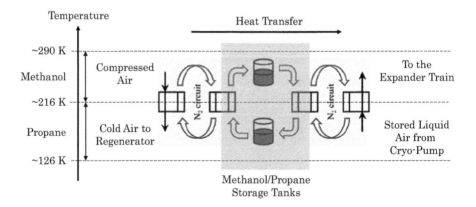

Figure 5.13 Indirect CTES utilizing methanol and propane in a multi-tank, multi-heat exchanger system with nitrogen as the intermediate heat transfer fluid.

methanol and then propane, which are stored in their respective tanks. In the discharge mode (on the right in the diagram in Figure 5.13), the process is reversed. This time, heated N_2 at near-ambient temperature regasifies the liquid air, which is pumped from the storage tank by the cryo-pump. Thereafter, air is sent to the expanders while N_2 at cryogenic temperature is reheated back to near-ambient temperature by cooling first propane and then methanol to cryogenic temperatures.

Utilizing the temperature limits in Figure 5.13 as a rough guide, the mean-effective temperature at which heat is transferred to the CTES from the compressed air can be estimated as

$$\overline{T}_L = \frac{126 - 290}{\ln\left(\frac{126}{290}\right)} = 197 \text{ K} = -76°\text{C}.$$

Obviously, a conceptual heat engine that receives heat from a reservoir at $-76°$C and rejects it to a reservoir at $15°$C while producing work violates the second law of thermodynamics, an impossibility. In that case, we have to look at the problem in a different way. Note that the work done by a Carnot engine receiving heat from a high temperature reservoir and rejecting heat to a low temperature reservoir can be expressed in two different ways. Namely,

$$w = q_H - q_L, \tag{5.16}$$

$$w = q_H \left(1 - \frac{q_L}{q_H}\right) = q_H \left(1 - \frac{\overline{T}_L}{\overline{T}_H}\right), \tag{5.17}$$

$$w = q_L \left(\frac{q_H}{q_L} - 1\right) = q_L \left(\frac{\overline{T}_H}{\overline{T}_L} - 1\right). \tag{5.18}$$

Equation (5.17) is the conventional version with the term in parentheses on the right-hand side of the equation representing the efficiency of the Carnot engine (the numerical value less than one). In the alternative version, Equation (5.18), the Carnot engine work is referenced to the heat rejected from the engine and the term in parentheses on the right-hand side of the equation represents a coefficient of performance, which is greater than one.

On the other hand, from a purely *mechanical* perspective, it is eminently possible to utilize a fluid at 200 bar to produce useful work via expansion to 1 bar. This is quantitatively reflected by the high exergy of pressurized and liquefied air at cryogenic temperatures. Using Equation (5.3) and the NIST REFPROP package (see Section 16.2), exergy of air at $-76°$C and 200 bar is calculated to be 464.3 kJ/kg with a dead state of 1 bar and $15°$C. In contrast, the enthalpy of air at the same conditions is -185.9 kJ/kg. For an in-depth understanding of this unique feature of liquefied air, one must separate the exergy into its mechanical and thermal parts. This is covered in detail in Kotas [17, pp. 40–41].

Hamdy et al. [16] analyzed different LH and Claude process variants (the latter includes *Kapitza* and *Heyland* processes) with and without HTES/CTES. Modifications to the basic LH process include precooling of air using a conventional refrigeration cycle and the introduction of a second pressure level (i.e., two throttling processes in series). As noted earlier, the Claude cycle differs from the basic LH process by the addition of an expander that is parallel to the throttle valve to reduce the exergy destruction in the latter and charging power consumption. The exhaust stream of the expander is directly utilized in the regenerator for cooling of air. The Kapitza variant of the Claude cycle mixes the cold expander exhaust stream with cold air vapor from the flash tank (stream 8 in Figure 5.12). The Heyland variation of the Claude cycle places the expander upstream of the regenerator so that the precooling improves the heat transfer process in the regenerator. Detailed cycle diagrams are provided in Ref. [16].

Based on the detailed heat and mass analysis in Ref. [16], Claude cycle variants are superior to the LH cycle variants in terms of liquid yield (about 0.61 on average vis-à-vis 0.3–0.45, respectively), exergetic efficiency (about 77% vis-à-vis 53–61%, respectively), and specific charging power (less than 1,100 kJ/kg vis-à-vis 1,500–2,000 kJ/kg).

Claude cycle performance is a function of the splitting ratio, r, which is the ratio of air flow rate through the expander to the compressed air flow rate. The optimum performance for Claude and Kapitza cycles is found at an r value between 0.35 and 0.4. The optimum for the Heyland cycle is at an r value of between 0.25 and 0.3. The roundtrip efficiency of the Claude and Kapitza cycles were calculated as 46.9%, and for the Heyland cycle as 49%. Roundtrip efficiencies for CES systems cited in the literature are as high as 60% with varying charge/discharge flow ratios and pressures and liquid yields (e.g., see Table 1 in Ref. [16]). The specific charging power calculated by Hamdy et al., of about 265 kWh/ton, was about half that found in the literature. The specific investment cost was evaluated at slightly above EUR 900 per discharge power output (in kW) vis-à-vis EUR 500–3,000 found in the literature.

In conclusion, due to the large variation in cycle configurations, design parameters, calculation models, and property packages, lack of experience base in terms of commercial construction and field operation experience, to pass judgment on the feasibility of CES in comparison to other bulk storage technologies such as CAES and pumped hydro is full of proverbial *booby traps*. Nevertheless, let us look at one example.

There are several commercial CES storage offerings out there from technology startups. One of them, *Highview Power*, offers a 50 MW discharge power and 5-hour system (250 MWh) at £85 million.[2] This translates to EUR 1,989 per kW ($2,278/kW). A bigger system at 200 MW and 10-hour duration (2,000 MWh) is stated to have a levelized cost of storage of £110/MWh ($147/MWh). The claim of the technology developer is that capacity can simply be increased by adding more liquid air storage tanks and using larger turbomachinery. The reported cost increase for doubling the capacity from 50 MW to 100 MW is 40%, which implies a scaling exponent of 0.49

---

[2] *Really cool storage*, The Energy Industry Times, September 2020 issue, p. 15.

(presumably the same duration). Thus, the cost of the 200 MW and 2,000 MWh system would be roughly £235 million or about EUR 1,100 per kW (~$1,250/kW). The RTE is claimed to be above 70% with hot and cold energy storage. So far, the company has built a 5 MW/15 MWh demonstration plant in the UK, which was commissioned in June 2018. Construction of a 50 MW in UK began in 2020.

The levelized cost of storage formula (LCOS) is given by

$$LCOS = \frac{CAPEX + \sum_{t=1}^{N} \frac{OPEX_t(1+i)^t + CCHG_t(1+e)^t}{(1+r)^t}}{\sum_{t=1}^{N} \frac{MWHd_t(1+e)^t}{(1+r)^t}}, \quad (5.19)$$

where CCHG is the cost of the electricity used for charging the system, MWHd is the electric power supply to the grid during discharge, r is the discount rate, i is the annual inflation rate, e is the annual cost of electricity increase rate, RTE = 70%, and N is the economic lifetime. Assuming that r = 8%, i = 2%, and e = 3%, the charging electricity price of 20/MWh, N = 20, and 350 cycles per year, the LCOS for the 50 MW/250 MWh system is found as £114/MWh ($153/MWh). The CAPEX is £85 million and OPEX is set to 1% of CAPEX. Note that

$$MWHd_t = 250 \text{ MWh} \times 350 = 87,500 \text{ MWh},$$

$$CCHG_t = 87,500/70\% \times £20/MWh = £2.5 \text{ million}.$$

If the discharge duration is changed to 9 hours, i.e., the capacity is increased from 250 MWh to 450 MWh, the CAPEX becomes £85 × (450/250)$^{0.49}$ = £113 million and LCOS is calculated as £92/MWh ($123/MWh).

The interested reader can compare these numbers to numbers cited for competing technologies available in the literature. Unfortunately, without a proper front-end engineering design (FEED) study for a *real* project by a credible engineering, procurement, and construction (EPC) contractor, there is no way to verify CAPEX, OPEX, and LCOS numbers for any of them. As far as the claimed RTE, there is no published heat and mass balance data for the technology to verify it independently. (The cycle, as depicted in the company brochure, is essentially identical to what is depicted in Figure 5.10.) Suffice that, at 70% or higher, it is quite higher than what is reported in the literature.

5.4   CO$_2$ Energy Storage

In a pumped hydro *battery*, the working fluid is water, charged by pumping water from a zero head reservoir to a high head reservoir and discharged by letting it flow down in the reverse direction. During the discharge phase, the *head difference* (potential energy) is converted into *velocity* (kinetic energy) and the difference is translated into useful work in a hydraulic turbine.

In CAES, the same principles are at work with two major differences. First, the working fluid is air with a much lower density than water. Second, there is only one reservoir, i.e., the high head (in this case, high pressure) reservoir. The zero head (i.e., zero gauge pressure) reservoir is the atmosphere. This technology is covered in detail in Chapter 6.

In LAES, not surprisingly, once again the same principles are at work. The difference between LAES and CAES is primarily the rather ingenious solution of the Achilles' heel of the latter, i.e., low density leading to the need for a large volume storage reservoir, via *cryogenics*. The density of high-pressure stored air is increased by orders of magnitude by cooling it to cryogenic temperatures. In this way, instead of finding (or excavating) a subterranean cavern that can hold the *Empire State Building*, one can resort to reasonably sized cryogenic holding tanks made from suitable materials. LAES technology was discussed in the preceding section.

At the risk of sounding like a parrot, yes, there is another way to approach the energy storage concept by using the same principles of charge/pump/compress and discharge/expand but with yet another working fluid, i.e., *carbon dioxide* (CO_2). Why? At this point, the reader probably knows the answer by heart: to solve the storage problem presented by a low-density (gaseous) working fluid like air without going to extreme solutions such as cryogenics. The advantage of CO_2 as a storage medium vis-à-vis air is readily explained by the comparison in Table 5.8. However, its disadvantage is rather obvious and difficult to overcome; there is not a readily available "reservoir" of low-pressure CO_2 whereas atmospheric air is practically an unlimited reservoir in that sense.

There are two proposed solutions in this regard. One is an extension of the original CAES approach, i.e., using *two* saline aquifers or caverns instead of *one*. The first reservoir is a low-pressure reservoir at low depth from the surface used to store CO_2 exhausted from the turbine during the discharger/generation phase, whereas the second, much deeper reservoir is at higher pressure to store CO_2 from the compressor during the charge phase. A conceptual schematic of this particular approach to CO_2 storage is provided in Figure 5.14. Simplified (idealized) thermodynamic and parametric analyses of the performance of this particular energy storage system under supercritical and trans-critical conditions is provided in paper by Liu et al. [18].

In the forecited paper, the authors consider two variants: *trans-critical* (TC) and *supercritical* (SC). The difference between the two variants is based on the turbine inlet and exit pressures. In particular,

Table 5.8 Compression of energy storage media

	CO_2	CAES	LAES
Volume at ambient conditions	Base	+50%	+50%
Storage pressure	Base	~ same	<10%
Storage temperature	Base	~ same	cryogenic
Storage volume	Base	10×	~ same
Storage energy density	Base	<10%	+60%

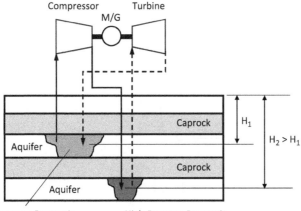

Figure 5.14 Simplified schematic of CO$_2$ energy storage system. $H_2 = 3{,}000$ m, $H_1 = 760$ m for the SC variant, 200 m for the TC variant (see the text). Storage pressure in the high-pressure reservoir is about 40 MPa. The storage pressure in the low-pressure reservoir is 8 MPa for the SC variant, 2 MPa for the TC variant.

- In the TC variant, the CO$_2$ turbine inlet pressure is supercritical (40 MPa) but turbine exit pressure is subcritical (2 MPa), whereas
- In the SC variant, the turbine inlet pressure is supercritical (but lower, i.e., 20 MPa) and turbine exit pressure is critical (8.04 MPa). In other words, turbine expansion takes place across the supercritical states.

A simplified flowsheet and the accompanying T-s diagram of the TC CO$_2$ energy storage system is shown in Figure 5.15. Its salient characteristics are as follows:

- Intercooled compressor train
- Reheat turbine (overall PR = 10:1, HP turbine PR = 1.6:1, LP turbine PR = 6.3:1)
- Natural gas fired reheater (873 K CO$_2$ exit temperature)
- TES to store the heat rejected by the compressor intercoolers to be used during discharge to heat stored CO$_2$ (not shown)

A simplified flowsheet and the accompanying T-s diagram of the SC CO$_2$ energy storage system is shown in Figure 5.16. Liu et al. defines the roundtrip efficiency as [18]

$$RTE = \frac{WT}{WC + 45.6\% \times HFC}, \tag{5.20}$$

Where WT is the turbine power output, WC is the compressor power consumption, and HFC is heater fuel consumption. The second term in the denominator represents the amount of electricity that could have been made from HFC if that fuel had been used to make electricity in a stand-alone power plant with an effective thermal efficiency of 45.6%. The calculation results summarized in Table 5 of Ref. [18] indicate RTE values in excess of 60% for SC and TC variants.

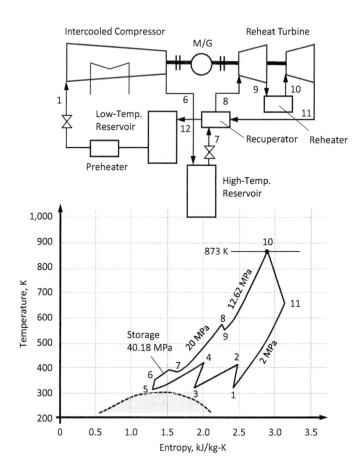

Figure 5.15 Schematic flowsheet and T-s diagram of the TC CO_2 energy storage.

These values are, unfortunately, highly unrealistic. The biggest source of error stems from the fact that the authors used the heat transferred to CO_2 in the heater for HFC. This, of course, is patently impossible in a fired heater. To perform a check on calculated numbers, the SC CO_2 energy storage calculation has been repeated by the author using Thermoflow, Inc.'s THERMOFLEX software (see Section 16.2). Carbon dioxide properties are calculated using the built-in REFPROP package (see Section 16.2). Cycle pressure and temperatures are taken from Table 5 of Ref. [18]. The same compressor and turbine isentropic efficiencies are adopted, i.e., 85% and 87%, respectively. A comparison of the results is summarized in Table 5.9. A natural gas (100% CH_4) fired heater diagram is presented in Figure 5.17.

Another approach to CO_2 energy storage is to utilize two man-made reservoirs. One example of this approach is the "CO_2 Battery" developed by an Italian company, *EnergyDome*.[3] It is very similar to LAES in its basic operating principle. In this system, CO_2 is liquefied and stored in carbon steel tanks at 65 bar and 25°C. Heat generated

[3] https://energydome.it/co2-battery/, last accessed on May 2, 2021.

Table 5.9 Comparison of SC CO₂ energy storage calculations

	Units	This Work	Liu et al. [18], Table 5
Compressor	kJ/kg	86.76	83.81
Turbine	kJ/kg	118.4	123.58
Fuel Input	kJ/kg	325.92	N/A
Heat Input	kJ/kg	222.25	240.73
Heater Air Fan	kJ/kg	4.78	0
Net Output	kJ/kg	113.62	123.58
RTE	%	47.0	62.3

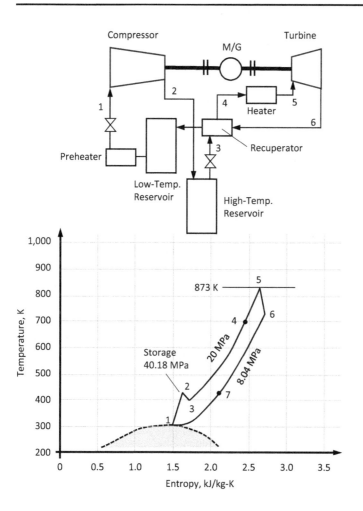

Figure 5.16 Schematic flowsheet and T-s diagram of the SC CO₂ energy storage.

during compression and liquefaction is stored in a packed bed TES (and hot water). In the discharge phase, liquid CO_2 is released from the tanks and heated utilizing heat from the stored hot water and TES and expands through a turbine to generate electric power. The turbine exhaust CO_2 is stored in a flexible membrane (the so-called dome). The

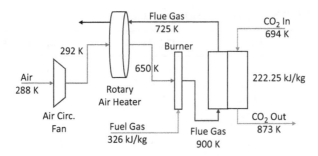

Figure 5.17 Natural gas fired CO_2 heater.

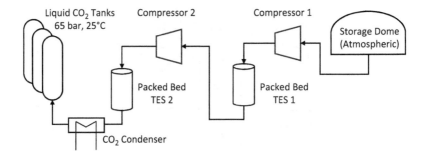

Figure 5.18 EnergyDome CO_2 storage – compression/charge phase schematic.

compression (charge) process is schematically described in Figure 5.18. The heart of the technology is the packed beds that act as compressor intercooler and aftercooler. Carbon dioxide from the flexible membrane dome at atmospheric conditions (gas) is compressed and liquefied in a water-cooled condenser. The discharge process takes placed in the reverse direction. There is no additional heat input (e.g., in gas fired heater).

On the website of the technology promoter, the RTE is claimed to be over 75%. No cycle information is available to assess the validity of this performance number. However, a quick check has been done in THERMOFLEX using two-stage compression and expansion as shown in Figure 5.19. Same compressor and turbine isentropic efficiencies as before are assumed, i.e., 85% and 87%, respectively. RTE is calculated as $26,122/(17,685 + 17,689) = 73.8\%$. While this is lower than the claimed RTE, it is close enough that the proposed storage system can be said to be of some merit. The critical component is the TES system. Further development of the packed bed TES technology and the overall CAPEX and OPEX of the proposed system are the key determinants of whether it can be a viable competitor to CAES or LAES.

5.5   Ancillary Services

Energy storage, especially utilizing batteries, can also provide ancillary grid services, e.g., frequency regulation, which refers to providing immediate (i.e., within 4 seconds)

Table 5.10 Ancillary service capabilities of selected energy storage technologies.

(RES: Renewable Energy Source)

	PHES	CAES LAES	Batteries	Hydrogen	Molten Salt
Power Quality			X		
Energy Arbitrage	X	X	X		X
RES Integration		X	X	X	X
Emergency Back-Up			X		
Peak Shaving	X	X	X		
Time Shifting	X	X	X		
Load Leveling	X	X	X		
Black Start			X		
Seasonal Storage				X	X
Spinning Reserve			X		
Network Expansion	X		X		
End User Services			X		

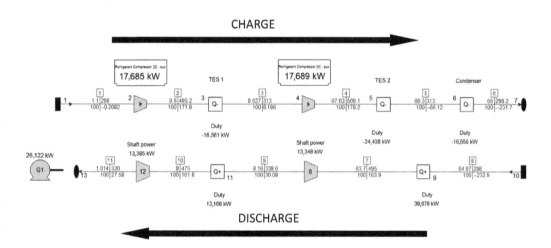

CHARGE

DISCHARGE

Figure 5.19 Approximate heat and mass balance of EnergyDome CO_2 battery.

power to maintain generation-load balance and prevent grid frequency fluctuations. Another important ancillary service is *spinning reserve*, which refers to maintaining electricity output during an unexpected contingency event (e.g., an outage) immediately. A brief overview of such capabilities for selected energy storage technologies is provided in Table 5.10 [2]. The reader can consult Ref. [2] for case studies of projects where these capabilities are utilized.

In Table 5.10,

- *Power quality* is the degree to which the voltage, frequency, and waveform of a power supply system conform to established specifications. It includes frequency and voltage regulation, which are essentially *network stabilization* actions.

- *Energy arbitrage* refers to charging the storage system with cheap, off-peak power and discharging it at peak times when electric power can be sold at much higher prices.
- *RES integration* is either using curtailed power from RES to charge the storage system or to respond to the fluctuations during RES operation (e.g., sudden cloud cover, drop in wind speed).
- Emergency back-up, also known as uninterruptible power supply or uninterruptible power source (UPS), is the ability to provide power to critical systems when the grid is down and no other resource (e.g., a diesel generator) is able to respond immediately.
- *Black start* is the process of restoring an electric power station without relying on the grid. For example, a small GT can be cranked using a battery. Once that GT is up and running, it can power the plant auxiliary systems and can be used to crank the larger, main gas turbines.
- *Peak shaving* refers to leveling out sudden increases in electricity demand on the grid. One way to do this is for the users to temporarily scale down their usage. Another way is to engage an energy storage system to take care of the temporary spike in power demand so that the remaining generating assets continue with their normal operation.
- Time shifting and load leveling examples were provided in the preceding paragraphs.
- The *expansion of transmission networks* is typically supported by the construction of new power lines to meet growing electric power demand. This is a costly and lengthy process. Thus, energy storage systems present an alternative by reducing the need to procure excess capacity to deal with demand peaks and, thus, avoiding unnecessary network expansion.

5.6   References

1. Akhil, A. A., Huff, G., Currier, A. B., et al. 2013. *DOE/EPRI 2013 Electricity Storage Handbook in Collaboration with NRECA*, Albuquerque, NM: Sandia National Laboratories.
2. World Energy Council. 2020. *Five Steps to Energy Storage, Innovation Insights Brief*, London, UK: World Energy Council.
3. U.S. DOE, Office of Fossil Energy. 2020. *Electricity Storage Technology Review*, Prepared by OnLocation. Available at www.energy.gov/sites/default/files/2020/10/f79/Electricity%20Storage%20Technologies%20%20Report.pdf
4. Sarbu, I. and Sebarchievici, C. 2018. A comprehensive review of thermal energy storage, *Sustainability*, 10, 191.
5. Ebeling, J., Balsbaugh, R., Blanchard, S. and Beaty, L. 1994. Thermal energy storage and inlet air cooling for combined cycle, International Gas Turbine and Aeroengine Congress and Exposition, June 13–16, The Hague, Netherlands.
6. Gkoutzamanis, V., Chatziangelidou, A., Efstathiadis, T., Kalfas, A., Traverso, A. and Chiu, J. N. W. 2019. Thermal energy storage for gas turbine power augmentation, *Journal of the Global Power and Propulsion Society*, 3: 592–608.

7. Tyner, C. E., Sutherland, J. P. and Gould, Jr., W.R. 1995. Solar Two: A molten salt power tower demonstration, SAND-95-1828C; CONF-951072-1, VDI-GET Meeting: Solar Thermal Power Plants II, 11–12 October, Stuttgart, Germany.

8. Relebohile, J. R. and Dinter, F. 2015. Evaluation of parasitic consumption for a CSP plant, *SolarPACES 2015, 13–16 October*, Cape Town, South Africa.

9. Laughlin, R. B. 2017. Pumped thermal grid storage with heat exchange, *Journal of Renewable and Sustainable Energy*, 9, 044103.

10. Kisi, E., Sugo, H., Cuskelly, D., et al. 2017. Miscibility gap alloys – A new thermal energy storage solution, World Renewable Energy Congress XVI, 5–9 February, Murdoch University, Australia.

11. Conlon, W. 2019. Decarbonizing with energy storage combined cycles, *POWER Magazine*, December, 36–39.

12. Manwaring, M., Mursch, D. and Tilford, K. 2012. Challenges and opportunities for new pumped storage development, A White Paper Developed by NHA's Pumped Storage Development Council. Available at www.hydro.org/wp-content/uploads/2017/08/NHA_PumpedStorage_071212b1.pdf

13. US Army Corps of Engineers, Northwest Division, Hydroelectric Design Center. 2009. Technical analysis of pumped storage and integration with wind power in the Pacific Northwest. Final report. Available at www.hydro.org/wp-content/uploads/2017/08/PS-Wind-Integration-Final-Report-without-Exhibits-MWH-3.pdf

14. Hüttermann, L., Span, R., Maas, P. and Scherer, V. 2019. Investigation of a liquid air energy storage (LAES) system with different cryogenic heat storage devices, *Energy Procedia*, 118, 4410–4415.

15. Damak, C., Leducq, D., Hoang, H.-M., Negro, D. and Delahaye, A. 2020. Liquid Air Energy Storage (LAES) as a large-scale storage technology for renewable energy integration – A review of investigation studies and near perspectives of LAES, *International Journal of Refrigeration*, 110, 208–218.

16. Hamdy, S., Moser, F., Morosuk, T., and Tsatsaronis, G., 2019. Exergy-based and economic evaluation of liquefaction processes for cryogenics energy storage, *Energies*, 12, 493. doi:10.3390/en12030493.

17. Kotas, T. J. 2012. *The Exergy Method of Thermal Plant Analysis*, London: Exergon Publishing Co. UK Ltd.

18. Liu, H., He, Q., Borgia, A., Pan, L. and Oldenburg, C. M., 2016. Thermodynamic analysis of a compressed carbon dioxide energy storage system using two saline aquifers at different depths as storage reservoirs. *Energy Conversion and Management*, 127, 149–159.

6   Compressed Air Energy Storage

Compressed air energy storage (CAES), with the exception of *pumped hydro energy storage* (PHES, see Section 5.2), is the only proven (i.e., TRL 9) large-scale, long-duration energy storage technology utilizing turbomachinery [1]. Therefore, it deserves a chapter of its own. The key idea behind the concept of CAES is to transfer the off-peak energy produced by base load units (e.g., coal or nuclear) or renewable sources (e.g., wind) to the high-demand peak periods while using only a fraction of the fuel that would be used by a standard peaking machine such as a heavy-duty industrial or aero-derivative gas turbine. A CAES plant comprises four main components: (1) compressor train, (2) motor-generator unit, (3) expander train, and (4) underground compressed air storage (see Figure 6.1).

During low-cost off-peak load periods (sometimes with *negative* electricity price, i.e., generators *pay* users to consume electricity!), a motor consumes power to compress and store air in the underground storage unit (the storage pressure increases). During peak load periods, the process is reversed; the compressed air is returned to the surface (the storage pressure decreases). The air from the storage is used to burn fuel (natural gas or distillate) in the combustors. The resulting combustion gas is then expanded in turbo-expanders to spin the generator and produce electricity. The process, commonly known as *time shifting*, is succinctly illustrated by the conceptual hourly load and cavern pressure profiles in Figure 6.2, which shows how the Huntorf CAES power station provided support to the NWK grid in Germany on a particular day [1]. Note that at around 5:00 pm and 5:30 pm, a fossil unit failed due to a fault in its electronics. In each case, the CAES plant was run up in only six minutes with a fast start and supplied power to the grid. The second interesting event took place at around 10:00 pm (22:00 hours) when the system was being pressurized. The system load started to increase at a fast rate so that the CAES plant was switched to *emergency power operation*, wherein the gas turbine quickly started during pressurization and automatically connected to the grid while compressor blow-off valves opened [2].

Time shifting enables CAES to provide peak shaving and load leveling services as well as ramping and load following, which are critical for the stable integration of renewable power resources with the existing generation portfolio. In addition, CAES can also provide *ancillary services* (see Section 5.5) such as black start (e.g., as Huntorf does for the German nuclear units located near the North Sea) and reactive power (reactive volt-amperes or VAR) compensation (which requires a synchronous, self-shifting [SSS] clutch between the turbomachinery and synchronous machines).

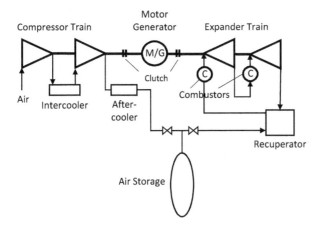

Figure 6.1 Conceptual representation of the CAES system.

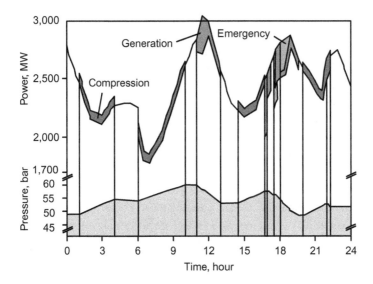

Figure 6.2 Sample daily profiles of power generation and cavern pressure [2]. This reflects the operation of the Huntorf CAES plant on January 9, 1983 (a Sunday). The light-gray shaded areas at the top denote the failure of a power station on the grid.

As shown in Figure 6.1, the turbomachinery and the synchronous alternating current (ac) machine (alternately run as motor or generator) can be arranged as a single train with SSS clutches. Alternatively, the compressor and expander trains can be separate units, each with a dedicated motor and generator, respectively. To minimize the compression power, the compressor train comprises intercoolers. The aftercooler ensures that the air pumped into the storage unit is at about the same temperature as the air therein. The heat rate of the expansion process is significantly improved by the use of a recuperator, which preheats the compressed air from the storage unit using hot expander exhaust gas.

For a complete review of CAES technology development and overview of major recent CAES projects, refer to the paper by King et al. [3]. At the time of writing, there are only *two* CAES plants in the world: the 290 MW plant in Huntorf, Germany (commissioned in 1978) and the 110 MW plant in McIntosh, Alabama, USA, which was commissioned in 1991. The general arrangement of each plant is similar to that shown in Figure 6.1 with the exception that the Huntorf plant does not have a recuperator. Both plants were, until recently, actively in operation with high reliability and availability. (The author has not been able to verify their status as of 2021.) The turbomachinery OEM for the Huntorf plant was the *Brown Boveri Company* (later ABB and then Alstom, now acquired by General Electric). The OEM for the US plant was *Dresser-Rand* (now a Siemens company), which is still active in CAES equipment development.

The underground storage in either plant is a solution-mined *salt cavern* (the Huntorf plant has two caverns).[1] Alternatives have been considered for many proposed CAES projects in the nearly four-decade period since the commissioning of the Huntorf plant. They include limestone mines, depleted natural gas reservoirs, aquifers (water-filled underground reservoirs), and porous rock formations. Each has its advantages and disadvantages as well as CAPEX implications.

This brief, high-level introduction is concluded by emphasizing that the CAES is a simple and field-proven concept, whose wide-spread adoption which was prevented primarily due to cost (especially that associated with storage cavern mining or construction) and the inability of developers to convince lenders of a firm revenue stream (bankability).

Nevertheless, increasing penetration of renewable technologies into the mix of generating assets, with the increased need for load following, VAR compensation, and other ancillary services, might finally tip the scale in favor of CAES deployment. In that regard, consider that CAES is analogous to PHES in principle – i.e., conceptually, just think of storage reservoir water depth instead of cavern pressure – and comparable to it in scale albeit at lower capital cost.

6.1   Huntorf and McIntosh CAES Plants

Worldwide, there are two CAES plants in commercial operation: Huntorf in Germany and McIntosh in Alabama, USA. A comparison of both plants is provided in Table 6.1.

Both plants have been consistently reported to operate reliably since commissioning (after the early operational problems have been successfully ironed out) [4]. Recent reliability numbers reported over a period of thirteen years are as follows [5]:

---

[1] The McIntosh CAES salt cavern took two years to mine. This is the longest work element in the construction schedule. McIntosh was able to dispose of the waste brine by paying Olin Chemical to take it. The resulting cavern is big enough to accommodate the Empire State Building.

Table 6.1 McIntosh and Huntorf CAES plant data

	McIntosh	Huntorf
Operating Since	1991	1978
Rating	110 MW @ 60 Hz	290 MW @ 50 Hz
Powertrain OEM	Dresser-Rand	Brown Boveri Co.
Air Storage	2,640 MWh (24 hours @ 110 MW)	1,160 MWh (4 hours @ 290 MW)
Cavern Volume	19.8 million cuft	10.6 million cuft
Cavern Type	Solution-Mined Salt Dome	Solution-Mined Salt Dome
Cavern Temperature	95°F	95°F
Type of Storage	Sliding Pressure	Sliding Pressure
Number of Caverns	1	2
Airflow Ratio (E/C)	1.7	4
Expander Airflow	340 pps (154 kg/s)	917 pps (416 kg/s)
Recuperator	YES	NO
Fuel	Gas	Gas or Oil
HP Expander Inlet	650 psig/1,000°F	667 psig/1,000°F
LP Expander Inlet	213 psig/1,600°F	160 psig/1,600°F
LHV Heat Rate	4,100 Btu/kWh	5,500 Btu/kWh

- Compression
  - Starting reliability 92.7%
  - Operating reliability 99.6%
- Generation
  - Starting reliability 91.6%
  - Operating reliability 96.7%

In the summer of 2016, the author was part of a team that visited the McIntosh CAES plant, which is now operated by the *PowerSouth Energy Cooperative*.[2] According to the plant manager and operations/maintenance superintendent, maintenance is minimal for the very robust system. Compressors have been rebuilt only once in the 24-year life of the plant. Temperatures are such that material degradation is minimal. According to the plant team, the largest issue with maintenance is that the plant is not dispatched frequently enough. The low number of cycles the plant sees increases maintenance over that for a more frequently operating machine (i.e., the minimal amount of time when the turbomachinery lies dormant). Specifically, compressor corrosion issues were encountered during non-compression periods if there were only one or two compression cycles per week. It may be possible to improve this by material selection or possibly by using of a small purge flow to warm/dry the compressor during the non-compression periods. Although originally designed for

---

[2] First known as Alabama Electric Cooperative (AEC), the name was changed to PowerSouth Energy Cooperative in 2008 to better reflect the cooperative's geographical service territory and to position the company for future growth opportunities. Headquartered in Andalusia, AL, PowerSouth is a generation and transmission cooperative providing the wholesale power needs of 20 distribution members – 16 electric cooperatives and 4 municipal electric systems – in Alabama and northwest Florida. It was formed in 1941 by 11 cooperatives to generate and sell electricity.

dual fuel operation, the plant no longer runs on fuel oil. One item mentioned by the plant team as an improvement item is that the lube oil system is currently common to both the compression and expansion trains. It would be better for maintenance if these were split systems (independent).

6.2   Basics

The essence of the CAES concept is the separation of the compressor of a gas turbine from its *hot gas path*, i.e., its combustor and expander (turbine). While this may sound like an insignificant modification, in practice, it is a tremendous enabler. Consider a gas turbine rated at 100 MW and 35% efficiency. Its expander generates 200 MW, of which 100 MW is consumed by the compressor. Once they are separated and run individually (for simplicity, ignore the practical details), the expander can generate 200 MW with the same heat input, i.e., 100 MW/35% = 286 MWth, so that its efficiency is doubled to 200/286 = 70%. Furthermore, the compressor can be replaced by an intercooled process machine to provide the same compression duty at, say, 50 MW power consumption at a time when electric power costs only 20% of the normal rate.

Thus, with the same fuel expenditure, the revenue, R, that can be generated is higher than that with the original gas turbine by an amount equal to

$$R = e \cdot t(200 - 0.2 \times 50 - 100) \approx 90 e \cdot t,$$

where e is the electricity sale price in $/MWh and t is the generation time in hours. In effect, the same technology is deployed in a manner equivalent to what can be achieved with the addition of 90 *free* megawatts (in economic terms that is)! As will be illustrated below, the *reality* is actually as impressive as predicted by this *back-of-the-envelope* exercise.

To start the discussion of a real CAES system, consider the heat and mass balance (HMB) diagram of the 110 MW CAES power plant in McIntosh, Alabama, USA, in Figure 6.3.

First note the ratio of expander and compressor airflows, which is equal to 340/197 = 1.726. Consequently, since the total amount of air pumped into the cavern must be equal to the total amount of air taken from the cavern, for 1 hour of generation at full load the compressor must be run for 1.73 hours.

The second item to note in Figure 6.3 is the difference between compressor discharge (i.e., air *into* the cavern) and recuperator inlet (i.e., air *from* the cavern) pressures. This is a result of the chosen operation philosophy for this plant (also for the other CAES plant in Huntorf, Germany).

There are two methods for CAES reservoir operation:

1. *Constant volume*: The output pressure of the reservoir varies over time as air is discharged from it. In this method, there are two design options to control the expander power output:

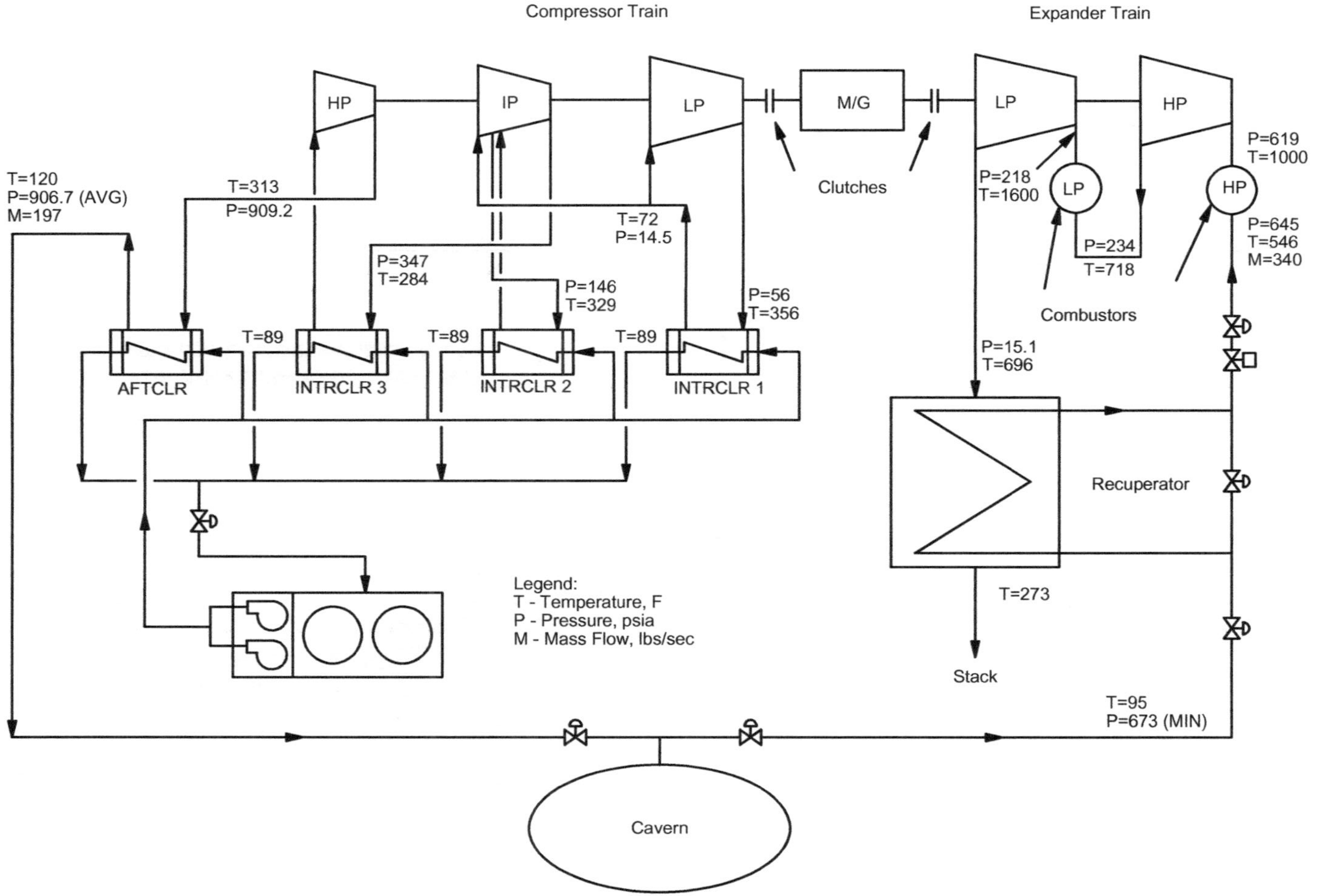

Figure 6.3 McIntosh CAES plant heat and mass balance (HMB) [5].

a. the high-pressure (HP) expander inlet pressure varies with the reservoir output pressure, and
b. the inlet pressure of the HP expander is kept constant by the throttle valve. Both existing CAES plants use this option, i.e.,
  i. the Huntorf CAES plant is designed to throttle the air to 46 bar at the HP expander inlet (with the cavern operating between 48 and 66 bar). whereas
  ii. in the McIntosh plant, the air is throttled to 45 bar (with the cavern operating between 45 and 74 bar).

2. *Constant pressure*: The storage reservoir is maintained at constant pressure throughout its operation via a compensation system by employing an aboveground water reservoir (see Figure 6.4).

The third item of interest in Figure 6.3 is the recuperator (75% effectiveness), which utilizes the exhaust of the low-pressure (LP) expander to heat the air from the cavern at 95°–546°F before the HP combustor. This reduces the fuel consumption in the HP combustor and significantly improves the overall expander heat rate (i.e., thermal efficiency).

The fourth item to note from Figure 6.3 is the air temperature downstream of the compressor intercooler (120°F – *into* the cavern) and the air temperature at the

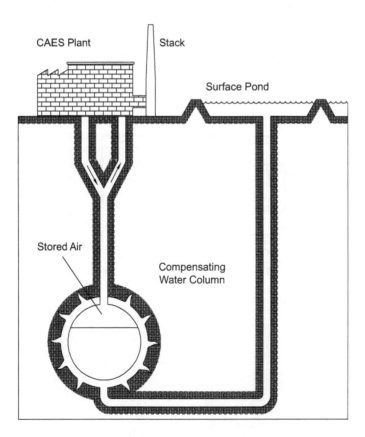

Figure 6.4 CAES plant with constant-pressure compensation [6].

recuperator inlet (95°F – *from* the cavern). According to the basic gas formula, for a cavern of a fixed volume, V, the amount (mass) of air that can be stored therein is directly proportional to the storage pressure and inversely proportional to the storage temperature, i.e.,

$$m \propto \frac{pV}{T}. \tag{6.1}$$

Thus, for maximum storage, air must be compressed to the highest possible pressure and cooled to a suitable temperature prior to injection into the cavern.

Based on the data in Figure 6.3, the following performance can be calculated using a flowsheet simulation tool (in this case THERMOFLEX by Thermoflow, Inc. – see Section 16.2):

- Net output during generation, 110 MW at nearly 85% thermal efficiency (about 4,025 Btu/kWh,[3] includes the fuel compressor power consumption)
- Recuperator effectiveness of 75%
- Fuel consumption of about 130 MWth (fuel-air ratio 0.0171, 362 Btu/lb of air)
- Total 51 MW of power consumption during compression (245 Btu/lb of air)
  - Compressor motor power consumption of 50.5 MW
  - Cooling tower fan and cooling water circulation pump power consumption of 425 kW

If one had to use a gas turbine rated at 35% efficiency, with a fuel consumption equal to that of CAES combustors, only 45.8 MWe would be generated. Thus, noting that 1.726 hours of compression is requisite for one hour of generation, the net revenue benefit per hour of generation is

$$R = e(110 - 0.2 \times 1.726 \times 51 - 45.8) \approx 47\,e.$$

Scaling this to the ideal example above,

$$R = \frac{100}{45.8}(47\,e) \approx 102\,e,$$

i.e., even better than predicted!

Alternatively, fuel consumption during generation is only a fraction of a comparable gas turbine with 35% efficiency, i.e., about 0.42. In new CAES designs, this fraction is claimed to be improved to about one-third by improved design and higher recuperator effectiveness (90% vis-à-vis 75% in this case).

6.2.1   Efficiency

The energy performance of a conventional fossil fuel power plant is easily described by a single efficiency: the ratio of the electrical energy generated to the thermal energy in the fuel. The situation is more complicated for CAES due to the presence of two very different energy inputs. On the one hand, electricity is used to drive the

---

[3] In SI units, ~4,250 kJ/kWh; cf. 4,330 kJ/kWh reported in the literature (e.g., *110 MW McIntosh CAES plant over 90% availability and 95% reliability*, Gas Turbine World, Vol. 28, pp. 26–8, 1998).

Table 6.2 Selected CAES efficiency expressions and values in the literature

Parameter	No Recuperator	With Recuperator
Heat Rate, kJ/kWh	5,500–6,000	4,200–4,500
CER $= E_T / E_M$	1.2–1.4	1.4–1.6
PEE – Nuclear ($\eta_T = 33\%$)	24.50%	29.7%
PEE – Fossil ($\eta_T = 42\%$)	28.20%	34.4%
PEE – CHP ($\eta_T = 35\%$)	N/A	35.1–41.8%
PEE – Avg. Grid ($\eta_T = 35\%$, CER $= 1.4$)	N/A	42–47%
RTE 1 (CER $= 1.5$, $\eta_{NG} = 47.6\%$)	N/A	81.7%
RTE 2 (CER $= 1.5$, $\eta_{NG} = 47.6\%$)	N/A	66.3%

compressors and on the other, natural gas or oil is burned to heat the air prior to expansion. This situation makes it difficult to describe CAES performance via a single index in a way that is universally useful – the most helpful single index depends on the application for CAES that one has in mind.

Several indices encountered in the literature are summarized in Table 6.2 (from Ref. [6]). Parameters used in the table are as follows:

- E_T is the turbine (expander) power output, MWe
- E_F is the turbine (expander) combustor fuel consumption, MWth
- E_M is the compressor (motor) power consumption, MWe
- η_T is the thermal efficiency of the (average) baseload power plant (i.e., nuclear, fossil, combined heat and power [CHP])
- η_{NG} is the thermal efficiency of a stand-alone power plant using E_F (instead of CAES)
- The *charging electricity ratio* (CER) is the ratio of generator output to compressor motor input, which is the inverse of the CEF (*compression energy factor*) used in this chapter
- Note that
  - *Primary energy efficiency* (PEE) in Table 6.2 is what is referred to as PEE′ in this chapter (see below)
  - *Roundtrip efficiency* (RTE) 1 in the table is what is referred to as PEE in this chapter (see below)

Formulas defining the parameters in Table 6.2 are:

$$HR = \frac{3600 E_F}{E_T},$$

$$PEE = \frac{E_T}{\dfrac{E_M}{\eta_T} + E_F},$$

$$RTE1 = \frac{E_T}{E_M + \eta_{NG} E_F},$$

$$RTE2 = \frac{E_T - \eta_{NG} E_F}{E_M}.$$

Arguably, the most useful measure of CAES performance is the PEE, which is the ratio of generated energy to consumed energy, i.e., using the definitions and nomenclature of this chapter,

$$\text{PEE} = \frac{t_g \dot{W}_E}{t_c \dot{W}_C + t_g \overline{\eta} \dot{Q}_F}, \tag{6.2a}$$

where

$\dot{W}_E$ = Expander net power output, MW or kW

$\dot{W}_C$ = Total power consumption during cavern charging, MW or kW

$\dot{Q}_F$ = Total fuel consumption of the combustors, MWth or kWth

t_g = Time of generation, hours

t_c = Time of charging, hours

$\overline{\eta}$ = Average power plant efficiency

(Refer to Figure 6.5 for a graphical depiction of Equation (6.2a).). Equation (6.2a) provides an electricity-for-electricity *roundtrip efficiency* that isolates the energy losses in the conversion of electricity to compressed air and back to electricity. The second term in the denominator of the PEE formula represents the electric energy that *could* have been generated with the *same* amount of fuel input with *average* power plant efficiency (typically ~50% for NG-fired plants). For the McIntosh CAES example above,

$$\text{PEE} = \frac{1 \cdot 110}{1.726 \cdot 51 + 1 \cdot 50\% \cdot 130} \approx 72\%.$$

In Equation (6.2a), both the denominator and the numerator of the fraction on the right-hand side of the PEE definition are in terms of electric energy, which is the correct way to look at the system efficiency (i.e., an *apples-to-apples* ratio). An alternative formulation for PEE is as follows:

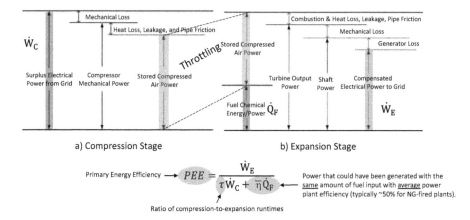

Figure 6.5 A graphical "anatomy" of PEE.

$$PEE' = \frac{t_g \dot{W}_E}{t_c \frac{\dot{W}_C}{\eta} + t_g \dot{Q}_F} \qquad (6.2b)$$

In Equation (6.2b), the compressor motor power consumption is replaced by an expression for the effective thermal energy input required to produce it. Thus, the overall efficiency value given by Equation (6.2b) reflects the system (grid + CAES) efficiency of converting primary (thermal) energy into electrical energy. For the McIntosh data above, PEE' is about 36%.

Frequently, a so-called *CAES efficiency* (or heat rate) is calculated and cited in the literature in proverbial *apples-to-oranges* terms, i.e.,

$$\eta_{CAES} = \frac{t_g \cdot \dot{W}_E}{t_c \cdot \dot{W}_C + t_g \cdot \dot{Q}_F},$$

$$HR_{CAES} = \frac{3,412}{\eta_{CAES}}.$$

The denominator of the fraction in the CAES efficiency equation above, i.e., the consumed energy, is a mix of electric energy (work) and fuel chemical energy (heat). (For the McIntosh data above, η_{CAES} is about 50%.) It is therefore imperative to understand the exact definition of the term "efficiency" within the context of the CAES operation. Simply stating that the "CAES efficiency is x%" does not convey meaningful information.

It is equally important to recognize the wide spectrum covered by the PEE definition even on a self-consistent, *apples-to-apples* basis. The critical assumption is the value of $\bar{\eta}$, which is a function of the power generation technology mix of the particular grid during off-peak times. The value used in the sample calculations above, 50%, is quite optimistic in the sense that it assumes a fully natural gas–fired generation mix. The reality can be quite different, i.e., substantially worse if the grid has many old coal-fired generators or substantially better if the grid comprises large nuclear units and wind farms (during off-peak generation times). For example, if the compression power is supplied *entirely* by a wind farm (which can be owned by the entity owning the CAES facility[4]), fuel consumption during air injection into the cavern is zero and PEE' becomes 84%, which is the thermal efficiency during generation!

Note that, via the mass conservation principle, for air injected into the storage cavern and withdrawn therefrom, one can write that

$$m_{charge} = t_c \dot{m}_C = m_{discharge} = t_g \dot{m}_E \qquad (6.3a)$$

The total amount of air (kg, lb, or tons) charged (injected) into the storage chamber and discharged (withdrawn) from it is referred to as the *cycled air*. From the equality above it follows that

---

[4] In fact, this is one lesson learned as cited in Ref. [6], i.e., the ideal is a storage owner who is both a load serving entity (e.g., a distribution utility with an obligation to serve its retail customers) and owns significant quantities of wind resources.

$$\tau = \frac{t_c}{t_g} = \frac{\dot{m}_E}{\dot{m}_C} \qquad (6.3b)$$

As calculated earlier, τ is 1.726. Furthermore, the key cycle parameters below can easily be determined from first principles:

$$\dot{W}_E = \eta_{th}\dot{Q}_F$$

$$\dot{Q}_F = \dot{m}_F LHV$$

$$f = \frac{\dot{m}_F}{\dot{m}_E}$$

$$q_f = fLHV$$

$$w_C = \frac{\dot{W}_C}{\dot{m}_C}$$

where η_{th} is the thermal efficiency during generation, f is the fuel-air ratio, LHV is the fuel lower heating value, q_f is the *specific fuel energy (heat) input*, and w_C is the specific charging power consumption. Note that w_C is a function of compressor design (e.g., stage efficiency, number of casings and intercoolers) and *compressor discharge pressure* (CDP). Substituting them into the PEE formula, the result is

$$PEE = \frac{\eta_{th}}{\frac{w_C}{q_f} + \bar{\eta}} \qquad (6.4)$$

With the existing technology and cavern pressure (which dictates the CDP), w_C is constant, i.e., around 240 Btu/lb (560 kJ/kg). Furthermore, $\bar{\eta}$, assumed to be 50%, is a fixed number (whatever it turns out to be for the given grid generation technology mix). Thus, Equation (6.4) suggests that PEE is primarily a function of the expander *thermal efficiency* and specific heat input, which, in turn, are functions of the *recuperator effectiveness* (for specified HP and LP combustor exit temperatures). However, a closer look will reveal that thermal efficiency is the only parameter of consequence.

Interestingly, PEE is *not* a function of τ. This, of course, is exactly as it should be because, one way or another, air mass taken from the cavern must be pumped back into it. This can be done (i) with a small compressor over a long time or (ii) with a large compressor over a short time. Equations (6.2) and (6.3) above clearly indicate that the total energy input, for a given total energy output, is not going to change at all.

Another way to look at the PEE relationship is to realize that, using the relationships enumerated above, the specific expander power output is given by

$$w_E = \eta_{th}q_f.$$

For the expander train in Figure 6.3, w_E is about 307 Btu/lb (714 kJ/kg); it is a function of expander design (i.e., stage efficiency, number of expanders and combustors, combustor exit temperatures) and inlet pressure. Substituting w_E into Equation (6.4) gives

$$PEE = \frac{\eta_{th}}{\frac{w_C}{w_E}\eta_{th} + \bar{\eta}} \tag{6.5a}$$

$$PEE = \frac{\eta_{th}}{CEF\,\eta_{th} + \bar{\eta}} \tag{6.5b}$$

In other words, PEE is primarily a function of expander thermal efficiency and the CEF, which is a ratio of compression energy input to expander energy output, i.e., making use of Equation (6.3),

$$CEF = \frac{\dot{W}_C \cdot t_c}{\dot{W}_E \cdot t_g} = \frac{w_C \cdot \dot{m}_C \cdot t_c}{w_E \cdot \dot{m}_E \cdot t_g} = \frac{w_C}{w_E}.$$

For recuperator effectiveness between 70% and 90%, however, calculations show that CEF is constant at about 0.80. This is as expected because the technology as represented by component efficiencies and combustor exit temperatures is the same. Thus, for a given turbomachinery technology level, both PEE and η_{th} are functions of recuperator effectiveness (see Figure 6.6).

The heat rate increase of a CAES plant (the expander/turbine during generation) at part load is small relative to a conventional gas turbine because of the way the turboexpander output is controlled (see Figure 6.7). Rather than changing the turbine inlet temperature as in a conventional turbine, the CAES output is controlled by adjusting the airflow rate with inlet temperatures kept constant at both expansion stages. This leads to better heat utilization and higher efficiency during part load operation. The high part load efficiency of CAES makes it well suited for balancing variable power sources such as wind.

The McIntosh CAES plant was reported to deliver power at heat rates of 4,330 kJ/kWh at full load and 4,750 kJ/kWh at 20% load [7]. This excellent part load

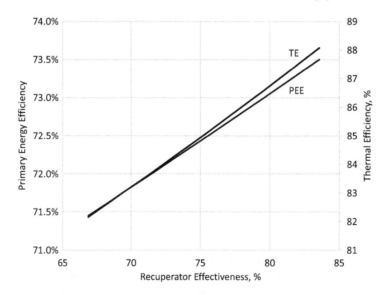

Figure 6.6 PEE and thermal efficiency as a function of recuperator effectiveness.

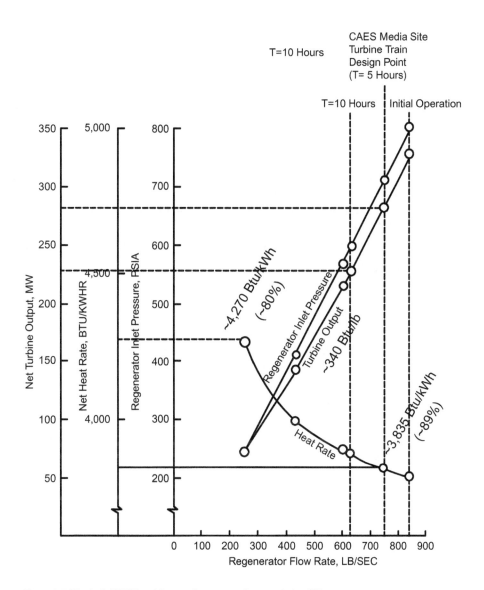

Figure 6.7 Typical CAES turbine performance characteristics [7].

behavior could be further enhanced in modular systems, where the full plant output would be delivered by multiple modules (e.g., several 100 MW CAES blocks). In this case, the system could ramp down to a fraction of the full load output (i.e., less than 10%) and still be within 10% of the full load output heat rate.

6.2.2 CO_2 Emissions

It is often claimed that CAES helps reduce CO_2 emissions. The statement is true only if CAES enables more renewables to be built than would otherwise be built without it.

However, the inherent veracity of the statement depends upon the resources used for the CAES unit's compression cycle, and the resources displaced by the CAES unit's output in generation mode. In an ISO (*independent system operator*) area, where the off-peak compression energy would be likely to come from coal resources, the situation would be less advantageous for CAES. Similarly, if the off-peak compression energy would be likely to come from wind farms, this would be a perfect "case for CAES" (no pun intended). (It is more likely, however, that the peak resources displaced by CAES would be coal or natural gas.)

Exhaust CO_2 flow for the 110 MW McIntosh CAES turbine (per the HMB in Figure 6.3) is 16 lb/s (7.3 kg/s), which corresponds to about 525 lb/MWh (238 kg/MWh) and, as such, about one-third less than that of a state of the art GTCC power plant.

Accounting for the CO_2 emissions during compression with $\bar{\eta}$ of 50% (gas-fired generation) and $\tau = 1.726$, the total becomes,

$$\text{Total CO}_2 = (525 + 875 \times 1.726)/(1 + 1.726) = 747 \text{ lb/MWh}(339 \text{ kg/MWh}),$$

which is more or less the same as that from a H or J Class GTCC at 60% efficiency. If the resources utilized during compression are coal-fired units with 1,750 lb/MWh (795 kg/MWh), total CAES CO_2 emissions would be a not-so-advantageous 1,300 lb/MWh (590 kg/MWh). An overall outlook of CAES CO_2 emissions is provided in Figure 6.8.

One interesting take-away from the figure is the effect of τ on total CAES CO_2 emissions. In general, as one would expect, higher τ is detrimental to specific CO_2 emissions (i.e., more time spent on flue gas generation). This holds true as long as

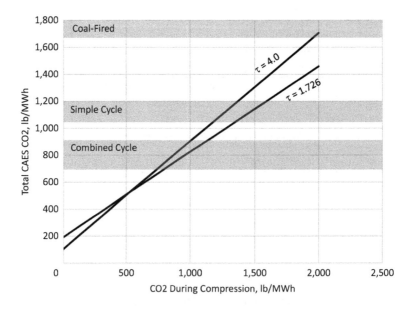

Figure 6.8 CO_2 emissions from a CAES power plant.

average specific emissions during compression (from the fossil fuel–fired plants on the grid) is above the *intrinsic* specific emissions of the CAES plant during generation (i.e., 525 lb/MWh). If a renewable resource such as a wind farm is used to drive the CAES compressor train during cavern charging (i.e., zero flue gas), a higher τ is beneficial due to lower overall specific emissions.

Specific CO_2 emissions during generation are a function of the turboexpander thermal efficiency. Any CAES technology that improves η_{th} is also going reduce specific CO_2 emissions.

Finally, utilization of hydrogen as fuel in CAES combustors is also going to help reduce specific CO_2 emissions. Hydrogen can be produced in situ utilizing excess power or transported to the site via pipeline. The feasibility of any such project should be evaluated on a case-by-case basis with accurate CAPEX and OPEX estimates. This would only be possible for a real project and after a diligent front-end engineering design (FEED) study by an experienced EPC contractor.

6.3   Economics

6.3.1   Peak and Off-Peak Price

The intrinsic value of a CAES facility is mainly derived from the *arbitrage* opportunity stemming from market prices during peak periods such as weekdays in winter and summer months when the off-peak and peak price spreads are greater. (Typically, during the spring and fall months, price spreads are lower.)

The *intrinsic* value represents the value of CAES (or any resource) calculated using *average* hourly prices [8]. In reality, the CAES unit will respond to real-time ISO price signals, which have significant uncertainty and price volatility.

The *extrinsic* value represents the *option value* of a resource to address future load quantity and price volatility, above and beyond the intrinsic value calculated using average hourly prices. The extrinsic value is calculated based on historical volatility in a particular ISO's (e.g., PJM) prices using well-known option valuation techniques such as Black–Scholes.

The variability of real-time prices can be easily appreciated by charting hourly PJM-reported *locational marginal prices*[5] at a particular location and a chosen period of time. The resulting *spaghetti diagram* of real-time prices provides a visual illustration of the variability of prices and the difficulty of defining in advance any model of *time-of-use* prices that will provide a good approximation of real-time prices.

---

[5] Locational marginal pricing reflects the value of the energy at the specific location and time it is delivered. When the lowest-priced electricity can reach all locations, prices are the same across the entire ISO grid. When there is congestion – heavy use of the transmission system – the lowest-priced energy cannot flow freely to some locations. In that case, more expensive electricity is ordered to meet that demand. As a result, the locational marginal prices are higher in those locations. Congestion generally raises the LMP in the receiving area of the congestion and lowers the LMP in the sending area. Operating conditions that limit the delivery capacity of specific transmission lines also can contribute to congestion and result in LMP changes.

The extrinsic value of CAES is significantly higher than the conventional alternatives such as combined or simple cycle gas turbine [8]. This is so because the CAES unit can ramp fast (in both generation and storage modes), accommodate multiple starts and stops in a day, store energy, and provide better *insurance* against future quantity and price volatility than the conventional alternatives, particularly in the storage mode not offered by the latter.

Extrinsic value is thus a potentially large component of a CAES unit's economic benefits, above and beyond the intrinsic value based on average hourly price off-peak to peak arbitrage, which is usually the primary focus of traditional storage economic studies. Unfortunately, monetizing it accurately and reliably during the project development stage to attract funding still remains an elusive goal.

6.3.2   Basic Considerations

The arbitrage revenue from time shifting of compression (cavern charge) and expansion (cavern discharge) is given by (again, assuming during charging/compression the tariff is a fraction, c, of the normal rate – which was 0.2 in the example above),

$$R = et_g(\dot{W}_E - c\tau\dot{W}_C),$$
$$R = et_g(\dot{m}_E w_E - c\tau\dot{m}_C w_C),$$
$$R = et_g\dot{W}_E(1 - c\,CEF).$$

As was shown earlier, for a given turbomachinery technology level, the term in the parentheses is constant and the arbitrage revenue is simply a function of the expander size, i.e.,

$$\frac{R}{et_g} \propto \dot{W}_E. \tag{6.6}$$

Thus, once again, the charge/discharge (also referred to as injection/withdrawal) time ratio, τ, does not play a role in the calculation. As explained above, this is exactly as it should be: air taken from the storage cavern must be put back in.

The equipment cost is, however, a different matter. For a given technology level, the equipment cost is a function of size, which can be measured by airflow, i.e.,

$$C \propto \dot{m}^\alpha.$$

A typical value of the exponent α for turbomachinery (compressors, expanders, etc.) is 0.6. For the CAES turbomachinery, the total equipment cost can be written as

$$C = k_E\dot{m}_E^\alpha + k_C\dot{m}_C^\alpha, \tag{6.7a}$$

$$C = k_C\dot{m}_C^\alpha \cdot \left(\frac{k_E}{k_C}\tau^\alpha + 1\right), \tag{6.7b}$$

$$C = k_E\dot{m}_E^\alpha \cdot \left(\frac{k_C}{k_E}\tau^{-\alpha} + 1\right), \tag{6.7c}$$

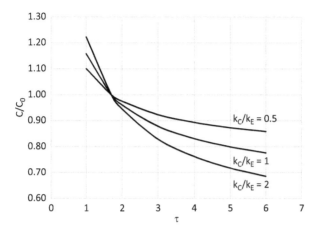

Figure 6.9 Turbomachinery equipment cost as a function of compression/expansion time ratio. Note that $k_C/k_E = 1$ means that the compressor and turboexpander train costs are *similar* if their sizes, as represented by mass flow rate, are similar. Thus, $k_C/k_E < 1$ means that the compressor train is *less expensive* than the expander train for similar size and vice versa for $k_C/k_E > 1$.

where k_E and k_C are proportionality factors for expander and compressor train costs, respectively. For a given expander size (i.e., $\dot{m}_E$), τ is inversely proportional to the compressor size, i.e.,

$$\tau \propto \frac{1}{\dot{m}_C}.$$

The relationship between the total equipment cost and τ is illustrated by the chart in Figure 6.9. Note that for the Huntorf CAES plant τ is 4. For the same expander size, changing τ from 1.7 (the value for the McIntosh CAES plant) to 4 should be expected to result in total equipment cost saving of 10–25%. The operational impact of increasing τ is the increasing time of cavern charging. In other words, for $\tau = 4$, each 1 hour of generation at full load requires 4 hours of charging whereas for $\tau = 1.7$ less than 2 hours is sufficient.

What is shown in Figure 6.9 can be described as follows: for a given generation capacity (i.e., $\dot{m}_E$) and cost (i.e., k_E),

- If τ increases, the total turbomachinery CAPEX decreases due to decreasing compressor train CAPEX
- The magnitude of the decrease is a function of the relative CAPEX of the compressor train vis-à-vis the expander train

In addition to the cavern volume and pressure (i.e., CDP), the trade-off between charge/discharge time and equipment CAPEX is the key optimization problem in CAES design. It should be done on a case-by-case basis due to myriad factors involved in the design and selection of the turbomachinery and the underground storage chamber.

The source of cavern charging energy is another consideration for determining the appropriate project τ. For example, charging with off-peak (say, overnight) natural gas–fired generation (or coal-fired or a combination thereof) for a few hours of discharge/generation makes a large τ reasonable. However, in the case of a renewable, say, solar source with a narrow generation window of less than 8 hours/day available for cavern charging using excess solar generation, one could be forced to go with a much smaller τ. Similarly, a wind source might mandate an intermediate ratio, depending on the expectations for prevailing wind duration and intensity.

A CAES plant comprises three major systems:

1. The turbomachinery trains
   a. Expander train
   b. Compressor train
2. The underground air storage chamber
3. The balance of plant (BOP)
   a. Myriad, pipes, valves, and auxiliary systems
   b. Pumps and compressors
   c. Inter- and aftercoolers
   d. Heat rejection system
   e. Recuperator

The important factors that go into the performance-cost trade-off for the optimal system design and equipment selection are:

1. The operating cycle
   a. Hours of generation (t_g)
   b. Equivalent hours of compression (t_c, τ)
2. Amount of energy generation (MWh), i.e., expander power output $(\dot{W}_E)$
   a. Airflow $(\dot{m}_E)$
   b. Expander inlet pressure
   c. Combustor exit temperature
3. Fuel consumption
   a. Combustor exit temperature
   b. Recuperator effectiveness

Note that net income from a CAES operation is the difference between arbitrage revenue and fuel expenditure (other items such as O&M expenses, etc., are ignored for this simple discussion). Thus, using the fuel price f in $/kWh (not to be confused with the fuel-air ratio f used above) and the revenue formula, the *net income* (NI) is found as

$$NI = e \, t_g \dot{W}_E \cdot (1 - c\,\mathrm{CEF}) - f \, h \, \dot{m}_F \mathrm{LHV}, \tag{6.8a}$$

$$NI = e \, t_g \dot{W}_E \left(1 - c\,\mathrm{CEF} - \frac{f/e}{\eta_{th}}\right), \tag{6.8b}$$

$$\frac{NI}{e \, t_g \dot{W}_E} = n' = 1 - c\,\mathrm{CEF} - \frac{f/e}{\eta_{th}}. \tag{6.8c}$$

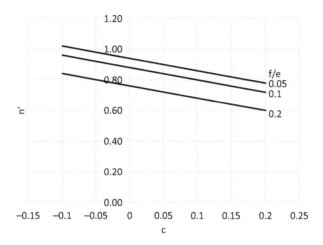

Figure 6.10 Normalized net income (Eq. [8c]) as a function of c (off-peak electricity tariff as a fraction of peak tariff) and f/e (ratio of fuel price to electricity price during generation, i.e., peak tariff).

The ratio f/e is a function of the *spark spread* at the time of generation (determined by market conditions and the power purchase agreement). Similar to the arbitrage revenue, R, the main driver of NI, is the expander power output, i.e., for a given turbomachinery technology level, expander airflow (size). For fuel prices $2–$6 per MMBtu(HHV) and US electricity price (about 10 cents per kWh), f/e is between 0.05 and 0.25. For CEF = 0.80 and η_{th} = 84%, the n′ plots in Figure 6.10 show that the latter is more sensitive to c. In other words, unless the fuel is extremely expensive, the economic feasibility of the CAES scheme has a stronger dependence on the *inexpensiveness* of the electricity price during off-peak compression hours (e.g., see Ref. [8] for typical spread between peak and off-peak prices).

Equation (6.8) suggests that, to maximize the NI from the CAES operating cycle, one should minimize CEF[6] and maximize η_{th}. In expensive fuel markets, the latter is more likely to have a stronger impact on n′ and *vice versa*. Either approach must be undertaken carefully to balance CAPEX and NI (via a proper *levelized cost of electricity* analysis with applicable financial parameters). Since reliability and availability are of utmost importance in a CAES operation (remember that a CAES plant is called upon to respond rapidly to an urgent grid need), a strictly conservative approach to expander and combustor technology development is highly advisable.

6.3.3 More Complex Issues

Almost all potential CAES projects in the last three decades failed to make it to construction and commercial operation stages due to the inability of developers to

---

[6] Note, however, that if c is *negative*, i.e., during off-peak hours the CAES facility is paid to run its compressors, a higher CEF is actually a benefit!

secure funds. The primary deficiency – as perceived by the lenders and insurers – was the lack of firmness in projected revenue stream. This is somewhat puzzling. In addition to its *raison d'être*, off-peak to peak price arbitrage, a CAES power plant offers a long list of ancillary services, which can be monetized – at least on paper.

A recent report by the Sandia Labs [9] provides an excellent coverage of economic and ownership issues bedeviling the CAES projects in the current power generation infrastructure. In a nutshell, the problem boils down to (i) the ability of the CAES plant owner/operator to take full advantage of all the unique capabilities of the technology in a contractually enforceable manner and (ii) to provide it at a competitive price.

As far as the second part is concerned, a typical utility-scale CAES plant is about 20% more expensive than a GTCC and about 80% more expensive than a simple cycle gas turbine [9]. Both technologies are direct alternatives to CAES because they can conceptually be classified as storage facilities in terms of energy stored chemically in the pipeline natural gas.[7] Either can be used, say, to "firm" wind resources by coming on-line and/or ramping up to base load fast when needed. The additional benefit of CAES is to prevent wind curtailment at off-peak times when the wind is still blowing hard.

Unfortunately, as discussed in detail in the aforementioned report [9], just based on intrinsic and extrinsic price arbitrage, low capacity factors and low fuel prices caused the CAES option to come in third in terms of NPV (although quite close to the simple cycle GT option). CAES did better under a high fuel price scenario but still could not beat the GTCC option. Only by the addition of other benefits such as lower CO_2 penalty, avoided wind curtailment, renewable energy credit, system baseload unit profit, and O&M savings, does the CAES option come ahead.

The problem is that, for the lone owner of the CAES facility, unless s/he owns other generating assets, especially wind farms, there is no current method for transferring most of these benefits to his/her bank account. Outside of a few unique items such as carbon penalty, the root cause can be tracked to the difficulty faced by the regulators in approving and rate-basing a CAES facility under a clear-cut category such as distri-bution, transmission, or generation. (The large PHES facilities in operation today were all justified and financed under the old vertically integrated investor-owned utility (IOU) business model [10].)

Deregulation and competition strategies pulled the component pieces apart, so that today there are a plethora of entities such as ISO/RTOs, independent and merchant generators, transmission companies, traditional integrated utilities, and utilities, which are largely distribution businesses, etc. [10]. An ISO could put a CAES facility to good use but (i) requisite ancillary services tariffs must be in place, (ii) resource planning models should be able to accurately calculate those tariffs, and (iii) since the

---

[7] Come to think of it, *all* fossil fuel–fired generation technologies can be classified as 'storage' technologies. In terms of oil, coal and gas, the storage time is measured in millennia whereas in CAES it is mere hours or days.

benefit would be shared by the ISO market *as a whole*, this should still incentivize the *individual* CAES owner.

Large IOUs can conceivably combine all CAES services and monetize them much more easily; especially when spurred by state renewable generation mandates. Otherwise, the most likely scenario under which a CAES project can be realized is an entity owning both a large wind farm and CAES facility or having bilateral agreements with relevant parties (quite a chore and a mess).

6.4   Air Cycle

The operating cycle and amount of energy generation determine the mass of cycled air (see Equation (6.8a)). The storage chamber pressure, temperature, and mass of cycled air determine the chamber volume, V. Ideally, the designer would desire a chamber pressure very close to the expander inlet pressure (accounting for losses in the piping, valves, and the recuperator as well as the HP combustor). This, however, is not practically possible (unless the designer is willing to live with progressively decreasing inlet pressure and power output during withdrawal). To see why, consider the derivative of Equation (6.1):

$$\frac{dp}{dm} \propto \frac{T}{V}. \tag{6.9}$$

For constant expander inlet pressure and no throttling between the wellhead and expander/recuperator inlet, dp/dm in the cavern should be nearly zero, which requires a very large and expensive storage chamber to the point of virtual impracticality (*unless* a water compensation system similar to that in Figure 6.4 or an *aquifer* is used). The compromise is to find an optimum cavern operating pressure range, p_{max} and p_{min}, corresponding to the amount of cycled air, i.e.,

$$m_{charge} = m_{discharge} = m_{max} - m_{min},$$

And storage chamber/cavern volume, V, to minimize the overall investment and operating costs of the CAES plant. For the requisite air flow rate and volume calculations to match the storage system to CAES powertrain requirements, see King and Moridis [11].

The maximum air pressure in a cavern, which is created via solution mining of a salt dome, is a linear function of the depth – the deeper the cavern the higher the p_{max}, which is typically around 70 bar (~1,000 psia), which corresponds to a minimum cavern roof depth of 1,400 ft (about 425 m). Note that the cavern temperatures at p_{min} and p_{max} are *not* the same (i.e., the *adiabatic* assumption is incorrect). Huntorf experience and computer simulations show a significant heat exchange with the cavern walls. In fact, when the air is expanded to atmospheric pressure, after reaching a minimum, the temperature rises again (see Figure 6.11). Representative calculations for Huntorf and McIntosh are summarized in Table 6.3.

Table 6.3 Cavern data

		McIntosh	Huntorf
Max. temperature	°C	35	35
Air inlet temperature	°C	15	15
Volume	m^3	540,000	160,000
Max. pressure	bar	70	66
Min. pressure	bar	45	46
$\Delta p/\Delta t$	bar/h	1.0	10.0
Max. air mass	kg	42.8 million	12 million
Min. air mass	kg	29.4 million	9 million
Cycled air mass	kg	13.4 million	3 million
Cycle time	Hours	24	2
Air flow rate	kg/s	154.8	422.9

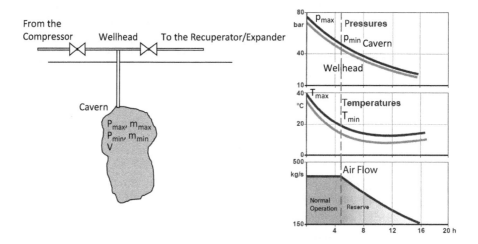

Figure 6.11 Cavern pressure, temperature, and airflow during discharge/withdrawal [12].

6.5   Turbomachinery

Although this section is intended to be a general coverage of CAES compressor and expander technology, because there are only two operating CAES plants worldwide to date, it is unavoidable that the discussion be dominated by the type of turbomachinery deployed in those two facilities.

The heart of the CAES plant is the turbomachinery, i.e., the compressor and expander trains. The former can be constructed using existing (off-the-shelf) components in myriad ways. As explained in Section 6.4, the maximum storage in a given chamber volume requires the maximum possible air pressure. Early technology and optimization studies ended up at around 70–80 bars of maximum cavern pressure (Huntorf and McIntosh CAES plants). For an expander train receiving compressed air directly from the storage chamber (cavern) wellhead, the existing combustor/expander technology precludes a pressure level higher than about 80 bars (about 1,200 psi).

However, some recent proposals (e.g., APEX Matagorda Energy Center, LLC (APEX) in Texas) are based on much higher storage pressures, i.e., 1,900–2,830 psia (about 130–195 bars) at the wellhead. This is enabled by a "very high pressure" unfired air expander, which generates additional power while reducing the air pressure to levels commensurate with the existing fired expander technology. For flow rates similar to that in McIntosh, this brings up the expander train rating up to around 160 MWe (same HP and LP combustor exit temperatures).

Compression to such high pressures can only be accomplished in a multi-casing, intercooled compressor train comprising axial and radial (centrifugal) units. Optimal design considerations and aerodynamic principles dictate that the last stage (high pressure and density, small airfoils) must rotate at a higher-than-synchronous speed (e.g., 6,000 rpm in McIntosh and 7,622 rpm in Huntorf). Thus, a speed-increasing gearbox is provided between the low and high pressure compressors. An artist's rendering of the compressor train in McIntosh CAES plant is shown in Figure 6.12. Note that the speed-increasing gear is between the LP (axial) and IP/HP (centrifugal, barrel-type) compressors.

While the CAES compressor train comprises proven, off-the-shelf components, the expander train and combustors required a careful and rigorous development stage. Especially during the 1970s and 1980s, when the first two CAES plants in Europe and North America were being designed, typical gas turbine pressure ratios (PR) were around 11–12:1 (i.e., then state of the art D/E Class machines). The air pressure at the inlet to the CAES air expander is above 40 bars, implying an overall PR of more than 40:1. Thus, a two-expander-in-series with two-stage combustion (for efficiency) configuration emerged as the optimal solution. To draw upon the existing steam turbine technology, the HP expander was designed with a PR of 4:1 and inlet temperature of ~1,000°F (~535°C). Consequently, there is no cooling requirement for the HP hot gas path [13]. The LP expander for the Huntorf CAES plant was based on the turbine section of BBC's Type 13 gas turbine with an inlet temperature of ~1,500°F (~815°C). The LP rotor is made of ferritic disks, which are welded together at their outer rims. This is the time-honored BBC technology, which is continued today by Alstom (now owned by GE and, partly, by Ansaldo). The rotor temperature is maintained by a cooling system comprising *heat shields* mounted on the rotor, whose inner sides are cooled by cooling air. A corresponding cooling system is also provided for the stator blade carriers [13].

As shown in Figure 6.13, the Huntorf gas turbine comprised two expanders on a *single* rotor between *two* bearings. Since the turbine operates while disconnected from the compressor, a *balance piston* is required to offset the axial thrust of the turbine. The rest of the axial load, especially during the operating transients, is carried by the thrust bearing, which is combined with the radial bearing on the HP end of the machine in a common bearing housing [13].

The two silo-type combustors were top-mounted similar to other BBC gas turbines of the day. The design afforded lower axial differential thermal expansion, reduced

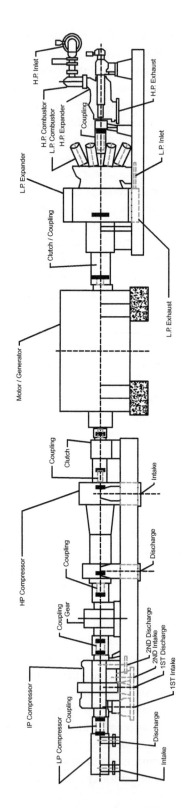

Figure 6.12 McIntosh, AL, CAES plant turbomachinery train (Dresser-Rand).

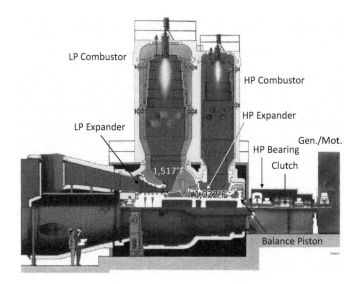

Figure 6.13 Huntorf CAES plant gas turbine (by BBC) [14].

bearing losses, easier shaft alignment, and ease of rotordynamic behavior monitoring and control in addition to overspeed protection (high inertia) in case of turbine trip [13].

The gas turbine in the McIntosh CAES plant is a Dresser-Rand (D-R) (now owned by Siemens) design and in many aspects similar to that in Huntorf, i.e., two expanders in series with reheat combustion. The HP expander is a steam turbine derivative with similar inlet pressure and temperature and two top-mounted combustors. The LP expander is developed specifically for the CAES by D-R; it is a six-stage gas turbine with eight circumferentially arranged combustor cans and a 1,600°F (870°C) inlet temperature (slightly higher than that in Huntorf). While no information can be found on the turboexpander hot gas path cooling system of the D-R unit, it is not too farfetched to assume that it is quite similar to the BBC machine in Huntorf (see above). The heat balance calculation in THERMOFLEX using the data in Figure 6.3 suggests that the LP turbine chargeable cooling flow is about 7% of the airflow.

In both CAES plants, the LP combustors operate in the exit stream of the HP expander and have to be capable of burning fuel properly in a partially vitiated airflow. Brown Boveri's experience in reheat gas turbines dating back to late 1940s and early 1950s (e.g., the Beznau GT in 1948) provided an advantage in that respect (and eventually led to GT24/26 sequential combustion gas turbines). Nevertheless, experience in gas turbine design, aside from the application of afterburners in jet engines and some laboratory installations, was scarce at the time the D-R system with eight cans was designed. In either plant design, the HP combustors operate at pressures that are well above the state of the art existing at the time of the conception. This required significant design and development effort including laboratory rig tests to the extent possible (e.g., test capability at pressures above 40 bars were not available back then).

For example, the desired temperature distribution at the inlet of the Huntorf HP turbine was investigated in a model combustion chamber.

Consequently, significant *in situ development* was involved in both CAES plants when field issues were resolved by design modifications over a period extending to nearly a decade following the commissioning. Details of combustor design and field issues encountered in either CAES plant are covered in several publications [15–16].

While modern aero-derivative gas turbines operate at similarly high PRs, the bottom line remains that the CAES combustor technology has matured to a proven and reliable status over a long and bumpy road with many lessons learned, which have been distilled into the current hardware. It is well known that, especially when it comes to combustor design, large-scale laboratory tests are very costly, have long time schedules, and can provide misleading results where lab simulations may not faithfully reproduce actual engine operation. As such, it is only prudent that new CAES plants should comprise incremental design improvement and upgrades, which are more likely to be cost-effective with maximized reliability and availability.

Both Huntorf and McIntosh turbomachinery were arranged as a single train with a synchronous ac machine (operated as a motor and generator, alternately, during charging and withdrawal, respectively) in the middle (e.g., see Figure 6.14). Each train is connected to the synchronous ac machine via a SSS clutch, which separates the idle train from it. This arrangement has several advantages, e.g.,

- Compact arrangement of machinery and auxiliaries
- Minimizes plot space requirements
- Modular design using existing turbomachinery components to meet specific requirements

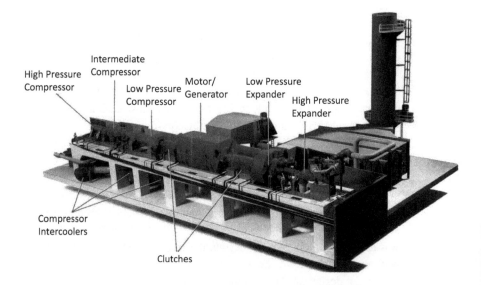

Figure 6.14 McIntosh CAES plant arrangement (Courtesy: Siemens Energy).

- Less investment in electrical infrastructure (transformers, switchyards, protective relays, etc.)

A small (0.5–1.5 pps airflow) atmospheric pressure *standby combustor* was incorporated into the McIntosh design to minimize the life reducing effects of low-cycle thermal stress commonly encountered in peaking power service. The standby combustor is located upstream of the HP expander and is used to keep both expanders warm when not in use, particularly vanes, buckets, and casings. The standby combustor is programmed within the distributed control system (DCS) to fire up when the HP expander external casing temperature falls below 500°F (260°C). If the casing temperature goes above 650°F (343°C), the standby combustor is automatically shut down. The standby combustor operation can also be initiated by the plant operator at any time provided that the expanders are on turning gear and the indicated HP expander case temperature is less than 650°F (343°C).

The startup behavior of the CAES plant in compression and generation modes are shown in Figure 6.15. The turbine is accelerated by the air coming from the cavern with the simultaneous ignition of the HP combustor. When the synchronous speed is reached (after about 3.5 minutes), the generator is synchronized to the grid and the LP combustor is ignited followed by the turbine loading. Normal startup to full load is about 10 minutes; emergency start takes about half of that time. A starting motor is *not* needed because the cavern has enough air to run up the turbine to full load.

The compressor train is run up to synchronous speed with the turbine, which is driven by the air in the cavern. After synchronization, the synchronous ac machine operates as a motor to drive the compressor and the turbine is run down. The SSS

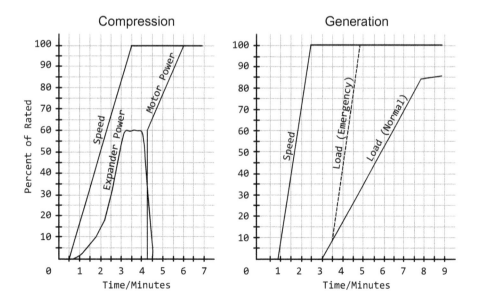

Figure 6.15 Typical CAES startup curves [17].

coupling between the turbine and the synchronous ac machine opens up when a sufficiently low turbine speed is reached.

In terms of operability, in a single train arrangement á la Huntorf or McIntosh CAES plants (i.e., only one synchronous ac machine), a critical consideration is switching from one mode of operation to the other, i.e.,

- from generation (discharge/withdrawal) mode to compression (charge/ injection) mode
- from compression (charge/injection) mode to generation (discharge/ withdrawal) mode

A third and somewhat trivial mode is a combination of the two main modes, i.e., simultaneous compression and generation (simultaneous charge/discharge or withdrawal/injection). In this mode, the net power supply to the grid is reduced by the amount of power absorbed by the compressor train. As discussed earlier, an example of this third mode of operation is the emergency power operation in the Huntorf CAES facility (see the discussion in conjunction with Figure 6.2). For the operability characteristics of the Huntorf CAES powertrain, the reader should consult the paper by Hoffeins et al. [18].

When operating in compression mode, the turboexpander can be started and the SSS will engage the synchronous ac machine, i.e., the motor/generator (MG), when the speeds of the prime mover and the MG match. The MG will shift from motor to generator mode when the torque supplied by the expander turbine matches the compressor mechanical load (i.e., unloading the motor). The power available to the grid will be based on the total turboexpander power minus the compressor mechanical shaft load. The compressor side SSS clutch will not disengage at synchronous speed, so the compression load would have to be controlled by the use of inlet guide vanes (IGVs) and/or variable guide vanes (VGVs) to throttle the airflow and/or unloader valves. In the Huntorf emergency power generation example, without reducing compression load the CAES plant would be able to supply 260 MW to the grid while still devoting 30 MW to compression.

When operating in generation mode with the compressor off-line, the SSS clutch for the compressor side is locked out and the speed differential is too high to allow engagement. Consequently, the turbine and the MG have to shut down and the rotor must coast down to essentially zero speed to pick up the compressor. Once at zero speed (or less than compressor turning gear speed), the SSS is released, and the compressor can be engaged and brought up to synchronous speed by the turboexpander. Once the generator is synchronized, the turbine can either keep running with excess power exported to the grid or the turbine is shut down with the MG running as a motor driving the compressor. In that case, the turbine-side SSS disengages as the turboexpander starts to coast down.

In the McIntosh CAES plant, the coast-down time for the turboexpander and the generator takes about 45 minutes to 1 hour. This is why a *fluid coupling* on the compressor side would be a good option for a future CAES plant (if a single train option is chosen), since at any time it can be engaged and perform a controlled roll-up of the compressor train, even if the turboexpander and the MG are already at full speed. The SSS clutch on the turboexpander is the best solution since it is always able to come up from a lower speed than the generator and engages when the speeds match.

6.6   Recuperator

The recuperator is vital to the performance of the CAES expander. It significantly improves the thermal efficiency during generation via the reduction of HP combustor fuel consumption. The impact is quantified as 20–25% reduction in heat rate.

The recuperator in the McIntosh CAES system diagram in Figure 6.3 increases the temperature of air from the wellhead by about 450°F (232°C) by extracting heat from the hot LPT exhaust gas and reducing its temperature by about 425°F (218°C).

The latter (hot) fluid is at approximately atmospheric pressure and generally has a significantly lower heat transfer coefficient than that of the cold fluid, i.e., compressed air from the cavern. Due to this dissimilarity, the tube temperature is much closer to the temperature of compressed air. As a result, in a conventional counter-flow recuperator, the tube temperature is typically below the acid dew point of the LPT exhaust gas, especially in the area where heat transfer takes place between cooled exhaust gas exiting the recuperator and cold compressed air.

Since the exhaust gas may contain SOx, this enters in reaction with the condensed water on the exterior surface of the tubes, forming sulfuric acid, which in turn causes severe corrosion and ultimately the failure of the tubes in the recuperator [19]. (Note that the McIntosh CAES plant was designed as a dual-fuel facility with number 2 fuel oil as a backup fuel.) This problem of corrosion is especially of concern in CAES systems due to compressed air temperatures as low as approximately 95°F (35°C), significantly less than the acid dew point of the exhaust gas, which can be as high as 250°F (121°C).

Various corrosion resistant materials have been suggested to combat this severe problem of corrosion. However, test results at the time of initial CAES development indicated that even the most expensive corrosion resistant materials would not last longer than 2,000 operating hours [20]. Another proposed solution required the use of additional equipment including a separate hot water storage device to initially preheat the compressed air prior to entry into the recuperator.

An innovative recuperator design (described in a US patent [21]) was eventually used in the McIntosh CAES plant. In particular, the temperature of the tubes is prevented from falling below the dew point of the exhaust gas utilized to heat the tubes containing compressed air through efficient utilization of a combination of parallel-flow and counter-flow sections. In a parallel-flow section, compressed air within the tubes flows in a direction parallel to that of the exhaust gas flow; in a counter-flow section, compressed air within the tubes flows in a direction opposite to that of the exhaust gas flow. The recuperator design philosophy is illustrated by the drawings in Figures 6.16 and 6.17.

The recuperator was eventually designed and fabricated by Struthers Wells Corporation under contract to the Electrical Power Research Institute (EPRI) in a

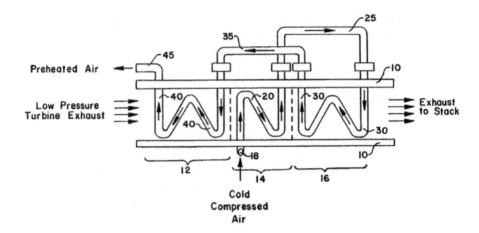

Figure 6.16 CAES recuperator arrangement (from US Patent 4,870,816).

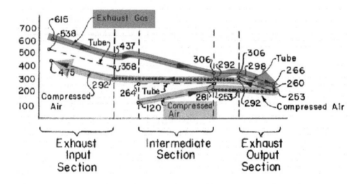

Temperature Profile, °F

Figure 6.17 Temperature of the compressed air (in °F), exhaust gas, and tubes at various locations within the CAES recuperator (from US Patent 4,870,816).

joint effort with Gibbs & Hill, Inc., the EPC contractor for the McIntosh project. Details of the recuperator design and cost-performance trade-off can be found in [20]. As told to the author on a site visit in July 2016, McIntosh had early issues with corrosion products from the recuperator depositing into the turbine. An upgrade of entire recuperator would likely be cost prohibitive since it includes almost five miles of five inch piping. Thus, it is imperative that recuperator corrosion issues be minimized by specifying improved materials in the low temperature stages of the recuperator at the design stage.

The recuperator design by Alstom for the Norton CAES project (which did not make it to construction and commercial operation) was more like a heat recovery steam generator (HRSG) with a duct burner and SCR. Each block in the project had a gross output of 297 MW at the generator terminals with a nominal heat rate of

3,900 Btu/kWh (4,115 kJ/kWh) and 3.5/15 ppmv NOx/CO emissions (with SCR and catalyst).

6.7   Air Storage Options

There are three main types of geological structures that are suitable as CAES storage chambers or reservoirs: *salt dome*, *hard rock* formation, and *soft rock* formation. Of those three, salt dome is the most favorable for solution mining a storage cavity (cavern). Both Huntorf and McIntosh CAES facilities (the only two CAES plants so far) use caverns mined into salt domes as storage reservoirs. The major hurdle in this type of cavern construction is disposing of the dissolved salt which would be in the form of brine that is many times saltier than seawater.

Hard rock formations are also suitable for use as a CAES storage reservoir, but at a much higher cost of mining vis-à-vis salt domes. Several CAES projects proposed to use existing mines, which can reduce the cost of cavern construction. However, there are only a limited number of such readily available hard rock caverns.

Porous rock formations are advantageous in terms of the lowest cost of cavern construction. However, a potential porous rock site for a CAES project requires extensive research and testing into its geological characteristics to determine its feasibility. One such project, *Iowa Stored Energy Park*, which aimed to use a *porous sandstone aquifer* for air storage, was ultimately canceled due to the low permeability of the sandstone and limited storage capability unable to support the intended plant size (270 MW). (Another problem with sandstone aquifers is the limited period of air storage due to the reaction of stored compressed air with local pyrites in the sandstone. Care must be taken to address impacts of mineralogical reactions resulting from introduction of air into reservoir by methods such as dehydration of injected air.)

Many suitable porous rock formations with higher porosity and permeability are currently being used for underground natural gas storage, which is a lucrative business in several places in the USA. In a sense, there is competition for the pore space. On the plus side, natural gas storage experience provides relevant tools for analyzing CAES storage site suitability. Further competition for both depleted reservoirs and potential brine injection targets might come from geological carbon sequestration where reservoirs or brine aquifers with similar characteristics may be needed to sequester CO_2 captured from power plants. However, the latter typically requires deeper formations and, thus, higher storage pressures.

The EPRI prepared a chart showing US geological formations potentially suitable as underground air storage reservoirs for CAES facilities.[8] A superimposition of geology suitable for CAES and Class 4+ wind resources (i.e., wind speed higher than 15–16 mph) on that chart shows that deploying CAES in a large scale for wind

---

[8] EPRI Journal, Oct/Nov 1992.

balancing implies a substantial role for aquifers. In particular, the footprint of aquifer needed to base load wind is ~15% of wind farmland area [22].

As discussed earlier, expander inlet pressure requirements for a feasible CAES unit dictate that the reservoir depth should be at least 1,400 feet below the surface. Ideal reservoir characteristics include the following:

- No faults or fractures
- Good vertical and lateral sealability
- High porosity (minimum 15%) and permeability (>300 md)[9]

For a given CAES expander airflow, the requisite storage volume is a function of porosity and permeability. Unfavorable porosity and/or permeability characteristics would result in a requirement for extremely large rock volumes for the planned operation. As such, the best candidates are salt caverns followed by depleted natural gas reservoirs. (The fact that both Huntorf and McIntosh CAES facilities are located adjacent to natural gas storage facilities mined from the same formation suggests that the conditions favorable for CAES development and natural gas development might often overlap.) As stated earlier, mining the caverns is not a problem. It is disposing of the brine that is produced during this process that presents the biggest challenge. (McIntosh was able to dispose of the waste brine by paying Olin Chemical to take it.) The injection of brine into deeper formations is the most appealing option but, unfortunately, there are not many porous and permeable formations to accommodate both the storage and disposal caverns. Note that significant volumes of brine are produced during solution mining (roughly *eight* volumes of brine to make *one* volume of cavern). Thus, a proper study of brine injection potential is the first thing to do (not the last!). Other options for brine disposal include salt mining, pipelines or trucking to the ocean, low-rate release into rivers during high-water events, and evaporation in ponds.

To determine whether a depleted natural gas reservoir is suitable as an air storage chamber for CAES, several preliminary studies are required. A preliminary assessment of the ignition and explosion potential in a depleted hydrocarbon reservoir from air cycling associated with CAES in geologic media can be found in a recent Sandia Labs report [23]. The study identifies issues associated with this phenomenon as well as possible mitigating measures that should be considered. Purging the reservoir before use, in situ gas monitors, prevention of surface breach, and venting of hot combustion gas to the surface equipment in the event of ignition are some of the measures. In any event, the natural gas content of withdrawn air entering the surface equipment should be monitored. The turbine control system will automatically compensate for entrained fuel components, which pass through the combustor since the

---

[9] Permeability in fluid mechanics and the earth sciences (commonly symbolized as κ, or k) is a measure of the ability of a porous material to allow fluids to pass through it. High permeability will allow fluids to move rapidly through rocks. Permeability is affected by the pressure in a rock. The unit of measure is called the darcy. Sandstones may vary in permeability from less than one to over 50,000 millidarcys (md). A rock with 25% porosity and a permeability of 1 md will not yield a significant flow of water.

"fuel" components will be thoroughly pre-mixed into the air stream and the control is simply demanding more or less fuel input to achieve a desired output or operating temperature setting. As long as there is not a sudden spike, the controller operation should not be affected.[10]

At all times, it must be ensured that the composition of natural gas and air remains outside the ignition envelope (about 4–16% by volume for typical natural gas). This concept is critical in maintaining safety. If the mixture of natural gas and air is either too rich or too lean combustion cannot occur.[11] Once the underground storage facility is filled to significantly above the lower flammability limit the mixture cannot burn. Nevertheless, residual intruding natural gas could act to bring the mixture back to an ignitable concentration and this condition should be avoided.

Finally, in addition to the concerns associated with flammability, residual hydrocarbons in the pore spaces of the formation might lead to the formation of permeability-reducing compounds and corrosive materials. For an expansive discussion of technical barriers for the CAES storage media discussed above, the reader is referred to the paper by King and Apps [24].

One obvious air storage option is, of course, utilizing tanks. Unfortunately, the low density of air renders this approach unfeasible as a means of storage. In fact, this is the reason behind the new technologies such as liquid air energy storage (LAES) and CES (e.g., see Section 5.3). Nevertheless, under the name *FastLight Storage Engine*, this is exactly the technology proposed by *Powerphase*,[12] the company that developed the modular air injection technology for gas turbine power augmentation, Turbophase. The latter comprises a natural gas or diesel fired reciprocating internal combustion engine (RICE) and a multi-stage intercooled centrifugal compressor, which injects hot, HP air directly into the compressor discharge plenum of the gas turbine, resulting in an increased mass flow through the combustor and turbine. The exhaust gas heat from the reciprocating engine is used in a recuperator (heat exchanger) to heat up the air to gas turbine compressor discharge temperature.

A schematic description of the system is shown in Figure 6.18. During the charge mode, an intercooled LP/HP process compressor train with electric motor drivers, powered, ideally, by renewable (i.e., zero carbon) or, alternatively, less expensive off-peak grid power, compresses air to roughly 100 bara (aftercooled to about 38°C) for storage in a "tank farm," which comprises a large number of carbon steel cylindrical tanks. During the discharge mode, compressed air is released from the storage tanks

---

[10] An area for additional consideration would be the presence of hydrocarbons in the HP expander cooling flow – chargeable flow in turbine terminology – which would bypass the HP combustor. These would ignite when they were exposed to a temperature higher than their auto-ignition temperature. With the relatively low 1,000°F HP turbine inlet temperature, the HP hot gas path is uncooled so that any residual hydrocarbons coming from the air storage volume would be burned out in the HP combustor.

[11] For example, if one assumes that the underground storage facility is filled with 100% natural gas, combustion cannot take place until enough air is admitted for the gas mixture to reach below the upper flammability limit. As air is added further, the ignitable range of air and natural gas mixture is traversed until, finally, with enough air the mixture becomes too lean to burn, i.e., below the lower flammability limit.

[12] https://powerphase.com/fastlight-energy-storage/

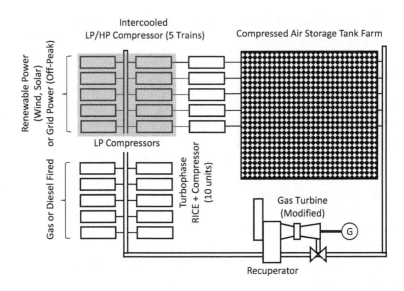

Figure 6.18 FastLight compressed air energy storage technology (by Powerphase).

and heated in the recuperator to the gas turbine compressor discharge temperature. Note that the gas turbine can be any industrial unit but with modifications for compressed air admission, i.e., debladed compressor rotor, compressor bell-mouth replaced with inlet manifold, and forward bearing compartment modified to take the thrust load. The hot gas path section (turbine), combustor, exhaust, and foundation are unchanged. Drive shaft, generator, and transformer changes may be needed.

In a variation of the conventional discharge mode above, compressed air from the storage tanks is combined with compressed air generated by the LP compressors and heated in the recuperator to the gas turbine compressor discharge conditions. This operating mode extends the discharge period.

Turbophase RICE-compressor modules enable operation of the gas turbine in a peaker mode even when the storage tanks are empty. In this operating mode, gas or diesel fired RICE powered compressors and LP compressors (driven by the renewable resources or the grid) generate air to deliver continuous peaking power.

In quantitative terms, let us consider a 166 MWe gas turbine with 935 lb/s airflow and about 450 MWth fuel consumption (i.e., vintage 60 Hz F Class). If the compressor is removed and air is supplied externally, during the discharge mode the output goes to approximately 320 MWe, i.e., it nearly doubles. During the charge mode, compression of 935 lb/s air to 1,500 psia can be done at a power consumption of about 245 MW, with two LP compressors (about 75 MW shaft power each, with intercooling) and one HP compressor (about 90 MW shaft power).

As a peaker this gas turbine would generate 166 MWe with an efficiency of about 37%. In the discharge mode, 320 MWe is generated with an efficiency of 320/450 = 0.71 or 71%. If the charge power is taken off the grid, assuming an effective grid fuel efficiency of 50%, PEE can be calculated as 320/(245/50% + 450) = 0.34 or 34%.

If charging is powered by a renewable resource, PEE becomes 320/(245 + 50% × 450) = 0.68 or 68%.

6.8   Advanced Concepts

While the last four decades have seen only two commercial CAES plants become a reality (albeit quite successfully), there has been a plethora of CAES projects that have ultimately failed to reach the *notice to proceed* stage. The CAES technology in those two facilities is generally referred to as *conventional* or Gen−1 and it has been the key building block of almost all proposed projects. Nevertheless, there has been a wide array of *advanced* or Gen-2 technologies, which have been the subject of DOE/EPRI and other studies and numerous articles in archival journals and trade publications (and patents). The paper by Nakhamkin et al. provides a reasonably detailed description of those advanced CAES variants [7], the highlights of which are summarized in Table 6.4. The numbers for construction costs in the table should be disregarded as "real" in absolute terms. They can easily be twice that shown or more when a bona fide FEED study is done. They should only be used as yardsticks to gauge the relative costliness of the listed technologies.

The advanced CAES concepts are based on the injection of the stored and pre-heated air directly into the compressor discharge plenum of a heavy-duty industrial gas turbine (e.g., GE's Frame 7 or 9 units) thus providing a power boost. The schematic of the CAES-AI concept with major performance characteristics is presented in Figure 6.19. Conceptually, CAES-AI is akin to using a turbine with an integrated gasification gas turbine (IGCC)-type hot gas path design. The IGCC design has the extra flow cross section nominally used for the IGCC unit's extra volume flow of the low Btu syngas and diluent flow (e.g., nitrogen from the air separation unit [ASU]). Typically, this results in 12–14% more gas flow through the turbine section. In CAES-AI, the enlarged stage 1 nozzle (S1N) inlet area can be used for passing the extra air flow coming from the storage volume.

All major gas turbine OEMs have variants of their most robust heavy-duty "frame" machines modified for IGCC applications (one recent example is the GE's 7FB gas turbine in the Duke Edwardsport IGCC power plant). In that sense, the design and development of a CAES-AI system as shown Figure 6.19 does not present a significant technology challenge. In all likelihood, it would be even less of a technology risk because the entire gas turbine, including its combustor, would be offered as a package by an OEM with the requisite performance guarantees and maintenance package.

In addition to Gen-2 CAES, there are *nine* different types of innovative CAES or CAES-like technologies (as identified by the Sandia Labs [8]):

1. Adiabatic CAES
2. Adsorption-enhanced CAES
3. Diabatic CAES
4. Hydrokinetic Energy

Table 6.4 CAES technology comparison

	CAES 1st Generation	CAES 2nd Generation w/Air Injection	CAES 2nd Generation w/Inlet Chilling*	2nd Generation CAES w/Above Ground Storage
Technology	Custom burners and equipment from Dresser-Rand on a single shaft	Uses off-the-shelf equipment including CT on separate shafts –Scalable 15–430 MW	Uses off-the-shelf equipment including CT on separate shafts –Scalable 15–430 MW	Uses off-the-shelf equipment including CT on separate shafts –Scalable 15–430 MW
Air Emissions (No SCR)	No dry-low NOx burner	CT burner technology	CT burner technology	CT burner technology
Working Air Pressure	~650–900 psi	400–2,000 psi+	400–2,000 psi+	400–1,200psi+
Total Power, MW	110	433	427	15
Off-peak Power, MW	85	318	313	12
Plant Heat Rate, Btu/kWh	~4,000	~3,800	~3,800	~4,000
Constructor Costs (2008 $)	~$1,200/kW	~$750/kW	~$750/kW	~$1,250/kW

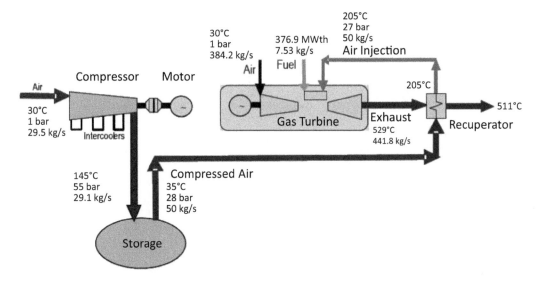

Figure 6.19 Advanced CAES with air injection (CAES-AI) – shown in power generation (discharge) mode, 137.4 mWe, 9,870 kJ/kWh net (LHV).

5. LAES (see Section 5.3)
6. Near-isothermal CAES
7. Transportable CAES
8. Underwater CAES
9. Vehicle Compression

The innovations in these technologies are in the storage vessel, the storage medium, the energy conversion process, or some other feature. Unlike conventional CAES covered herein, many of these technologies do not rely on underground geologic formations to store compressed air. Some technologies, such as near-isothermal and underwater CAES, can store compressed air in transportable vessels or underwater bladders. Other technologies, such as adiabatic and near-isothermal, are considered innovative for their theoretical improvement in the efficiency of the energy conversion process.

The *adiabatic CAES* attempts to capture the heat produced by the compressed air, store it using liquid or solid thermal energy systems (e.g., mineral oil, molten salt, and ceramics), and recycle it to reheat stored compressed air before it enters the expander for power production. One such technology is developed by a company in Canada, Hydrostor under the name A-CAES (Advanced CAES).[13] The A-CAES technology is schematically described in Figure 6.20. In addition to the conventional adiabatic CAES feature, i.e., storage of heat rejected by compressor inter- and aftercoolers during charging for use during discharging and power generation, the A-CAES includes hydrostatically compensated air storage. As described earlier in Section 6.2,

---

[13] https://www.hydrostor.ca/company/, last accessed on June 26, 2021.

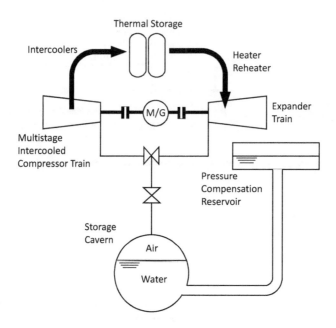

Figure 6.20 Hydrostor's A-CAES system with hydrostatic pressure compensation.

this enables a near constant pressure regardless of charge state and a reduced storage volume. According to Hydrostor, this was demonstrated in 2015. In particular, the air was stored in underwater air storage vessels approximately 55 m below the surface of Lake Ontario. A small facility in Canada (1.75 MW discharge, 2.2 MW charge (CEF = 1.25) and 10+ MWh) has been (according to the company website) in commercial operation since 2019.

One interesting possibility for deploying CAES is in combination with PHES. The basic idea has been proposed under different names by different investigators, e.g., ground-level integrated diverse energy storage (GLIDES) [25] and pumped hydro compressed air (PHCA) [28]. It is developed commercially by the Energy Internet Corporation (https://energyinternetcorporation.com/) under the name isothermal CAES with a liquid piston heat engine.

The heart of the system is a "liquid piston," i.e., a pressure vessel initially filled with air at a certain pressure. During charging, a pump driven by grid power (ideally, from renewable resources such as wind or solar) pumps water from a reservoir into the bottom of the pressure vessel (i.e., the piston). As the water level rises in the vessel, air at the top is compressed. Compressed air is cooled by water spray from the top of the vessel (from the pump discharge) to cool the air heated via compression. During discharge, compressed air in the vessel pushes water out, which is connected to the reservoir via a hydraulic turbine for electric power generation. The basic scheme and calculations can be found in Refs. [25–26].

The system is relatively simple and modular. It can be used in conjunction with an underground cavern as the HP reservoir (if one is readily available) [25]. The predicted RTE is above 80% [25]. Until an actual project has been designed, engineered, and built, it is not possible to gauge the veracity of claims made on CAPEX and the performance of the PHCA technology.

6.9   CAES with Hydrogen

Hydrogen is a key enabler of carbon-free power generation because the combustion product is water vapor (H_2O). However, unique properties of H_2 such as high flammability, high ignition temperature but low ignition energy (vis-à-vis methane) and high flame speed present significant design challenges such as flashback, high pressure drop, and unstable combustion if H_2 content in the fuel gas exceeds a threshold. Without development, the current capability of gas-fired spark-ignition engines in terms of maximum allowable H_2 content in fuel gas is about the same as that in gas turbine dry-low-NOx (DLN) combustors, i.e., 5%(v). Typical NOx control techniques include equivalence ratios of 0.65 or lower (known as *lean burn*) and water injection (same drawback as in the case of gas turbine diffusion flame combustors, i.e., consumption of scarce water resources). A technology concept introduced below utilizes *exhaust gas recirculation* (EGR) and is described in Figure 6.21.

The way EGR works is via reduced O_2 and increased inert content (i.e., CO_2) in combustion air. This helps to curtail NOx production via reduced peak cylinder temperature. In addition, EGR slows down the heat release rate (which becomes faster roughly by 0.15 ms for each 10%(v) H_2 in fuel gas) and flame speed to keep peak cylinder pressures down and ensure stable combustion. For optimal EGR (i.e., as a percent of exhaust gas flow redirected to the cylinder inlet), spark timing and equivalence ratio adjustments for specified levels of H_2 content in fuel gas are necessary. The start of combustion is likely to be earlier than that of the stock engine

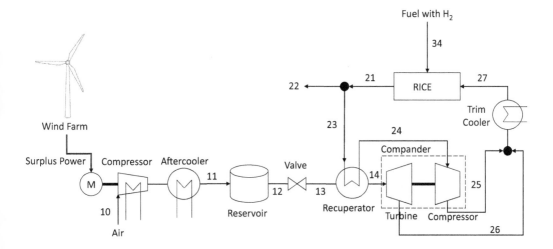

Figure 6.21 Proposed technology, CAES with EGR – schematic description.

because of the low H_2 ignition energy, which may lead to knocking. As a first guess, 40% EGR is used herein for 100%(v) H_2 fuel gas to demonstrate the concept. Actual levels must be determined by the engine OEM.

As shown in Figure 6.21, surplus power from a carbon-free generation resource such as solar or wind (or nuclear) is utilized to compress air in an intercooled, multi-stage process compressor to the storage pressure (150 bar), which is then cooled in an aftercooler with moisture removal to the storage temperature (25°C). Pressurized and cooled air is delivered to the storage reservoir. This completes the charge phase of the energy storage process.

During the discharge phase of the process, compressed air is sent through a pressure control valve to a recuperating heat exchanger and heated to a temperature of 240°C by cooling the recirculated portion of the RICE exhaust gas from the discharge temperature (327°C) to 62°C. It should be emphasized that the numbers cited herein are meant to be representative of the process based on a particular RICE for illustration purposes and are thus subject to change and optimization, particularly by the engine OEM, for actual field deployment. Compressed air at 50 bar and 240°C is expanded through the expander section of a "compander," which drives the compressor of the same. The *compander* (a word contraction for *combined compressor and expander*) increases the pressure of the recirculated portion of the exhaust gas from 1 bar to 6.5 bar, which then mixes with the expanded air stream and cooled in the trim cooler to a temperature of 50°C. Mixed air-gas stream at 5 bar and 50°C comprises the charge air for the RICE, which is ideally but not necessarily designed to burn 100% H_2 fuel gas.

Exhaust gas stream 21 of the RICE in Figure 6.21, at 1 bar and 327°C, is separated into two streams, 22 and 23. Stream 22 is sent to the stack whereas stream 23 is sent to the recuperator. Based on the assumption that recirculated gas stream 23 flow rate is 40% of the RICE exhaust flow stream 21, selected stream properties are summarized in Table 6.5.

The engine selected for the sample calculation is loosely based on MAN 18V51/60 and has the following performance: 18.9 MW generator output at 48% net LHV efficiency with 1,180 kg/h of H_2 consumption (LHV of 120,068 kJ/kg). The single-stage charge compressor power consumption is estimated as 7.5 MW for 30 kg/s air

Table 6.5 Key air and gas stream data (see Figure 6.21 for streams)

	10	21	23	27	34
	Air	Gas	Gas	Air + Gas	Fuel
Flow rate, kg/h	64,760	109,300	43,720	108,100	1,180
Temperature, °C	15.0	327.0	327.0	50.0	25.0
Pressure, bar	1.0	1.0	1.0	1.0	20.7
O_2, %(v)	20.74	6.86	6.86	14.88	0.00
H_2O, %(v)	1.01	23.37	23.37	10.12	0.00
H_2, %(v)					100

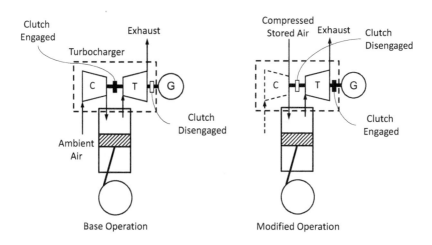

Figure 6.22 RICE operation with a turbocharger module equipped with a generator.

flow and 5 bara delivery pressure, which is supplied by the exhaust gas turbine of the turbocharger unit. As illustrated in Figure 6.22, the turbocharger is modified to include (i) a clutch/coupling between the compressor and turbine and (ii) a generator attached to the turbine by another clutch/coupling. Thus, during the discharge phase of the energy storage process, the compressor can be deactivated (i.e., the clutch disengaged), and 5 bar charge air is directly supplied to the engine. At the same time, the generator is activated (i.e., the clutch is engaged) and electric power is supplied to the grid. As a result, with the same amount of fuel consumption, the net output increases to $18.9 + 7.5 = 26.4$ MW and the engine efficiency becomes $26.4/18.9 \times 48\% = 67\%$. Another possible modification is to transmit exhaust turbine shaft power output to the main engine shaft and generator via a gearbox (instead of a separate generator). The final configuration is subject to design optimization by the engine OEM. Alternatively, if the RICE is going to be deployed exclusively in an energy storage mode, the turbocharger module can be modified by permanently removing the compressor and adding a generator (Figure 6.23a) or connecting the exhaust turbine to the main shaft and generator through a gearbox (Figure 6.23b).

The power consumption of the intercooled compressor during the charge phase of the process is calculated as 12.1 MW (including the power consumed by the fin-fan coolers for inter- and aftercoolers – not shown in Figure 6.21). Engine fuel consumption during the discharge phase is

$$(1,180/3,600)\text{kg/s} \times 120,068 \text{ kJ/kg} = 39.4 \text{ MWth}.$$

Arguably, the most useful measure of CAES performance is the PEE, which is defined in Equation (6.2a). The equation provides an electricity-to-electricity "roundtrip efficiency" that isolates the energy losses in the conversion of electricity to compressed air and back to electricity. The second term in the denominator of the PEE formula represents the electric energy that could have been generated with the same amount of fuel input with average power plant efficiency (45.5% for natural gas–fired,

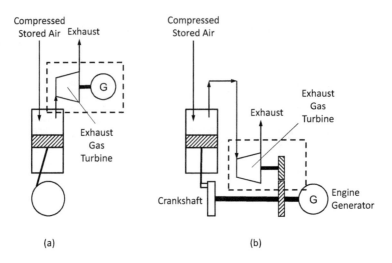

Figure 6.23 RICE with exhaust turbine and generator.

32.7% coal-fired power plants in the USA in 2015 based on the US EIA data). Assuming $t_g = t_c$ and $\bar{\eta} = 50\%$, the PEE of the proposed energy storage concept becomes

$$\text{PEE} = \frac{26.4}{12.1 + 50\% \cdot 39.4} = 83\%.$$

The underground storage in Huntorf and McIntosh CAES plants is a solution-mined salt cavern (the Huntorf plant has two caverns). Alternatives have been considered for many proposed CAES projects in the nearly four-decade period since the commissioning of the Huntorf plant. They include limestone mines, depleted natural gas reservoirs, aquifers (water-filled underground reservoirs) and porous rock formations. Each have their advantages and disadvantages as well as cost impact.

For the engine considered herein, 4 hours' worth air storage at 150 bar and 25°C can be accomplished either in a spherical tank of 8.1 m diameter or twenty tanks with a diameter of 12 ft (3.6576 m) and length/height of 10.5 m, which can be shop-fabricated and transported to the site. The specified storage volume holds 6 hours' worth of air but at the engine's rate of air consumption (30 kg/s), in 4 hours, the storage pressure drops from 150 bar to 50 bar.

The critical question is whether the system can be cost effective. For the example above, the energy storage capacity is $4 \times 26.4 = 105.6$ MWh. If the capex bogey is set to \$250/kWh to be competitive with Li-Ion battery technology, the maximum allowable installed cost is \$250/kWh $\times$ 105,600 kWh = \$26.4 million. For an 8-hour system, i.e., 211.2 MWh storage capacity, the allowable installed cost target will double to \$52.8 million but the actual installed cost increase will be much smaller, i.e., determined by the size and cost of the storage reservoir. All components are off-the-shelf, engineered products widely used in power and chemical process industries. The critical item from a cost-effectiveness perspective is the final design, sizing, and costing of the storage tank (or tanks) or identification of a suitable natural reservoir (readily available or to be mined).

In conclusion, the CAES system with a modified RICE as described above

- increases the base engine output by $7.5/18.9 = 40\%$ in energy storage mode at the same fuel consumption as in base operation mode
- improves the heat rate (efficiency) by the same factor
- has better than 80% roundtrip efficiency, and
- enables (or, at the very least, contributes significantly to enabling) 100% H_2 combustion for carbon-free power generation.

It should be noted that the concept can also be applied to operation with natural gas or a mixture of H_2 and methane (CH_4) or a syngas (mainly, a mixture of H_2 and CO). EGR is beneficial to reduction of NOx emissions with any gaseous fuel by the same mechanism, i.e., reduction of flame temperature by dilution with an inert gas component (e.g., H_2O and/or CO_2). However, unlike the case with combustion of 100% H_2, the exhaust gas will include CO_2.

Another "green" possibility for combining CAES with hydrogen storage is to locate H_2 production (via electrolysis powered by carbon-free resources, e.g., curtailed solar or wind energy) and storage with CAES. During the charge period, the electrolyzer and H_2/air compressors are powered by, say, a renewable energy resource (RES). Compressed H_2 and air (at, say, 100 bara) are stored in their respective reservoirs. Depending on the design size of the system, H_2 storage can be accommodated aboveground in pressure vessels. This is a question of cost–performance trade-off to be done for a given project in a FEED study.

During the discharge period, the CAES gas turbine generates power using the compressed air (transmitted to the combustor through a recuperator) in a combustor utilizing the stored H_2. The system offers high flexibility such that, when the RES is down, the electrolyzer can be run at minimum turndown (say, 20% load) using the power generated by the CAES turbine. Preventing frequent start-stop cycling of the electrolyzer is key to equipment life. During the discharge, extra power can be generated by a H_2 expander operating between the H_2 reservoir and the gas turbine and by an air expander operating between the air reservoir and the CAES turbine.

6.10   Compressed Gas Energy Storage

The compressed gas energy storage (CGES) concept is based on an analogy to the CAES technology described above. The idea originated from the owner of a gas storage facility in Madera, California, which has two of the major CAES components already in place, namely the compressor train (comprising five units with variable frequency drives (VFD), inter- and aftercoolers – basic fin-fan air coolers) and the storage space (natural gas reservoirs) [27]. Thus, the present concept simply comprises the addition of a turboexpander train to transfer the existing natural gas storage facility

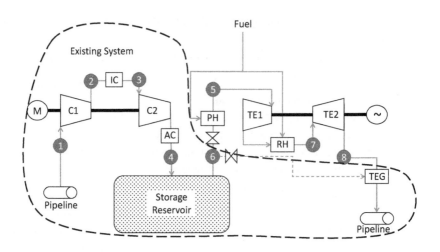

Figure 6.24 Compressed gas energy storage (CGES) – schematic diagram.

into a CGES power plant. The resulting power plant is analogous to CAES with the main difference being the storage medium (i.e., natural gas instead of air).

The heart of the CGES concept, schematically described in Figure 6.24, is a turboexpander with gas heaters. A two-stage turboexpander (TE1 and TE2) with a preheater (PH) and reheater (RH) generates power by expanding the gas from the reservoir to the entrance of triethylene glycol dehydration unit. The variation between different implementation options is in the manner of gas heating and the temperature to which it is heated. The primary reason for gas heating is to prevent formation of liquids within the expander. The natural gas from the reservoir is at 2,500± psia (170+ bara) and approximately 100°F (38°C). If the expansion from the storage to the pipeline pressure (around 900± psia [62 bara]) were done without any preheating, the gas temperature at the end of the expansion would be −18°F (−28°C, very common in refrigeration plants); this is not desirable in this case due to liquid water and hydrate formation. An additional benefit of gas preheating upstream of each turboexpander stage is the increased power output due to increased availability. Without any preheating, the turboexpander would generate about 8 MWe (320 MMSCFD gas flow, about 9 million standard m^3 per day); with preheating to 200°F (93°C) at each stage, the power generation is about 11 MW with about 130°F (54°C) at the end of the expansion. For a detailed description of the system and an in-depth cost-performance study, the reader is referred to the article by Gülen et al. (Ref. [16] in Chapter 2).

6.11   References

1. Nakhamkin, M. and van der Linden, S. 2000. Integration of a gas turbine (GT) with a compressed air storage (CAES) plant provides the best alternative for mid-range and daily cyclic generation needs, ASME Turbo Expo 2000: Power for Land, Sea, and Air. DOI:10.1115/2000-GT-0182.

2. Brown-Boveri Company (BBC). 1979. Huntorf air storage gas turbine power plant, Report publication no. D GK 90202 E.

3. King, M., Jain, A., Bhakar, R., Mathur, J. and Wang, J., 2021. Overview of current compressed air energy storage projects and analysis of the potential underground storage capacity in India and the UK, *Renewable and Sustainable Energy Reviews*, 139, 110705.

4. Daly, J., Loughlin, R. M., DeCorso, M., Moen, D. and Davis, L. 2001. CAES – reduced to practice, ASME Turbo Expo 2001: Power for Land, Sea, and Air. American Society of Mechanical Engineers.

5. Holden, P., Moen, D., DeCorso, M. and Howard, J. 2000. Alabama Electric Cooperative Compressed air energy storage (CAES) plant improvements, Proceedings of the ASME Turbo Expo 2000: Power for Land, Sea, and Air. Volume 3: Heat Transfer; Electric Power; Industrial and Cogeneration, May 8–11, Munich, Germany.

6. Succar, S. and Williams, R. H. 2008. *Compressed Air Energy Storage: Theory, Resources and Applications for Wind Power*, Princeton Environmental Institute, Princeton University.

7. Nakhamkin, M. and Chiruvolu, M. 2007. Available compressed air energy storage (CAES) plant concepts, Power-Gen International.

8. Agrawal, P., Markel, L., Gordon, P. et al. 2011. Characterization and assessment of novel bulk storage technologies: A study for the DOE energy storage systems program, Report SAND 2011-370, Sandia National Laboratories.

9. Schulte, R. H., Critelli, Jr., N., Holst, K. and Huff, G. 2012. Lessons from Iowa: Development of a 270 Megawatt compressed air energy storage project in Midwest Independent System Operator, Report SAND 2012-0388, Sandia National Laboratories.

10. Makansi, J. 2014. ENERGY STORAGE: Does grid-scale storage threaten gas turbine plants? *Combined Cycle Journal*, (Q4). Available at www.ccj-online.com/4q-2014/energy-storage-does-grid-scale-storage-threaten-gas-turbine-plants/

11. King, M. J. and Moridis, G. 2022. Compressed Air Energy Storage in Aquifer and Depleted Gas Storage Reservoirs, In A. Hauer (ed.), *Handbook of Energy Storage*, London, UK: Wiley Press.

12. Crotogino, F., Mohmeyer, K.-U. and Scharf, R. 2001. Huntorf CAES: More than 20 years of successful operation, In *Proceedings of the Solution Mining Research Institute (SMRI) Spring Meeting, 15–18 April*, Orlando, FL.

13. Haselbacher, H., Stys, Z. S., Karalis, A. J. and Sosnowicz, E. J. 1985. Design and expected performance of 45 MWe to 100 MWe CAES turbine machinery, *The American Society of Mechanical Engineers (ASME)*, Paper 85-JPGC-GT-10.

14. Brown-Boveri Company (BBC). 1979. Operating Experience with the Huntorf Air Storage Gas Turbine Power Station, Publication No. D GK 1274 86 E.

15. Hounslow, D. R., Grindley, W., Loughlin, R. M. and Daly, J. 1998. The development of a combustion system for a 110 MW CAES plant, *Journal of Engineering for Gas Turbines and Power – Transactions of the ASME*, 120, 875–883.

16. Karalis, A. J., Sosnowicz, E. J., Haselbacher, H. and Istvan, J. 1987. Optimizing the design conditions of a 100 MWe CAES plant with salt dome air storage, *The American Society of Mechanical Engineers (ASME)*, Paper 87-GT-42.

17. Nakhamkin, M., Schainker, R. B., Hutchinson, F. D., Stange, J. R. and Canova, F. 1985. Compressed air energy storage: Plant integration, turbomachinery development, *The Americal Society of Mechanic Engineers*, Paper 85-IGT-4.

18. Hoffeins, H., Romeyke, N., Hebel, D. and Sütterlin, F. 1980. Die Inbetriebnahme der ersten Luftspeicher-Gasturbinengruppe, *Brown Boveri Mitteilungen*, 67(8), 465–473.

19. Lukas, H. and Mehta, B. 1986. Acid corrosion in a CAES recuperator, *The Americal Society of Mechanic Engineers*, Paper 86-GT-83.

20. Nakhamkin, N., Stange, J. R., Marshall, R., Pelini, R. and Schainker, R. B. 1986, Advanced Recuperator for Compressed Air Energy Storage Plants, In *Proceedings of the 1986 Joint Power Generation Conference: GT Papers, October 19–23*, Portland, OR.

21. Nakhamkin, N. 1989. Advanced Recuperator, US Patent 4,870,816.

22. Succar, S. 2008. Wind-CAES integration, *CAES Scoping Workshop, Center for Energy & Life Cycle Analysis*, New York, NY.

23. Grubelich, M. C., Bauer, S. J. and Cooper, P. W. 2011. Potential hazards of compressed air energy storage in depleted natural gas reservoirs, Report SAND 2011-5930, Sandia National Laboratories.

24. King, M. and Apps, J. 2013. Compressed air energy storage: Matching the earth to the turbomachinery – no small task, *Electrical Energy Storage Applications and Technologies (EESAT) October 20–23*, San Diego, CA.

25. Kassaee, S., Abu-Heiba, A., Raza Ally, M. et al. 2019. Part 1– Techno-economic analysis of a grid scale ground-level integrated diverse energy storage (GLIDES) technology, *Journal of Energy Storage*, 25, 100792.

26. Wang, H., Wang, L., Wang, X. and Yao, E. 2013. A Novel Pumped Hydro Combined with Compressed Air Energy Storage System, *Energies*, 6, 1554–1567.

27. Adams, S. S., Gülen, S. C., Haley, R. M., Weber, D. A., and White, J. K. 2017. System for Compressed Gas Energy Storage, US Patent 9,803,803.

7   Hybrid Systems

The term hybrid is used for a combination of three different systems:

- Fossil fired, e.g., gas turbine or gas fired recip engine
- Renewable generation system, e.g., solar or wind
- Energy storage system, e.g., lithium-ion (Li-Ion) batteries

  The combination can be of different types, e.g.,

- Gas turbine and *battery energy storage system* (BESS)
- Solar photovoltaic (PV) and BESS
- Solar PV and gas turbine (or gas engine)
- Solar PV plus gas turbine/engine plus BESS

The main motivation underlying the design of a hybrid electric power system is *decarbonization*, which is the heart of the *energy transition*, which refers to the global energy sector's shift from fossil fuel–based energy production and consumption to renewable energy sources like wind and solar supplemented by energy storage. Decarbonization is the reduction of carbon dioxide emissions by different means. Carbon dioxide (CO_2) is a potent *greenhouse gas* contributing significantly to global warming and thus plays a major role in climate change. In the USA, transportation, industry, and electricity are three economic sectors with roughly equal shares in GHG emissions. By far the dominant mechanism of CO_2 generation is combustion of hydrocarbon (fossil) fuels, i.e., coal, petroleum derived liquid fuels such as diesel and gasoline, and natural gas. Thus, as dictated by simple logic, the first and foremost action to be taken is elimination of fossil fuels from the picture. There are several solutions in this respect:

- Increase the share of carbon-free resources such as wind, solar, nuclear, and hydro in electricity generation portfolio
- Switch from coal to natural gas and, eventually, hydrogen

A key component of decarbonization is *electrification* of the industrial (e.g., utilization of electric motor drives for large compressors in LNG plants), residential (e.g., use of heat pumps in the buildings), and transportation sectors (e.g., electric vehicles). An integral part of this endeavor is improving, updating, and extending the transmission infrastructure (accompanied by energy storage) so that carbon-free and low-carbon generation use in electrification is maximized.

Well-known intermittency and low capacity factors of solar and wind resources prevent these technologies from fulfilling the demands of the *energy transition* on their own – at least in the near future. They require backup in the form of dispatchable resources, e.g., fossil-fired power plants and energy storage systems. Such systems must be nimble enough to address the short-term fluctuations and maintain the grid stability in addition to taking over the base load generation when the renewable resources are not available. Aeroderivative gas turbines, small industrial gas turbines, gas fired recip engines, and energy storage systems such as compressed-air energy storage (CAES), liquid air energy storage (LAES), pumped hydro (PHES), and electric batteries are readily available technologies that can accomplish these tasks. Large-scale, long-duration systems such as CAES and PHES are discussed elsewhere in this book. Herein, the focus is on BESS and its integration with gas turbines and solar PV.

7.1   Gas Turbine Plus BESS

Gas turbines are characterized by their fast startup and shutdown as well as fast load ramp, up or down, capabilities. Even the advanced class, heavy-duty industrial gas turbines rated at several hundred megawatts can roll from cold iron to full speed, no load (FSNL), synchronize to the grid, and ramp up to full speed, full load (FSFL) rapidly, e.g., in twenty minutes or less (in simple cycle). Even in combined cycle, OEMs are advertising hot start times to full load in 30 minutes or slightly less. (Typically, these times are subject to several caveats, e.g., from ignition to all steam bypass valves fully closed. Due to the significant thermal inertia of the heat recovery steam generator (HRSG) and the steam turbine, the actual time to true 100% rated load in megawatts is much longer.)

Nevertheless, "lighter" and even nimbler small industrial gas turbines, especially the aeroderivative gas turbines, are much more suitable to this role with hundreds of annual start/stop cycles and many more load up-down ramp cycles (when they are used for frequency response). Examples of such gas turbines are General Electric's (GE's) LM series (Land and Marine), Rolls-Royce Trent (under Siemens Energy now), and Pratt & Whitney's FT8 machines (owned by Mitsubishi Power). The reader can refer to Chapter 23 in **GTFEPG** for detailed information on the performance and operability characteristics of aeroderivative gas turbines.

In terms of installed capacity, not surprisingly, aeroderivative gas turbines larger than 10 MWe typically constitute less than 20% of the worldwide market for *all* gas turbines. In the more focused segment of the market however, i.e., small gas turbines rated between 18 MW and 65 MW, aeroderivatives account for two-thirds of the installed capacity.[1] Note that, even outside the USA, GE's LM2500 and LM6000

---

[1] As presented by Mark Axford in Western Turbine Users Inc's (WTUI) 2013 Conference
(March 10–13, San Diego). The numbers may be somewhat different at the time of writing (2021).
The interested reader can check the internet for more up-to-date information. However, the distribution is
not expected to have changed appreciably.

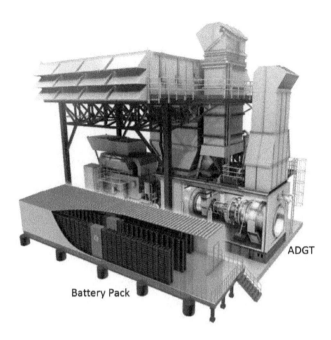

Figure 7.1 Hybrid gas turbine plus battery storage (Courtesy: General Electric).

products hold nearly a 70% market share (in 2012).[2] The cited numbers include *all* aeroderivatives with applications ranging from marine propulsion to oil and gas and pipeline industry to electric power generation (about one-third of the total). (In passing, naval applications for ship propulsion is dominated by GE's LM2500 gas turbines.) There is no reason to expect that the future role of aeroderivative gas turbines in electric power generation will differ significantly from their current role: peaking (emergency) power, renewable support, and reactive power in addition to focused industrial power generation with cogeneration (combined with battery storage).

In conjunction with the last item, note that GE's *LM6000 Hybrid EGT*, which combines an aeroderivative gas turbine with battery storage (see Figure 7.1). The new system was introduced in 2016 to provide 50 MW of spinning reserve (without burning fuel), flexible capacity, and peaking energy; 25 MW of high-quality regulation; and 10 MVAR of reactive voltage support and primary frequency response when not online (i.e., as *synchronous condenser*). The way the hybrid system works is as follows:

A standard LM6000 can start in less than 5 minutes in fast response mode. The battery included with the LM6000 provides electricity for the first 4.5 minutes. Consequently, the combined LM6000 plus battery Hybrid EGT can function as a single generating asset and bid into *spinning reserve*, while the gas turbine is switched off. When the request for spinning reserve comes in, the battery will provide

---

[2] Ibid.

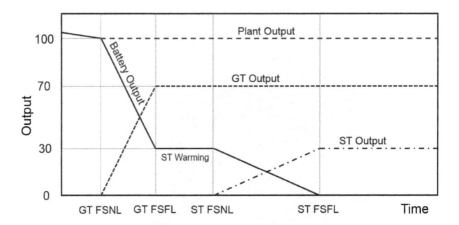

Figure 7.2 Gas turbine and battery hybrid combined cycle power plant startup.

electricity for the first 4.5 minutes and the gas turbine will start up in that time and provide seamless electricity after the 4.5 minutes. This is how it delivers *green* 50 MW of spinning reserve without burning fuel. The battery can be charged later when the gas turbine has taken over. The first two LM6000 Hybrid EGT units went commercial in California in March 2017. Each unit integrates a 10 MW/4.3 MWh BESS with the 50 MW aeroderivative gas turbine. They are owned and operated by *Southern California Edison*.

Siemens offers the Hybrid EGT capability under the trade names SIESTART and SIESTORAGE. Unlike GE, Siemens does not limit the concept to a simple cycle aeroderivative gas turbine (GT) (at least not on a product basis). How the system operates is conceptually demonstrated in Figure 7.2. When the power plant is dispatched from "cold iron" in the event of an unpredicted generation loss, the battery engages almost instantaneously and provides a full power grid while powering the load commutating inverter (LCI) for the gas turbine start. Once the gas turbine rolls up to FSNL and starts its load ramp, the battery contribution goes down in lock-step so that full power is still delivered to the grid. After a warming period, during which the battery contributes to the plant output steadily, the steam turbine rolls up to FSNL and starts its load ramp. This time, the battery contribution goes down in lock-step with the steam turbine load ramp until the steam turbine and, consequently, plant full load is reached. The sequence of events shown in Figure 7.2, minus the steam turbine, would be conceptually the same for the GE Hybrid EGT.

As described above, the usual application of the hybrid system is *gas turbine primary*, i.e., the GT is the main source of power and the battery is used to assist with startup, ramp rates, and low-level output. The other possibility is *battery primary*, i.e., the battery is the primary source of power, and the GT is used to provide additional power when the capacity of the battery is exceeded. This configuration is advantageous for making use of older machines. (Apparently, in Germany alone, there are close to 400 such gas turbines, nearly 300 of which have a nameplate power rating below 15 MW.) As an example, see the article by Fuhs that describes an installed

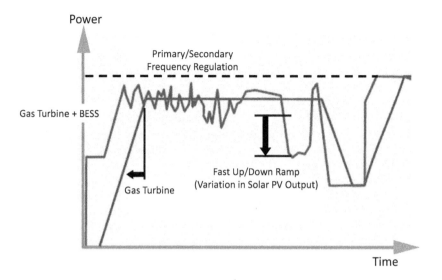

Figure 7.3 Conceptual operation of gas turbine plus BESS hybrid.

system comprising a 4 MW gas turbine with a 9 MW/6.5 MWh battery.[3] The combined power plant was designed to compete in both the primary and the secondary frequency regulation. Additional potential applications include utilization as a district PV storage and balancing unit, balancing group compensation, network peak avoidance or black start capability after a power failure, and participation in blockchain-based energy trading.

The operation of the hybrid system is illustrated in Figure 7.3. Key features are as follows:

- BESS can supply the necessary electrical power and possibly the frequency reference to enable the gas turbine to start when the grid is down (*black* start). The battery needs to have the capacity to power the turbine starter as well as the auxiliary equipment. Once the GT is up and running, the BESS absorbs the energy produced until customers can be re-connected to their supply from the grid.
- BESS can both absorb energy from the grid as well as feed energy back into it. The fast response capability of BESS technology (of the order of milliseconds) enables it to act as a primary reserve to stabilize grid frequency fluctuations while the GT continues to run at a stable load. The BESS-GT hybrid is capable of providing primary and secondary reserve services. BESS assumes the task of frequency regulation, while the gas turbine supplies the energy to deliver extended service requirements and balance the battery charge level. This mode of operation significantly reduces the low cycle fatigue on the gas turbine. (The BESS can provide frequency response services even when the GT is not running.)

---

[3] Michael Fuhs, "Combined gas and battery grid services power plant by Technische Werke Ludwigshafen and Younicos," PV Magazine, March 2018.

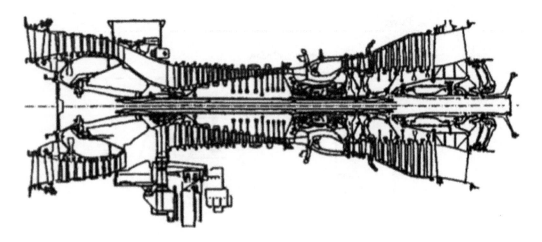

Figure 7.4 LM6000 aeroderivative gas turbine cross section (reprinted by permission of the American Society of Mechanical Engineers[4]).

- BESS can provide load ramp up/down support by either injecting additional power into the grid or absorbing power from the GT during ramp down. In the case of sudden loss load events, BESS can assist the GT by absorbing power and providing a soft shutdown.

7.1.1   LM6000 Gas Turbine

The LM6000 gas turbine is derived from GE's CF6–80C2 engine for wide-body aircrafts. It is a two-shaft design (see Figure 7.4). The low pressure (LP) rotor consists of a 5-stage LP compressor and a 5-stage LP turbine connected by means of a mid-shaft, which extends through the center of the engine. The high pressure (HP) rotor consists of a 14-stage HP compressor and a direct coupled two-stage HP turbine. It is available in different configurations with standard and dry-low-emissions (DLE) combustors as well as a SPRINT option. In its latest DLE version, LM6000 is a 45 MW – 42% net lower heating value (LHV) gas turbine. Key off-design performance curves for the LM6000 are shown in Figures 7.5–7.7.

7.1.2   Gas Fired Recip Engines

An alternative to aeroderivative and small industrial gas turbines are gas-fired *reciprocating internal combustion engines* (RICE) or, in short, recip engines. These machines are essentially car engines on steroids. They are classified in speed (rpm) and output as shown in Figure 7.8. They are either *spark ignited* (*Otto* cycle) or *compression ignited* (*Diesel* cycle) engines utilizing *premix* method (*lean burn* for emissions control) for

---

[4] Casper, R.L., Application of the LM6000 for Power Generation and Cogeneration, 93-GT-278, International Gas Turbine and Aeroengine Congress and Exposition, Cincinnati, Ohio, May 24–27, 1993.

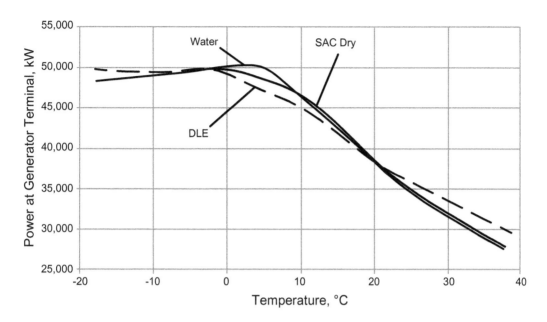

Figure 7.5 LM6000 aeroderivative gas turbine – output variation with inlet air temperature (SAC: Standard Annular Combustor, DLE: Dry Low Emissions).

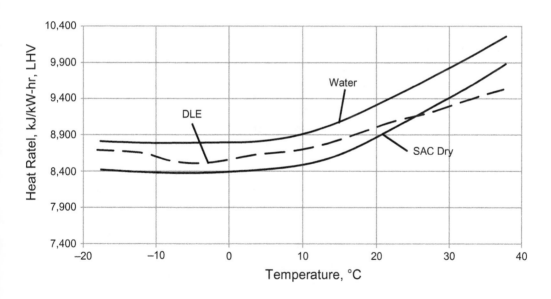

Figure 7.6 LM6000 aeroderivative gas turbine – heat rate variation with inlet air temperature.

gas fuels and *diffusion* burn for liquid fuels. They are either single fuel (gas or liquid) or dual fuel.

Prima facie, as touted widely in the trade literature, gas-fired RICE offer significant advantages over simple cycle gas turbines in that their fast start capabilities and

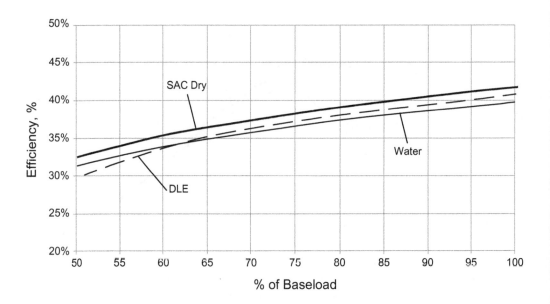

Figure 7.7 LM6000 aeroderivative gas turbine – part load efficiency.

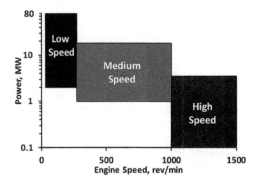

Figure 7.8 RICE landscape, Low: 60–275 rpm, Medium: Up to 1,000 rpm, High: >1,000 rpm.

broader load ranges constitute a better match to changing grid needs. In particular, they can reach full load within five minutes, and, depending on the number of units installed in the field, they can operate from 10% to 100% of the total plant load at their base efficiency (by turning units on or off as needed).

Unlike their air-breathing cousins, piston-cylinder engines are less susceptible to changes in site ambient conditions, i.e., high temperature and altitude (e.g., see Figure 7.9). This basic manifestation of their piston-cylinder construction, which restricts the amount of charge air via limited physical volume, is not surprising. Furthermore, due to their cycle advantage (namely, *constant volume combustion*), their simple cycle efficiencies can be very high, i.e., 48% net LHV or even higher in some cases (as quoted by the OEMs). There is, however, a big, proverbial *asterisk*

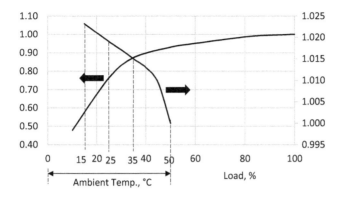

Figure 7.9 Normalized efficiency of RICE as a function of load and ambient temperature.

associated with the latter. In particular, RICE are allowed a +5% *fuel flow tolerance*, which can reduce the quoted engine efficiency by about 2.5 percentage points when measured at the generator terminals of the actual engine running in the field. Furthermore, if RICE are run on heavy fuel oil or LPG, their efficiency can be hit by about 2–4% (points).

The downside of the high simple cycle efficiency (even after making a discount for the 5% heat rate tolerance) is low combined cycle efficiency. This is not a defect per se but a direct result of the engine thermodynamic cycle, which has a very high effective cycle pressure ratio (via *constant volume* combustion inside the cylinders), i.e., the key enabler of high cycle efficiency. This, of course, results in low exhaust gas temperatures and low exhaust gas *exergy*, which, combined with the low gas flow rate, leaves little opportunity to generate additional power from the bottoming cycle (steam turbine generator).

Admittedly, to a certain extent the same problem is also present in aeroderivative gas turbines due to their high cycle pressure ratios vis-à-vis those in advanced class, large gas turbines. Nevertheless, an aeroderivative gas turbine with 41–43% simple cycle efficiency can readily reach above 50% net LHV in combined cycle. In comparison, a combined cycle with multiple RICE (45% net, 0% tolerance simple cycle efficiency) cannot go much further than 47–48% net LHV.

Exhaust gas emissions is another area where gas turbines and RICE comparisons are difficult to make on a truly even playing field. Aeroderivatives equipped with DLE combustors can guarantee NOx and CO emissions as low as 15 ppmvd down to 15% to 35% load (depending on the OEM and model). This is not only a question of technology. It involves different environmental regulations imposed on each technology and different emissions quoting criteria (not to mention the discrepancies in definition of *standard temperature and pressure* (STP) values requisite for some calculations).

Regardless of the architecture, i.e., heavy-duty industrial or aeroderivative, gas turbines equipped with dry low NOx (DLN) combustors (the term is DLE in aeroderivatives), can get down to 15–25 ppmvd NOx in their exhaust gas. In the USA,

Table 7.1 Full load NOx emissions from typical RICE and aeroderivative gas turbine

	RICE		Aero GT
Output, MW		20.4	42.7
Efficiency, %	46.6	lower	41.6
NOx @ 5% O_2, mg Nm^3	600	200	NA
NOx @ 5% O_2, ppmvd	292.4	97.5	NA
NOx @ 15% O_2, mg/Nm^3	225	75	51.3
NOx @ 15% O_2, ppmvd	109.6	36.5	25.0
CO_2, lb/MWh	716	722	807

this number is typically quoted as "dry" on a volume basis, hence the units "ppmvd," at 15% O_2 content. Without selective catalytic reduction (SCR), even with lean burn combustion technology, RICE cannot match this performance. A comparison is provided in Table 7.1. The fuel used is natural gas with a methane number of 80. The higher efficiency of RICE is reflected by better CO_2 emission numbers. Reduced NOx is achievable by engine tuning, but it comes at the cost of slightly lower efficiency by up to 0.5 percentage points and higher CO and other emissions. In both cases, the units can be equipped with downstream SCR systems to meet extremely stringent emissions requirements (e.g., single digit NOx). The bottom line, however, is that, in terms of criteria pollutants, RICE is at a disadvantage vis-à-vis gas turbines equipped with DLN/DLE combustors. This is so because, unlike the gas turbines, they also emit unburned hydrocarbons (volatile organic compounds, up to 700 mg/Nm^3) and particulate matter (e.g., PM10, about 10 mg/Nm^3).

On the plus side, vis-à-vis gas turbines, regulatory agencies usually impose higher NOx emission limits on RICE. For example, the US EPA limit on gas engines with lean burn technology is 0.1 g/hp, which translates into about 168 mg/Nm^3 at 15% O_2. World Bank 2007/2008 emissions guidelines specified 200 mg/Nm^3 at 15% O_2 for gas fired, spark-ignition engines. The number from the Integrated Pollution Prevention and Control (IPPC) directive of the European Union for the same is 75 mg/Nm^3 at 15% O_2. (You can divide the listed numbers by two to estimate the ppmvd equivalent.) In contrast, in some localities, e.g., California in the USA, single digit NOx limits (in ppmvd) are imposed on gas turbine exhaust gas emissions.

It is not surprising then to run into situations where gas fired, medium speed RICE can have higher full load emissions than gas turbines due to different regulations, e.g., 3–6 times more NOx, 10–17 times more CO, 6 times more VOC, and 3 times more PM. On top of this, depending on the type/model, aeroderivates have a minimum emissions compliance load (MECL) as low as 15–35%. It is not clear whether RICE have similar capability. (The author does not know about MECL guarantees from RICE, but this may not be the case.)

One widely advertised capability of multi-unit installations, with gas turbines or gas engines, is the ability to maintain base load efficiency at lower loads by turning off individual units successively. Much smaller gas engines can do this in a manner that approximates a straight line load-efficiency curve more closely than the larger gas

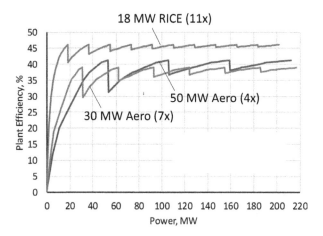

Figure 7.10 Comparison of multi-unit plant efficiency at different plant loads.

turbines can do. This is illustrated for a nominal 200 MW power plant in Figure 7.10. The situation depicted in the figure is certainly accurate – on paper, that is. In the field, as mentioned above, the emissions characteristics of RICE can be a severe impediment. The reason for that is that being off-line but ready for fast restart does not come inexpensively or free for RICE, which uses pre-warming, pre-lubrication, and intermittent slow turning to prepare for fast start. This imposes a parasitic load on RICE plants (estimated as 650 kWe for a 100 MW plant – should be verified by the OEM). Gas turbines do not have that limitation and parasitic load penalty. They can be shut down and restarted – in 10 minutes (normal) or 5 minutes (fast) – from cold iron as many times as necessary. When this is taken into account and gas engines are kept online at very low load (and high emissions) at low plant loads, the efficiency discrepancy, while not totally absent, can be not as drastic as shown in Figure 7.10. A typical comparison of RICE and aeroderivative startup speed and load characteristics is presented in Figure 7.11.

Gas-fired RICE and BESS hybrid systems are offered by OEMs like Wärtsila (under the name *Engine+ Hybrid Energy*). The system can also include a solar PV component. A conceptual system configuration is shown in Figure 7.12. Envisioned system operation in an islanded mode is depicted in Figure 7.13.

As shown in Figure 7.13, the hybrid power plant comprising the three technologies provides the load demand of a particular customer. The main generation source is the gas-fired engine component. In the middle of the day, when demand as well as solar irradiation are at their highest, solar PV contributes as much as it can and reduces the fuel burn, e.g., by turning engines off as necessary. Some of the solar power is used to charge the BESS. When there is a sudden loss of solar power due to a weather event such as cloud cover, BESS responds fast and covers up for the lost megawatts until the idle engines are restarted and brought online.

It is worth noting that the same response could be delivered by RICE without a need for BESS. This, however, requires an N + 1 or N + 2 configuration (depending on load

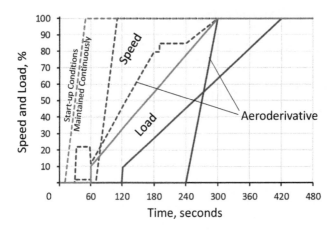

Figure 7.11 Comparison of RICE and aeroderivative GT startup speed and load.

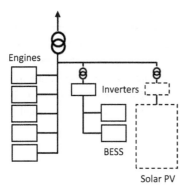

Figure 7.12 Conceptual engine plus BESS (plus solar PV) hybrid system configuration.

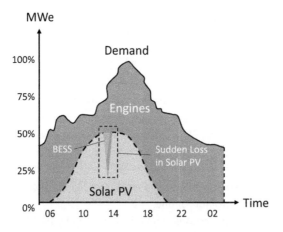

Figure 7.13 Conceptual operation of a triple hybrid system.

demand size and other needs). As an example, say, there is a 300 MW industrial facility served by a dedicated RICE plant (simple cycle). Each engine can deliver 20 MW at 100% load. Thus, nominally 300/20 = 15 engines are needed. Usually, however, one engine will off-line for maintenance at any given time. Thus, at least, 15 + 1 units must be installed in the field. Now, consider that this facility is combined with a solar PV rated at 100 MW (midday peak). At that time, then, 200/20 = 10 engines must be running at 100% load to meet the load demand.

However, as is well known, solar irradiation is finicky. Spinning reserve must be available when, suddenly, 20 MW solar power is lost. (To be fair, with advanced warning from weather stations and other forecasting methods, this magnitude of loss cannot be that sudden or unexpected.) One way to accomplish this is to run 11 engines at 200/11 = 18.2 MW or 90.9% load each. When 20 MW from solar PV is lost, they ramp up to 100% load very fast and deliver 220 MW to maintain the 300 MW total plant output.

8　Hydrogen

8.1　Basics

Hydrogen as a carbon-free fuel is amenable to utilization in all heat engines, including gas turbines and reciprocating internal combustion engines, which are the most efficient technologies for electric power generation from fossil fuels. If readily available, existing combustion technology can make use of H_2 today with no difficulty. (In fact, there are hundreds of E and F Class gas turbines in refinery and steel mill applications burning H_2 blended with natural gas at fractions varying from a few percent (by volume) to up to 80–90%.)

Alas, H_2 is not readily available; to be precise, it is not an energy *resource* – it is an energy *carrier*. As such, it must be produced first, then transported or stored to be used as a primary fuel for power generation. As will be dramatically clear from the short discussion below, the price of achieving this goal at an industrial scale is exorbitant, physically and economically. For instance, for each 1 MWh of electricity generated by a state of the art gas turbine burning 100% H_2 fuel, 4.5 MWh electricity consumption is required to generate the requisite amount of H_2 via electrolysis. A brief look at the generation portfolio of Germany in 2016 revealed that curtailment resulting from increasing wind power share to 50% overnight (from the current 17%) can only provide energy for H_2 production to supply only *two* 900 MWe gas turbine combined cycle plants running at a 40% capacity factor.

Production of hydrogen via *steam methane reforming* (SMR) with *carbon capture and storage* (CCS) is another technologically mature option widely used in the chemical process industry. In this scenario, CCS is pushed upstream (from the power plant stack end) to the fuel production step. Natural gas remains the primary fossil fuel and H_2 is relegated to the role of energy carrier. In a more challenging variant, SMR is replaced by coal gasification to produce H_2. Either variant is exorbitantly expensive (without even accounting for transmission and distribution pipeline and storage) and highly unlikely to compete with the best-available alternative today: advanced class *gas turbine combined cycle* (GTCC) with post-combustion CCS using amine-based chemical absorption technology.

Hydrogen is the lightest element in the periodic table and the most abundant chemical substance in the universe, constituting roughly 75% of all normal atomic matter. Furthermore, hydrogen fuel (H_2) is a zero-emission fuel when burned with oxygen, i.e.,

$$2H_2(g) + O_2(g) \rightarrow 2H_2O(g) + \text{energy}.$$

Consequently, if used as a fuel in heat engines such as gas turbines, the combustion of hydrogen does *not* produce carbon dioxide! In fact, if combusted with pure oxygen, the only combustion product is water vapor. It is noteworthy that, at nearly 120,000 kJ/kg, the lower heating value (LHV) of H_2 as a fuel is more than twice that of natural gas.

Perfect! Well, not exactly because of the following inconvenient facts:

- Pure H_2 and pure O_2 do not occur naturally in large quantities.
- A primary energy input is required to produce both on an industrial scale.
- The combustion of hydrogen with ambient air generates a significant amount of *nitrous oxides* (NOx), a *criteria pollutant*.

As already stated, H_2 is not an energy *resource*, it is an energy *carrier*. Prior to its use as a fuel, it must be produced, stored, or transported. There are significant problems associated with all three phases of the hydrogen fuel chain. These aspects will be discussed qualitatively and quantitatively in the remainder of this chapter.

8.2   Production Technologies

The "color coding" of hydrogen production technologies is as follows:

- **Green hydrogen** is produced by water electrolysis: water is split into hydrogen and oxygen by an electric current and with the help of an electrolyte. If the electricity required for electrolysis comes exclusively from renewable, CO_2-free sources, the entire production process is completely CO_2-free.
- **Blue hydrogen** is generated from fossil fuels and is CO_2-neutral. The CO_2 is separated and stored or reused (CCUS).
- **Gray hydrogen** is obtained from fossil fuels. For example, natural gas is converted under heat into hydrogen and CO_2 (SMR). Approximately nine tons of CO_2 are generated to produce one ton of hydrogen from methane.

There are two technologies for producing H_2 that are commercially proven: *electrolysis* and *SMR*. In electrolysis, electricity is run through water to separate the hydrogen and oxygen atoms. The electricity requirement for H_2 production via electrolysis is very high, in the neighborhood of 50–70 kWh/kg. At an average value of 60 kWh/kg, this translates to 216 MWe per kg/s of H_2. Consider a modern gas turbine rated at 300 MWe and 40% of net LHV efficiency. The hydrogen consumption of this gas turbine would amount to about 6 kg/s, which would require 1,300 MWe of electricity to produce it from water (consumed at a rate of about 50,000 gallons per hour)! For a complete picture covering the full range of gas turbine output capacities, refer to the chart in Figure 8.1. Clearly, this is not a feasible scenario if the requisite electricity is to be supplied by the grid. Apart from the significantly negative power (and revenue) balance and water consumption, there is the problem of emissions (NOx, SOx, CO_2, etc.)

from the fossil fuel–fired generation resources supplying the grid. Even if one totally disregards CAPEX and size concerns, straightforward electrolysis is seemingly a proverbial long shot for a utility-scale electric power perspective.

Nevertheless, it behooves this introductory discourse to provide an idea about the cost and size of electrolysis technologies. Three main electrolyzer technologies are used or are being developed today. Of those three, only *alkaline* electrolyzers (ALK), which have been used by industry for nearly a century, can be considered fully mature. *Proton exchange membrane* electrolyzers (PEM) are commercially available today and are rapidly gaining market traction as, along with other factors, they are more flexible and tend to have a smaller footprint. *Solid oxide electrolyzers* hold the potential of improved energy efficiency but are still in the development phase and, unlike ALK and PEM, work at high temperatures. The techno-economic characteristics of ALK and PEM electrolyzers are summarized in Table 8.1.

Table 8.1 Techno-economic characteristics of ALK and PEM electrolyzers[1]

	ALK		PEM	
	2017	2025	2017	2025
Efficiency, kWh/kg	51	49	58	52
Efficiency (LHV), %	65	68	57	64
Lifetime stack, op. hrs	80,000	90,000	40,000	50,000
CAPEX, EUR/kW	750	480	1,200	700
OPEX, % CAPEX/yr		2		
Stack replacement, EUR/kW	340	215	420	210
Output pressure, bar	1	15	30	60
System lifetime, years	20		20	

[1] Hydrogen from Renewable Power – Technology Outlook for the Energy Transition, International Renewable Energy Agency (IRENA), September 2018.

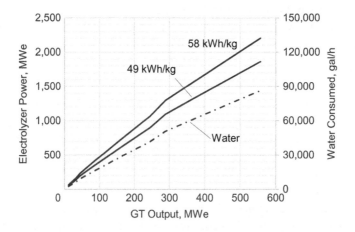

Figure 8.1 Electricity and water consumption for H_2 production via electrolysis (at 49 and 58 kWh/kg) as a function of gas turbine output (100% H_2 fuel).

A recent idea in vogue, under the name *power-to-hydrogen*, is to utilize carbon-free renewable electricity to produce hydrogen. In this scenario, hydrogen becomes an *energy storage* medium. Recall that solar and wind are not *dispatchable* generation technologies. In other words, even when there is no demand for their full capacity, without curtailment they would simply keep going and generating electric power. The logical solution to this problem is the storage of "free" electricity generated by such resources for future use. One obvious and straightforward option is using lithium-ion batteries – apart from their own cost and size/capacity issues, of course. Another option is to utilize the "free" renewable electric power to produce H_2 via electrolysis, which will be looked more closely below.

8.2.1 Thermodynamics of Electrolysis

Electrolysis is a technique that uses direct electric current (DC) to drive an otherwise non-spontaneous chemical reaction. The electrolysis of one mole of water produces one mole of hydrogen gas and half a mole of oxygen gas in their normal diatomic forms. The reaction formula is as follows:

$$H_2O(l) \rightarrow H_2(g) + 0.5\,O_2(g).$$

If the process takes place at *standard state* (SS), i.e., 298.15 K (25°C) and one bar (100 kPa), the process and the pertinent thermodynamic data for the chemical reaction is as shown in Figure 8.2.

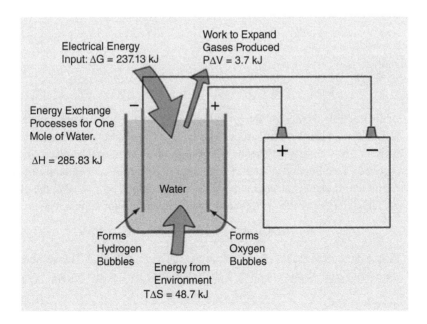

Figure 8.2 Schematic description of the simplified water electrolyzer.

The reactions taking place at the *anode* (oxidation) and *cathode* (reduction) at SS are as follows:

$$2\,H_2O(l) \rightarrow O_2\,(g) + 4\,H^+(aq) + 4e^- \qquad E^0 = +1.23\,V,$$

$$2\,H^+(aq) + 2e^- \rightarrow H_2(g) \qquad\qquad E^0 = 0.00\,V.$$

Consequently, for the electrolysis cell, one can write the following equality:

$$\text{cell } E^0 = \text{cathode } E^0 - \text{anode } E^0 = -1.23\,V.$$

The *Gibbs free energy* for the electrolysis reaction is given by

$$\Delta G^0 = -n \cdot F \cdot E^0,$$

where the *Faraday constant*, $F = 96,485.3321233$ C/mol, and $n = 4$ (i.e., two H_2 molecules formed from two H_2O molecules). Thus, $\Delta G^0 = 4 \times 96.485 \times 1.23 = 474.48$ kJ per two moles of water, or 237.24 kJ/mol of water. The enthalpy of water at SS is -285.83 kJ/mol so that the enthalpy change of the reaction is $\Delta H^0 = 285.83$ kJ/mol. (Note that, by definition, the enthalpies of H_2 and O_2 at SS are 0 kJ/mol.) From the definition of the Gibbs free energy,

$$\Delta G^0 = \Delta H^0 - T\Delta S^0,$$

$$237.24 = 285.83 - T\Delta S^0,$$

$$T\Delta S^0 = 285.83 - 237.24 = 48.59\,kJ/mol.$$

Note that the entropy of H_2O, H_2, and O_2 at SS is 69.91 J/K-mol, 130.68 J/K-mol, and 205.14 J/K-mol, respectively. Thus, the entropy change for the electrolysis reaction is

$$\Delta S^0 = 130.68 + 205.14/2 - 69.91 = 163.64\,J/K\text{-mol},$$

so that

$$T\Delta S^0 = 298.15\,K \times 163.64\,J/K\text{-mol}/1,000 = 48.7\,kJ/mol.$$

In the electrolysis process shown in Figure 8.2, two things happen: (1) dissociation of H_2O and (2) expansion of gaseous H_2 and O_2. The electrolysis does not happen on its own, i.e., it is not a spontaneous reaction. In order for it to take place, energy input is required. This energy input translates into a change in the enthalpy of the system. Using the fundamental thermodynamic relationship $H = U + PV$, the change in the enthalpy can be written as (ignoring the superscript 0 for simplicity)

$$\Delta H = \Delta U + \Delta PV + P\Delta V = \Delta U + P\Delta V.$$

(Since the process takes place at constant pressure, $\Delta P = 0$.) The second term on the right-hand side of this equation is the work done by the system, i.e.,

$$W = P\Delta V$$

$$= 101,325\,Pa \times 1.5\,mol \times 22.4 \cdot 10^{-3}\,m^3/mol \times (298.15\,K/273.15\,K)$$

$$= 3,715\,J = 3.715\,kJ.$$

(Note that the electrolysis process generates 1 mol H_2 and 0.5 mol O_2 gas, i.e., a total of 1.5 mol of gas from 1 mol of H_2O. At 273.15 K and 1 atm, 1 mol of gas occupies a volume of 22.4 liter.) Consequently, the change in the internal energy of the system is found from

$$\Delta U = \Delta H - P\Delta V = 285.83 \text{ kJ} - 3.72 \text{ kJ} = 282.1 \text{ kJ (per mol of } H_2O).$$

In conclusion, the change in enthalpy represents the necessary energy to accomplish the electrolysis. It is equal to the change in the internal energy of the system *after* subtracting the work done during the expansion of the gases produced. Where does this energy, i.e., $\Delta H = 285.83$ kJ/mol, come from? As shown in Figure 8.2, it comes in two forms: electrical energy and heat. The latter, $\Delta Q = T\Delta S = 48.7$ kJ/mol, is provided from the environment at temperature $T = 298.15$ K. The remainder, i.e., the electrical energy, is supplied by the battery and is equal to the change in the Gibbs free energy:

$$\Delta G = \Delta H - T\Delta S = 285.83 - 48.7 = 237.1 \text{ kJ/mol}.$$

This is the minimum theoretical energy required for the electrolysis of water to take place. Thus, 237.1 MJ is required to produce 1 kmol, i.e., 2 kg, of H_2 from 1 kmol, i.e., 18 kg, of H_2O. Alternatively, 118.55 MW power is required to produce H_2 at a rate of 1 kg/s, which translates into 118.55/3.6 = 32.93 ~ 33 kWh/kg of H_2. This is the theoretical yardstick that one must use to gauge the claims made by electrolyzer manufacturers. For instance, claiming to have an electrolyzer that can electrolyze water to make hydrogen with a 33 kWh/kg energy consumption is akin to having built a heat engine with the efficiency of a Carnot engine, i.e., an impossibility. In terms of heat engines, e.g., the advanced class gas turbines, the state of the art in terms of a "technology factor" is about 0.70–0.75. In other words, at the present stage of human technological development, we can make "real" heat engines, whose thermal efficiency is 70–75% of that achievable in a theoretical Carnot engine operating between the same temperature "reservoirs." Thus, what one should expect from a "mature" electrolyzer technology is at best 33/0.75 = 44 kWh/kg. A more likely number is most likely to be around 0.33/0.7 = 47 kWh/kg. It is encouraging to note that the numbers listed in Table 8.1 are in concurrence with this assertion.

8.2.2 Sulfur-Iodine Cycle

There is another method for producing hydrogen and oxygen from water. Water thermally dissociates at significant rates into hydrogen and oxygen at temperatures approaching 4,000°C. Trying to do it this way is obviously not a practical proposition. The sulfur-iodine (SI) process is developed to achieve the same result at much lower temperatures. The SI cycle consists of three chemical reactions that result in the dissociation of water into hydrogen at oxygen:

Bunsen Reaction (T ~ 120°C): $I_2 + SO_2 + 2H_2O \rightarrow 2HI + H_2SO_4,$

H_2SO_4 Decomposition (T > 800°C): $H_2SO_4 \rightarrow SO_2 + H_2O + \frac{1}{2}O_2,$

HI Decomposition (T > 350°C): $2HI \rightarrow I_2 + H_2,$

Net: $H_2O \rightarrow H_2 + \frac{1}{2}O_2.$

All three reactions are operated under conditions of chemical equilibrium. *Sulfuric acid* (H_2SO_4) and *hydrogen iodide* (HI) decomposition reactions are *endothermic* and thus require heat input. Sulfuric acid decomposition takes place in two steps and the second step requires a catalyst. Heat at about 120°C is rejected from the exothermic Bunsen reaction. Except for water, all reactants are regenerated and recycled.

The cycle chemistry is well understood. In addition to the three primary reactions described above, the actual process/cycle requires a significant number of separation and distillation steps. They are described in detail in a report by Russ [1]. The estimated technology readiness level (TRL) for this technology is probably not higher than 5 at the time of writing (TRL was stated to be 4 in Ref. [1]). The predicted efficiency of this process is about 45% (LHV basis), which is worse than electrolysis (about 60% LHV at 55 kWh/kg). This efficiency is only possible at a large scale because increased thermal integration and lower heat losses from larger equipment can be readily provided by large power plants, e.g., nuclear power plants with high-temperature reactors. In such a plant, the production of 760 mt/day of H_2 could be possible with 2,400 MWth heat from the reactor [1]. The same production rate would require roughly 1,750 MWe of electricity with electrolysis.

In February 2010, the *Japan Atomic Energy Agency* (JAEA) established the HTGR *Hydrogen and Heat Application Research Center* at Oarai to progress operational technology for hydrogen production via the SI cycle. The facility has demonstrated laboratory-scale and bench-scale hydrogen production via the SI process of up to 30 liters per hour. A pilot plant test project producing hydrogen at 30 m^3/h utilizing a 400 kWth heat exchanger with helium (reactor coolant) was used to test the engineering feasibility of the SI cycle. An SI cycle plant producing 1,000 m^3/h of H_2 was to be linked to the JAEA's *High Temperature Engineering Test Reactor* (HTTR) to confirm the performance of an integrated production system, envisaged for the 2020s. In 2014, H_2 production at up to 20 liters per hour was demonstrated. In January 2019, it used the HTTR to produce H_2 using the SI cycle over 150 hours of continuous operation. JAEA aims to produce hydrogen at less than $3/kg by about 2030 with very high temperature reactors.[1]

Another hydrogen production technology closely associated with nuclear energy is *methane pyrolysis*, which is covered in Section 9.7.

8.2.3   Steam Methane Reforming (SMR)

The largest share of global hydrogen demand is from the chemicals sector for production of ammonia and in refining for hydrocracking and desulfurization of fuels.

---

[1] From the website of World Nuclear Organization.

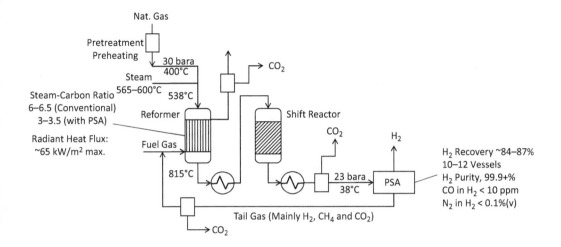

Figure 8.3 Steam methane reforming (SMR) process with pressure swing adsorption (PSA).

Currently, SMR is the leading technology for producing hydrogen in large enough quantities to meet that demand. The SMR process is more than a century old and quite simple.[2] At high temperatures (700–1,100°C) and in the presence of a metal-based catalyst (nickel), steam reacts with methane (CH_4) to yield carbon monoxide and hydrogen according to the (endothermic) chemical reaction

$$CH_4 + H_2O \rightleftharpoons CO + 3\,H_2, \qquad (\Delta H° = 206.1 \text{ kJ/mol}).$$

Additional H_2 is obtained by reacting CO with water via the (exothermic) *water-gas shift reaction:*

$$CO + H_2O \rightleftharpoons CO_2 + H_2, \qquad (\Delta H° = -41.2 \text{ kJ/mol}).$$

It should be noted that steam reforming is not limited to methane (natural gas). The process can be applied to LPG or naphta as well.

A typical progress diagram is shown in Figure 8.3. Natural gas (free from sulfur and other contaminants) is fed to the reformer, which converts the feed and steam to syngas (mainly hydrogen and carbon monoxide) in an endothermic reaction at high temperature (~1,500–1,600°F) and moderate pressure (~300–400 psia). The requisite energy is provided by additional fuel gas (natural gas or refinery gas) combustion. Hot syngas from the reformer is cooled in a heat recovery heat exchanger and fed to the CO shift reactor. In earlier versions, the shift reaction took place in a two-stage process (high and low temperature shift reactors at 650–700°F and 450°F, respectively). The exothermic water-gas shift reaction converts CO and steam to CO_2 and H_2. In modern systems, hydrogen from the shift reactor(s) is purified in a *pressure swing adsorption*

---

[2] See US Patent 1,128,804 (February 16, 1915), *Process of Producing Hydrogen*, by Alwin Mittasch and Christian Schneider, assigned to Badische Anilin & Soda Fabrik (BASF) in Germany.

(PSA) unit to achieve the final product purity (99+% in newer designs).[3] Tail gas from the PSA is recycled back to the SMR and can provide about 90% of the fuel requirement for the reformer furnace. Older systems used amine-based CO_2 stripping. Residual CO and CO_2 were converted to methane in a fixed-bed, catalytic methanation reactor (700–800°F).

As shown in Figure 8.3, CO_2 could be captured at three locations [2]: from the PSA tail gas or from the reformer flue gas with about 90% efficiency (~45% (v) and ~20% (v) concentration, respectively, and less than 1 bar partial pressure) or from the raw H_2 at the shift reactor exit with 99+% efficiency (~15% (v) concentration and ~3.5 bar partial pressure). The removal technologies include amine-based scrubbing, physical solvents, and membranes and there is widespread experience in chemical process industry in CO_2 removal from raw hydrogen at high pressure. CO_2 scrubbing from the flue gas at low partial pressures and high-volume flows requires larger and more expensive equipment and consumes more parasitic power (e.g., see the discussion of post-combustion capture in Section 11.1).

As the chemical reaction formulae clearly show, using SMR for large-scale H_2 production to fuel carbon-free electric power generation is counterproductive. It generates a significant amount of CO_2, which must be captured and stored and, prima facie, defeats the purpose in making hydrogen in the first place. To provide a quantitative picture, consider the 300 MWe gas turbine example above. From the chemical reaction formulae, on a theoretical basis, each kilogram of H_2 produced comes with 5.5 kg of CO_2 production. In the actual SMR plant, one must include the fuel burn in the reformer as well. This can raise the CO_2/H_2 ratio to about 9.2. With the latter number as a benchmark, using SMR to supply the hydrogen fuel to that unit's combustor would generate about 200,000 kg/h of CO_2 corresponding to *1.6 million tonnes* per year (at 8,000 hours per year). On a simple cycle basis, this corresponds to about 675 kg/MWh (about 1,500 lb/MWh) CO_2 generation. If the gas turbine is in a combined cycle configuration, CO_2 emissions are about 50% lower, i.e., about 450 kg/MWh.

This last number deserves closer scrutiny, which will be done with the help of recent data. Based on the US EIA statistics, average CO_2 emissions from natural gas-fired electricity production in the USA were about 400 kg/MWh (vis-à-vis roughly 1,000 kg/MWh from coal) in 2015. For the most advanced H or J Class gas turbine combined cycles with 60–61% net LHV efficiency, the CO_2 emission number is around 330 kg/MWh. Consequently, at about 450 kg/MWh, CO_2 emission from SMR is about 35% higher on a comparable technology basis. On average, carbon capture from SMR in refinery applications costs (in $/tonne) 30% less vis-à-vis post-combustion capture from natural gas–fired combined cycle power plants. Combining the two deltas, hydrogen from SMR with carbon capture can be at a par with

---

[3] PSA is based on the physical binding of gas molecules to a solid absorbent such as activated carbon, silica gel, zeolites, or a combination thereof. Separation is based on the fact that highly volatile compounds with low polarity such as H_2 are not adsorbed at all. The process takes place at constant temperature (no heating or cooling) because adsorption and desorption are driven by the effect of alternating pressure and partial pressure.

post-combustion carbon capture (everything else being equal). Based on these assumptions, a quick calculation shows that for 100%(v) H_2 firing, carbon capture from the SMR plant is about 6–10% cheaper in CAPEX vis-à-vis post-combustion capture from natural gas–fired GTCC.

This is demonstrated in a detailed study using an advanced class gas turbine in simple and combined cycle modeled in Thermoflow, Inc.'s GTPRO software. The base case is firing with typical natural gas. Hydrogen blending with natural gas fuel is carried out in steps until it is 100% H_2 firing. The gas turbine output is kept the same as in the base case. The results are summarized in Table 8.2. SMR CO_2 generation is estimated as a function of H_2 consumption of the gas turbine (9.17:1 mass ratio). The capture cost is summarized in Table 8.3.

The results are slightly different when the gas turbine firing temperature is kept constant and the gas turbine is run at full load. The results from that series of runs are summarized in Table 8.4. The capture cost is summarized in Table 8.5.

From a performance perspective, the reader is cautioned that these calculations are to be treated as guesstimates. Of the two approaches, the most likely one is the constant GT output case with the turbine inlet temperature (TIT) being reduced with increasing fuel H_2 content. It is even likelier that, depending on an OEM's particular combustor design, a further reduction in TIT is required to prevent flashback with a safety margin. In that case, the gas turbine output with H_2 firing will be lower than the base 100% NG–fired case. Furthermore, with reduced TIT and exhaust temperature, the steam turbine output will suffer as well so that the combined cycle output will also be affected negatively. The exact nature of the gas turbine performance variation with increasing levels of fuel H_2 content can only be obtained from the OEM.

The cost calculations presented above do not include the capital investment required for constructing the SMR facilities and pipelines to supply H_2 fuel to the GTCC power plants. In a 2017 study,[4] the CAPEX for SMR with 90% CCS was estimated at \$3,000 per Nm^3/h of H_2, which comes to about \$750 million for the example gas turbine burning 750 MWth H_2 fuel. For a 450 MWe GTCC power plant, the budgetary price from the *Gas Turbine World 2019 Handbook* is \$300 million. In other words, the total cost of the GTCC+SMR+CCS is about \$1 billion, or 200% higher than that for the base GTCC with no CCS.

One can certainly envision a few cases where the fuel production is in situ and pipeline cost can be ignored. This is unlikely to be the case for a widespread deployment of the technology for (almost) carbon-free power generation from natural gas (via H_2 as the energy *carrier*).

8.2.4   Autothermal Reforming

Autothermal reforming (ATR) is an oxy-combustion process, which combines steam reforming (endothermic) with *partial oxidation* (POX), which is exothermic. In other

---

[4] IEAGHG Technical Report 2017-02, *Techno-Economic Evaluation of SMR Based Standalone (Merchant) Hydrogen Plant With CCS*, February 2017.

Table 8.2 Hydrogen co-firing (constant gas turbine output)[1]

		BASE		H₂ Co-Firing Cases						
Fuel NG	%(v)	100	85	70	50	30	20	10	0
Fuel H₂	%(v)	0	15	30	50	70	80	90	100
H₂ Flow	kg/h	0	1,063	2,410	4,885	8,728	11,589	15,568	21,478
Fuel Cons. (LHV)	kWth	742,296	740,762	739,096	736,024	731,316	727,891	723,198	716,323
GT Output	kWe	312,358	312,358	312,358	312,358	312,358	312,358	312,358	312,358
GT Efficiency	%	42.08	42.17	42.26	42.44	42.71	42.91	43.19	43.61
GT CO₂	lb/MWh	1,064	1,011	944	820	628	488	294	7
	lb/hr	332,208	315,846	294,906	256,114	196,246	152,372	91,815	2,183
GT Exhaust Flow	lb/s	1,359	1,358	1,357	1,354	1,351	1,348	1,344	1,338
GT Exhaust Temp	F	1,156	1,154	1,150	1,144	1,134	1,127	1,118	1,104
STG Output	kW	155,840	155,191	154,231	152,469	149,790	147,856	145,228	141,428
CC Gross Output	kW	468,198	467,549	466,589	464,827	462,148	460,214	457,586	453,786
Aux. Power	kW	7,491	7,481	7,465	7,437	7,394	7,363	7,321	7,261
CC Output	kWe	460,707	460,069	459,124	457,390	454,753	452,850	450,265	446,525
CC Efficiency	%	62.07	62.11	62.12	62.14	62.18	62.21	62.26	62.34
GTCC CO₂	lb/MWh	721.1	686.5	642.3	559.9	431.5	336.5	203.9	4.9
GTCC CO₂	kg/MWh	327.1	311.4	291.4	254.0	195.7	152.6	92.5	2.2
SMR CO₂	kg/h	150,687	143,265	133,767	116,171	89,016	69,115	41,646	990
	kg/h	0	9,749	22,103	44,792	80,037	106,271	142,762	196,949
	kg/MWh	0.0	21.2	48.1	97.9	176.0	234.7	317.1	441.1
SMR+GTCC CO₂	kg/MWh	327.1	332.6	339.5	351.9	371.7	387.3	409.6	443.3
SMR+GTCC CO₂	lb/MWh	721.1	733.2	748.5	775.8	819.6	853.8	902.9	977.3
	kg/h	150.7	153.0	155.9	161.0	169.1	175.4	184.4	197.9

[1] Note that for the case with 100% hydrogen fuel, CO_2 in the gas turbine exhaust (990 kg/h) is nonzero because it is the CO_2 that comes in with the ambient air and, thus, strictly speaking it should not be considered as CO_2 "emission," i.e., CO_2 created by the combustion process. Thus, for a precise accounting of combustion-created CO_2 emissions, that number should be subtracted from all the cases. This is done in Table 8.4.

Table 8.3 Capture cost (5,000 hours per year, $110/tonne GTCC PCC, $75/tonne SMR capture) – constant GT output (MWe)

Fuel H$_2$, % (v)	0	15	30	50	70	80	90	100
Annual CO$_2$ Emissions, mt	753,436	765,074	779,351	804,818	845,261	876,929	922,044	989,696
Capture CO$_2$, mt	678,092	688,566	701,416	724,336	760,735	789,236	829,840	890,726
Capture Cost, $/mt	$110	$108	$105	$100	$93	$89	$83	$75
Capture CAPEX, MM$	$74.59	$74.21	$73.67	$72.62	$71.08	$70.08	$68.80	$66.96
Delta, %		−0.51	−1.23	−2.64	−4.71	−6.05	−7.77	−10.2

Table 8.4 Hydrogen co-firing (constant firing temperature at 100% load)

See the footnote to Table 8.3

		BASE		H₂ Co-Firing Cases					
Fuel NG	%(v)	100	85	70	50	30	20	10	0
Fuel H₂	%(v)	0	15	30	50	70	80	90	100
H₂ Flow	kg/h	0	1,066	2,424	4,940	8,901	11,893	16,115	22,518
Fuel Cons. (LHV)	kWth	742,296	742,870	743,387	744,354	745,871	747,008	748,601	751,020
GT Output	kWe	312,358	313,358	314,405	316,360	319,418	321,690	324,871	329,678
GT Efficiency	%	42.08	42.18	42.29	42.50	42.82	43.06	43.40	43.90
CO_2	lb/MWh	1,057	1,004	936	812	620	479	286	0
	lb/hr	330,025	314,554	294,420	256,787	197,893	154,109	92,773	0
GT Exhaust Flow	lb/s	1,359	1,358	1,357	1,355	1,351	1,348	1,345	1,339
GT Exhaust Temp	F	1,156	1,156	1,155	1,154	1,151	1,149	1,147	1,143
STG Output	kW	155,840	155,787	155,440	154,795	153,794	153,057	152,033	150,506
CC Gross Output	kW	468,198	469,146	469,845	471,155	473,212	474,747	476,904	480,184
Aux. Power	kW	7,491	7,506	7,518	7,538	7,571	7,596	7,630	7,683
CC Output	kWe	460,707	461,639	462,327	463,616	465,640	467,151	469,273	472,501
CC Efficiency	%	62.07	62.14	62.19	62.28	62.43	62.54	62.69	62.91
GTCC CO_2	lb/MWh	716.3	681.4	636.8	553.9	425.0	329.9	197.7	0.0
GTCC CO_2	kg/MWh	324.9	309.1	288.9	251.2	192.8	149.6	89.7	0.0
	kg/h	149,697	142,679	133,547	116,477	89,763	69,903	42,081	0
SMR CO_2	kg/h	0	9,777	22,231	45,298	81,623	109,055	147,774	206,487
	kg/MWh	0.0	21.2	48.1	97.7	175.3	233.4	314.9	437.0
SMR+GTCC CO_2	kg/MWh	324.9	330.3	336.9	348.9	368.1	383.1	404.6	437.0
	lb/MWh	716.3	728.1	742.8	769.3	811.4	844.6	891.9	963.4
	kg/h	149.7	152.5	155.8	161.8	171.4	179.0	189.9	206.5

Table 8.5 Capture cost (5,000 hours per year, $110/tonne GTCC PCC, $75/tonne SMR capture) – constant firing temperature at 100% load

Fuel H_2, %(v)	0	15	30	50	70	80	90	100
Annual CO_2 Emissions, mt	748,485	762,283	778,891	808,872	856,930	894,789	949,278	1,032,436
Captured CO_2, mt	673,637	686,055	701,002	727,985	771,237	805,310	854,351	929,192
Capture Cost, $/mt	110	108	105	100	93	89	83	75
Capture CAPEX, MM$	74.10	73.93	73.61	72.94	71.98	71.41	70.70	69.69
Delta, %		−0.23	−0.66	−1.56	−2.86	−3.63	−4.58	−6.0

words, unlike SMR where the oxidant O_2 for the combustion process (to generate steam) comes from air, in ATR, the oxidant is pure O_2. In POX, the amount of O_2 is not enough to completely oxidize the feedstock, e.g., methane, completely to CO_2 and H_2O. With less than the stoichiometric amount of oxygen available, the reaction products contain primarily hydrogen and carbon monoxide. Steam (H_2O) is used as moderator. The reaction equations are given as

$$POX : CH_4 + 0.5\,O_2 \rightarrow 2\,H_2 + CO, \Delta H = -35.2\ kJ/mol,$$

$$SMR : CH_4 + H_2O \rightarrow 2\,H_2 + CO, \Delta H = 206.1\ kJ/mol.$$

The H_2:CO ratio in the syngas is 2.5:1. The syngas temperature is $950-1{,}100°C$. The syngas pressure is 30–50 bar but can be as high as 100 bar.

The main advantages of ATR vis-à-vis SMR are the increased energy efficiency, faster startup times, and faster response times to transient events. Developed in 1950s, ATR is used to generate syngas for ammonia (NH_3) and methanol synthesis. In the case of NH_3 production, where high H_2/CO ratios are needed, the autothermal reforming process is operated with steam. In the case of methanol synthesis, the required H_2/CO ratio is provided by manipulating the carbon dioxide recycle.

The typical ATR process comprises three steps:

1. First, the feed gas reacts with oxygen (partial oxidation) and steam to produce syngas.
2. Second, the syngas enters a catalyst bed for further reforming to achieve a high yield reaching thermodynamic equilibrium.
3. Finally, the syngas stream is cooled in a process gas boiler, generating high-pressure steam that can be exported to neighboring units or used for power generation.

For hydrogen production, just like in the SMR process, water-gas shift (WGS) reaction is added to convert CO to CO_2 followed by H_2 purification. Prima facie, POX and WGS reactions should be sufficient for H_2 production. For one, the POX reactor is more compact than an SMR reactor, in which heat must be added indirectly via a heat exchanger. Furthermore, the efficiency of the POX unit is relatively high (70–80%). However, POX systems are typically less energy efficient than SMR because of the higher temperatures involved (which exacerbates heat losses) and the problem of heat recovery. As discussed in the preceding section, in SMR, heat can be recovered from the flue gas to raise steam for the reaction, and the PSA purge gas can be used as a reformer burner fuel to help provide heat for the endothermic steam reforming reaction. In a POX reactor, in which the reaction is exothermic, the energy in the PSA purge gas cannot be recovered effectively to the same extent.

This is why ATR, combining the advantages of SMR and POX, can be the better route to H_2 production. With the right mixture of input fuel, air, and steam, the POX reaction can supply all the heat needed to drive the catalytic steam reforming reaction. In addition, excluding the air separation unit (ASU) to manufacture the requisite O_2, ATR is likely to be smaller in size and complexity with a lower CAPEX

Table 8.6 Comparison of SMR and ATR vis-à-vis gas-fired GTCC

		GTCC	SMR	ATR
CH_4	lb/h	1	1	1
CO_2	lb/h	2.75	4.58	2.75
H_2	lb/h	NA	0.5	0.375
CO_2/H_2		NA	9.17	7.3
Heat Consumption	kWth	6.31	7.56	5.67
Specific CO_2 (Heat Cons.)	kg/MWh	198	275	220
GTCC Efficiency	%	61	62	62
GTCC Output	kWe	3.86	4.71	3.53
Specific CO_2 (Power Out.)	lb/MWh	713	973	778
	kg/MWh	324	441	353

(vis-à-vis SMR). This also explains why ATR is more efficient than POX, where excess heat is not easily recovered. In terms of CO_2 generation, ATR is also likely to be advantageous with 7.3 kg of CO_2 for each kg of H_2 produced (per the theoretical reaction formulas). Unlike the SMR process, since no extra gas-fired furnace is needed, the actual ratio may not be too far from this value.

To have a side-by-side comparison, a basic table has been constructed with the assumption of a unit flow rate of methane feedstock, i.e., 1 lb/h. Gas-fired GTCC performance and specific CO_2 emission provides the base. Using the data from Table 8.2, SMR-produced H_2-fired GTCC performance is compared with ATR-produced H_2-fired GTCC performance and CO_2 emissions. The results are summarized in Table 8.6. The objective is to provide a simple, clean picture devoid of obfuscating assumptions and caveats. Clearly, from an electric power generation perspective, *blue* hydrogen schemes are unlikely to be a match for natural gas–fired GTCC with post-combustion capture.

For an in-depth technical and economic comparison of SMR, ATR, and POX for hydrogen production, the reader is pointed to *Concepts for Large Scale Hydrogen Production*, by Jakobsen and Åtlar (MS, thesis, Norwegian University of Science and Technology, 2016).[5] In addition to the three basic technologies, the authors also considered what they called SMR+ (using H_2 as fuel in the furnace), electrolysis, and combined ATR and electrolysis as H_2 production options (CRE). In CRE, O_2 produced by the electrolysis of water to make H_2 replaces O_2 from the ASU in the ATR process. The thesis study defined large-scale H_2 production as 500 tons/day, which translates into 5.8 kg/s or about 21,000 kg/h. From Table 8.2, this is roughly equivalent to the amount of fuel consumed by a 300 MW (nominal) gas turbine if it burns 100%(v) H_2. Basically, all options ended up with a hydrogen price between 1.5 and 2 EURO per kg (including capture). The conclusion of the study was that SMR was the most cost-effective solution for large-scale, "carbon-lean" hydrogen production (at least in Norway). However, since SMR still emitted three times as much CO_2

---

[5] https://ntnuopen.ntnu.no/ntnu-xmlui/handle/11250/2402554

as ATR (in tons/day), with a sufficiently high carbon tax, the situation can be easily reversed in favor of the latter.

8.2.5   Coal Gasification

Hydrogen can also be produced via coal gasification. About 95% of America's hydrogen is produced from natural gas via SMR (of the order of 10 million tonnes) and the rest from gasification (mostly) and electrolysis. Globally, hydrogen production is about 70 million metric tonnes with about 75% from SMR and the rest mainly from gasification. In the gasification process, the first step is reacting coal with oxygen and steam under high pressures and temperatures to form *synthesis gas* (syngas) a mixture consisting primarily of carbon monoxide and hydrogen via the following generic reaction (unbalanced):

$$CH_{0.8} + O_2 + H_2O \rightarrow CO + CO_2 + H_2 + \text{other species.}$$

After the impurities are removed from the syngas, the carbon monoxide in the gas mixture is reacted with steam through the water-gas shift reaction (see above) to produce additional hydrogen and carbon dioxide. Hydrogen is removed by a separation system, and the highly concentrated carbon dioxide stream can subsequently be captured and stored. In a 2018 study,[6] CAPEX for this technology was estimated at three times that for SMR with CCS.

The economic feasibility of this route to hydrogen production is rather iffy. Nevertheless, the USA has an abundant, domestic resource in coal. The use of coal to produce hydrogen, especially for the transportation sector, can reduce America's total energy use and its reliance on imported petroleum while helping create jobs through the creation of a domestic industry. This idea has strong political appeal. In a future "hydrogen infrastructure" with large central coal gasification plants and a pipeline network, it is conceivable that some of the hydrogen thus produced will be used for power generation.

The cost of hydrogen production and CO_2 intensity is summarized in the chart in Figure 8.4, which is taken from a report published by the US DOE's Office of Fossil Energy. (The source is listed as *IEA Roadmap for Hydrogen and Fuel Cell and DOE Baseline Studies*.) At $2/kg, the hydrogen production cost is equivalent to $15.5/MMBtu for natural gas, which is roughly five times the Henry Hub price of the latter. Thus, as a fuel, for green hydrogen from electrolysis to be competitive with natural gas, the latter's price should be $39/MMBtu to $46.5/MMBtu. At the end of the day, the whole matter of trying to decide which one, i.e., SMR or coal gasification, makes more sense economically is highly likely to be moot. Neither can compete with the best-available alternative today: advanced class *GTCC* with post-combustion CCS using amine-based chemical absorption technology.

The reader may be skeptical about the short shrift given to the H_2 production technologies discussed above, i.e., SMR, ATR, and gasification. Note that *gray*

---

[6] *Hydrogen as an energy carrier – An evaluation of emerging hydrogen value chains*, DNV-GL Position Paper, 2018.

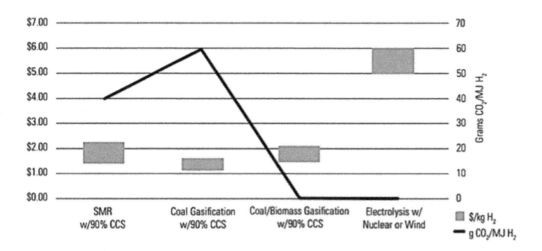

Figure 8.4 Cost of hydrogen production and CO_2 intensity for major technology options.

variants of SMR (with methane or naphta) and gasification (coal or petroleum coke) are TRL 9 technologies currently in use for commercial H_2 production. POX and ATR, at least for large commercial scale production, are still under development. Transforming these technologies into *blue* hydrogen producers is going to be quite expensive. Experience from projects such as Petra Nova, Boundary Dam, and Kemper IGCC amply illustrated that CAPEX estimates found in published academic studies are nowhere near reality. The author's experience from actual EPC FEED studies of similar projects suggests that some of those numbers should be multiplied by at least two. Even if one can make a plausible economic case, simple calculations (results summarized in Table 8.6) show that, if the goal is electric power generation with minimal CO_2 emissions, it is patently impossible to do better than an advanced GTCC with post-combustion capture.

This is not to say that there is not a place for a given technology in a diverse portfolio of technologies. On a case-by-case basis, under certain site criteria (economic, legislative, etc.), a particular technology can be a suitable choice. For example, an existing SMR facility can be repurposed as a *blue* hydrogen fuel producer (for whatever reason) by the addition of a carbon capture plant utilizing chemical absorption technology with a generic amine solvent. Captured CO_2 can be sequestered permanently or utilized for enhanced oil recovery (EOR), thus generating a steady, additional income stream, and produced hydrogen can be injected into an existing natural gas pipeline. (Granted, using the captured CO_2 for pulling another fossil fuel from underground defeats the purpose in capturing CO_2 in the first place.)

8.2.6   A Critical Look

The preceding paragraphs contain a lot of information, assumptions (especially pertaining to CAPEX), and modeling techniques. Combined with the lack of clear-cut assessment of OPEX, it is difficult to rank hydrogen production technologies and decide whether they make sense or not. However, it is possible to eliminate the sources of unnecessary obfuscation and have a good idea about the feasibility of "green," "gray," and "blue" technologies by resorting to first principles. Let us start with "Green H_2."

At the time of writing, for purposes of discussion, an advanced class GTCC (60 Hz) in $1 \times 1 \times 1$ is rated at 600 MWe. For ease of calculation, let us assume that the effective efficiency is 60% net LHV. Thus, it consumes $600/0.6 = 1,000$ MWth of fuel. At 120 MJ/kg LHV for H_2, this comes to $1,000/120 \times 3,600 = 30,000$ kg/h H_2 if this gas turbine burns 100% H_2 fuel.

Let us assume that green H_2 from electrolysis requires 55 kWh/kg energy (see Table 8.1). Ignoring storage, transport, losses, etc., one needs

$$55 \times 30,000/1,000 = 1,650 \text{ MWe} = 1.65 \text{ GWe}$$

carbon-free (wind, solar, nuke, etc.) surplus power to sustain that single (just one) GTCC. (In other words, the round-trip efficiency (RTE) is $0.6/1.65 = 36\%$. To have a benchmark, as of January 2021, the total installed wind power nameplate generating capacity in the USA is 122.5 GWe.)

The total wind power generation in the USA in 2020 was 338,000 GWh – for convenience, say, 350,000 GWh. Assuming a 5% curtailment (in fact, the average was only 3.4%), 17,500 GWh can be diverted to H_2 production. This can support about two (correct, just *two*) 600 MWe GTCC plants for 5,000 hours/year, i.e., $17,500/(1.65 \times 5,000) = 2.12$. In short, the infeasibility of green H_2 from utility-scale electric power generation is obvious.

Since, electrolysis for large-scale green H_2 production is not ready for prime time, pretty much the only game in town today is SMR, which is a thermal process, i.e., from Section 8.2.3,

$$CH_4 + 2H_2O \rightarrow 4H_2 + CO_2 \ (\Delta H° = 165 \text{ kJ/mol}).$$

In other words, from 1 kg of methane and 2.25 kg steam (H_2O), one can make 0.5 kg of H_2. Unfortunately, one also generates 2.75 kg of CO_2. This is called "Gray H_2". Even more unfortunately, if one accounts for the fuel burn in the reformer, CO_2 production can reach approximately 5 kg (per kg of CH_4 feedstock).

Let us do a trade-off analysis with CH_4 LHV = 50 MJ/kg and H_2 LHV = 120 MJ/kg. Assuming a 60% net LHV GTCC efficiency and ignoring all other stuff (H_2 storage, transport, losses, etc.),

CH_4 electricity generation:	$1 \times 50 \times 0.6 = 30$ MJe/kg,
H_2 electricity generation:	$0.5 \times 120 \times 0.6 = 36$ MJe/kg,
CH_4 CO_2 generation (GTCC stack):	2.75 kg/s or 329.9 kg/MWh,
H_2 CO_2 generation (SMR stacks):	5.0 kg/s or $5 \times 3,600/36 = 500$ kg/MWh.

It is eminently clear that gray H_2 is a futile proposition. Now, however, if one adds 90% carbon capture (CC) to the SMR, the picture changes, i.e., CO_2 generation is reduced to only 50 kg/MWh. This is the premise behind the famous "Blue H_2". But the thing is that you can match that with only 85% post-combustion capture (PCC) added to the GTCC, i.e., $0.15 \times 329.9 = 49.5$ kg/MWh.

Let us now do some thinking. Which one makes more sense:

- GTCC + PCC (85%) for < 30 MJe/kg electricity and 49.5 kg/MWh CO_2 emission, or
- SMR + CC (90%) (+ H_2 Storage + H_2 Compression/Piping) + GTCC for < 36 MJe/kg electricity and > 50 kg/MWh CO_2 emission?

(Note that, everything else being the same, 85% capture has much less parasitic power consumption vis-à-vis 90% capture. On the other hand, and as discussed in Section 8.2.3, at least some of the SMR CO_2 streams are more amenable to capture, e.g., higher concentration, higher pressure). It is left to the reader to draw his or her own conclusions.

A third possibility is gasification of coal, refinery residue, or other problematic hydrocarbons. Unlike the two prior examples, in this case, the feedstock, in contrast to surplus electric power plus water or natural gas (mostly methane, that is), is typically not a good candidate for "clean" electric power generation. An example calculation for this process is provided in Section 13.4. Another possibility is to gasify a mix of biomass (e.g., wood pellets, corn stover) and coal as the feedstock. In this case, due to the carbon credit of biomass, hydrogen can be manufactured in a *net negative* carbon process. One such project funded by the US DOE is *Gasification of Coal and Biomass: The Route to Net-Negative-Carbon Power and Hydrogen* led by the Electric Power Research Institute, Inc. (Palo Alto, CA) and partners, who plan to perform a system integrated design study on an oxygen-blown gasification system coupled with water-gas shift, pre-combustion CO_2 capture, and pressure-swing adsorption working off a coal/biomass mix to yield high-purity hydrogen (8,500 kg/h) and a fuel off-gas that can generate export power (50 MWe net). The plant will be hosted at one of two Nebraska Public Power District sites, where opportunities for enhanced oil recovery and sequestration have been investigated and where the need for low-carbon power and hydrogen is imminent. The principal biomass to be used is corn stover, prevalent in Nebraska where the plant will be located. It will be mixed with Powder River Basin coal, necessitating a gasifier that can use this feedstock and be flexible to allow other types.[7]

8.3   Power-to-Hydrogen

Having thus downplayed the significance of SMR and electrolysis (using power from the grid) as viable means of large-scale hydrogen fuel production for electric power generation, let us return to the concept of generating hydrogen with electrolysis using renewable power. The key to the feasibility of this power-to-hydrogen route is having

---

[7] https://www.energy.gov/fecm/project-descriptions-coal-first-initiative-invests-80-million-net-zero-carbon-electricity-and. Last accessed November 20, 2022.

Table 8.7 Curtailed wind power (GWh)[1]

	2012	2013	2014	2015	2016
Germany	410	358	480	3,743	4,722
Ireland	103	171	236		
Italy	164	106	119		
United Kingdom	45	380	659	1,277	

[1] Data in Table 8.7 and Figure 8.5 is from the white paper by GE's Jeff Goldmeer, Power to Gas: Hydrogen for Power Generation, GEA33861, February 2019.

readily available excess power, above and beyond that needed to meet the electricity demand. One way to gauge this capability is to look at the curtailment of renewable sources. Table 8.7 shows wind curtailment for Germany, Ireland, Italy, and the UK between 2012 and 2016.[8]

The largest curtailment in Table 8.7 is close to 5,000 GWh and took place in Germany in 2016. Let us look at our 300 MWe gas turbine example again and recall that it consumed nearly 6 kg/s of hydrogen, which would require 1.3 GWe of electricity to produce it from water. Thus, the entire national wind curtailment in Germany in 2016 would be enough to sustain the full load operation of this *one* gas turbine for about 3,500 hours (equivalent to a capacity factor of 40%).

In 2016, at 79.8 TWh, wind power constituted 12.3% of all German power production (see Figure 8.5). Let us assume that, by the wave of a "magic wand," we increase that share to 50% *overnight*. Let us be pessimistic and assume that the curtailment (which one would logically expect to be minimized) also increases by the same factor. The result is to sustain the full load operation of *four* (4) 300 MWe gas turbines for about 3,500 hours each. Assuming that these gas turbines are in combined cycle configuration, this is equivalent to two $2 \times 2 \times 1$ combined cycle power plants rated at 900 MWe each operating at 40% capacity factor. In other words, this would be a perfect replacement for two (2) coal-fired power plants (40% share in German power production in 2016). To be sure, this is nothing to sneer at, but it is hardly a game changer.

There are multiple studies evaluating the cost of hydrogen generated via electrolysis and renewable power. Figure 8.6 summarizes some of the published price forecasts.

To make a comparison with the numbers shown in Figure 8.6, Henry Hub natural gas spot prices in the US have been around $3/MMBtu in 2018–2019 but dropped significantly in 2020 below $2/MMBtu, mainly because of the economic downturn caused by the Covid-19 pandemic. At the end of the 2020, the price levels, while still below the $3 benchmark, started a recovery. At the time of writing in February 2021, the average LNG price for March delivery into Northeast Asia LNG-AS was estimated

[8] Curtailment may occur for reasons other than excess supply, e.g., transmission congestion, insufficient reserves, maintenance of system stability during voltage dips, or tripping of wind generators due to high wind speeds.

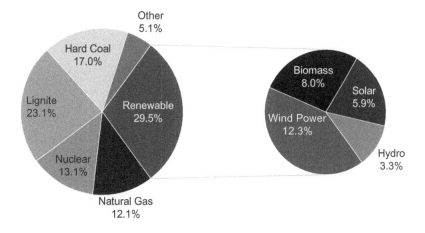

Figure 8.5 Gross power generation mix in Germany (2016) – total 648 TWh.

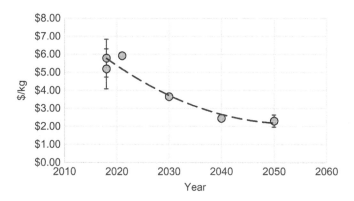

Figure 8.6 Green hydrogen (from renewable power generation via electrolysis) price estimates and forecasts. To convert to $/MMBtu, multiply by 7.44.

at around \$8.90/MMBtu. However, a record high of \$32.50/MMBtu in mid-January 2021, according to the Japan-Korea-Marker, which is used as a reference for spot markets in Asia.[9] At the same time, due to the extreme cold weather in the USA, the Henry Hub natural gas spot price hit a 20-year record high of \$23.86/MMBtu. Operating a power plant on a fuel whose cost ranges from 3 to 10 times the current cost of natural gas will increase the cost of electricity by similar factors.

At this point, it is rather obvious that 100% H_2 as a standard, carbon-free heat engine fuel is unlikely to be a viable proposition, from a physical as well as an economic perspective. Let us then direct our attention to a *middle-of-the-road*

---

[9] While the final draft of the book was reviewed by the author in mid-October 2021, the average LNG price for November delivery into Northeast Asia was estimated at about \$38.50 per MMBtu. Henry Hub price of natural gas was \$5.42 per MMBtu.

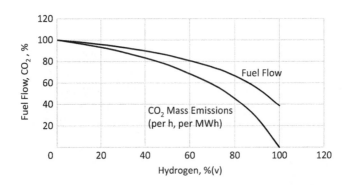

Figure 8.7 Relationship between CO_2 emissions and hydrogen/methane fuel blends.

solution, i.e., blending natural gas with hydrogen. Consequently, the amount of CO_2 reduction will be a function of the percentage of H_2 in the fuel. As a percentage of gas turbine fuel gas composition, the relationship between H_2 content of the fuel gas and CO_2 reduction is as shown in Figure 8.7. Thus, to achieve a 50% reduction in CO_2 emissions, a blend that is roughly 75%(v) hydrogen and 25%(v) methane is required. (It is worth noting that, if the correlation is based on a heat content basis, the requisite fuel gas blend is 50:50 in hydrogen and methane.)

This is small consolation. A 50:50 (in heat content) H_2-CH_4 blend to reduce CO_2 emissions (from gas turbines burning natural gas fuel, that is) by 50% would increase the capacity factor of our example gas turbine from 40% to 80%, while making use of the entire wind curtailment in Germany in 2016 (remember: only *one* gas turbine!). The equivalent of this is two (2) gas turbines at 40% capacity factor each, which translates into one combined cycle power plant rated at 900 MWe. Repeating the exercise for the USA, the findings are not much more encouraging, as was explained with simple calculations in Section 8.2.6.

It should, however, be noted that there are circumstances when hydrogen is available as a byproduct of an industrial or petrochemical process. In such cases, there is usually not enough H_2 to fully load a gas turbine so that a blend of hydrogen and natural gas is used as fuel. This may be just as well because, as will be touched upon briefly below, burning H_2 in gas turbine combustors has its own unique problems.

8.4   Combustion Technologies

The ability of a heat engine (e.g., gas turbine, reciprocating internal combustion engine, boiler-steam turbine) to operate on a high-H_2 fuel requires a combustion system that can deal with the specific nature of this fuel. Key considerations are listed below:

- Low energy density on a volume basis (vis-à-vis natural gas)
- High flame speed (danger of flashback)
- High stoichiometric combustion temperature (high NOx production)
- Safety (low luminosity and tendency to leak)

Consequently, state of the art "Dry-Low-NOx" (DLN) combustors cannot handle high-H_2 fuel without significant modifications. Vintage diffusion flame combustors can but they require diluent injection for NOx control. These combustion systems have been installed on many older E and F Class gas turbines by different OEMs, and have accumulated millions of fired operating hours on a variety of low calorific value fuels, including syngas, steel mill gases, and refinery gases. The downside is that the most suitable diluent is steam, which exacerbates the water consumption problem.

DLN combustors and their fuel handling systems are designed for a maximum of 5%(v) H_2 in the fuel gas. Some OEMs already claim up to 30%(v) H_2 capability for their combustors but the fuel handling systems must also be upgraded accordingly. There are ongoing research efforts (supported by the US DOE and other governmental organizations elsewhere) toward 100%-H_2 capable DLN combustion technology.

Hydrogen can be burned in *diffusion flame* type combustors easily up to 100%(v) without any difficulty. The main drawback is excessive NOx production (due to the high flame temperature), which is typically controlled by diluent injection and post-combustion treatment (e.g., SCR in the heat recovery steam generator [HRSG]). Diluent options are water or steam in GTCC natural gas and nitrogen in IGCC syngas, blast furnace gas, coke oven gas, or refinery gas combustion. All major OEMs have significant experience with vintage E and F Class gas turbines equipped with diffusion flame combustors in such applications.

Burning hydrogen in *premix flame* combustors, i.e., DLN or *Dry Low Emissions* (DLE) combustors, is an altogether different matter. Unique properties of H_2 such as high flammability, low ignition energy (vis-à-vis methane), and high flame speed present significant design challenges such as flashback, high pressure drop, and unstable combustion if H_2 content in the fuel gas exceeds 5%(v). The difference between the two types of combustors is schematically explained in Figure 8.8. Because of the recently heightened focus on decarbonization, all major OEMs,

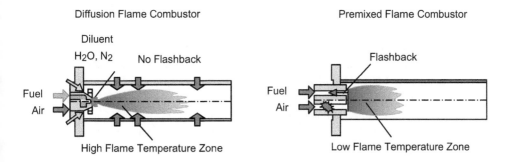

Figure 8.8 Diffusion and premixed flame (DLN/DLE) combustor comparison.

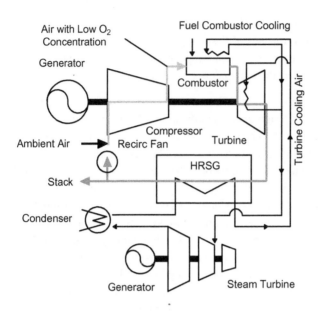

Figure 8.9 Schematic diagram of Mitsubishi Power's EGR system [3].

including Mitsubishi, are actively developing DLN combustors that can ultimately handle 100% H_2 fuel (the combustion products are practically free of CO_2).

One technology that can enable burning high H_2 fuels with reduced NOx emissions in DLN/DLE combustors (or in diffusion flame combustors *without* diluent injection) is *exhaust gas recirculation* (EGR). The EGR method is based on a semi-closed arrangement in which some of the exhaust gas is mixed with air and reduces the O_2 concentration in the combustion air (see Figure 8.9). The exhaust gas is divided downstream of the HRSG and is mixed with ambient air through a cooler, and then introduced into the compressor. One major OEM, Mitsubishi Power (MP, then MHI) considered EGR for its 1,700°C TIT Class gas turbine target under the national project (in Japan). At the end, however, it seems that this rather complex and costly technology was dropped in favor of redesigned DLN combustor nozzles (verified for 30%(v) H_2 content in fuel gas) and *multi-tube, micro-mixer* (*multi-cluster* in MP parlance) combustors (for 100%(v) H_2 fuel).

In the USA, *General Electric* (GE) carried out hydrogen combustion R&D under the sponsorship of DOE's *Advanced IGCC/Hydrogen Gas Turbine Program.* Their report covering the work done in the 2005–2015 period (Phases I and II) is available on the DOE National Energy Technology Laboratory (NETL) website [4]. The key component developed by GE for high-hydrogen DLN combustion is the *multi-tube, micromixer.* (*Siemens* also participated in the same DOE program.) Mitsubishi's 100% H_2 combustion-enabling *multi-cluster* technology is similar to that of GE as shown in Figure 8.10. The essence of the multi-hole nozzle technology is two-fold: (1) rapid mixing of fuel and air for low NOx and (2) flame lifting to prevent flashback (see Figure 8.11).

The technology shown in Figure 8.11 comprises converging and diverging swirl flows. As described by Asai et al. [5], the air passing through the air holes develops

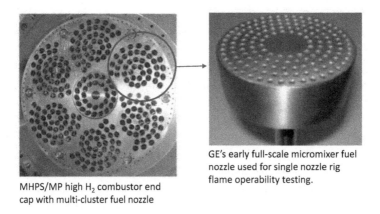

MHPS/MP high H_2 combustor end
cap with multi-cluster fuel nozzle

GE's early full-scale micromixer fuel
nozzle used for single nozzle rig
flame operability testing.

Figure 8.10 Comparison of Mitsubishi and GE high H_2 DLN combustor technologies.

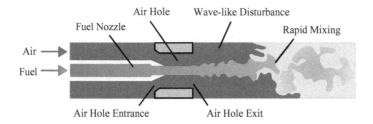

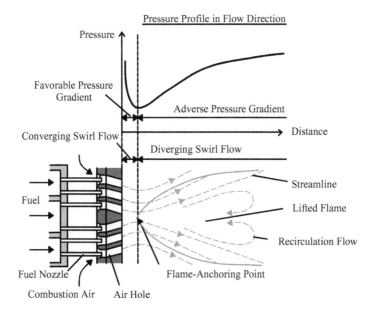

Figure 8.11 Multi-cluster combustor operating principles [5].

into a swirling flow that first converges toward and then diverges from the *flame-anchoring point* (a little ahead of the burner). The converging-diverging swirling flow induces a negative-positive (in the flow direction) pressure gradient. The positive (adverse to the flow) pressure gradient leads to vortex breakdown and generates a recirculation flow that stabilizes the flame by providing a stable heat source of combustion gas to sustain the ignition of fresh reactants. At the same time, the negative part of the pressure gradient acts as an adverse gradient to the combustion gas and prevents it from entering the fuel nozzle channels. This process of flashback prevention is referred to by the designers as "flame lifting."

There is no question that combustion of H_2 can produce more NOx than natural gas (mostly methane, CH_4). However, using the conventional reporting system, i.e., "parts per million by volume (ppmv) on a dry basis at the reference condition of 15% O_2" (ppmvd at 15% O_2 or ppmvdr) artificially exaggerates the difference between the two fuels. The validity of this assertion has been demonstrated by Douglas et al. in a recent paper [6]. The authors compared four different NOx emission metrics, i.e., ppmvdr, ppmv (without correction), and emitted NOx mass per unit of work output (ng/Je) and heat input (ng/Jth), across the full CH_4-H_2 fuel blend range.[10] A bar graph of each approach shows that the percentage difference in emissions across the range of CH_4–H_2 fuel blends relative to the 100% CH_4 case (under conditions where the emitted mass per unit of work output, ng/Je, is constant) as represented by the two measures, ppmvd and ng/Jth, shows little variation. However, the percentage difference represented by ppmvdr increases dramatically as the H_2 content increases, e.g., about 115% at 75%(v) H_2 and nearly 140% at 100% H_2.

As an example, consider an advanced class gas turbine with an ISO base load performance of 346 MWe and 42.3% net LHV efficiency. Let us assume that with 100% CH_4 fuel, NOx emissions are reported to be 25 ppmvdr. This number translates to 102.3 ng/Je and 43.2 ng/Jth. For this machine, with 100% H_2 fuel and at the same emitted NOx mass per unit of work, 102.3 ng/Je, the other emissions metrics are calculated as 44.3 ng/Jth (2.5% higher) and 35.1 ppmvdr (40% higher). (This is a purely thermodynamic exercise. On a real hardware basis, the numbers can be quite different depending on the OEM's changes to the machine to enable hydrogen combustion.) The discrepancy stems from the changing proportions of H_2O and O_2 in the exhaust gas as the fuel's H_2 content is changed. Specifically, increased concentrations of those two constituents in the exhaust gas with high-H_2 fuels lead to artificially high ppmvdr values even without a physical increase in NOx production.

8.5   Transportation

Currently, most hydrogen production (primarily via SMR as mentioned above) is in situ. For widespread deployment of power-to-hydrogen (using renewable energy), this is unlikely to be the typical case. The construction of new hydrogen pipelines will be

---

[10] The units are nanograms per Joule, electric and thermal, respectively.

required. Even though this is not an insurmountable technical hurdle, there are some significant issues.[11] In particular [7],

- Existing steel pipelines are subject to *hydrogen embrittlement* and are inadequate for widespread H_2 distribution.[12]
- Current joining technology (welding) for steel pipelines is a major cost factor and can exacerbate hydrogen embrittlement issues.
- New H_2 pipelines will require large capital investments for materials, installation, and right-of-way costs.
- H_2 leakage and permeation pose significant challenges for designing pipeline equipment, materials, seals, valves, and fittings.
- H_2 delivery infrastructure will rely heavily on sensors and robust designs and engineering.

NIST researchers calculated that hydrogen-specific steel pipelines can cost as much as 68% more than natural gas pipelines, depending on the pipe diameter and operating pressure [8]. Alternatives to metallic pipelines, i.e., pipelines constructed entirely from polymeric composites and engineered plastics, can enable reductions in capital costs and provide safer, more reliable H_2 delivery. For an overview of hydrogen piping design considerations from embrittlement, leakage, and safety (e.g., explosion limits) perspectives, the reader is referred to a recent article by Huitt [9]. For problems associated with use of hydrogen in chemical and refining processes, see API Recommended Practice (RP) 941, *Steels for Hydrogen Service at Elevated Temperatures and Pressures in Petroleum Refineries and Petrochemical Plants.*

There are efforts to investigate the use of *fiber-reinforced polymer* (FRP) pipeline technology for the transmission and distribution of hydrogen to achieve reduced installation costs, improved reliability, and safer operation of hydrogen pipelines. The development of *polymeric nanocomposites* with dramatically reduced hydrogen permeance for use as the barrier/liner in nonmetallic hydrogen pipelines is another possible remedy.

The hydrogen pipeline construction cost can be a significant factor in project economics. In a 2005 US Department of Energy (DOE) study, the 2015 cost for hydrogen pipeline (transmission and distribution) was estimated at $1 million per mile. Yet, in 2021, the cost of new onshore transmission pipeline was estimated as $2.2–4.5 million per kilometer (1 mile ~ 1.6 km). The number for the new subsea transmission pipeline was $4.7–7.1 million per kilometer.[13]

---

[11] Per US DOE, about 1,600 miles of hydrogen pipelines operate in the USA, generally near large users such as oil refineries (accounting for 70% of hydrogen use). The longest in the world is the 600 miles of hydrogen pipeline along America's Gulf Coast built by Air Products.

[12] Hydrogen embrittlement is caused by the penetration of atomic hydrogen into a metal matrix, which results in a loss of ductility and tensile strength. The use of more resistant materials or vented linings of low permeation rate metals are mitigating solutions.

[13] https://www.statista.com/statistics/1220856/capex-new-retrofitted-h2-pipelines-by-type/. Last accessed November 20, 2022.

In February 2022, Kawasaki Heavy Industries (KHI) ship *Suiso Frontier* delivered the world's first cargo of LH2 to the country (suiso means hydrogen in Japanese). The ship's vacuum-insulated, double-shell-structure stainless steel tank has a 1,250 m_3 capacity (75 tons of LH2). The support structure for the tank is made of highly durable FRP to further ensure a reduction in heat transfer. The hydrogen carried by the ship is gray hydrogen manufactured from the gasification of brown coal (lignite) in Victoria, Australia. (Future implementation of carbon capture is planned.) In April 2022, S&P Global Commodity Insights assessed the price of Victorian hydrogen (produced via lignite gasification with CCS, including CAPEX) at $3.6/kg.

8.6   Storage

Hydrogen can be stored physically as either a gas or a liquid. The storage of hydrogen as a gas typically requires high-pressure tanks (350–700 bar [5,000–10,000 psi] tank pressure). The storage of hydrogen as a liquid requires cryogenic temperatures[14] because the boiling point of hydrogen at one atmosphere pressure is −252.8°C. Hydrogen can also be stored on the surfaces of solids (by adsorption) or within solids (by absorption).

A comparison of specific energy (energy per mass or gravimetric density) and energy density (energy per volume or volumetric density) for several fuels based on lower heating values is provided in Figure 8.12. The significant disadvantage of H_2 from a volume perspective is unmistakable.

Resorting again to our gas turbine example, its heat consumption is 750,000 kWth. Two days' worth of hydrogen fuel is 36 million kWh in heat content. At 700 bar of pressure, from Figure 8.12, the volumetric density of H_2 is about 1.2 kWh/L (L is liter). A tank that could store two days' worth of H_2 for this gas turbine would have a volume of 30 million liters, i.e., 30,000 cubic meters! Furthermore, it should be constructed from suitable materials such that it could withstand 700 bar of pressure inside.

When it comes to the industrial storage of hydrogen, salt caverns, exhausted oil, and gas fields or aquifers can be used as underground stores. Although more expensive, as demonstrated by the quick calculation above, cavern storage facilities are most suitable for large-scale hydrogen storage. Underground cavities have been used for many years for natural gas and crude oil/oil products, which are stored in bulk to balance seasonal supply/demand fluctuations or for crisis preparedness. Salt-mined caverns represent the most economical solution (up to 6,600 tons of H_2 per cavern) based on a proven technology (but geographically limited), e.g., $10–$36 per kg of stored hydrogen. Appalachia salt caverns represent the high end of the cost range. Gulf Coast salt caverns are in middle range. Another underground storage option is a purpose-built steel- or concrete-lined vertical underground shaft or "silo." The silo can be "dry" or water-compensated (for pressure). The shaft depth changes between 750 and 2,000 ft.

[14] The cryogenic temperature range has been defined as from −150°C (−238°F) to absolute zero (−273.15°C or −459.67°F), the temperature at which molecular motion comes as close as theoretically possible to ceasing.

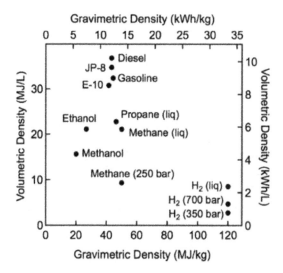

Figure 8.12 Comparison of specific energy (energy per mass or gravimetric density) and energy density (energy per volume or volumetric density) for several fuels based on lower heating values.
Source: www.energy.gov/eere/fuelcells/hydrogen-storage

Diameter, depth, and number of silos can be determined to fit the storage need. This option is suitable to 20 metric tons equivalent of storage volume and multiples thereof at a cost anywhere between $1,000 and $2,000 per kg of stored hydrogen,

To date, the operational experience of hydrogen storage caverns exists only in a few locations in the USA and Europe. Only a relatively small proportion of all potential locations are natural caverns; the most prominent and common form of underground storage consists of depleted gas reservoirs. In particular, the latter could potentially be used as large reservoirs for hydrogen generated from surplus renewable energies. It should also be noted that the natural gas stores are unevenly distributed at a regional level.

There are also aboveground hydrogen storage methods, e.g., utilizing horizontal steel pipeline segments. Pipeline storage is highly sensitive to site and distance (about 4 metric tons of H_2 storage per mile of 24 in. pipeline). The cost can be anywhere between $1,000 and $2,000 per kg of stored H_2. It should be noted that storing H_2 in the pipeline at 200 bara has not been done before. Thus, it is critical to develop a comprehensive pipeline design and construction specification (including welding) that meets all code and industrial standard requirements.

Another option is to store hydrogen in pressure vessels. They can be steel or steel-concrete hybrid vessels in horizontal or vertical alignment. For smaller storage volumes, hydril tube modules can be used as well. Typical installations currently in practice consist of 3–18 tubes with total capacities up to about 4,250 m^3 of hydrogen. (Hydril is the name of the company who manufactures these tubes.) Pressure vessels and hydril tubes can cost around $2,000 per kg of stored hydrogen.

Hydrogen storage energy consumption estimates are as follows:[15]

- Compression energy requirements from on-site production range from approximately 5–20% of LHV (33.3 kWh/kg)
- Liquefaction of hydrogen (including conversion to *para* LH2[16]) with today's processes requires 30–40% of LHV

Assuming electrolysis takes place at 1 bar and 300 K, the upper end of the range would be a good first guess for storage energy consumption. This would correspond to 40% × 33.3 = 13.3 kWh/kg of H_2 for storage in liquid form at 700 bar and –40°C.

The reader is cautioned that these estimates are subject to wide variation based on the assumptions made in the analysis, specifically pressures and temperatures of the individual processes, e.g., the pressure of hydrogen at the discharge of the electrolyzer and the basis of the numbers, i.e., LHV or higher heating value (HHV) (39.5 kWh/kg for H_2). Thus, it is imperative to ascertain whether the electrolysis energy cited for a particular technology includes compression from atmospheric pressure to a higher pressure, say, 20 or 30 bar (e.g., as in the case of the PEM technology). This would make a significant difference in the energy accounting. As an example, consider compression of H_2 from atmospheric pressure to 700 bar with an intercooled compressor (70% polytropic efficiency). If the suction pressure is atmospheric, the compressor power consumption is about 4.5 kWh/kg (auxiliary power consumption of the heat rejection equipment is ignored), which is 13.7% of the LHV. If the suction pressure is 20.7 bar, the power consumption is about 2.7 kWh/kg (8% of LHV).

For a review and discussion of underlying assumptions applied in H_2 value chain analysis, including electrolysis and liquefaction, the reader is referred to a recent paper by Berstad et al. [10]. In that reference, liquefaction energy is cited as about 6.5 kWh/kg (~20% of LHV). For a comprehensive source of information on myriad aspects of hydrogen as a key player in the energy transition, the reader should consult the superb handbook edited by Brun and Allison [11].

8.7   Ammonia

One proposed solution to the H_2 storage and transportation problem is utilizing *ammonia* (NH_3) as a hydrogen-carrying energy vector. The primary reason is ammonia's lower cost of storage, i.e., roughly 5% or less vis-à-vis hydrogen on a per

---

[15] DOE Hydrogen and Fuel Cells Program Record # 9013, July 7, 2009, Energy requirements for hydrogen gas compression and liquefaction as related to vehicle storage needs.

[16] There are two spin isomers of hydrogen: para, in which the two nuclear spins are antiparallel, and ortho, in which the two are parallel. Liquefied H_2 typically comprises 99.79% para H_2 and 0.21% ortho H_2 because, apparently, the para isomer is more stable than the ortho isomer. At room temperature, however, gaseous H_2 is mostly in the ortho isomeric form due to thermal energy. In this form, when liquified, H_2 will be metastable and (slowly) undergo an exothermic reaction to become the para isomer. This conversion will release enough heat to cause some of the liquid to boil and be lost. To prevent such a loss during long-term storage, H_2 is first converted to the para isomer as part of the liquefaction process by the use of a catalyst such as activated carbon or some nickel compounds.

kilogram basis, which is a direct result of its higher density. Ammonia has a molecular weight of 17 kg/kmol, its H_2 weight fraction is about 18%, and its H_2 volume density is about 100 g/L vis-à-vis about 70 g/L for liquefied H_2 (LH2). It is worth noting that, at STP, H_2 is a gas with a density of less than 0.1 g/L and must be liquefied by expending energy, anywhere from 6.5 to 13 kWh per kg of LH2, depending on the type and size of the liquefaction operation, whereas NH_3 is a liquid at less than 10 bar and atmospheric temperature.

Ammonia is a critical fertilizer component and one of the largest produced synthetic chemicals in the world. Ammonia production in the USA amounted to about 17 million metric tons (mt) in 2021 (US Geological Survey estimate). The global production capacity in 2019 was about 235 million mt (with about 180 million mt of actual production) and is expected to increase to 290 million mt by 2030. The growth is largely expected in Asia and the Middle East with natural gas being by far the dominant feedstock. Natural gas is advantageous due to its low energy consumption, i.e., 7.8 MWh per mt of NH_3 produced vis-à-vis more than 10 MWh for coal or fuel oil, and low CO_2 production, i.e., 1.6 mt of CO_2 per mt of NH_3 vis-à-vis 3 mt or more for coal and fuel oil.

The heart of the ammonia production is the *Haber-Bosch* process that converts atmospheric nitrogen to ammonia by a reaction with hydrogen using a metal catalyst at about 100 bar pressure and 400–500°C:

$$N_2 + 3H_2 \rightarrow 2\,NH_3 \;(\Delta H° = -46 \text{ kJ per mol of } NH_3).$$

The catalyst (typically multi-promoted magnetite) compensates for the low reactivity of N_2 (because nitrogen atoms are held together by strong triple bonds). Nitrogen in the reaction comes from the atmosphere but hydrogen must be produced by steam reforming of a hydrocarbon feedstock. The most commonly used feedstock is natural gas (methane) or naphta. Thus, SMR (see Section 8.2.2) is a key step in ammonia production. This can be referred to as *brown* ammonia production. If the CO_2 generated by the shift reaction during SMR is captured and stored, one ends up with *blue* ammonia production. Obviously, H_2 can also be produced by a *green* process such as electrolysis so that ammonia production itself becomes green (provided that a zero-carbon power source is employed in the process, of course, e.g., wind, solar, or nuclear). Due to the nature of the catalyst used in the Haber–Bosch reaction, only very low levels of O_2 and/or O_2-containing compounds can be tolerated in the synthesis gas. Consequently, in this case N_2 must be produced by a cryogenic *ASU* and supplied to the reactor. The energy required is estimated in the range of 10–12 MWh per ton of NH_3 (vis-à-vis 8–12 MWh via SMR) [12]. While the Haber–Bosch technology is at its limit at 7.8 MWh/mt (from natural gas), it is projected that, in theory, the electrolysis process can be improved to 7 MWh/mt [12].

Refrigerated liquid ammonia at –33°C is stored in atmospheric storage tanks (e.g., see API Standard 650, Appendix R). Liquid NH_3 for the fertilizer industry has been transported in pipelines for a long time. They are also used in the oil and gas industry to transport hydrogen. At the delivery point, H_2 gas is produced via NH_3 *cracking* (see below). There are close to 5,000 kilometers of 150–200 mm diameter carbon steel pipelines in the USA. Annually, close to 2 million tons of NH_3 (equivalent to about 350,000 tons of H_2) are transported in those pipelines. Ammonia pipelines normally operate at about 17 bar pressure (NH_3 becomes liquid at 8.6 bar).

Ammonia has an LHV of 18.6 MJ/kg (5.17 MWh/kg, cf. 120 MJ/kg or 33.3 MWh/kg for H_2) and can be combusted with zero CO_2 emissions. The drawback, of course, is nitrogen in the fuel itself, significantly boosting NOx production. Unlike H_2 with high reactivity and flame speed, NH_3 has a narrow flammability limit (15.5% to 27%(v) in air), high autoignition temperature, 630°C, and low flame speed, i.e., about 6–8 cm/s vis-à-vis 140–150 cm/s for H_2 or 40 cm/s for CH_4. These characteristics make combustion of 100% NH_3 in a DLN combustor very challenging, if not impossible, such that significant R&D in this area is required.

One possibility is co-firing NH_3 with H_2 or natural gas. The combustion of a NH_3-H_2 mixture does not produce CO_2 either and can, in fact, be beneficial in the sense that the combustor design challenges of burning each fuel alone are mitigated by their counteracting characteristics. For example, the presence of NH_3 dampens the thermal NOx mechanism dominant in H_2 combustion via lowering the flame temperature while H_2 compensates for the propensity of NH_3 for early blowoff. Nevertheless, this is another area still in need of further research including reaction modeling and bench-scale experiments.

The obvious alternative to burning NH_3 alone or in a mixture is *cracking* it to make H_2 and N_2. The process is the exact reverse of the Haber–Bosch synthesis, i.e.,

$$2NH_3 \rightarrow N_2 + 3H_2 \left(\Delta H° = +46 \text{ kJ per mole of } NH_3 \right).$$

Unlike the *exothermic* synthesis reaction, the cracking reaction is *endothermic*, i.e., it requires energy input. Depending on the catalyst used, the reaction may require temperatures as high as 1,000°C, certainly no lower than 500–600°C. According to the reaction formula, one kmol of NH_3 produces three kilograms of hydrogen at the theoretical heat of reaction of 46 MJ = 12.8 kWh, i.e., 12.8/3 = 4.26 kWh per kg of H_2, which is 4.26/33.3 = 12.8% of hydrogen LHV. On top of this, one should add thermal losses and the heat required to raise the temperature of the reactor to as high as 1,000°C. In Ref. [12], the total energy requirement is estimated as 1.41 MWh per ton of ammonia, i.e., 1.41 kWh/kg or 1.41×17 = 24 kWh/kmol or 86.4 kJ/mol, which is 86.4/46 = 188% of the theoretical heat of reaction.

A comparison of electricity-to-hydrogen and electricity-to-ammonia processes is provided in Table 8.8. For H_2, PEM with 30 bar discharge pressure is assumed. For

Table 8.8 Electricity to H_2 or NH_3 – energy budgets (H_2 LHV = 33.3 MWh per ton)

	Production	Storage	Cracking	Total	
	MWh	MWh	MWh	MWh	per
H_2	55.0	7.0	0.0	62.0	ton of H_2
NH_3	11.0	1.1	1.4	13.5	ton of NH_3
	62.3	6.2	8.0	76.6	ton of H_2

ammonia storage 10% of production energy is assumed. On an efficiency basis, excluding transportation, ammonia seems to be at a disadvantage. When transport and regasification energy costs are included, the picture may change in favor of ammonia.

Performance implications of different fuel gas combinations have been explored using Thermoflow, Inc.'s GT PRO software. A J Class gas turbine is used as an example (engine model #540 from the software library, Mitsubishi Power's M501 JAC, 2015 vintage). Base performance is at ISO base load with 100% CH_4 fuel (heated to 440°F). Other cases are run with different fuel compositions to the same generator output as the base case, i.e., 312.4 MWe. The TIT is adjusted by the software using the choked stage 1 nozzle assumption so that the turbine flow parameter

$$TFP = \frac{TIF}{TIP} \sqrt{\frac{RUNV}{MW}} \, TIT,$$

with turbine inlet flow rate and pressure, TIF and TIP, respectively, are constant for the simulation of fixed turbine hardware. (The universal gas constant is RUNV = 8.3145 kJ/kmol-K and the hot combustion gas molecular weight is MW.) The results are summarized in Table 8.9. The key takeaway from the table is that 50:50 (by volume) ammonia-methane fuel provides a 30% reduction in CO_2 emissions, whereas the reduction is 24% for 50:50 hydrogen-methane fuel. The reason for the difference is higher fuel and turbine inlet gas flow (due to lower LHV of the fuel gas mixture) in the former case, which leads to a lower TIT (at constant TFP) for the same net output. Consequently, simple cycle efficiency is 0.44 percentage points higher. It is worth noting, however, that the combined cycle efficiency is *lower* by 0.15 percentage points as a result of 35°C lower exhaust gas temperature (i.e., lower steam turbine output).

What is not considered in this analysis is the increase in H_2O content of the combustion gas and its impact on hot gas path component life (due to the increased specific heat of the hot gas and heat flux to the components). In the case of 50:50 ammonia-methane firing, H_2O in the exhaust gas is 1.7%(v) higher (absolute basis) than the base 100% CH_4 case. On the other hand, the TIT is already lower by 76°C and this could be enough to make up for the higher heat transfer rate. A caveat regarding TIT, similar to the one stated earlier in Section 8.2.3, is valid here, too. The exact determination of TIT and GT performance is a function of the respective OEM's combustor design and can be lower than estimated herein.

In conclusion, while recognizing the preliminary nature of the calculations, or, more aptly, *estimates*, presented above, it is not possible to prefer one fuel, i.e., NH_3 or H_2, over the other as gas turbine fuel in a definitive manner. In all likelihood, however, if the distance between fuel production (e.g., H_2 or NH_3 from electrolysis) and the point of end use is great, direct combustion of NH_3 in a DLN combustor is (probably) the more cost-effective option.

In fact, Japan, in accordance with the country's *Basic Hydrogen Strategy*, embarked on delivery of H_2 converted from coal gasification from Latrobe Valley

Table 8.9 Simple and combined cycle performance with different fuel gas combinations (in co-fired cases, each constituent is 50%(v))

		CH4	H₂	H2+CH4	NH₃	NH3+CH4	NH3+H₂
Fuel MW	lb/lbmol	16.0	2.0	9.0	17.0	16.6	9.6
GT Fuel Flow	lb/s	32.68	13.15	28.03	83.42	47.28	53.14
TIT	°F	2,748	2,674	2,730	2,616	2,672	2,619
TIP	psia	317.8	316.0	317.4	309.2	320.3	315.5
TIF	lb/s	1,208	1,188	1,203	1,167	1,222	1,194
Turbine Inlet Gas MW	lb/lbmol	28.246	27.015	27.947	26.722	27.817	26.877
Fuel Input (LHV)	kWth	741,962	716,330	735,821	705,478	728,324	708,772
Simple Cycle Output	kWe	312,360	312,360	312,357	312,419	312,400	312,418
Simple Cycle Efficiency	%	42.10	43.61	42.45	44.28	42.89	44.08
Simple Cycle CO₂	lb/MWh	1033.4	0	787.4	0	727.5	0
	lb/hr	322,799	0	245941	0	227256	0
GT Exhaust Temp.	°F	1,156	1,104	1,143	1,092	1,109	1,080
GT Exhaust Flow	lb/s	1,358	1,338	1,353	1,306	1,373	1,339
Combined Cycle Output		453,490	440,017	450,166	434,175	444,482	434,930
Combined Cycle Eff.	%	61.12	61.43	61.18	61.54	61.03	61.36
Combined Cycle CO₂	lb/MWh	712	0	546	0	511	0

in Victoria, Australia. The first *blue ammonia* (NH_3 from SMR with CC) shipment arrived in Japan in the fall of 2020. Since 2017, as part of the Japanese Cabinet's *Strategic Innovative Promotion Program*, and Japan's *New Energy and Industrial Technology Development Organization* (NEDO), Mitsubishi Power has focused efforts on developing a system that thermally cracks (using waste heat) NH_3 into H_2 and N_2, and then combusts that H_2 in a gas turbine.[17] At the same time, in May 2021, the company also announced the development of the 40 MW 100% NH_3 capable gas turbine. (According to the company spokesmen, "larger gas turbines posed many technical problems, such as upsizing and complication of the combustor.")

8.8   Hydrogen Fuel Cells

Strictly speaking, since they are neither "turning" nor "burning," fuel cells do not belong to this treatise. Nevertheless, in the context of hydrogen being widely touted as the "savior of the globe" (green) fuel, a few words on this technology are in order.

A fuel cell converts the chemical energy of a fuel directly into electrical energy. It is different from a battery in the sense that, rather than being a *storage* device, it is a *conversion* device to which fuel and oxidant are supplied continuously. The intermediate processes of heat generation, transfer, and conversion to shaft power are avoided so that fuel cells are *not* subject to the Carnot limit. Since there is no combustion, the fuel cell itself generates power with no pollutants. Of course, from a holistic perspective, this is strictly true only if the fuel is green H_2; otherwise, pollutant and greenhouse gas emissions during fuel production must be accounted for. For the details of the fuel cell technology the reader is referred to Chapter 25 in Ref. [6] in Chapter 2 of the present book (on which the coverage below is based).

Fuel cells, as stand-alone individual power generation systems, are ideally suited to *distributed generation*. From a gas or steam turbine perspective, especially for utility-scale electric power generation, the application of most interest is a *fuel cell hybrid system*, wherein the fuel cell acts as the *combustor* of a gas turbine [12]. From a different perspective, the hybrid system is a *combined* cycle with the fuel cell as the *topping* cycle. (Another variant has the fuel cell as the *bottoming* cycle.) A hybrid gas turbine fuel cell (as the topping cycle) system is shown in Figure 8.13. Fuel cell hybrid systems are based on *molten carbonate* and *solid oxide* fuel cells (MCFC and SOFC) due to their high operating temperatures (650–900°C). The typical fuel cell fuel is hydrogen (H_2), carbon monoxide (CO), or methane.

A detailed study of fuel cell GT hybrid systems was done by Bhargava et al. [14]. Their findings are summarized in Table 8.10. (Note that, at the time of writing, modern GTCC is represented by 25:1 cycle PR, close to 1,700°C TIT, and 64% net LHV efficiency.) For a detailed analysis, the reader is pointed to the cited work. The fuel cell technology is characterized by cell voltage (0.70 V), fuel cell *utilization factor*

---

[17] Article by Sonal Patel in POWER online, March 3, 2021, https://www.powermag.com/mitsubishi-power-developing-100-ammonia-capable-gas-turbine/. Last accessed May 17, 2021.

Table 8.10 Comparison of optimum performance data for fuel cell GT hybrid cycles [14]

(REC: recuperated, ICR: intercooled and recuperated)

Technology	Cycle PR	TIT, °C	Eff., %	Specific Output, kJ/kg
SOFC+GT	30:1	1,300	65.0	588
SOFC+REC	3.5:1	1,300	68.7	620
SOFC+ICR	4:1	1,300	67.8	614
MCFC+GT	3.5:1	1,300	63.5	464
Modern GTCC [15]	23:1	1,600	62.0	650–700

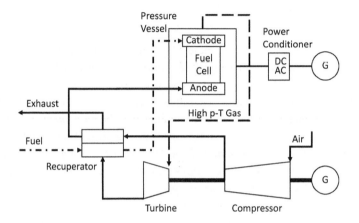

Figure 8.13 Basic fuel cell – gas turbine hybrid power generation system.

(UF, with a typical value of 0.80), and the air utilization factor (0.25).[18] Fuel cell UF (for MCFC and SOFC) is defined as the ratio between oxidized hydrogen mass flow and the equivalent hydrogen mass flow available at the cell inlet, accounting for the internal reforming reactions. Air UF (for MCFC) is defined as a ratio between oxygen mass flow consumed by the fuel cell electrochemical reactions and the oxygen total mass flow available. Gas turbine technology was defined by two cooled stages and compressor/turbine polytropic efficiencies (0.90/0.89).

The US DOE, NETL conducts a SOFC program under the *Clean Coal Research Program*, specifically within the *CCS and Power Systems* program. Solid oxide fuel

---

[18] If the fuel is H_2 and the cell product is liquid water, the ideal cell emf is 1.229 V (at 25C and 1 atm). Maximum possible ideal (reversible) cell efficiency is 83% and is the ratio of the Gibbs energy of H_2 (237.3 J/mol) to its enthalpy (286 J/mol). In that sense, it is analogous to the efficiency of an ideal (Carnot) heat engine and the ideal emf is equal to the exergy of the heat source. If the product is H_2O in vapor form, the ideal cell emf is 1.185 V. The emf increases with increasing pressure and decreases with increasing temperature. Power generation requires when a reasonably large current is drawn from the cell, which introduces irreversibilities into the process. Thus, in practice, the cell emf is limited to 0.6–0.8 V, corresponding to exergetic efficiencies (or technology factors if you will) of 0.49–0.65.

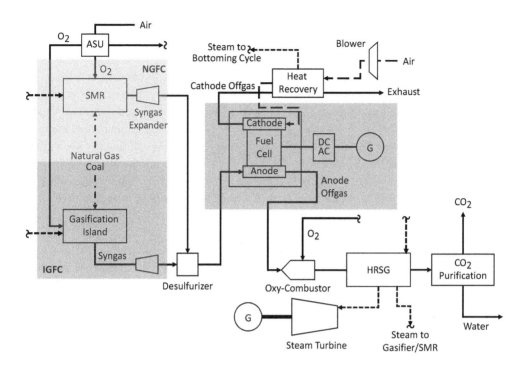

Figure 8.14 IGFC and NGFC Power Systems [15] (condenser and feed water BOP not shown for simplicity).

cells constitute one of the four areas within the *Advanced Energy Systems* subprogram.[19]

The research and development of key technologies within the SOFC program leading to electric power generation systems are coordinated through the *Solid State Energy Conversion Alliance* (SECA). The primary objective of the SECA program is utility-scale power generation with a coal feedstock that generates cost-effective electricity, with near-zero levels of air pollutants, facilitates >97 percent CO_2 capture, and has an efficiency of $60+\%$ in HHV (which corresponds to $63+\%$ net LHV efficiency for subbituminous coal feedstock), with minimal raw water consumption [15]. This is more than *twice* that for a typical pulverized coal power plant with CCS.

Solid oxide fuel cell can also use syngas generated in a gasification plant from a solid feedstock, e.g., coal, in which case the overall system is referred to as an *integrated gasification fuel cell* (IGFC). Another option is to use syngas generated by SMR, which is known as the *natural gas fuel cell* (NGFC). The basic IGFC and NGFC configuration is shown in Figure 8.14. The gasification island of an IGFC is similar to that of an IGCC power plant (see the synthetic fuels section below). There are two system configurations: *atmospheric* (as shown in Figure 8.14) and *pressurized*.

---

[19] The other three are (i) Gasification Systems, (ii) Advanced Combustion System, and (iii) Advanced Turbines.

In the atmospheric IGFC, SOFC modules are operated at near-atmospheric pressure. Coal is fed to the gasifier where it is converted to syngas (comprising CO and H_2). Contaminants are removed from the syngas via conventional gas cleanup technology (e.g., Selexol™ or Rectisol®). Pressurized, clean syngas is expanded to near-atmospheric pressure in a syngas expander. Syngas exiting the expander is processed to reduce sulfur to acceptable levels, and subsequently delivered to the SOFC power island. The SOFC electrochemically utilizes 85–90% of the incoming fuel to produce electric power. The anode off-gas (mainly depleted fuel in addition to CO_2 and H_2O) is combusted in an oxy-combustor. Heat is recovered from this process and used to generate steam for use in a bottoming cycle and for the gasification process. Process air for the electrochemical reaction and for module cooling is delivered by an air blower. Heat is recovered from the cathode off-gas (mainly vitiated air) and the process air to generate steam for use in the bottoming cycle.

In the pressurized IGFC, syngas from the gasification block or the SMR is supplied to the anode of the fuel cell. Cathode off-gas can be utilized to generate power in a turbine. Anode off-gas is sent to an oxy-combustor for combustion with oxygen supplied by the ASU. The combustion products (mainly CO_2 and H_2O) are used to generate power in a turbine with the exhaust treated in a heat recovery unit to generate CO_2 for storage or EOR (about 97% is captured). Condensed water is recycled to the gasifier or the SMR.

The pressurized IGFC is more efficient vis-à-vis the atmospheric variant due to the increase in cell voltage (e.g., at 40 bars pressure, cell voltage is about 90 mV higher than that at 1 bar), but it also has a more complex system configuration. The air blower is replaced by a compressor, and both the anode and cathode off-gas streams utilize a turboexpander generator. Along with increased system complexity, pressurization also presents operational challenges, particularly during startup, shutdown, and during transients. High-integrity seals are required to keep the anode and cathode streams separated. The steam bottoming cycle is eliminated since there is insufficient heat remaining in the off-gas streams to raise steam. In addition to reducing system complexity and cost, eliminating the steam bottoming cycle also reduces the pressurized IGFC power-system water requirements.

IGFC systems with existing gasifiers are projected to have efficiencies in the 45–50% range (in HHV). With a *catalytic gasifier* (lower process temperature, less oxygen input, product syngas with a significantly higher methane concentration, 15–30% (v), in addition to H_2 and CO), 5–10 percentage points of efficiency increase is projected [15]. Similar efficiencies can be achieved by enriching the syngas produced by a conventional gasifier (typically 4–5% (v) methane) with natural gas.

The ultimate SOFC-based power plant envisioned by SECA is an NGFC system that features complete *internal reformation* (IR) of the natural gas within the SOFC to utilize its ability to internally reform methane, i.e., IR-NGFC [16]. Since its inception in 2001, the SECA program has been reported to manage to drop the SOFC cost by a factor of ten in 2010 while maintaining the power density [15]. The goal as recently stated in 2013 was to begin IGFC commercial operation by 2035.

Fuel cell technology is a long shot for utility-scale electric power generation. For *distributed generation*, e.g., instead of being captive to the grid, each house in a

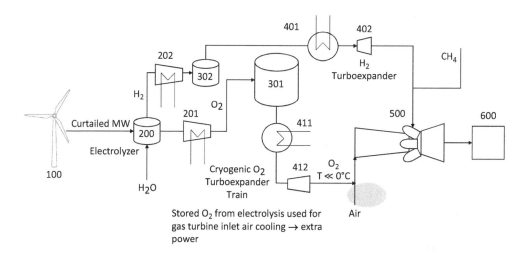

Figure 8.15 Hydrogen and oxygen stored in separate reservoirs; oxygen used to cool gas turbine compressor inlet air via mixing.

neighborhood is equipped with its own fuel cell generator, it may be a suitable option. Obviously, one area of promise for fuel cell application is *fuel cell electric vehicles* (FCEVs) that use electricity generated by a hydrogen fuel cell to power an electric motor. (Hydrogen is considered an alternative fuel under the *Energy Policy Act of 1992* and qualifies for alternative fuel vehicle tax credits.) According to the *International Energy Agency* (IEA), *Clean Energy Ministerial, and Electric Vehicles Initiative*, by the end of 2020, about 31,000 passenger FCEVs powered with hydrogen had been sold worldwide, with South Korea leading the pack with nearly 10,000 vehicles, followed by the USA, China, and Japan. The key challenge for FCEVs to become competitive with other technologies is the lack of an extensive hydrogen infrastructure.

8.9   What about Oxygen?

Electrolysis produces 8 kg of O_2 for each kg of H_2. Prima facie, O_2 produced in the electrolysis plant can be sold as a byproduct. This may, however, require additional storage and transportation infrastructure and may or may not be economical. One option is combined reforming (ATR) and electrolysis (CRE) mentioned at the end of Section 8.2.4, where O_2 from electrolysis is used in the ATR process in lieu of O_2 from the ASU. The author also came up with several alternatives to make use of O_2 produced during electrolysis of water to produce H_2. The basic concept for utilizing O_2 from electrolysis in power generation is shown in Figure 8.15.[20]

---

[20] Although a patent disclosure was prepared by the author, ultimately, he decided not to file it because the novelty angle did not seem sufficiently promising to incur tens of thousand dollars in lawyer fees. Furthermore, while there is absolutely no doubt that there will be a thermodynamic benefit, its cost-effectiveness would require a detailed FEED study to ascertain and is by no means a foregone conclusion.

As shown in Figure 8.15, during the charging phase of the storage process, O_2 generated by the electrolyzer (200) is compressed in an intercooled compressor (201) to the storage pressure. The storage reservoir (301) can be a cavern (natural or man-made) or a vessel (man-made). During the discharge phase, O_2 stored in the reservoir (301) is expanded through a turboexpander train (412) from the storage pressure to the atmospheric pressure. Prior to entry into the turboexpander (412), O_2 is heated in a heat exchanger (411) to a suitable temperature (e.g., 300°C) to optimize turboexpander power output and subzero O_2 temperatures at the turboexpander discharge. The heat exchanger (411) can utilize waste heat recovered from the gas turbine exhaust or from another source. At subzero temperatures, expanded O_2 mixes with the compressor inlet air of the gas turbine (500) and cools it below the atmospheric temperature. The reduced air temperature increases the air mass flow rate and power output of the gas turbine. This is especially beneficial at high ambient temperatures when the power output of the gas turbine drops significantly.

Another variation of the concept described in Figure 8.15 is depicted in Figure 8.16. In this case, warm, pressurized O_2 gas from the turboexpander (412) discharge is injected into the gas turbine (500) compressor discharge. The temperature of O_2 at the gas turbine boundary is controlled via heat transfer in heat exchanger (411) to adjust the O_2 temperature at the turboexpander (412) inlet. The net power output is increased via an additional turbine mass flow rate. The extra gas turbine power output in this variant is 12% of the base value vis-à-vis 6% in the original variant depicted in Figure 8.15 via inlet airflow chilling. However, the applicability of this embodiment is subject to the approval gas turbine OEM.

To assess the feasibility of the concepts presented above, sample calculations were done. Hydrogen-fired gas turbine efficiency is assumed to be 40%. Oxygen storage pressure is assumed to be 700 bara. For a range of base gas turbine outputs, results are

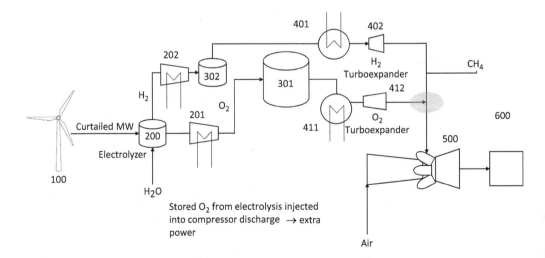

Figure 8.16 Hydrogen and oxygen stored in separate reservoirs; oxygen injected into the gas turbine compressor discharge.

Table 8.11 Charge-discharge performance for increasing gas turbine power output

Gas Turbine Base MWe	15	50	75	100	125	150	200
Gas Turbine Heat Input, MWth	37.5	125	187.5	250	312.5	375	500
Electrolyzer Power Input, MWe	62.5	208.3	312.5	416.7	520.8	625	833.3
H_2 Flow, t/h	1.12	3.75	5.62	7.5	9.37	11.24	14.99
O_2 Flow, t/h	9	30	45	60	75	89.9	119.9
O_2 Compression	2,384	7,945	11,918	15,890	19,863	23,835	31,781
O_2 Expansion	1,001	3,336	5,004	6,672	8,340	10,008	13,345
GT Added kW	1,301	4,335	6,503	8,670	10,838	13,005	17,340
Extra kW (Discharge)	2,301	7,671	11,507	15,342	19,178	23,013	30,685
Extra kW (Charge)	5,959	19,863	29,794	39,726	49,657	59,589	79,451
8-hour Storage Volume, m^3	124	413	619	826	1,032	1,239	1,652

summarized in Table 8.11. Note that the marginal RTE of O_2 storage is 38.6%. The charge and discharge times are 8 hours each. The RTE of the base process is increased by about two percentage points, i.e., from about 24.3% to 26.4%. (It should be emphasized that only electrolyzer and GT powers are used, other parasitic losses are ignored.)

As an example, consider a gas turbine rated at 125 MWe with 700 bar O_2 storage. Extra generation during discharge is roughly 20 MWe (see Table 8.11) and enables storage of 160 MWh for an 8-hour charge-discharge cycle. On paper, this is equivalent to a $40 million investment in Li-I batteries at $250/kWh. In practical terms, whether this magnitude of power/energy is realistic to expect from batteries over a reasonably long plant life without significant performance deterioration and exorbitant operating and maintenance costs is debatable. Equally debatable is, of course, the realism in an energy storage facility with an electrolyzer rated at more than 500 MWe and concomitant H_2 (and O_2) compression and storage equipment and/or facilities. The determination of the feasibility of such an installation requires a bona fide FEED study in addition to a realistic pro forma accounting for all pertinent regulatory requirements, pricing projections, power purchase agreements, and financial criteria.

8.10   References

1. Russ, B. 2009. Sulfur iodine process summary for the hydrogen technology down-selection, General Atomics.
2. Collodi, G.. 2010. Hydrogen production via steam reforming with CO2 capture, 4th International Conference on Safety and Environment in the Process Industry (CISAP4), 14–17 March, Florence, Italy.
3. Tanaka, Y., Nose, M., Nakao, M., et al. 2013. Development of low NOx combustion system with EGR for 1700°C-class gas turbine, *Mitsubishi Heavy Industries Review*, 50(1), 1–6.

4. York, W., Hughes, M., Berry, J., et al. 2015. Advanced IGCC/hydrogen gas turbine development, Final Technical Report, DE-FC26–05NT42643, GE Power and Water.

5. Asai, T., Akiyama, Y., and Dodo, S. (2017). Development of a State-of-the-Art Dry Low NOx Gas Turbine Combustor for IGCC with CCS, In Y. Yun (ed.), *Recent Advances in Carbon Capture and Storage*, London: IntechOpen. DOI: 10.5772/66742.

6. Douglas, C., Emerson, B., Noble, D., et al. 2022. Nitrogen oxide corrections, emissions reporting, and performance considerations for hydrogen-hydrocarbon fuels in gas turbines, *GT2022-80971, ASME Turbo Expo 2022. June 13–17*, Rotterdam, The Netherlands.

7. Smith, B., Frame, B., Eberle, C., et al. 2005. New materials for hydrogen pipelines, Hydrogen Pipeline Working Group Meeting, August 30–31, Oak Ridge National Laboratory, Augusta, GA.

8. Fekete, J. R., Sowards, J. W. and Amaro, R. L. 2015. Economic impact of applying high strength steels in hydrogen gas pipelines. *International Journal of Hydrogen Energy*. 40 (330), 10547–10558.

9. Huitt, W. M., 2021, Hydrogen Piping Systems: Mitigating Pitfalls by Design, *Chemical Engineering*, August, 34–40.

10. Berstad, D., Gardarsdottir, S., Roussanaly, S., Voldsund, M., Ishimoto, Y., Neksa, P. 2022. Liquid hydrogen as prospective energy carrier: A brief review and discussion of underlying assumptions applied in value chain analysis, *Renewable and Sustainable Energy Reviews*, 154, 111772.

11. Brun, K., Allison, T. (eds.). 2022. *Machinery and Energy Systems for the Hydrogen Economy*. Amsterdam, The Netherlands: Elsevier.

12. Giddey, S., Badwal, S. P. S., Munnings, C. and Dolan, M. 2017. Ammonia as a renewable energy transportation media, *ACS Sustainable Chemistry & Engineering*, 5(11), 10231–10239.

13. Bhargava, R. K., Bianchi, M., Campanari, S., De Pascale, A., Negri di Montenegro, G. and Peretto, A. 2010. A parametric thermodynamic evaluation of high performance gas turbine based power cycles, *Journal of Engineering for Gas Turbines and Power*, 132(2), 022001.

14. Campanari, S. and Macchi, E., 1998, Thermodynamic analysis of advanced power cycles based upon solid oxide fuel cells, *Proceedings of the ASME 1998 International Gas Turbine and Aeroengine Congress and Exhibition. Volume 3: Coal, Biomass and Alternative Fuels; Combustion and Fuels; Oil and Gas Applications; Cycle Innovations, June 2–5*. Stockholm, Sweden. https://doi.org/10.1115/98-GT-585.

15. Gas Turbine World. 2015. *Gas Turbine World 2014–15 Handbook*, 31, 66, Fairfield, CT: Pequot Publishing Inc.

16. U. S. Department of Energy. 2013. *Solid Oxide Fuel Cells Program*, Morgantown, WV: National Energy Technology Laboratory (NETL).

9 Nuclear Power

From the perspective of the current book, nuclear reactors are *boilers*. They either act as steam boilers for Rankine steam cycle power plants (conventional deployment) or as heat exchangers to increase the temperature of the power cycle's working fluid. As far as the second variant is concerned, it has not progressed from paper to practice. The power cycle in question is a *closed-cycle* gas turbine. There are several candidates for the working fluid in such a cycle with *supercritical CO_2* being a prime candidate.

This is not the place to enter a futile discussion on the merits or demerits of nuclear power. Suffice to say that, at least in this author's opinion, it is impossible to envision a "net zero carbon" electric power generation portfolio without nuclear power plants playing a prime role in it. Whether they will be in the form of large base loaded utility plants or small facilities amenable to distributed power generation remains to be seen.

9.1 Reactor Basics

Nuclear reactor technologies for land-based electric power production can be classified into four major groups or *generations*:

- Generation (*Gen* henceforth) I and II utilize regular water, H_2O, (usually referred to as *light* water in contrast to the *heavy* water, D_2O) as both coolant and moderator.[1] There are two major variants in this group: (i) *pressurized water reactor* (PWR) and (ii) *boiling water reactor* (BWR). By far the largest number of reactors built in the first few decades of nuclear power were one of these two types. CANDU (Canada Deuterium Uranium) is a *pressurized heavy water reactor* (PHWR), which uses D_2O as coolant/moderator. The *gas-cooled reactor* (GCR) and the *advanced GCR* (AGR) are primarily British designs, which can be counted among Gen I and Gen II technologies (more on GCR in the historical perspective section).
- Gen III reactors incorporate evolutionary improvements to Gen I and II reactors such as improved fuel technology and enhanced (*passive*) safety systems. Examples are *advanced BWR* and Westinghouse AP-600 (AP for *advanced passive*). *High*

---

[1] Gen I refers to the early prototypes such as Shippingport (PWR) or Dresden (BWR); Gen II refers to the commercial power reactors.

Table 9.1 List of reactors in operation (as of 2016) [1]

PHWR: Pressurized Heavy Water Reactor, LWGR: Light Water Graphite Reactor

	BWR	Fast Reactor	GCR	LWGR	PHWR	PWR	Total
Operable	78	3	14	15	49	289	448 (+7)
Africa						2	2
Asia	28	1			25	83 (+7)	137 (+7)
East Europe & Russia		2		15		33	50
North America	36				19	65	120
South America					3	2	5
West & Central Europe	14		14		2	104	134

temperature gas-cooled reactors (HTGR) such as the *pebble bed modular reactor* and the *gas turbine modular helium reactor* (GT-MHR) are also in this group.

- Gen IV reactors include *liquid metal fast breeder reactors* (LMFBR) and *very high temperature gas-cooled reactors* (VHTR). The majority of LMFBRs on the drafting board are cooled by liquid sodium (Na). VHTRs are also envisioned to use helium (He) as coolant. Since neither has moderating capability, a separate solid moderator such as graphite is employed. Another Gen IV reactor is the *supercritical water reactor* (SCWR), in which the reactor coolant H_2O is at a supercritical pressure (i.e., above 221 bar).

All three *light water reactors*, PWR, BWR and SCWR are also referred to collectively as LWR. (Note that SCWR is still on the drawing board.) Gen III and Gen IV reactors, which operate in the "fast neutron" spectrum with (mostly but not necessarily) high coolant temperatures are collectively referred to as *advanced nuclear reactors* (ANR).

At the end of 2016 there were a total of 448 reactors in operation [1]. In terms of reactor types used, the PWR remains predominant (see Table 9.1). Of the 39 reactors that were connected to the grid between 2011 and 2016, all except four have been PWRs, with the remainder consisting of two PHWRs and two fast breeder reactors [1].[2] Typical reactor types, in operation or on the drawing board, and their main characteristics are summarized in Table 9.2. A representative selection of ANR technologies is provided in Table 9.3.

In a thermal power plant, the energy source is combustion of the fossil fuel (e.g., coal), which releases the fuel's chemical energy in the form of heat, which is then transferred to the power cycle working fluid (e.g., water/steam in the Rankine steam turbine cycle). In the nuclear power plant, the energy source is the nuclear fission

---

[2] In the 1970s, fast breeder designs were being widely experimented on, especially in the USA, France, and the USSR. When uranium prices dropped to less than $20 (by 1984) and breeder reactor-produced fuel was of the order of $100–$160, the few units that had reached commercial operation became economically infeasible. Jimmy Carter's April 1977 decision to defer construction of breeders in the US due to proliferation concerns and the terrible operating record of France's Superphénix reactor pretty much killed FBR technology in the West.

Table 9.2 Fission-power reactor types and characteristics [2,3]

GCFR: Gas-cooled fast reactor

Neutron Energy	Reactor Type	Fuel (Enrichment)	Coolant	Moderator	Typical Fuel Assembly	Cladding
Thermal (<0.1eV)	**PWR (LWR)**	UO_2 (up to ~5%)	H_2O	H_2O	Square array of cylindrical rods	Zircaloy-4[3]
	BWR (LWR)	UO_2 (up to ~5%)	H_2O	H_2O	Square array of cylindrical rods	Zircaloy-2
	GCR (AGR)	UC (GCR), UO_2 (AGR – 2–4%)	CO_2, He	Graphite	Circular array of pins in graphite sleeve	Magnox[4] (GCR), SS (AGR)
	CANDU (PHWR)	UO_2 (natural, ~0.7%)	D_2O	D_2O	Circular bundle of tubes	Zircaloy-4
Fast (>10 eV)	**LMFBR**	$UO_2 + PuO_2$	Na	None	Hexagonal array of cylindrical rods	Stainless Steel (SS)
	GCFR	$UO_2 + PuO_2$, Thorium	CO_2, He	None	Plates within hexagonal tube	Ceramic
	HTGR	TRISO[5]	MS, He	Graphite	Prismatic block or pebble bed	SS

reaction, in which the "fuel," e.g., U^{235}, is consumed (by a neutron striking it), releasing fission products, i.e., two neutrons and energy. In other words, neutrons are consumed *and* created during the reaction. The reactor is said to become *critical* when neutrons consumed are equal to neutrons created. Clearly, neutrons are needed to sustain a fission chain reaction but *not* every neutron does lead to fission. Thus, reactor design evolves around the highest probability of a fuel nuclei-neutron collision event (*cross section* in nuclear jargon with the units of *barns*), which is a function of neutron speed and energy, as tabulated in Table 9.4.

Neutrons are classified into three categories according to energy: *thermal* (less than 1 eV), intermediate, and *fast* (greater than 0.1 MeV). Note that fission neutrons are "born" fast (about 2 MeV on average but most with less than 1 MeV) but they slow down via scattering collisions. Nuclear cross section is inversely proportional to neutron speed. For thermal (slow) neutrons it is high, e.g., about 600 barns at 0.025 eV for U^{235} whereas it is low, i.e., O(1) barns, in the fast neutron regime [2].

Thermal neutron fission reactors (PWR, BWR, PHWR) are designed to use enriched fuel (up to ~5% U^{235}) and slow down neutrons into the thermal region (i.e., high cross section) via design geometry and moderators so that a chain reaction

---

[3] Zircaloy-2 and -4 are alloys of zirconium with about 1.5% tin as the main alloying element.

[4] Magnox alloy is magnesium with about 1% aluminum or zirconium.

[5] Tristructural-isotropic (TRISO) fuel is a type of micro-fuel particle. It consists of a fuel kernel composed of UO_X (sometimes UC or UCO) in the center, coated with four layers of three isotropic materials.

Table 9.3 Advanced nuclear reactor (Gen IV) technologies [4]

HTMSR: High temperature molten salt-cooled reactor; HTR-PM: High temperature reactor – Pebble Bed Module; THTR: Thorium high temperature reactor

Reactor	Coolant	MWth	bar	TCIN, °C	TCOUT, °C	ΔTC, °C	Power Cycle	Examples
N/A	MS	1,475		360	510	150	Indirect	TerrapowerTravelling Wave Reactor
HTMSR	MS	80		600	650	50	Indirect	Terrestrial Energy IMSR
	MS	80		650	700	50	Indirect	
HTMSR	MS	750		550	750	200	Indirect	Moltex Enerav SSR
HTMSR	MS	1,250		500	650	150	Indirect	Transatomic MSR
LMFR	Pb	300		400	480	80	Indirect	ALFRED Lead Fast Reactor
LMFBR	Na	1,500		400	550	150	Indirect	Super Phenix Mk 1 (LMFBR. [2, sec 5]), ASTRID by CEA (France)
GCFR	He	75	90	490	850	360	Direct	Allegro GCFR
HTGR	He	125	40	260	750	490	Indirect	THTR-300 Pebble Bed Reactor [2, sec 10–13], X-Energy Xe-100
HTGR	He	250	77.3	520	900	380	Direct	MIT Modular Pebble Bed Reactor
HTGR	He	250	70	250	750	500	Indirect	GA Technologies HTGR [2, sec. 10–12]. INET (China) HTR-PM
GCFR	He	500	133	550	850	300	Direct	General Atomics EM^2 GCFR
HTGR	He	625	60	325	750	425	Direct	Areva HTGR

Table 9.4 Neutron speeds and energies [2]

°C	°F	eV	m/s
20	68	0.025	2,198
250	482	0.045	2,937
500	932	0.067	3,570
1,000	1,832	0.11	4,581
1,500	2,732	0.15	5,406
2,000	3,632	0.20	6,121

can be sustained. However, fuel consumption (*burnup* in nuclear jargon) takes place at low temperatures (less than 540°C) so that high coolant and power cycle temperatures are precluded.

Fast neutron fission reactors have no moderator, but the coolant and other reactor materials slow down the neutrons so that the average is around 2 MeV, but it can go down to 0.05 or 0.1 MeV. In reactors with liquid metal coolant and oxide fuels, the distribution extends to lower energies. In fast reactors, fuel burnup takes place at about 1,200–1,300°C.[6] This is what allows coolant temperatures as high as 760°C or higher.

If the ratio of final to initial fissile fuel content is less than unity, the fast neutron reactors are referred to as "burners," i.e., consuming more fissile material (U^{235}, Pu and minor actinides[7]) than they produce (fissile Pu). If the ratio is more than unity, they are referred to as "breeders" or FBR (fast breeder reactor). If the ratio is unity, they are *isobreeders*, producing the same amount of fuel as they consume during operation.

In almost all nuclear power plants in operation (see Table 9.2), reactors are based on thermal neutron fission technology. In all *existing* nuclear power plants, electric power is generated by a steam turbine generator. Steam is generated in a boiler (*steam generator* in nuclear jargon) that utilizes the primary reactor coolant, e.g., regular or heavy water, in PWR and PHWR, respectively, as the heat source. Thus, the Rankine steam turbine cycle is commonly referred to as the "secondary" loop. In BWR and SCWR, there is no secondary loop. In either variant, steam is generated in the reactor core while the moderator/coolant H_2O performs its cooling duty. A schematic description of SCWR, which would be identical in its general functional form to a BWR, is shown in Figure 9.1. The GCR and advanced gas-cooled reactor (AGR) cycle is illustrated in Figure 9.2. Similar to PWR and PHWR, the Rankine steam turbine cycle constitutes a secondary loop.

From a power cycle perspective, neither of the water-cooled Gen II reactors, which operate in the thermal neutron region, has sufficient temperature to generate high-temperature and/or high-pressure steam. Typical steam outlet temperatures are [2]

---

[6] Experimental fuel has operated successfully at up to 2,900°F (1,600°C) and some tests have been conducted at 3,500°F (1,925°C) [5].

[7] The term actinide refers to any of the series of fifteen metallic elements from actinium (atomic number 89) to lawrencium (atomic number 103) in the periodic table. They are all radioactive, the heavier members being extremely unstable and not of natural occurrence.

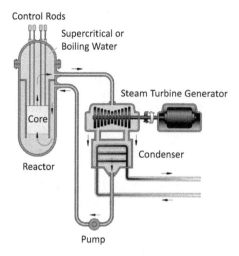

Figure 9.1 SCWR or BWR power generation scheme (single loop).

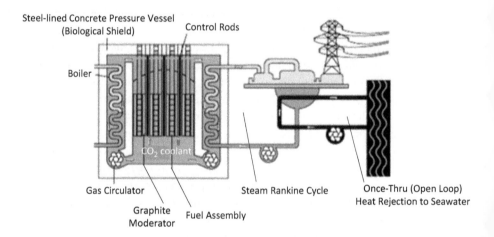

Figure 9.2 GCR and AGR power cycle.

- Up to 325°C (~620°F) saturated or slightly superheated steam for PWR
- 288°C (~550°F) saturated steam (~1,050 psi) for BWR

With non-reheat steam cycles,[8] their efficiencies are in the low 30s (as a percentage).

Early 1950s British GCR power plants with Magnox cladding were not much better than their water-cooled cousins in terms of source temperature and steam cycle. On the

---

[8] Note that PWR and BWR steam cycles *do* have a moisture separator-cum-reheater downstream of the HP turbine to remove moisture from the HP exhaust steam prior to its entry into the LP turbine. The "reheat" energy is supplied from the extraction ports in the HP turbine. This is not to be confused with the "true" reheat steam cycle feature, where the HP exhaust steam is reheated by using the energy in the boiler flue gas or the reactor coolant.

other hand, AGR power plants with stainless steel cladding and CO_2 coolant, achieved coolant gas outlet temperatures up to 665°C (1,230°F). This enabled high pressure and temperature reheat steam cycles (e.g., 2,400 psig and 1,005°F high pressure or main steam with 565 psig and 1,005°F hot reheat steam) at 40% cycle thermal efficiency [2].

In HTGRs, the heat source for the steam boiler is helium, which is the reactor coolant. Other than that, the generic power generation arrangement is similar to that in PWR and PHWR with a primary coolant loop and the secondary steam Rankine cycle loop. The technology was extensively investigated but eventually dropped in the USA after a 40 MWe prototype, Peach Bottom Unit 1 in Pennsylvania (1966–1974), and a 330 MWe commercial unit between 1981 and 1989 in Fort Saint Vrain, Denver, CO. In its ultimate Gen IV ANR version, however, the HTGR is envisioned to have a single coolant loop with a closed Brayton cycle helium gas turbine (see Figure 9.3). It is this particular technology, i.e., closed-cycle gas turbine, which is going to be covered in detail later in the chapter.

The GCFR is, in essence, very similar to a HTGR. It differs from the HTGR design in that the core has a higher fissile fuel content as well as a non-fissile and fertile

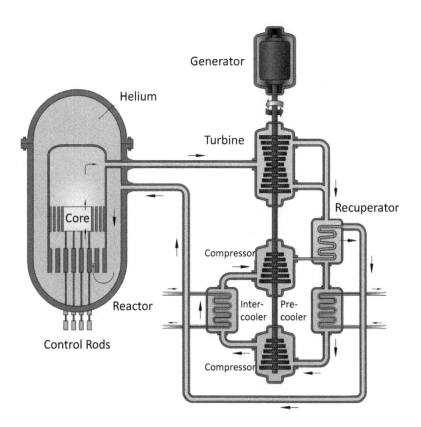

Figure 9.3 HTGR reactor with closed-cycle helium gas turbine.

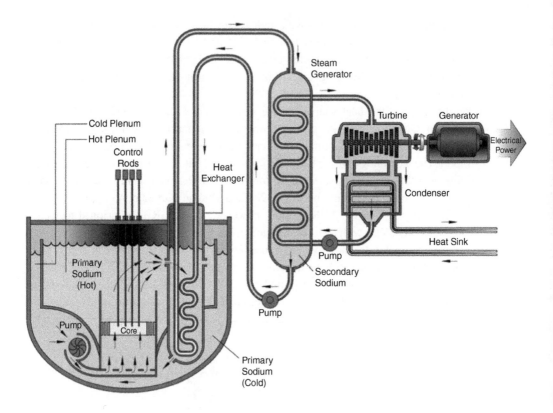

Figure 9.4 Liquid metal fast breeder reactor with steam turbine generator.

breeding component, i.e., there is no neutron moderator.[9] Due to the higher fissile fuel content, the design has a higher power density than the HTGR. From a power generation perspective, the GCFR arrangement is the same as that shown in Figure 9.3 for the HTGR.

In LMFBRs, there is an intermediary coolant loop between the primary reactor core coolant loop (both using liquid Na) and the secondary loop, i.e., the steam cycle. The reason for this is the high induced radioactivity of Na (and other liquid metals), which, in general, is also chemically active. The intermediate loop prevents possible reactions between the radioactive primary coolant and the water/steam working fluid of the power conversion unit (PCU). A generic representation is shown in Figure 9.4, whose primary coolant loop is of *pool* type (also referred to as tank or *pot* type). The other type is a conventional one with circulating liquid Na. The intermediate loop also uses liquid Na as the heat transfer fluid [2].

The Russian BN-600 FBR – Beloyarsk Unit 3 of 600 MWe gross, 560 MWe net – has been supplying electricity to the grid since 1980 (reportedly, the best of all Russia's nuclear power units in terms of its operating and production record). It

---

[9] U^{238} and Pu^{240} are "fertile," i.e., by capturing a neutron they become "fissile" Pu^{239} and Pu^{241}, respectively. They later undergo fission in the same way as U^{235} and produce heat energy.

chiefly uses uranium oxide fuel, enriched to 17, 21, and 26%, with some mixed-oxide fuel in recent years.[10] It has a pool-type coolant configuration, with heat exchangers for three secondary coolant loops inside a pool of sodium around the reactor vessel and three steam generators outside the pool, supplying three 200 MWe steam turbine generators. The sodium coolant temperature is 525–550°C (at little more than atmospheric pressure).

From this very brief introductory summary, the following picture emerges:

- In all existing nuclear reactor power plants with *thermal neutron* energy, i.e., PWR, PHWR, BWR, LMFBR (only three are operational – see Table 9.1) and British GCRs and AGRs, the electric power generation system is based on conventional steam Rankine cycle.
- In Gen IV (ANR) HTGR power plants with *fast neutron* energy, the envisioned electric power generation system is the closed-cycle Brayton gas turbine with helium as the working fluid (also the reactor coolant).

The following questions arise immediately:

1. Why is a helium gas turbine not an option for LMFBR technology?
2. Is another Rankine cycle working fluid (other than H_2O) possible? Under what conditions?
3. What working fluids other than He are possible for use as the HTGR coolant and power cycle working fluid? Under what conditions?

The answer to the third question requires detailed reactor design analysis and is beyond the scope of this discussion. The only viable alternative to helium so far, as coolant-*cum*-direct power cycle working fluid, is *supercritical carbon dioxide* (sCO2). For a detailed description of an sCO2-cooled fast reactor and its direct sCO2 Brayton power cycle, the reader is advised to refer to a recent Sandia report [6]. An in-depth coverage of the sCO2 technology and cycles is provided in Chapter 10.

Answers to other questions can be provided by fundamental cycle thermodynamics. This will be the subject of the following sections.

9.1.1   Fusion Reactor

Fusion reactors are the holy grail of electric power generation. In fusion reaction, mass is transferred into energy as predicted by the famous correlation, $E = mc^2$. Very high temperatures are required to trigger and sustain a fusion reaction. This is the root cause of the myriad difficulties encountered in designing and constructing a practically feasible fusion reactor. For conversion of heat energy from the fusion reactor into electric power, however, one must apply the same principles outlined above for the

---

[10] High-enriched uranium (over 20% U^{235}) would fission, too. At this concentration of U^{235}, the low cross-section (i.e., low probability) for fission with fast neutrons (see Table 9.4) is offset by more neutrons being released per fission above about 0.1 MeV. Up to 20% U is actually defined as "low-enriched" uranium.

fission reactors (e.g., heating of a working fluid to drive turbomachinery operating in a Rankine or Brayton cycle). Therefore, no specific consideration of fusion reactors is provided in the rest of the chapter.

9.2   Available Alternatives

Regardless of the type of nuclear reactor technology, the basic governing principle of nuclear electric power generation is to utilize the heat generated by the nuclear fission reaction and absorbed by the reactor coolant in a *heat engine*. The heat engine in question (the PCU, in nuclear parlance) can operate in one of the *two* thermodynamic cycles:

1. Brayton or gas turbine cycle (the working fluid does *not* change phase),
2. Rankine or steam turbine cycle (the working fluid does change phase).

Both are *closed* cycles, i.e., the mass (and composition) of the cycle working fluid does *not* change (ignoring leaks).

This basic concept can be implemented in *two* ways:

1. Direct – the reactor coolant is also the heat engine cycle working fluid (i.e., H_2O in BWR or SCWR, He in HTGR),
2. Indirect – the reactor coolant is *not* the heat engine cycle working fluid (i.e., He in HTGR or Na in LMFBR).

In the direct heat engine cycle, the reactor itself is the cycle *heat adder*. In the indirect heat engine cycle, the cycle heat adder is an *intermediate* heat exchanger where heat transfer between the reactor coolant and the cycle working fluid takes place.

Based on this high-level classification, there are two key questions facing the designer:

1. Which cycle?
2. Which working fluid?

To answer these two questions in a rational manner, the designer should look at the problem by utilizing the fundamental principles of the following engineering disciplines:

- Thermodynamics (cycle selection and conceptual design)
- Aerothermodynamics (compressor/pump and turbine design)
- Heat transfer and fluid mechanics (heat exchanger and piping design and sizing)

During each design activity, the designer must have readily accessible data for working fluid properties (equation of state, molecular weight, specific heat, thermal conductivity, speed of sound, dielectric strength, etc.)

Direct versus indirect cycle selection revolves around the requirement for minimum (ideally, zero) contamination of the PCU turbomachinery. From that perspective, the indirect cycle is preferable. The disadvantage is the added cost and complexity of the *intermediate heat exchanger* (IHX), where heat is transferred from the reactor coolant to the cycle working fluid. This detrimental impact is counterbalanced by reduced containment cost because the power generation equipment can be located outside the containment structure. Furthermore, the indirect cycle is the only choice when the reactor coolant is not suitable as working fluid, e.g., molten salt or liquid sodium.

From a thermodynamic perspective, on paper at least, *any* pure substance in a fluid state is suitable as cycle working fluid. The only requirement for the Brayton (gas turbine) cycle working fluid candidates is that the cycle state points *all* fall into the *supercritical fluid* or *superheated vapor* regions (i.e., no change of phase from vapor to liquid or vice versa). Practical considerations, alas, strongly limit the selection space, for example:

- thermal stability
- chemical inertness
- inflammability
- toxicity

As a result, practically all organic fluids are eliminated. Over the years, helium (He) and supercritical carbon dioxide (CO_2) have emerged as the most suitable candidates for closed Brayton (gas turbine) cycles in utility-scale, nuclear electric power generation applications in direct or indirect power conversion. Nitrogen (N_2) is also a viable candidate for indirect power conversion (it was considered for the PCU of French CEA's sodium-cooled ASTRID fast reactor). Steam (H_2O) is the only alternative for the Rankine (steam turbine) cycle for nuclear power generation with decades of design, development, and operational experience under the belt of industry participants all over the world. Nevertheless, in the last decade or so, supercritical CO_2 has emerged as a candidate for the Rankine power cycle of nuclear reactors. Physical properties of working fluids, which have been considered for closed-cycle gas turbine applications and steam are summarized in Table 9.5.

Table 9.5 Properties of typical cycle working fluids [7]

k is thermal conductivity, γ is specific heat ratio

Gas	Atomic #	MW	R (kJ/kg-K)	k (W/m-K)	r (kg/m³)	c_p (kJ/kg-K)	g
Helium	1	4.0	2.08	0.1462	0.179	5.190	1.660
Argon	1	39.5	0.21	0.0163	1.784	0.522	1.660
Hydrogen	2	2.0	4.12	0.1683	0.090	14.188	1.409
Oxygen	2	32.0	0.26	0.0266	1.429	0.917	1.399
Nitrogen	2	28.0	0.30	0.0242	1.250	1.041	1.400
Air	–	29.0	0.29	0.0242	1.293	1.006	1.402
Steam	3	18.0	0.46	–	–	–	–
Carbon Dioxide	3	44.0	0.19	0.0146	1.977	0.826	1.301

Two parameters are critical in selecting a working fluid for a closed-cycle gas turbine: specific heat (a fluid property) and the *heat transfer coefficient* (HTC). As shown by Lee et al. [8], the two can be correlated to each other via HTC $\propto c_p^{0.453}$ for constant tube (i.e., the flow channel) diameter and pressure drop parameter per unit tube length,

$$\frac{\Delta p}{pL}.$$

As will be clear from the thermodynamic cycle analysis in Section 9.4, to obtain the best possible performance from a closed-cycle gas turbine (as measured by thermal efficiency) there are two critical parameters:

- Recuperator effectiveness
- Pressure loss in pipes, ducts, and heat exchangers

The reason for this is the low cycle pressure ratio (not more than 3:1), which is optimal for a recuperated cycle but is very sensitive to pressure losses. Consequently, achieving high heat exchange effectiveness in a cost-effective manner while not incurring high pressure losses is strongly dependent on high HTC. The superiority of helium vis-à-vis other working fluid candidates in that sense is clearly illustrated in Figure 9.5.

When all is said and done, there are *four* gaseous reactor coolant candidates:

- air
- nitrogen
- helium
- carbon dioxide

Air is free but it comes with moisture and impurities, which would foul up the reactor chemically and nuclearly (i.e., about 0.85%(v) or 8,500 ppm argon in the air[11]). Nitrogen is better but it becomes slightly radioactive.[12] Both carbon dioxide and helium have been utilized as gas-cooled fast reactor coolants. At reactor temperatures of around 600°C, however, CO_2 dissociates into carbon monoxide and oxygen under the combined action of heat and radiation (radiolytic dissociation). The free oxygen resulting from dissociation oxidizes the graphite and metallic structures. The rate of oxidation becomes unacceptable at higher temperatures. Note that, in an emergency, N_2 or CO_2 as working fluids can be replaced by air without mechanical problems. In the case of CO_2, however, the output would be limited.

---

[11] The isotope A^{41} can be formed. It has strong gamma radiation (1.4 MeV) and a half-life of 100 minutes. This would prevent access to machinery for several hours after shutdown.

[12] Commercial nitrogen can contain argon, which, even at 10 ppm, would require lead shielding to reduce the gamma radiation. Leakage of such a radioactive gas into a confined space is a significant health hazard and precludes human occupancy.

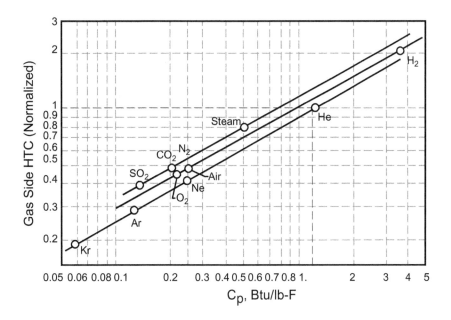

Figure 9.5 Comparative heat transfer coefficients for various working gases (based on equal pressure, temperature, length, and pressure loss ratio in the heat exchanger) [8].

Using the property data in Table 9.5, it can be shown that, for a given reactor pressure, a CO_2 cooled core will require less pumping power than a helium-cooled core. On the other hand, helium's HTC is higher by a factor of about two (see Figure 9.5). Overall, helium is a superior gas from a thermal-hydraulic perspective, but it costs more to pump at a given pressure, or costs more to contain if used at a higher pressure.

Helium is inert and does not run into the dissociation problem mentioned above. Nevertheless, even with He, there is still the possibility of chemical attack of the structural materials. The reason for that is quite intriguing. With most common structural metals, they are protected by a thin, self-repairing, oxide layer that forms naturally in an oxygen containing atmosphere. In an inert atmosphere, if the oxide layer is damaged, there is no oxygen available to repair the layer.

An associated problem in a helium environment is rooted in tribology. Surfaces that slide against each other, e.g., bearings, valves and valve seats, and screw threads can effectively weld themselves together (diffusion bonding). In an inert atmosphere, this occurs through of the exchange of metal atoms, via diffusion, through the oxide-free surfaces under the action of contact pressure, heat, and time. This is a problem especially for safety systems, such as decay heat removal system valves.

Helium is not cheap. Its price is set by the US government.[13] (The USA produces 75% of the world's helium.) In 2016, the price was about \$85–\$105 per thousand

---

[13] A policy artifact from World War I, when the USA assumed national control over the material in case it was needed for lighter-than-air ships (i.e., Zeppelins) in a time of war.

cubic feet. This corresponds to roughly \$3.5 per cubic meter or \$20,000 per tonne (compared with around \$500 for N_2 and less than \$50 for CO_2). Considering that approximately one tonne of helium is required per 100 MWth reactor (i.e., 25 tonnes per one GWe at an average 40% thermal efficiency), the inventory for a 1,000 MWe helium GCFR is about \$500K. The cost issue is aggravated by the propensity of helium to leak easily through even microscopic crevices (similar problem in hydrogen). Hermetic sealing is a must because making up for leaks is expensive.

9.3   A Historical Perspective

Before moving on, a very quick look at the history of cycle and working fluid selection is in order. This will provide a foundation for understanding the choices made in the 1940s and 1950s and the existing state of the art in nuclear reactor power cycles and working fluids.

At the dawn of the *nuclear age*, the primary goal was production of plutonium for atomic bombs. In 1945, it was decided that plutonium produced in a natural uranium fueled, graphite-moderated reactor would be the most economic nuclear explosive for manufacture in the USA [9].

Around the same time, nuclear power was also actively pursued by the US Navy for submarine and surface ship propulsion. Due to the unique nature of submarine warfare, the advantages of nuclear reactor–powered propulsion (i.e., practically unlimited range under water) put the submarine reactor design at the top of the priority list. There were two options at the beginning [9]:

1. Liquid metal-cooled, beryllium-moderated breeder reactor
2. Pressurized water-cooled/moderated fission reactor

In either case, coolant heat was utilized to make steam for a steam turbine prime mover. Practical considerations and the stage of development in each technology dictated that the latter was selected for the first naval propulsion application (SSN Nautilus) in 1948.[14]

In the late 1940s and early 1950s, the gas turbine as a prime mover, primarily for aircraft propulsion, was in its infancy. As a land-based electric power generation technology, it was too inefficient (cycle efficiency less than 20%) and too complicated. Consequently, the steam turbine was the natural choice available to the designers.

The first Brayton open-cycle gas turbine for electric power generation, designed by the venerable Swiss company Brown-Boveri (BBC), went operational in Neuchâtel, Switzerland in 1939 (rated at 4 MWe). This gas turbine did not exactly look like a "jet

---

[14] A General Electric liquid metal–cooled breeder reactor for submarine propulsion was developed and deployed in the second US nuclear sub, SSN Seawolf, commissioned in 1957. In 1959, although the reactor itself functioned without problems, primarily due to the problems caused by leaks in the secondary loop (specifically, the superheater, which was bypassed), its reactor was replaced by a PWR. This was the only instance of a non-PWR reactor in US naval propulsion [9].

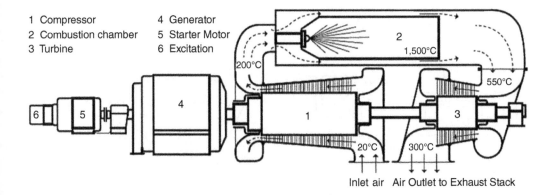

1 Compressor 4 Generator
2 Combustion chamber 5 Starter Motor
3 Turbine 6 Excitation

Inlet air   Air Outlet to Exhaust Stack

Figure 9.6 BBC's Neuchâtel gas turbine (1939–2002).

engine on steroids" like the heavy-duty industrial gas turbines of today (see Figure 9.6). With a cycle pressure ratio of 4.4:1 and turbine inlet temperature (TIT) of 537°C, it generated about 4 MWe with a thermal efficiency of 17.4%. It was operational until 2002, mostly as a peaker (on September 2, 1988, it was designated a *Historic Mechanical Engineering Landmark* by the ASME) [10]. More details on this gas turbine and others of historical significance can be found in chapter 4 of **GTFEPG**.

Around the same time as the commissioning of the BBC Neuchâtel gas turbine, in 1939, another Swiss company, Escher-Wyss (E-W), built a 2 MWe test installation in their factory in Zurich. The power cycle, with air as the working fluid in a closed loop, was named after its inventors, J. Ackeret and C. Keller, as the *AK cycle*. Eventually, between 1950 and 1981, several experimental and commercial closed-cycle gas turbine power plants were built in Europe, Japan, and the USA [11].

Of those power plants, only one, ML-1, a 400-kWe unit designed and built for the US Army, had a nuclear reactor as the heat source (the reactor coolant and the cycle working fluid was *nitrogen*; the moderator was water – see Figure 9.7) – all others were fossil-fueled. At the end, although the ML-1 worked and generated power (but never reached its design specs), it had numerous major issues (e.g., rapid shutdowns were commonplace) and, combined with the budget pressure imposed by the Vietnam War, it was canceled in 1965 [11]. A detailed description of the system and its testing can be found in Ref. [12].

Gas turbines for marine propulsion were considered in the USA in the 1950s (much earlier in Europe). In 1954, a WWII Liberty ship, *John Sargeant*, was equipped with two 3,000 high pressure (HP) recuperated gas turbines (by General Electric) burning marine fuel. With a cycle PR of 5:1 and TIT of 1,450°F (~790°C), each gas turbine had an efficiency of about 26.7% [13]. Compared to the marine boiler-steam turbines at the time, this was about the point of equivalency, in terms of thermal efficiency, between the two types of prime movers.

The attractiveness of using closed-cycle gas turbines with ship nuclear reactors in a direct conversion system (i.e., helium as reactor coolant *and* the gas turbine cycle

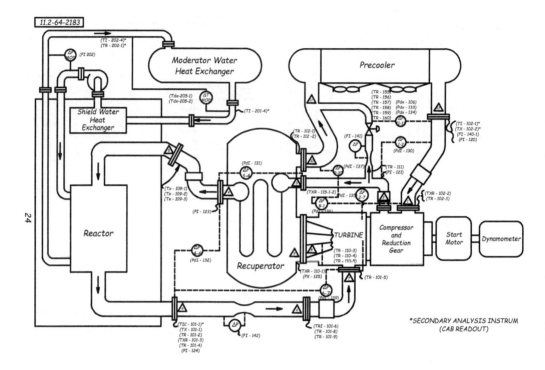

Figure 9.7 US Army's ML-1 nuclear reactor closed-cycle gas turbine [12].

working fluid) was recognized around the same time.[15] However, the high temperatures required for gas turbine thermal efficiency pushed nuclear fuel burnup temperatures too high – beyond the capability of thermal neutron technologies at the time. For a high performing gas turbine with a TIT of, say, 1,500–1,600°F (815–870°C), the fuel burnup temperature should be more than 2,000°F. This·would only be possible with gas-cooled fast reactors, which is a Gen IV technology.

Initially, air was the working fluid for closed-cycle gas turbine power plants. The first operational closed-cycle gas turbine with helium as the working fluid was designed by James La Fleur for an *air separation unit* in 1962 [11]. Note that, as shown in Figure 9.8, it did *not* generate electric power, i.e., the gas turbine shaft was free-running. Its useful output was the helium bled for the lower, cryogenic part of the cycle, where it cools and liquefies air (after being cooled to cryogenic temperature in an aftercooler, in a recuperator, and via expansion through the refrigeration turbine).

In 1974, a 50 MW-34.5% (design intent, only 30 MWe-23% was achieved) closed-cycle helium gas turbine power plant entered service in Oberhausen, Germany. This was a straight-out-of-the-textbook, intercooled-recuperated closed-cycle machine with a cycle PR of 27:1 and a TIT of 750°C (85 kg/s helium mass flow rate). The reason

---

[15] A 20,000 hp (15 MWe) maritime gas-cooled reactor (MGCR) with a direct closed-cycle gas turbine was studied between 1958 and 1962 with joint support from the Maritime Administration and the Atomic Energy Commission.

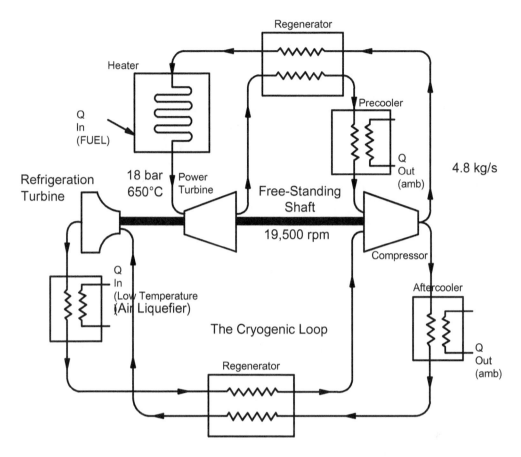

Figure 9.8 Cycle diagram of the La Fleur closed-cycle helium gas turbine plant for air liquefaction [11].

that its field performance fell short of the design intent could be traced to the breakup of the long-standing cooperation between *E-W* and *MAN Gutehoffnungshütte* (GHH) for developing and manufacturing these gas turbines. *Brown Boveri-Sulzer Turbomaschinen AG* (BST) took over from E-W, which left the GHH without a turbine designer. (The task was eventually relegated to a technical university.) Not surprisingly, as a result of myriad design shortcomings, the turbomachinery could not reach its design performance [11].

The reason that the customer, which already had an operating unit with air as the working fluid, i.e., a "tried and tested" system, on hand opted for the helium variant was the subsidy provided by the German–Swiss project for a high-temperature GCR with a helium turbine, HHT (for *high-power helium turbine*). In 1968, the Swiss Federal Institute for Reactor Research began a major study (in cooperation with General Atomics in the USA) into helium-cooled fast breeder reactors with direct (i.e., closed-cycle helium gas turbine) and indirect (i.e., Rankine steam turbine cycle) power generation systems. The rating was specified as 1,000 MWe with 650°C-90 bar

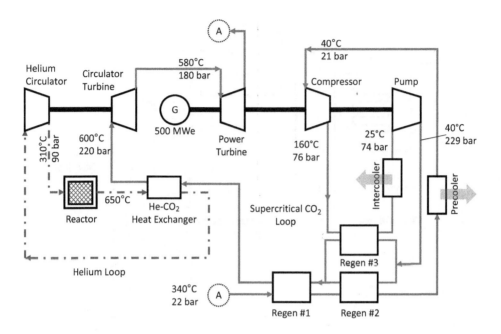

Figure 9.9 Cycle diagram of an indirect supercritical CO_2 turbine power cycle (there are two parallel 500 MWe cycles for 1,000 MWe gas-cooled reactor) [11].

(maximum) helium conditions. Since the indirect version with the steam turbine cycle presented some problems (e.g., risk of water penetrating the primarily cooling cycle), an option with supercritical CO_2 (sCO2) as the working fluid of the secondary power cycle was considered (see Figure 9.9). As shown in the figure, the sCO2 power cycle is also an intercooled-recuperated closed-cycle (textbook version) with a separate turbine to drive the helium circulator (blower).

The culmination of German research in HTGR technology was the 15 MWe (40 MWth) AVR (the acronym for *Arbeitsgemeinschaft Versuchsreaktor*[16]) research reactor at the nuclear research center in Jülich. The AVR, which was a pebble bed reactor, was constructed in 1960, connected to the grid in 1967, and shut down in 1988. During its initial years, the AVR was operated with helium outlet temperatures of 650–850°C. In 1974, the cooling gas outlet temperature was raised to 950° C. Design changes made to the AVR resulting from operating experience were incorporated in the design of the 300 MWe (750 MWth) thorium high-temperature reactor (THTR), which operated between 1985 and 1988. The AVR was also the basis of the technology licensed to China to build HTR-10 and the HTR-PM (see Table 9.3).

The coupling of a high-temperature, gas-cooled nuclear reactor with a closed-cycle gas turbine PCU using helium as the working fluid was first proposed by C. Keller (one of the inventors of the AK cycle) in 1945 [14]. However, in the 1940s the

---

[16] In German, the Experimental Reactor Consortium.

technology was in its infancy – clearly, Keller was way ahead of his time. Interestingly, though, the world's first commercial nuclear power plant at Calder Hall in West Cumbria, England, which became operational in 1956, was cooled with supercritical CO_2 (moderated with graphite), which was used as the heat source in a steam generator.[17]

Advanced, gas-cooled nuclear reactor technology and direct closed-cycle gas turbine power conversion emerged as a natural choice in the 1970s (in the USA, somewhat earlier in Europe – see above). Several historical developments played a role in this:

1. Experience gained in HTGR technology with helium coolant in Peach Bottom Unit 1 (demonstration) and Fort Saint Vrain (commercial) nuclear power plants in the USA.[18]
2. Experience gained in fossil fuel–fired closed-cycle gas turbines with helium as the working fluid, especially in Oberhausen II in Germany (see above).
3. Synergy between the two technologies to eliminate the cycle heat addition component (replaced by the GCR itself, which has a graphite core), which limited TITs due to heat exchanger material limitations.
4. Development of industrial gas turbine technology to a point where its efficiency matched fossil fuel–fired steam turbine power plants (mainly driven by aircraft jet engine technology and materials) and transfer of technology from industrial gas turbines to their closed-cycle cousins.

As a combination of these developments, up to 850°C (maybe somewhat higher) helium turbine TIT is possible with nickel-based superalloys in the hot gas path without cooling. This makes efficiencies well above 40% possible. As outlined by McDonald in a 1978 paper, without increasing the maximum fuel temperature in the HTGR, the coolant outlet temperature can be increased up to 982°C (1,800°F) [16]. An excellent historical perspective is provided by the graphic in Figure 3 of the same paper by McDonald. Even though it is four decades old, its predictions are uncannily accurate (see Figure 9.10).

---

[17] The first nuclear reactor in the world for power generation was the Argonne-Westinghouse Submarine Thermal Reactor, STR 1, which was taken critical in 1953 in Arco, Idaho. The first commercial nuclear power plant in the USA, in Shippingport, Ohio, which went critical in 1957, used the technology originally developed for aircraft carriers (CVR), which was similar to STR 1 [9].

[18] Fort Saint Vrain (FSV) was a commercial nuclear power plant owned and operated by the Public Service Company of Colorado (PSC). FSV was granted an operating license by the Atomic Energy Commission (AEC) on December 21, 1973; became critical on January 31, 1974; and went into commercial operation on July 1, 1979. The power plant remained in commercial operation for a little more than 10 years. On August 18, 1989, the plant was shut down to repair a stuck control rod pair. During the shutdown, numerous cracks were discovered in several steam generator main steam ring-headers. The required repairs were determined by the PSC board of directors to be too extensive to justify continued operation, so they decided to permanently terminate nuclear operations on August 29, 1989.

The reactor at FSV was helium cooled, graphite moderated, and utilized a U^{235}-thorium fuel cycle. The reactor's design employed many of the same fundamental principles that formed the basis for the prototype HTGR at Peach Bottom Unit 1, in Pennsylvania [15].

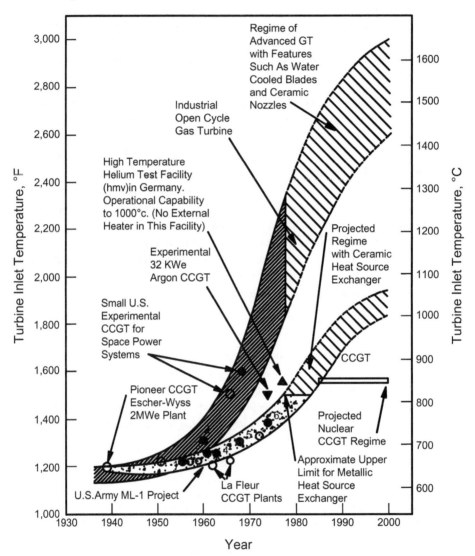

Figure 9.10 Turbine inlet temperature trends for closed-cycle gas turbines and open-cycle industrial gas turbines [16].

9.4   Thermodynamic Cycles

The starting point for *PCU* or heat engine cycle selection is the reactor design and size. The reactor design is characterized by reactor core coolant inlet and outlet temperatures. As discussed in the preceding sections, coolant temperature magnitudes are determined by the reactor type (i.e., whether it operates in a thermal or fast neutron spectrum) and the concomitant fuel burnup temperature. The reactor size is characterized by the cooling load. These three parameters, combined with the specific heat of the coolant, determine the coolant mass flow rate. High specific heat of helium (see Table 9.5) makes it an ideal candidate to minimize the coolant flow and the parasitic power consumption of the circulator blower.

The logarithmic mean of coolant inlet and outlet temperatures, TCIN and TCOUT, respectively, sets the mean-effective heat *addition* temperature (METH) of the heat engine cycle, i.e.,

$$\text{METH} = \frac{\text{TCOUT} - \text{TCIN}}{\ln\left(\frac{\text{TCOUT}}{\text{TCIN}}\right)}. \tag{9.1}$$

The lowest possible mean-effective cycle heat *rejection* temperature (METL) is the ambient temperature, which is typically set at the ISO value, i.e., 15°C or 59°F, i.e.,

$$\text{METL} = \text{TAMB}.$$

The two mean-effective temperatures set the *theoretical* maximum for the direct heat engine cycle efficiency, which is the efficiency of a Carnot cycle operating between the two temperature *reservoirs* at METH and METL, i.e.,

$$\text{EFFMAX} = 1 - \frac{\text{METL}}{\text{METH}} = 1 - \frac{\text{TAMB}}{\text{METH}}. \tag{9.2}$$

For an indirect cycle, one must account for the *approach temperature deltas* (ΔTA) between the reactor coolant and the cycle working fluid. Thus, the working fluid temperature at the IHX exit, TCWFH, is

$$\text{TCWFH} = \text{TCOUT} - \Delta\text{TA}. \tag{9.3}$$

Similarly, the working fluid temperature at the IHX inlet, TCWFL is

$$\text{TCWFL} = \text{TCIN} - \Delta\text{TA}, \tag{9.4}$$

so that, for the indirect heat engine cycle, we can write that

$$\text{METH}' = \frac{\text{TCOUT} - \text{TCIN}}{\ln\left(\frac{\text{TCOUT} - \Delta\text{TA}}{\text{TCIN} - \Delta\text{TA}}\right)}. \tag{9.5}$$

The theoretical maximum for the *indirect* heat engine cycle efficiency is given by

$$\text{EFFMAX} = 1 - \frac{\text{TAMB}}{\text{METH}'}. \tag{9.6}$$

It can be shown that (if ΔTA is the same at both ends of the IHX)

$$\text{METH} - \text{METH}' \cong \Delta\text{TA}. \tag{9.7}$$

9.4.1   Brayton Cycle

Basic Brayton cycle thermodynamics were covered in Chapter 3. Herein, we will address a significant deficiency of the direct closed-cycle helium gas turbine with low *cycle pressure ratio* (PRC): low T_2 and low METH, i.e., low cycle efficiency. (It will be obvious below why increasing PRC is *not* a feasible option. In fact, even a PRC of 4:1 is quite a stretch.) On top of this, the required reactor coolant temperature, 550°C, is much higher than the compressor exit temperature of 254.7°C. The ideal solution to this double-edged problem is *regeneration*, i.e., utilizing the heat in helium exhausting from the turbine (expander) for preheating the compressed helium in a heat exchanger. The heat exchanger in question is called a *recuperator*. A closed-cycle, regenerative gas turbine T-s diagram is shown in Figure 9.11.

In this example, with an approach temperature delta of $\Delta TA = 20°C$, compressed helium can be heated to 380°C, which is still lower than the specified 550°C. The only option available to the designer is to reduce the PRC to increase the expander exit temperature, T_4, so that the recuperator exit temperature of compressed helium, T_5, can be increased. Note that for the ideal cycle

$$T_5 - T_2 = T_4 - T_6. \tag{9.8}$$

Going through the requisite calculations, it is found that for PRC = 2.28:1 and $T_3 = 900°C$, $T_5 = 550°C$, $T_6 = 441.8$ K (168.6°C) and $T_4 = 570°C$. METH and METL for the recuperated cycle are now given by

$$METH_R = \frac{T_3 - T_5}{\ln\left(\frac{T_3}{T_5}\right)}, \tag{9.9}$$

$$METL_R = \frac{T_6 - T_1}{\ln\left(\frac{T_6}{T_1}\right)}. \tag{9.10}$$

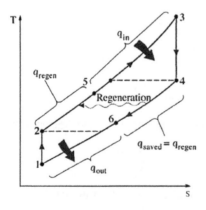

Figure 9.11 Brayton gas turbine cycle with regeneration.

Substituting the cycle temperatures into these expressions, it is found that

$$METH_R = 987.8K > METH = 790.5K \text{ and}$$

$$METL_R = 368.1K < METL = 454K, \text{ so that}$$

$$\eta_{ID,BR} = 1 - \frac{METL_R}{METH_R} = 1 - \frac{368.1}{987.8} = 62.7\%.$$

Key findings from this exercise, which are instrumental in closed-cycle gas turbine design for nuclear electric power generation can be summarized as follows:

1. Recuperation significantly improves cycle efficiency by
   a. increasing cycle METH and
   b. decreasing cycle METL, *simultaneously.*
2. A low cycle pressure ratio (PRC), less than 3:1, is requisite for
   a. effective recuperation and
   b. meeting the reactor coolant inlet temperature requirement.

Another Brayton cycle improvement is compressor *intercooling*, which reduces compression work and cycle heat rejection. A closed-cycle, intercooled regenerative gas turbine T-s diagram is shown in Figure 9.12.

For the same cycle PR of 2.28:1, each compression stage, 1-2A and 1A-2, is calculated to have PR = 1.51:1. Consequently, with $T_{1A} = T_1 = 30°C$, $T_{2A} = T_2 = 84.4°C$ (357.6 K), and $T_6 = 377.6$ K (104.4°C) the mean-effective heat addition temperature does not change, because T_5 and T_3 do not change, i.e.,

$$METH_{IR} = METH_R = 987.8K > METH = 790.5K,$$

but the mean-effective heat rejection temperature is lower (because T_2 and T_6 are lower):

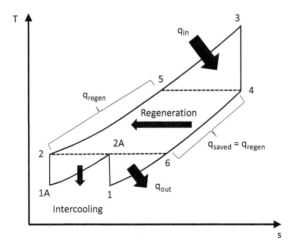

Figure 9.12 Brayton gas turbine cycle with intercooling and regeneration.

$$\text{METL}_{\text{IR}} = 339\text{K} < \text{METL}_{\text{R}} = 368.1\text{K} < \text{METL} = 454\text{K}.$$

The cycle efficiency improves by three percentage points (lower METL), i.e.,

$$\eta_{\text{ID,BIR}} = 1 - \frac{\text{METL}_{\text{IR}}}{\text{METH}_{\text{IR}}} = 1 - \frac{339}{987.8} = 65.7\%.$$

Thus, for the closed-cycle helium gas turbine with intercooling and regeneration, the *Carnot Factor*, CF, is quite high, i.e.,

$$\text{CF} = \frac{65.7}{70.8} = 0.93.$$

The efficiency of an ideal Brayton cycle is a function of PRC only. The efficiency of an ideal Brayton cycle with intercooling and regeneration is a function of PRC *and* T_3. For $\Delta\text{TA} = 0$, i.e., $T_5 = T_4$ and $T_6 = T_2$, it is given by

$$\eta_{\text{ID,BIR}} = 1 - \frac{2}{\tau_3} \cdot \frac{\sqrt{\text{PRC}^k} - 1}{1 - \text{PRC}^{-k}}, \tag{9.11}$$

with $\tau_3 = T_3/T_1$. For nonzero ΔTA, i.e., $T_5 = T_4 - \Delta\text{TA}$ and $T_6 = T_2 + \Delta\text{TA}$, the efficiency of an ideal Brayton cycle with intercooling and recuperation is given by

$$\eta_{\text{ID,BIR}} \simeq 1 - 2 \cdot \frac{\text{PRC}^{\frac{3k}{2}}}{\tau_3} \cdot \frac{-\sqrt{\text{PRC}^{-k}} + 1 + \Delta''}{\text{PRC}^k - 1 + \Delta'} \cdot \frac{\ln \text{PRC}^k + \Delta'}{\ln \text{PRC}^k + 2\Delta''}, \tag{9.12}$$

$$\Delta' = \frac{\Delta\text{TA}}{T_4}, \Delta'' = \frac{\Delta\text{TA}}{T_2}.$$

The cycle pressure ratio for state of the art, advanced class heavy-duty gas turbines is above 20:1 as calculated by using the compressor discharge pressure (see **GTFEPG** or Chapter 3 in this book for more on this subject). (The compressor inlet pressure is, of course, atmospheric.) Such high values are not realistic for closed Brayton (gas turbine) cycles considered for nuclear power generation. The reason for that is two-fold:

1. One cannot have a very low working fluid pressure at the compressor inlet. In other words, the cycle must be *charged*. This ensures that the volumetric flow rates in the pipes upstream of the compressor are low enough to result in reasonable size and cost. At the same time, high pressure is advantageous for high HTCs in the precooler and recuperator, which keeps the size and cost of these critical heat exchangers low. (For supercritical CO_2, it must be above the critical pressure of CO_2, which is 73.8 bar.)
2. The working fluid pressure at the turbine inlet is limited by material availability. Pipe wall thickness and stresses are functions of working fluid pressure and temperature. At high temperatures, expensive alloys are needed to accommodate the thermal stress imposed on the pipe wall. Existing experience with *supercritical* (SC) and *ultra-supercritical* (USC) coal-fired steam turbine power plants sets the upper limit for turbine inlet pressure.

While there is no definitive standard for USC steam conditions, it typically refers to steam temperature of greater than 600°C and steam pressure in excess of 240 bar. Modern supercritical units are being designed to produce main steam at 620°C and 300 bar and reheat steam at temperatures exceeding 620°C. Thus, for a closed-cycle gas turbine with supercritical CO_2, it would be unrealistic to envision a PRC greater than $300/75 = 4:1$ (assuming no pressure losses, which, of course, is quite unrealistic). The typical PRC for existing designs (all conceptual at this stage) is around 3:1.

For helium, there is no hard limit at the compressor inlet imposed by the critical pressure. The typical compressor inlet pressure is set to about 25 bar. As mentioned earlier, the higher value is advantageous for high HTC and optimal precooler and recuperator sizing. While metallurgical considerations allow a relatively high PRC, i.e., $300/30 \sim 10:1$, a PRC higher than about 3:1 is not realistic due to (i) compressor and turbine aerothermodynamic design requirements and (ii) optimum PRC for the recuperative cycle. The low molecular weight and high specific heat of helium translates into a large number of compressor and turbine stages at synchronous speed (i.e., 3,000 or 3,600 rpm depending on grid frequency), e.g., close to 20 stages for the compressor and eight stages for the turbine [17]. Increasing the rotational speed would limit the stage number at the additional cost and complexity of a gearbox and increased blade stresses.

For He and N_2, the ideal compressor inlet temperature is 30°C, as dictated by readily available cooling water temperature at ISO conditions (see above). For supercritical CO_2, the compressor inlet temperature is dictated by the critical temperature of CO_2, which is 304.19 K ($\sim$ 31°C).

9.4.2 Rankine Cycle

The key advantage of the Rankine cycle is its low METL. The reason for that is constant pressure (and temperature) heat rejection to the atmosphere (via the cooling water), which is the Carnot ideal, i.e., *isothermal* heat rejection. Typical condenser conditions are listed in Table 9.6. Note that, unless there is a naturally available cooling water source such as a river, lake, or ocean, a mechanical or natural draft cooling tower is required to achieve low condenser pressures and temperatures. This can become an onerous requirement especially in hot climates due to the parasitic power consumption of the mechanical cooling tower fans. (Natural draft cooling towers operate on the principle of the *chimney effect* – no fans are required.) Even in seemingly ideal

Table 9.6 Steam turbine condenser pressure and temperature

in. Hg	psi	mbar	°F	°C	°K
1.2	0.59	40.6	84.7	29.3	302.4
1.5	0.74	50.8	91.7	33.2	306.3
2	0.98	67.7	101.1	38.4	311.6
3	1.47	101.6	115.1	46.1	319.3

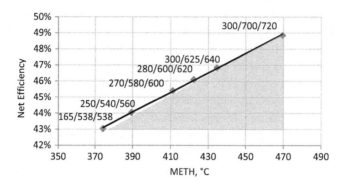

Figure 9.13 Steam cycle net thermal efficiency as a function of METH [18] (cycle conditions are in bara/°C/°C – HP steam pressure, temperature, and HRH temperature, respectively).

locations, environmental regulations can prohibit the dumping of warm cooling water back into the natural reservoir (because it is harmful to the ecosystem).

In summary, the disadvantage of the Rankine steam cycle vis-à-vis the Brayton gas turbine cycle in terms of lower METH is more than compensated by its advantage in terms of lower METL. This is so because, as dictated by fundamental thermodynamics, i.e., Equation (2.8), *both* are equally important to achieving highest possible cycle thermal efficiency. This is the reason why: advanced Rankine steam cycle efficiencies in Figure 9.13 are comparable to or higher than Brayton cycle efficiencies of advanced class gas turbines, which are in the low 40s.

Typical steam cycle conditions for coal-fired (subcritical) steam turbine power plants are 165 bar and 565°C steam temperatures. Supercritical power plants can go up to 250 bar steam pressure. Ultra-supercritical power plants have steam pressures of up to 300 bar and steam temperatures up to 700°C. To stay within proven technology (including boiler tube, steam pipe, and steam turbine materials) and a cost-effective design range, it is recommended not to exceed 600°C and to stay within the subcritical pressure range, i.e., 165–180 bar.

There are Rankine cycle working fluids other than H_2O. *Organic* Rankine cycles have found a certain measure of commercial success in low grade (i.e., low temperature) waste heat recovery and geothermal applications (more on this particular variant below). Another example is the *Kalina* cycle, where the working fluid is a mixture of ammonia (NH_3) and water. Due to the variable pressure and the temperature boiling characteristic of the binary working fluid, the Kalina heat recovery process attacks the key exergy destruction mechanism in the conventional Rankine cycle with pure working fluids. The concept did not go further than a small demo plant due to myriad reliability problems [18].

For liquid metal or molten salt (MS) cooled FBR or other reactors with indirect power cycles, a potentially feasible option is an sCO2 Rankine cycle. Due to its unique properties at the relatively low coolant temperature, this cycle can deliver higher efficiencies than the Rankine steam cycle with low pressure and temperature saturated steam. This cycle is discussed at great length in Chapter 10.

9.4.3   Rankine Bottoming Cycle

Heat rejection from the intercooled-recuperated Brayton cycle to the cooling water in the precooler (i.e., state-point 6 to state-point 1 in Figure 9.12) can be replaced by a waste heat recovery *bottoming* cycle. Since the temperatures are too low for a steam Rankine bottoming cycle (similar to those in modern heavy-duty industrial gas turbine combined cycles), different working fluids are required. The most suitable candidates are the organic fluids such as isopentane or ammonia. Thus, the bottoming cycle is commonly referred to as an *organic rankine cycle* (ORC). The ideal working fluid candidate for the bottoming ORC of a closed-cycle gas turbine in an HTGR power plant must have (i) the ability to absorb the heat rejected from the *topping* cycle helium in the precooler at supercritical pressure and (ii) the ability to reject heat in a condenser to the cooling water.

One such candidate working fluid, which was considered by General Atomics in 1970s, is ammonia (NH_3). A cycle diagram for an HTGR power plant with a recuperated helium gas turbine (no intercooling) and a supercritical ammonia bottoming cycle is shown in Figure 9.14 [19].

As shown in Figure 9.14, the precooler of the helium gas turbine is the ammonia *boiler*. The helium gas turbine cycle has no intercooling to keep T_6 (see Figure 9.12) high enough for a viable bottoming ORC. Heat rejection takes place from 280°C to

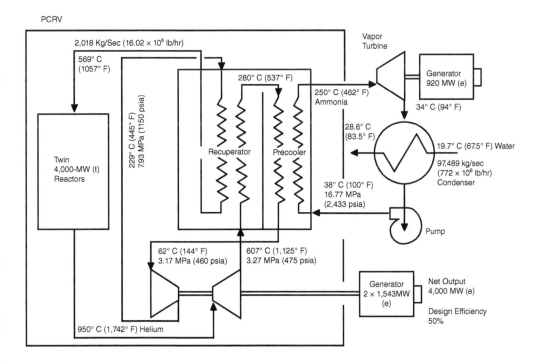

Figure 9.14 Cycle diagram for HTGR gas turbine power plant with ammonia bottoming cycle [19].

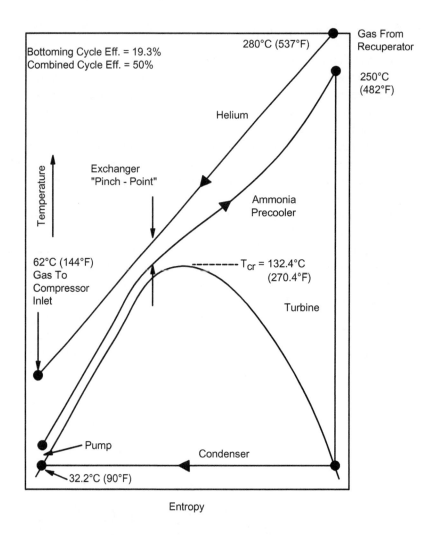

Figure 9.15 Supercritical ammonia bottoming cycle T-s diagram [19].

62°C to ammonia at 167.7 bar. In the supercritical ORC, heat is rejected to the cooling water 19.7°C. A supercritical ammonia bottoming ORC T-s diagram is shown in Figure 9.15. Detailed design information can be found in Ref. [19].

Whether a combined cycle approach to nuclear reactor power plant design is cost-effective or not should be evaluated on a case-by-case basis. (It should be noted that General Atomics' *Energy Multiplier Module* (EM^2), a helium-cooled fast reactor with a net unit output of 265 MWe, has a combined cycle PCU with a direct helium Brayton cycle and a Rankine bottoming cycle.)

For the ORC in Figure 9.15, METH can be calculated as

$$METH_{ORC} = \frac{280 - 62}{\ln\left(\frac{280 + 273.15}{62 + 273.15}\right)} = 435.1 \text{ K},$$

Table 9.7 Comparison of simple and combined cycle PCU performances

		Simple	Combined	Combined (Sized Equipment)
Reactor Duty	kWth	250,000	250,000	250,000
TCIN	C	520	520	520
TCOUT	C	900	900	900
He Flow Rate	kg/s	126.8	126.8	126.8
NH_3 Flow Rate	kg/s		103.8	103.8
He Compressor (LP)	kW	54,738	49,942	49,949
He Compressor (HP)	kW	54,941	79,026	79,037
Total Compressors	kW	109,679	128,968	128,986
He Turbine	kW	229,989	229,293	229,493
NH_3 Turbine	kW		36,338	36,355
Total Turbines	kW	229,989	265,631	265,848
He Generator	kW	118,647	99,104	99,087
NH_3 Generator	kW		35,584	35,600
Total Gross Output	kW	118,647	134,688	134,687
CT Fan	kW	1,260	2,235	3,049
Circ Pump	kW	1,212	5,687	6,578
NH_3 Pump	kW		3,947	4,005
Total Aux Load	kW	2,473	11,869	13,632
Total Net Output	kW	116,174	122,819	121,055
Net Efficiency	%	46.47	49.13	48.42

with METL = TAMB = 288.2 K, and the maximum ORC efficiency is 33.8%. According to Ref. [19], the ORC efficiency is 19.3%, which corresponds to a technology factor of 19.3/33.8 = 0.57.

A comparison of simple and combined helium–NH_3 cycles is presented in Table 9.7. From a purely thermodynamic cycle perspective, a combined cycle version can have about a 2.5 percentage point advantage. This is, however, somewhat misleading. When key pumps and the cooling tower (a dry cooling tower in the example) are actually sized according to the real (physical) hardware requirements (e.g., head and flow margins in the pump curve), the auxiliary power consumption of the combined cycle variant can be inflated. (Note that cooling water consumption of the NH_3 condenser is much higher than that of the He precooler.)

Ammonia bottoming cycle performance is summarized in Table 9.8. The technology factor is very close to that calculated earlier. In conceptual studies, a more conservative factor of 0.51 can be used (along with precooler helium inlet and exit temperatures, T_6 and T_1) to estimate the combined cycle efficiency without resorting to detailed ORC heat and mass balance calculations.

As noted earlier, General Atomics' EM^2 is the only PCU based on a direct helium closed-cycle gas turbine and a Rankine bottoming cycle. The helium Brayton cycle is

Table 9.8 Ammonia bottoming cycle performance

NH₃ Bottoming Cycle		Combined Cycle (Thermodynamic)	Combined Cycle (Sized Equipment)
METH	K	432	432
METL	K	306	306
Max Efficiency	%	29.2	29.2
NH₃ Net Output	kW	23,715	21,968
NH₃ Heat Input	kWth	147,156	147,173
Actual Efficiency	%	16.1	14.9
Technology Factor		0.55	0.51

located in the PCU vessel, while the Rankine cycle is located *outside* the reactor building. The Brayton cycle incorporates the turbomachinery train and the generator, which are mounted on an in-line vertical shaft suspended by active magnetic bearings inside the vessel [4]. This implies an additional construction cost on top of extra equipment.

The question is whether the performance gain is cost effective or not. For example, consider that the target EM^2 plant overnight capital unit cost of a 4×265 MWe installation is about \$4,300/kWe (in 2014 prices) plus an additional first-core cost of \$240/kWe [4]. A +5% output advantage, i.e., 50 MWe, is thus worth about \$230 million, i.e., roughly \$57 million per 265 MWe module. Is this amount (combined with the CAPEX saving from the elimination of the helium compressor intercooler) likely to pay for the extra equipment, especially for the large NH₃ condenser cooling water circ pump and the cooling tower? A detailed analysis during a full-blown FEED study is required for a reliable answer. However, note that a 50 MWe steam Rankine bottoming cycle is roughly \$1,500–\$2,000 per kWe (i.e., \$75–\$100 million). Thus, the likelihood of a positive answer is not promising.

9.5   Cycle Selection Guidelines

It is now time to put the learning in the preceding sections to use. A high-level selection algorithm to identify the best possible cycle type and working fluid for given Gen IV or ANR technology, coolant type, and coolant temperatures is provided in Figure 9.16. Note that the selection process described herein applies to ANR (Gen IV) technologies with reactor coolant exit temperatures of 500°C or higher. Small modular reactors with PWR technology can only support indirect Rankine cycle with steam as the working fluid. (Temperatures are not high enough for efficient sCO2 Rankine cycles.)

As described in Section 9.4.1, Brayton cycles with helium and sCO2 working fluids are limited to a cycle PR of about 3:1. To be reasonably efficient at such a low cycle

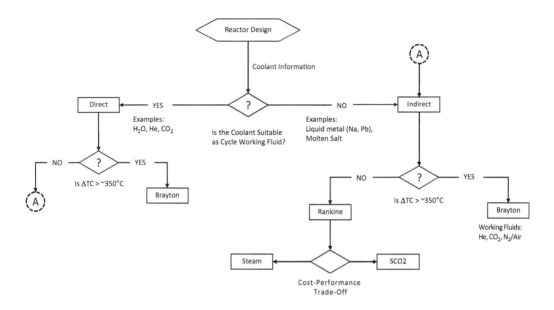

Figure 9.16 Cycle selection flow chart.

PR, the cycle must be regenerative. Coolant TCIN and TCOUT (i.e., T_5 and T_3 in Figure 9.12, respectively), with applicable approach temperature deltas, essentially set the temperature drop available to the Brayton cycle expander. If that drop (i.e., $T_3 - T_4$ in Figure 9.12) is too small, the cycle PR becomes too small and renders the cycle inefficient. By the same token, a very low coolant core inlet temperature (i.e., $T_5 \ll T_4$ in Figure 9.12) and very high temperature range can also be detrimental because it eliminates the possibility of effective recuperation.

Even if the reactor coolant is, say, helium and thus points to a direct Brayton gas turbine cycle as a feasible option (at a first glance), limited or excessive coolant temperature range can render this option infeasible. This can be numerically highlighted by considering a Gen IV ANR technology with helium as the coolant (see Table 9.3).

As an example, let us consider the 250 MWth, helium-cooled HTGR by INET (China) with coolant core inlet and exit temperatures of 250°C and 750°C, respectively. Even with a cycle PR of 4:1, with a 70 bar-750°C turbine inlet, the cycle net efficiency is around 31% net. In comparison, utilizing the heat from the coolant helium in a heat recovery boiler, with a 160 bar/600°C reheat steam cycle and 60 mbar condenser pressure, about 36% net efficiency is possible.

In steam cycle selection and specification, the following items should be considered:

• Condenser pressure and heat rejection system. This should be determined on a case-by-case basis based on site ambient conditions and the applicable environmental regulations. The mechanical cooling tower and water-cooled condenser with 60 mbar back-pressure is adequate for front-end conceptual studies.

- Reheat or no reheat with requisite steam temperature. This is also a case-by-case cost–performance trade-off study subject. However, for front-end conceptual studies
  - Reheat cycle for thermal capacity 200 MWth or higher and TCOUT 550°C or higher,
  - Main steam temperature [°C] = MAX(TCOUT – 25, 600),
  - Reheat steam temperature [°C] = MAX(TCOUT – 50, 600).
- Main steam pressure. High TCOUT and thermal capacity (which determines steam mass flow rate) can render SC or USC steam cycles feasible – on paper. A rigorous cost–performance trade-off study is required on a case-by-case basis. For front-end conceptual studies, it is highly advisable to stay within mature technology parameters, i.e., 160 bar and 600°C maximum. The relationship between high steam mass flow rate (i.e., high reactor thermal rating) and high steam pressure is a function of steam volumetric flow rate (i.e., steam density), which determines the steam-path aero design and geometry. In general,
  - For 600 MWth or higher, 160 bar,
  - For 300 MWth to 600 MWth, 125 bar,
  - For 200–300 MWth, 100 bar,
  - For less than 100 MWth, 80 bar.
- Steam turbine casing configuration (see Figure 9.17). Here, the key decision is to determine the number of double-flow LP casings. This is a function of steam mass flow rate, i.e., the thermal rating of the reactor. The limiting factor is the exhaust annulus area per LP end, which sets the length of the last stage bucket. The steam mass flow rate is estimated from the estimated steam turbine rating, which, of course, is a direct function of the reactor thermal rating.
- Number of feed water heaters. Note that from the ANR coolant temperature data in Table 9.3, it is quite clear that a steam turbine cycle à la gas turbine combined cycle bottoming cycle (i.e., multiple pressure levels and no feed water heating) is not a

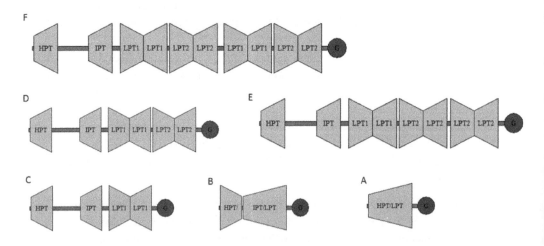

Figure 9.17 Steam turbine casing configurations.

feasible choice. The level of feed water heating and number of feed water heaters (which is a direct function of the former) is dictated by TCIN. The higher the TCIN, the higher the heat recovery boiler feed water inlet temperature. A good rule of thumb for front-end conceptual studies is as follows:

- If TCIN > 350°C, five feed water heaters
- Otherwise, four feed water heaters
- If thermal rating is 100 MWth or less, no feed water heating

The rules outlined above are summarized in Table 9.9.

For direct or indirect Brayton gas turbine cycles, the most feasible option is an intercooled-recuperated cycle to deliver the maximum possible efficiency (e.g., see Figure 9.12 for the cycle T-s diagram). In the case of sCO2, due to the unique properties of the supercritical fluid, a "split recompression" (also known as "part flow") cycle is a candidate for maximum cycle efficiency albeit at the expense of increased complexity and cost ([20]; also see Chapter 10).

Helium [17] and sCO2 [6] have been considered for GCFRs as coolant and direct power (Brayton gas turbine) cycle working fluid. Nitrogen has been considered as the (indirect) Brayton gas turbine cycle working fluid in an HTGR with helium as the coolant [21].

In cycle pressure/temperature selection, the critical consideration is heat exchanger and pipe materials to be used in the IHX and the fluid connection between the IHX and the turbine inlet. In general,

$$\text{PMAX} \propto \frac{S}{D},$$

where PMAX is the allowable pressure, S is the endurance limit for the material under consideration (e.g., average stress to produce a creep rate of 1% at 100,000 hours), and D is the channel diameter. Utilizing the relationship between the fluid mass flow rate and the channel (e.g., pipe) diameter, we end up with

$$\text{PMAX} \propto \frac{S}{\sqrt{\dot{m}}}.$$

The working fluid mass flow rate is directly proportional to the reactor thermal rating

$$\text{PMAX} \propto \frac{S}{\sqrt{\text{MW}_{\text{th}}}}.$$

The endurance limit S is inversely proportional to the temperature of the working fluid, which is determined by the reactor core coolant exit temperature. Thus, for a given material,

$$\text{PMAX} \propto \frac{1}{\text{TCOUT}\sqrt{\text{MW}_{\text{th}}}}.$$

Consequently, for coolant temperatures, which are characteristic of Gen IV HTGR and GCFR, and higher ratings, allowable pressures are limited by material capability. The

Table 9.9 Steam turbine selection matrix

Type designations are based on casing configuration letter in Figure 9.17 and the number of feed water heaters.

Capacity	Expected Expected	No Reheat Max Stm T	Reheat Max Stm T	Reheat	FWHTRs	#LP Casings	LP Ends/Casing	ST Type
MWth	Net MWe	C	C					
2,400	1,160	600	600	Y	5	4	2	F5
2,300	915	525	500	Y	5	3	2	E5
2,200	845	485	460	Y	5	3	2	E5
2,100	895	600	600	Y	5	3	2	E5
2,000	920	600	600	Y	5	3	2	E5
1,900	865	600	600	Y	4	3	2	E4
1,800	810	600	600	Y	5	3	2	E5
1,700	625	455	430	Y	5	2	2	D5
1,600	705	600	600	Y	5	3	2	E5
1,500	655	600	600	Y	4	3	2	E4
1,400	605	600	600	Y	5	2	2	D5
1,300	515	525	500	Y	5	2	2	D5
1,200	460	485	460	Y	5	2	2	D5
1,100	465	600	600	Y	5	2	2	D5
1,000	420	600	600	Y	5	2	2	D5
900	375	600	600	Y	4	2	2	D4
800	335	600	600	Y	5	1	2	C5
700	260	455	430	Y	5	1	2	C5
600	255	600	600	Y	5	1	2	C5
500	220	600	600	Y	4	1	2	C4
400	180	565	565	Y	5	1	1	B5
300	120	525	500	Y	5	1	1	B5
200	80	525	500	Y	4	1	1	B4
100	40	525	500	N	0	1	1	A
80	30	525	500	N	0	1	1	A
50	20	525	500	N	0	1	1	A

solution is to utilize expensive nickel-based alloys, which significantly adds to the plant cost.

In conventional LWRs, core pressures range between 1,000 psi (~70 bar) and 2,250 psi (~155 bar). As was noted in Section 9.4.1, for sCO2, turbine inlet pressure is around 200 bar for a viable cycle. Thus, sCO2 is more likely to be a candidate for an

Table 9.10 Closed-cycle gas turbine pressure and temperatures

			He or N_2 Direct/Indirect Brayton	sCO2 Direct/Indirect Rankine
Compressor/Pump Inlet	T_1	°C	30	27
	P_1	bar	25	~68
Turbine Inlet	T_3	°C	550–850	<600
	P_3	bar	70–80	~200

indirect Brayton gas turbine cycle. (This issue is discussed at length in Ref. [6]. One option is to limit the reactor pressure to 130 bar and use a double-sCO2 turbine cycle. Such schemes, not to mention the basic sCO2 turbine itself, are still at the early R&D stage.)

Up to about 600°C, stainless steel, e.g., 316 SS, is a viable option for sCO2 IHX and pipes (e.g., P91). Strictly speaking, temperatures higher than 600°C should not be considered for sCO2 turbines (see Section 9.2). For helium and nitrogen, whether in a direct or indirect Brayton gas turbine cycle, the turbine inlet pressure is capped around 70 bar. With these working fluids TITs above 600°C are possible. Above 600°C, up to about 750°C, an advanced nickel-based alloy such as Incoloy Alloy 800 is requisite [22]. At still higher temperatures, even more advanced (i.e., more expensive) nickel-based superalloys or cooling of the hottest-running turbine stages is necessary. Stainless steel such as 316 SS is adequate for recuperators [17]. (Medium grade carbon steel is adequate for the precooler.)

Under the light of the aforementioned thermodynamic, heat transfer, and metallurgical considerations (also see the discussion at the end of Section 9.2), the recommended cycle pressure and temperature requirements are summarized in Table 9.10. Note that

- For He and N_2 gas turbines, direct or indirect, the Brayton cycle is intercooled with regeneration
- For direct or indirect sCO2 gas turbines, the Rankine cycle is of split recompression (part flow) type

(Note that N_2 is unlikely to be a reactor coolant – see Section 9.3.)

As will be seen in Chapter 10, there is not a whole lot of difference between sCO2 Brayton and Rankine cycle performances. The latter is always slightly more efficient.

9.6   Nuclear Repowering

As stated in the beginning of the chapter, in a conventional nuclear power plant, the nuclear reactor serves as the steam boiler. A natural extension of this basic fact is the amenability of retired coal-fired power plants to repowering with a nuclear reactor (or multiple nuclear reactors). On paper, this is a trivial exercise, i.e., remove the fossil-fired boiler and its accessories and substitute them with a nuclear reactor. It goes without saying that translating this from the drawing board to the field application is not so easy. Nevertheless, let us start with the paper exercise anyway. Consider the typical coal-fired boiler-turbine power plant in Figure 3.20. In a nuclear repowering scheme using three *small modular reactors* (SMR), the simplified system diagram would look like that shown in Figure 9.18.

As shown in the figure, the heat source is three SMRs. Reactor heat from the SMRs is removed by the primary coolant. The removed heat is transferred from the primary coolant to a heat transfer fluid (e.g., MS) in three intermediate heat exchangers. The heat transfer fluid (HTF) is used in a heat recovery boiler (similar to the heat recovery steam generator in a gas turbine combined cycle) to make HP, from feed water) and hot reheat steam (HRH, from the cold reheat steam) to be used in the steam turbine for electric power generation.

There are several requirements to enable a brownfield retrofit of this type:

1. The steam cycle should be of recent technology, i.e., supercritical or ultra-supercritical, with good efficiency (i.e., at least 40% net low heating value basis), in good condition with reasonably long remaining life (at least 15–20 years).

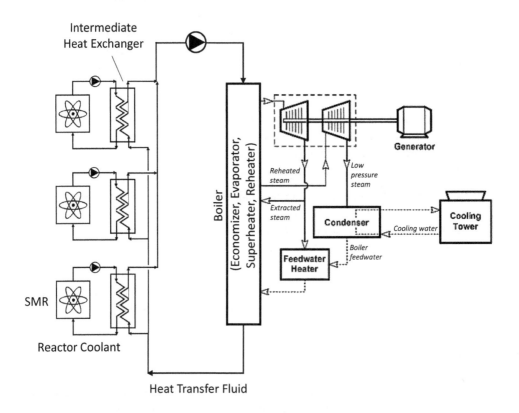

Figure 9.18 Repowering with nuclear reactors.

2. Steam and feed water temperatures should match the nuclear reactor coolant parameters closely (with allowance for the intermediate heat transfer loop).
3. Plant layout and site availability should accommodate the nuclear reactors and additional balance of plant equipment (including a seismic hazard assessment).

The last requirement is critical. Demolition and removal of the pulverized coal boiler, air quality control system (AQCS), and coal preparation and delivery equipment to make room for the new system would render such an undertaking economically infeasible. The ideal site would enable the construction of the new facility adjacent to the existing one and decommission the existing equipment without demolition and/ or removal. The reader can consult a recent paper by Bartela et al. [23] for a detailed techno-economic analysis of the nuclear repowering of a fossil power plant in Poland. The power plant in question is a 460 MW unit with 275 bar/560°C/580°C supercritical steam cycle (290°C feed water temperature at the boiler entry). The SMR considered for the repowering was a Gen IV design with a eutectic salt mixture used as a coolant with the following nominal design data:

- 320 MWth total thermal power
- Coolant inlet/exit temperatures 550/650°C
- Coolant flow rate 1,200–1,400 kg/s

The intermediate heat exchange loop is assumed to use MS with an approach temperature delta of 50°C, i.e., hot and cold MS temperatures are 600°C and 500°C, respectively. The MS flow rate is 2,080 kg/s (per SMR). Three SMRs are required to supply the requisite steam cycle heat input (see Figure 9.19). Note that this calculation must be taken with a grain of salt. As discussed in Section 5.1, the typical MS used in solar applications, i.e., sodium/potassium nitrate 60%/40%(w), has an upper stability limit of 565°C. This implies that the original steam cycle temperatures, 560°C/580°C for HP and HRH steam, respectively, cannot be achieved in this repowering scheme, at least, not with this particular HTF. Thus, there will be a corresponding steam turbine performance penalty.

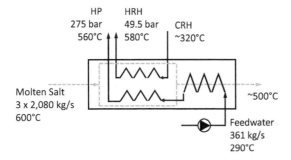

Figure 9.19 Molten salt supercritical boiler (total heat transfer, 998.4 MWth).

9.7   Methane Pyrolysis

Methane pyrolysis or cracking is an endothermic reaction that takes place at high temperatures:

$$CH_4(g) \rightarrow C(s) + 2H_2(g) \; \Delta H \text{ at } 25°C = 74.85 \text{ kJ/mol of } CH_4.$$

Without the use of catalysts, a practically high reaction requires temperatures at 1,200°C or higher due to the high activation energy required to break the stable C–H bonds of methane molecules. In contrast, catalytic methane decomposition occurs at a temperature in the range of 600–900°C, which is comparable to temperatures requisite for SMR. For the details of the chemical principles underlying this process, the reader is referred to the review article by Msheik et al. [24]. The estimated technology readiness level of this technology is between 4 and 5.[19] There are several options for catalyst selection. One of them is a nickel catalyst that can reduce the energy requirement down to a temperature range of 500–700°C. Another candidate is iron, which has the potential to work well at 700–900°C.

The advantage of methane pyrolysis vis-à-vis SMR is obvious, i.e., there is no need for expensive carbon capture. The carbon produced in a methane cracking reaction is pure solid carbon, i.e., what is colloquially referred to as *black carbon*. Solid carbon is a valuable by-product and much easier and cheaper to "capture" than CO_2. As an example, it can be used to produce carbon fiber, a material with highly desirable properties such as high stiffness, high tensile strength, low weight to strength ratio, high chemical resistance, high temperature tolerance, and low thermal expansion. These properties have made carbon fiber very popular in aerospace, civil engineering, military, and motorsports, along with other competition sports.

Reactor temperatures in Gen IV nuclear technologies range from 500 to 1,000°C, which make them attractive as carbon-free heat source to be used in methane pyrolysis. The concept has been explored by Serban et al. as described in their paper [25]. The researchers specifically considered a fast neutron spectrum, fissile self-sufficient converter reactor using a heavy liquid metal coolant (e.g., lead, lead-bismuth, or tin). High-temperature gas (e.g., He, N_2, or CO_2) is used for heat transport from the nuclear reactor to the thermal cracking unit. An experimental study was carried out in a microreactor with methane being bubbled through a bed of either low-melting-point metals (e.g., lead, Pb, or tin, Sn) or through a mechanical mixture of metal and solid media (e.g., silicon carbide, SiC). The conversion rate was found to be dependent on the contact time between the methane and the heat transfer media, as well as on the methane bubble size. The most efficient systems used for the pyrolysis process were the ones using Mott porous filters as feed tubes, with almost 57 and 51% methane conversions obtained when passing natural gas through a bed of 4-in. Sn + SiC and Sn, respectively [25].

---

[19] 2019 CEFIC (European Chemical Industry Council) Vision on Hydrogen, https://cefic.org/app/uploads/2019/11/Cefic-position-on-Hydrogen-1.pdf, last accessed December 3, 2021.

In passing, there is another, novel, method for methane cracking that has been recently proposed and elaborated upon in a paper, i.e., *shock wave heating* [26]. The underlying principle is quite simple. A shock wave propagating through natural gas (basically, methane) creates a high pressure and temperature zone downstream of the wave. This can be easily seen from the gas-dynamic equation for static temperature rise across a normal shock wave, i.e.

$$\frac{T_1}{T_0} = \frac{(2\gamma Ma^2 - \gamma + 1)((\gamma - 1)Ma^2 + 2)}{(\gamma + 1)^2 Ma^2},$$

where the subscripts 0 and 1 denote upstream and downstream states, respectively, and γ is the specific heat ratio of the gas (1.3 for methane). Easier said than done, of course. For methane at 25°C (298 K), a shock wave with a Mach number (Ma) of 5 is needed to bring the gas temperature above 1,100°C. To minimize the reaction time, much higher temperatures are needed, e.g., 1,500°C, which requires a shock wave with $Ma = 6$. It is practically impossible to design a continuous process to translate this theoretical requirement from paper to practice with a single, normal shock wave. Therefore, complex devices are necessary to generate multiple (reflected) oblique shock waves in a confined space in a continuous manner. One such device is the *wave rotor*, which was originally developed as an internal combustion engine supercharger by Brown-Boveri Co. under the commercial name *Comprex*. For more information and details of a wave rotor device for cracking methane (natural gas) to produce hydrogen, the reader can consult [26].

9.8   References

1. World Nuclear Association. 2017. World Nuclear Performance Report, 2nd ed., Report No. 2017/004.
2. El-Wakil, M. M. 1984. *Powerplant Technology*, New York, NY: McGraw Hill.
3. Krymm, R., Lane, J. A. and Zheludev, I. S. 1977. Future trends in nuclear fuel for power reactors, *International Atomic Energy Agency Bulletin*, 19(4), 47–54.
4. International Atomic Energy Agency. 2016. *Advances in Small Modular Reactor Technology Developments, A Supplement to: IAEA Advanced Reactors Information System (ARIS)*. Available at https://aris.iaea.org/Publications/SMR-Book_2016.pdf, accessed July 24, 2018.
5. Kasten, P. R. and Trauger, D. B. 1972. Gas-Cooled Power Reactors and Their Thermal Features, 8th Annual Southeastern Seminar on Thermal Sciences, Vanderbilt University, March 23.
6. Wright, S. A., Parma, E. J. Jr., Suo-Antilla, A. J., et al. 2011. Supercritical CO2 direct cycle gas fast reactor (SC-GFR) concept, ASME Small Modular Reactors Symposium. DOI: 10.2172/1013226.
7. Matsuo, E., Tsutumi, M., Ogata, K. and Nomura, S. 1995. Conceptual design of helium gas turbine for MHTGR-GT, Proceedings of IAEA Technical Committee Meeting, Beijing, China, 30 October – 2 November.
8. Lee, J. C., Campbell, Jr., J. and Wright, D. E. 1980. Closed-cycle gas turbine working fluids, ASME paper 80-GT-135, *Gas Turbine Conference & Products Show, New Orleans, La., March 10–13*.

9. Pocock, R. F. 1970. *Nuclear Ship Propulsion*, London, UK: Ian Allan Ltd.

10. Eckardt, D. 2014. *Gas Turbine Powerhouse*, Munich, Germany: Oldenbourg Verlag.

11. Frutschi, H. U. 2005. *Closed-Cycle Gas Turbines: Operating Experience and Future Potential*, New York, NY, USA: ASME Press.

12. Aerojet-General Nucleonics. 1964. Army gas-cooled reactor systems program, Final Report IDO-28634, https://doi.org/10.2172/4681181.

13. Crouch, H. F. 1960. *Nuclear Ship Propulsion*, Cambridge, MD: Cornell Maritime Press.

14. Keller, C. 1946. The Escher Wyss-AK closed-cycle turbines. Its actual development and future prospects, *Transactions of the ASME*, 68, 791–882.

15. Copinger, D. A., Moses, D. L., and Cupidon, L. R. 2003. Fort Saint Vrain gas cooled reactor operational experience, Oak Ridge National Laboratory, ORNL/TM-2003/223. https://www.osti.gov/biblio/1495207

16. McDonald, C. F. 1978. The closed-cycle turbine: Present and future prospectives for fossil and nuclear heat sources, *ASME Paper 78-GT-102, International Gas Turbine Conference and Products Show, London, England, April 9–13.*

17. McDonald, C. F., 1995, Helium and combustion gas turbine power conversion systems comparison, *ASME Paper 95-GT-263, International Gas Turbine and Aeroengine Congress and Exposition, June 5–8, Houston, TX, USA.*

18. Gülen, S. C. 2017. Advanced Fossil Fuel Power Systems, In D. Y. Goswami and F. Kreith (eds.), *Energy Conversion*, 2nd ed., Boca Raton, FL: CRC Press.

19. McDonald, C. F. and Vepa, K. 1977. Ammonia turbomachinery design considerations for the direct cycle nuclear gas turbine waste heat power plant, *ASME paper 77-GT-75, International Gas Turbine Conference and Products Show, Philadelphia, PA, USA, March 27–31.*

20. Utamura, M. 2010. Thermodynamic analysis of part-flow cycle supercritical CO_2 gas turbines, *ASME Journal of Engineering for Gas Turbines and Power*, 132(11), 111701.

21. Zhang, Z. and Jiang, Z. 1995. Design of indirect gas turbine cycle for a modular high-temperature gas-cooled reactor, *Proceedings of IAEA Technical Committee Meeting, Beijing, China, 30 October – 2 November* .

22. Hejzlar, P., Dostal, V., Driscoll, M. J., et al. 2005. Assessment of gas-cooled fast reactor with indirect supercritical CO_2 cycle, *Nuclear Engineering and Technology*, 38, 109–118.

23. Bartela, Ł., Gładysz, P., Andreades, C., Qvist, S. and Zdeb, J. 2021. Techno- Economic Assessment of Coal Fired Power Unit Decarbonization Retrofit with KP-FHR Small Modular Reactors, *Energies*, 14, 2557.

24. Msheik, M., Rodat, S. and Abanades, S. 2021. Methane cracking for hydrogen production: A review of catalytic and molten media pyrolysis, *Energies*, 14, 3107.

25. Serban, M., Lewis, M. A., Marshall, C. L. and Doctor, R. D. 2003. Hydrogen production by direct contact pyrolysis of natural gas, *Energy & Fuels*, 17, 705–713.

26. Akbari, P., Copeland, C. D., Tuchler, S., Davidson, M. and Mahmoodi-Jezeh, S. V. 2021. Shock wave heating: A novel method for low-cost hydrogen production, *IMECE2021–69775, International Mechanical Engineering Congress and Exposition, November 1–5.*

10 Supercritical CO$_2$

Supercritical CO$_2$ is the working fluid of a *closed-cycle* heat engine. The term supercritical refers to the fact that the cycle state-points are above the critical point of carbon dioxide (73.8 bar, 31°C). At such conditions, the phase of the substance in question is referred to as "fluid," i.e., not "gas" nor "liquid." Thus, a supercritical CO$_2$ (sCO2 henceforth) heat engine operating in a Brayton cycle can indeed be called a closed-cycle "dense gas turbine."

Readers interested in the theory and history of closed-cycle gas turbines are referred to the one and only source on the subject matter, namely, the monograph by Hans Ulrich Frutschi [1]. Several closed-cycle concepts with helium as the working fluid were developed primarily with nuclear power plant applications in mind. Helium, with the lowest neutron capture cross section (it is chemically and neutronically inert or *non-fertile*[1]), was considered an ideal coolant for high-temperature nuclear reactors and fast breeders. sCO2 and its mixture with helium were also considered as working fluids for similar applications and many studies were done on "direct" cycles with either working fluid [2]. This subject is covered in more detail in Chapter 9.

Around 2010 and thereafter, trade publications and archival journals were inundated with articles and papers filled with hyperbole and lofty claims about closed-cycle sCO2 turbines and their merits. In particular, the sCO2 cycle/turbine was/is touted as a technology that can replace Rankine (steam) cycles and steam turbines in conventional fossil fuel–fired power generation, as a stand-alone or as the bottoming cycle of a gas turbine combined cycle (GTCC). The author dissected the thermodynamics of sCO2 power cycles in an earlier article and showed that the claims have no basis in utility-scale power generation [3]. The reader can also refer to Chapter 22 in **GTFEPG** for a formal mathematical derivation of why an sCO2 bottoming cycle cannot be a realistic replacement for a Rankine (steam) bottoming cycle. For more information on the realistic description and assessment of supercritical CO$_2$ technology, the reader is referred to Refs. [4–9].

In this chapter, the performance of sCO2 in power generation applications is rigorously assessed with in-depth thermodynamic analysis and cycle data. Furthermore, we will look at the operability challenges presented by the unique structure of the sCO2 powertrain and heat exchangers. Note that sCO2 has no realistic

---

[1] A fertile material can be converted into a fissile material by neutron absorption and subsequent nuclei conversions. An example is U-234, which can be converted into U-235.

chance to become a topping cycle à la the gas turbine Brayton cycle in a GTCC configuration. Why this is so will be amply clear in the discussion below. Furthermore, sCO2 has no realistic chance of becoming a stand-alone electric power generation technology either – in utility scale, i.e., several hundred-megawatt ratings – with fossil fuels, coal, or natural gas. Thus, this option will not be given too much attention other than demonstrating this assertion dramatically with a sample cycle analysis. The fact that the sCO2 bottoming cycle cannot be a feasible alternative to the steam Rankine bottoming cycle in a GTCC has been elucidated with rigorous theoretical analysis in **GTCCPP**. The reader is pointed to that book or to the article by the author cited earlier [3].

10.1   sCO₂ Power Cycles

Supercritical CO_2 heat engines can operate in a Brayton or Rankine cycle. Both cycles are shown on a T-s diagram in Figure 10.1. In passing, it should be emphasized that the only difference between the sCO2 Rankine and Brayton cycles in Figure 10.1 is in the heat rejection part of the cycles. The phase change of the working fluid from vapor to liquid is present in the low-pressure region of the cycle labeled as *Rankine* but absent in the high-pressure region. To be precise, the cycle should be called "half-Rankine" or "Rankine–Brayton hybrid." Instead of this awkward phrasing, it is referred to as the Rankine cycle. However, the reader should be cognizant of the difference between the two Rankine cycles, i.e., sCO2 and steam.

In their most basic embodiment, both cycles incorporate *recuperation* of heat from the exhaust stream of the turbine to heat the compressed working fluid from the discharge of the compressor/pump. Due to its very low cycle pressure ratio (PR), i.e., about 3:1, without recuperation the sCO2 cycle efficiency would be very low because cycle mean-effective heat addition and rejection temperatures, METH and METL, respectively, would be very close to each other. Graphically, this situation can be readily recognized by the "thinness" of the Brayton and Rankine cycles in Figure 10.1. Without recuperation, logarithmic means of temperatures in the heat addition process $(2 \rightarrow 3)$ and the heat rejection process $(4 \rightarrow 1)$ would be quite close to each other. (For verification, it is easy to visualize that, in the limit of cycle PR $\rightarrow$ 1:1, METH and METL would be exactly equal.)

The sCO2 Rankine cycle variant in Figure 10.1 is referred to as a *split-flow recompression* cycle. Using the cycle diagram in the upper left corner of Figure 10.1, let us describe the cycle processes. The heat exchanger between state-points 6 and 3 is where heat is added to the cycle. The logarithmic mean of the two temperatures defines the cycle's mean-effective heat addition temperature, i.e.,

$$\text{METH} = \frac{h_3 - h_6}{s_3 - s_6} \approx \frac{T_3 - T_6}{\ln\left(\frac{T_3}{T_6}\right)}. \tag{10.1}$$

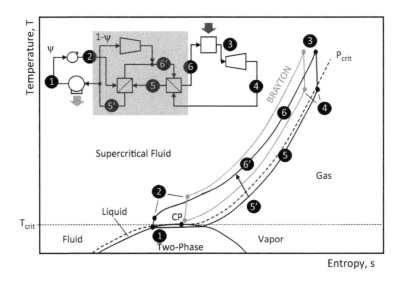

Figure 10.1 Temperature-entropy (T-s) diagram of supercritical CO_2 Rankine cycle with split-flow and recompression.

The heat exchanger can one be of two types:

- Fired (analogous to the boiler of a conventional steam power plant),
- Unfired (analogous to a heat recovery steam generator [HRSG], utilizing the exhaust gas of a gas turbine).

Hot sCO2 (at 250–300 bara and 500–760°C, depending on the heat source) is expanded in the sCO2 turbine to state-point 4 (about 65–85 bara depending on the heat rejection type and coolant source). Expanded sCO2 goes through a *high-temperature recuperator* (HTR) to preheat compressed "cold" sCO2 coming from the cycle compressor discharge (state-point 6′) and cools down to state-point 5. After the HTR, the sCO2 at state-point 5 goes through a second, *low temperature recuperator* (LTR) and preheats compressed sCO2 from the cycle pump/compressor discharge (state-point 2). Downstream of the LTR, sCO2 (state-point 5′) splits into two branches: Only a fraction, ψ, is sent through the condenser and the main cycle pump; the remaining fraction, $1 - ψ$, is directed to the "recompression" or "bypass" compressor (recompressor). (One can easily verify that for $ψ = 1.0$, the split-flow cycle reverts to the basic recuperated cycle.) Due to the reduction in cycle heat rejection ($ψ < 1$), recuperation effectiveness and cycle thermal efficiency improve significantly. Cycle studies have shown that the optimal value of ψ is around 0.7. The two branches merge again upstream of the cold inlet of the HTR.

As was shown in Chapter 2, for an ideal cycle, the efficiency can be written as

$$EFFID = 1 - \frac{METL}{METH},\tag{10.2}$$

where METL is the mean-effective heat rejection temperature of the cycle. In this case, for the cycle shown in Figure 10.1,

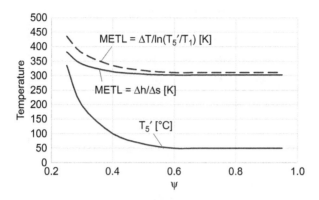

Figure 10.2 sCO2 split-flow recompression cycle mean-effective heat rejection temperature.

$$\text{METL} = \frac{h_{5'} - h_1}{s_{5'} - s_1} \approx \frac{T_{5'} - T_1}{\ln\left(\frac{T_{5'}}{T_1}\right)}. \qquad (10.3)$$

Before moving on, let us spend some time on METL. The approximation in Equation (10.3) is not as solid as the approximation in Equation (10.1). Calculations using the REFPROP property package in THERMOFLEX (see Section 16.2) confirm that, as illustrated in Figure 10.2. As ψ decreases, state-point 5′ moves up along the lower cycle isobar in Figure 10.1, i.e., deeper into the supercritical fluid region, whereas state-point 1 is constant (saturated liquid). The approximation in Equations (10.1) and (10.3) is dependent on the weak nonlinearity of enthalpy and entropy as a function of temperature. This is indeed the case for state-points 6 and 3 (and the error is less than 1 K) but not so for state-points 5′ and 1. The error is about 8 K for $\psi > 0.6$ when state-point 5′ is closest to the two-phase dome. It becomes as high as about 54 K when $\psi = 0.25$ and state-point 5′ is farthest from the two-phase dome.

We can now return our attention to the cycle efficiency. It will be shown below that, in the case of split-flow cycle, one cannot use Equation (10.2). Why? Because the implicit assumption leading to Equation (10.2) is that the working fluid flow rate is the same during the heat addition and heat rejection processes. In other words, the derivation steps of Equation (10.2) are as follows:

$$\text{EFFID} = \frac{W}{QH} = \frac{OH - QL}{QH} = 1 - \frac{QL}{QH},$$

$$\text{EFFID} = 1 - \frac{QL}{QH} = 1 - \frac{ML \times CP \times DTL}{MH \times CP \times DTH},$$

$$MH = ML \ (\psi = 1.0),$$

$$\text{EFFID} = 1 - \frac{CP \times DTL}{CP \times DTH},$$

$$CP \times DTH = METH \times DSH,$$

$$CP \times DTL = METL \times DSL,$$

$$DSH = DSL \text{ (always, for the } ideal \text{ cycle)},$$

$$EFFID = 1 - \frac{METL \times DSL}{METH \times DSH} = 1 - \frac{METL}{METH}. \tag{10.4}$$

In the formulae above, individual parameters are defined as follows:

W = Cycle net work output [kW]
QH = Heat added to the cycle [kWth]
QL = Heat rejected from the cycle [kWth]
MH = Working fluid flow rate during heat addition [kg/s]
ML = Working fluid flow rate during heat rejection [kg/s]
DTH = Temperature change during heat addition [K]
DTL = Temperature change during heat rejection [K]
DSH = Entropy change during heat addition [kJ/kg-K]
DSL = Entropy change during heat rejection [kJ/kg-K]

What if $MH \neq ML$? In that case, Equation (10.4) becomes

$$EFFID = 1 - \frac{ML \times METL}{MH \times METH},$$

$$\psi = \frac{ML}{MH},$$

$$EFFID = 1 - \psi \frac{METL}{METH}. \tag{10.5}$$

For $\psi = 1$, there is no flow split and $ML = MH$ so that Equation (10.5) reverts to Equation (10.4). For $\psi = 0$, Equation (10.5) returns $EFFID = 1$, which implies a violation of the Kelvin–Planck statement of the second law of thermodynamics, i.e., $W = QH$.

Ideal cycle calculations are done in THERMOFLEX with the REFPROP property package (see Section 16.2). Cycle assumptions are as follows:

- Cycle high pressure 205.6 bar, low pressure 65.88 bar (PR = 3.12:1)
- Turbine inlet temperature 600°C
- Zero pressure and heat losses
- 100% recuperator effectiveness
- 100% component isentropic efficiency
- Flow split ψ changes between 0.25 and 0.95

The calculation results are summarized in Figure 10.3 and Table 10.1. Figure 10.3 contains three plots: cycle efficiency from Equation (10.5) with METH from Equation (10.1) and METL from Equation (10.3) (using state-point enthalpies and entropies from the heat and mass balance output), cycle efficiency as the ratio of net cycle shaft

Table 10.1 Ideal sCO2 split-flow recompression cycle – key quantities

Q_HPR: High-pressure recuperator duty; Q_LPR: Low-pressure recuperator duty

ψ	SHAFT, kJ/kg	PUMP, kJ/kg	COMP, kJ/kg	QL, kJ/kg	QH, kJ/kg	NET, W kJ/kg	Q_HPR, kJ/kg	Q_LPR, kJ/kg	Total REC, kJ/kg
0.25	65	5	109	−129	189	60	33	160	193
0.30	98	6	76	−109	201	92	144	136	280
0.40	127	8	47	−101	220	119	267	124	392
0.50	142	10	32	−102	234	132	315	124	439
0.60	151	12	23	−109	248	139	346	117	463
0.68	156	14	18	−123	265	142	374	89	463
0.75	160	16	14	−136	280	144	392	71	463
0.80	162	17	11	−145	291	146	404	59	463
0.90	168	19	6	−163	313	150	427	36	463
0.95	171	20	3	−172	324	152	441	21	463

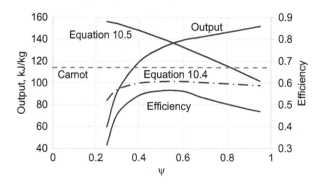

Figure 10.3 Ideal sCO2 split-flow recompression cycle performance.

output and heat input, and net cycle shaft output. Table 10.1 provides details about the performance of individual cycle components with changing ψ to help with the analysis of the findings from Figure 10.3. State-point thermodynamic properties are listed in Table 10.2.

Takeaways from the chart in Figure 10.3 are as follows:

- Ideal efficiency from Equation (10.4) displays the expected trend with varying ψ
- Actual cycle efficiency is significantly lower and goes through a maximum at ψ ~ 0.5
- Net cycle output decreases with decreasing ψ monotonically ntil ψ ~ 0.5 and exponentially thereafter

This finding is somewhat surprising because when we construct an ideal cycle with zero losses and isentropic components, the expectation is that the efficiency from the cycle heat and mass balance would be the same as that given by Equation (10.5). Before delving into this deeper, let us look at another interesting finding from this exercise.

Table 10.2 Ideal sCO2 split-flow recompression cycle – state-point data

State Point	Flow Fraction	p Bar	T °C	h kJ/kg	s kJ/kg-K
1	ψ	65.9	26.0	−226.5	−1.4735
2	ψ	205.6	50.2	−206.3	−1.4735
3	1.0	205.6	600.0	591.3	0.0719
4	1.0	65.9	446.4	417.5	0.0719
5	1.0	65.9	116.2	44.0	−0.6223
5'	1.0	65.9	50.2	−45.3	−0.8748
6	1.0	205.6	385.5	326.2	−0.2763
6'	1.0	205.6	116.1	−47.4	−1.0205

In the example we looked at, the cycle maximum temperature is 600°C. For a Carnot engine operating between a hot temperature reservoir at 600°C and a cold temperature reservoir at 15°C (i.e., ISO ambient), the efficiency is $1 - (15 + 273.15)/(600 + 273.15) = 0.670$, or 67%. This value is lower than the ideal cycle efficiency given by Equation (10.5) for $\psi < 0.8$ Prima facie, this seems like a violation of the second law of thermodynamics. In fact, however, what we see here is another manifestation of the split-flow feature of the sCO2 cycle. The theoretical Carnot efficiency implicitly assumes $\psi = 1$. Thus, a consistent comparison with the Carnot efficiency is the efficiency predicted by Equation (10.2).

Coming back to the original question, the discrepancy between the prediction of Equation (10.5) and the actual ideal cycle efficiency is due to the non-ideality of 100% effective recuperators. This can be readily seen in the heat transfer diagram of the high-pressure (HP) recuperator (for $\psi = 0.68$) in Figure 10.4. The exergy destruction ("lost work") in this heat exchanger is quantified by the area between the hot and cold fluid heat transfer lines. It can be calculated using the steady-state, steady-flow exergy (availability) balance for the heat exchanger control volume, i.e.,

$$\dot{W}_{lost} = \sum_{i=1}^{4} \dot{m}_i (h_i - T_0 s_i),$$

where the subscript i denotes cold and hot fluid inlet and exit state-points, h is the fluid enthalpy (in kJ/kg), s is fluid entropy (in kJ/kg-K), and T_0 is the "dead state" temperature (i.e., the ambient temperature, in K). Based on the state-point property data in Table 10.2, lost work in the HP recuperator in Figure 10.4 is 14.4 kJ/kg.

The other impact of "imperfect" heat exchange on cycle efficiency is via increased heat addition (by about 75 kJ/kg). When a hypothetical case is run with "perfect" recuperators (recognizing that this is a practical impossibility), the ideal cycle efficiency is calculated to be 74.34% (for $\psi = 0.68$), which is the same as predicted by Equation (10.5), 74.08%. The small difference can be attributed to the real gas property package used in the simulation software and the small lost work due to mixing downstream of the bypass compressor (about 0.75 kJ/kg).

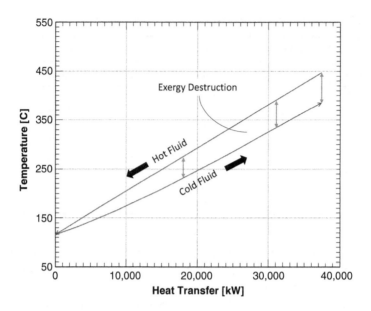

Figure 10.4 High-pressure recuperator (100% effectiveness) heat transfer diagram.

Finally, let us answer the most important question: What performance should one expect from a "real" sCO2 power plant in the field? As was shown in detail in Chapter 6 in **GTFEPG** and in Chapter 2 of the present book (see Figure 2.4 in particular), the state of the art in gas turbine technology at present is quantified by a technology factor value of 0.7–0.75. It is unreasonable to expect to do better than this with sCO2 technology. In Figure 10.3, the maximum ideal cycle efficiency is about 56%. Thus, what we could realistically expect from an sCO2 cycle with a turbine inlet temperature (TIT) of 600°C and cycle PR of about 3:1 is in the range of 39.2–42.1%. For 700°C TIT, the incremental cycle efficiency is 0.6 percentage points.

10.1.1   Partial Cooling Cycle

There are three basic types of supercritical CO_2 cycles, i.e.,

- Simple recuperation cycle
- Split-flow recompression cycle
- Partial cooling cycle

All three cycles incorporate recuperation, without which, due to the low cycle PR (about 3:1), cycle efficiency would be extremely poor. (In other words, METH and METL would be quite close to each other.) Recuperation fixes this basic cycle deficiency by simultaneously increasing METH and decreasing METL. The split-flow recompression feature further improves the efficiency by reducing the cycle heat rejection. The basic thermodynamic mechanism was explained in detail above (i.e., Equation (10.5) vis-à-vis Equation (10.2)).

The partial cooling cycle is a modification of the split-flow recompression cycle. The modification comprises the addition of a cooler and compressor upstream of the recompressor. The flow split takes place at the discharge of the precompressor. The net effect of the precooler and precompressor is to reduce the temperature of sCO2 at the discharge of the recompressor. One effect of this modification is to increase the temperature range of the cycle heat addition, which makes it more suitable to bottoming cycle, i.e., heat recovery, applications. The downside is, of course, a reduction in METH, which is detrimental to cycle efficiency (i.e., see Equations (10.2) and (10.5)).

Fortunately, the efficiency penalty due to low METH is somewhat offset by lower METL due to the low discharge temperature of the pre-compressor vis-à-vis the recompressor in a split-flow recompression cycle. The other effect of partial cooling modification is increase in cycle PR, i.e., from 3:1 to about 4:1 by enabling lower turbine exhaust pressure. Overall, though, with everything else being the same, split-flow recompression has comparable efficiency but the partial cooling cycle boosts plant net output significantly. For a cost and performance compression of the three cycles in a concentrated solar power application, see Neises and Turchi [10]. For the application of the partial cooling cycle to a coal-fired application (with a circulating fluidized bed sCO2 heater), refer to Pidaparti et al. [11], who found higher efficiency and lower COE for the partial cooling cycle with reheat.

10.1.2   Real Cycle Calculation

Published performances of sCO2 cycles depict an unrealistic picture and create expectations that are never going to be fulfilled. In this section, we will look at the real potential of this closed-cycle gas turbine. Before jumping into the analysis, let us first state a few things that are usually neglected (deliberately or not) in marketing hyperbole:

- The end goal is *net electric power* delivered to the grid.
- This number is the power measured at the high voltage terminals of the step-up transformer at the switchyard *after* subtracting the power consumed by the power plant itself.
- The thermal efficiency is the ratio of the net power to the fuel consumption (LHV). In fossil fuel applications, this requires accounting for the fired heat exchanger efficiency and auxiliary power consumption.

The last item is of extreme importance. It will be shown that, in a realistic, cost-effective design, the heat picked up by the cycle working fluid or the intermediate *heat transfer fluid* (HTF) in the fired heater is *much less* than the fuel input.

In the example below, we will look at the performance of a nominal 300 MW (net) sCO2 power plant under two assumptions:

- Natural gas fired heater (760°C sCO2 TIT)
- Nuclear plant with an intermediate heat transfer loop (580°C sCO2 TIT)

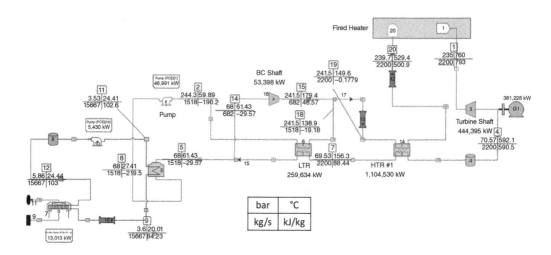

Figure 10.5 Split-flow recompression sCO2 cycle power plant with natural gas fired heater.

Key cycle assumptions are as follows:

- 235 bara sCO2 turbine inlet pressure
- 68 bara sCO2 condenser pressure
- Split-flow recompression cycle with $\psi = 0.69$
- Turbine isentropic efficiency 92%
- Compressor isentropic efficiency 90%
- Main CO₂ pump/compressor isentropic efficiency 90%
- Low temperature recuperator effectiveness 98%
- High-temperature recuperator effectiveness 98.5%
- ISO ambient conditions

The cycle is modeled in THERMOFLEX. Supercritical CO_2 properties are calculated with the REFPROP package (see Section 16.2). The sCO2 turbine and bypass compressor are on the same shaft with the synchronous ac generator and the gearbox. The shaft speed is assumed to be 27,000 rpm; the generator synchronous speed is 3,600 rpm (60 Hz grid). The gearbox efficiency is assumed to be 98.5%. The overall generator plus gearbox efficiency is 97.5%. The main compressor/pump is driven by its own motor. The cycle flowsheet is shown in Figure 10.5 (sCO2 flow rate is 2,200 kg/s). A schematic diagram of the (100% CH_4) natural gas–fired heater is shown in Figure 10.6. The plant and cycle performance is summarized in Table 10.3.

From the stream data in Figure 10.5, the following cycle characteristics are obtained:

- METH is 902.4 K
- METL is 306.5 K
- Thus, from Equation (10.5), the equivalent Carnot efficiency of the cycle is $1 - 0.69 \times 306.5/902.4 = 76.6\%$; as discussed in the previous section, the actual ideal cycle is lower, i.e., 55.4%

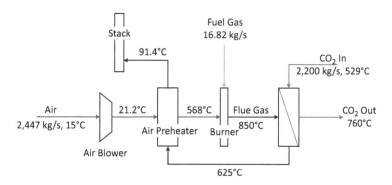

Figure 10.6 Natural gas fired sCO2 heater.

- Carnot efficiency with cycle maximum temperature of 760°C and 15°C ambient temperature is 72.11%
- Thus, the ideal cycle Carnot factor is 55.4/72.1 = 0.79

Plant/cycle gross output is the net output found by subtracting the main pump/compressor motor power from the generator output. Miscellaneous losses are set to 0.1% of the generator output. Transformer losses are estimated as 0.35% of the generator output. Plant performance is calculated for two different heat rejection systems:

- sCO2 condenser cooling water heat rejection through fin-fan coolers
- Once-through (open loop) water-cooled condenser

The first one is the most likely option in many sites due to strict environmental protection regulations and/or water scarcity. The second one is the bare bones minimum for rating performance. The efficiency of the fired heater in Figure 10.6 is 76.3% (the ratio of sCO2 heat pickup to fuel input in LHV). As an entitlement, the performance with a water-cooled condenser is recalculated with an assumed "boiler efficiency" of 90%, typical of modern coal-fired utility boilers with 20% excess air and 60 mbar total draft loss.

There is a fourth case in Table 10.3 with partial cooling cycle described in Section 10.1.1. A precooler and precompressor have been added to the cycle upstream of the recompressor (bypass compressor). Turbine PR is increased to 4:1 by reduction of the discharge pressure to 58.2 bar. In the precooler, sCO2 from the LT recuperator is cooled to 29°C and compressed to 68 bar in the precompressor. Flow split is optimized at $\psi = 0.56$.

The key takeaways from the data presented in Table 10.3 for the split-flow recompression cycle are as follows:

- Best possible efficiency (akin to ISO base load rating numbers published in the GTW Handbook) is 46% net LHV (equivalent to a technology factor, TF, of 0.67), i.e., not too far from a modern, supercritical, pulverized coal power (SCPC) plant
- The efficiency can be as low as about 35% (TF = 0.61) if the open-loop water-cooled heat rejection option is not available and the fired heater efficiency is low (76% in the example)

Table 10.3 Natural gas-fired sCO2 power plant performance

	Case 1	Case 2	Case 3	Case 4
Cycle Type	Split-Flow Recompression	Split-Flow Recompression	Split-Flow Recompression	Partial Cooling
Heat Rejection	Air-Cooled	Water-Cooled	Water-Cooled	Water-Cooled
Turbine Pressure Ratio	3.3:1	3.3:1	3.3:1	4.0:1
Turbine Shaft Output, kW	444,395	444,395	444,395	507,110
Re-compressor Shaft Power, kW	53,398	53,398	53,398	65,878
Pre-compressor Shaft Power, kW	NA	NA	NA	19,859
Net Shaft Output, kW	390,997	390,997	390,997	410,977
Generator + Gearbox Efficiency, %	97.5	97.5	97.5	97.5
Generator Output, kW	381,228	381,228	381,228	381,228
Fuel Consumption (LHV), kWth	842,046	842,046	714,086	784,422
Plant Gross Thermal Efficiency, %	39.7	39.7	46.8	47.5
Heat Added to sCO2, kWth	642,677	642,677	642,677	705,980
"Boiler" Efficiency, %	76.3	76.3	90.0	90.0
Cycle Gross Efficiency, %	52.0	52.0	52.0	52.8
Main Pump/Compressor Motor, kW	46,991	46,991	46,991	38,150
Fin-Fan Cooler Motor, kW	13,010	0	0	0
Circ Pump Motor, kW	5,429	2,292	2,292	2,526
Air Fan, kW	15,929	15,929	2,000	2,196
Miscellaneous, kW	381	381	381	411
Transformer Loss, kW	1,334	1,334	1,334	1,438
Total Auxiliary Power, kW	83,075	66,927	52,998	44,721
Plant Net Output, kW	298,153	314,301	328,330	366,256
Plant Net Thermal Efficiency, %	35.4	37.3	46.0	46.7
Cycle Net Efficiency, %	46.4	48.9	51.1	51.9
Specific Power, kJ/kg	136	143	149	166

- Low power density of the power plant, i.e., 314 MW/2,200 kg/s = 143 kJ/kg (vis-à-vis around 1,200 kJ/kg for an SCPC power plant with an advanced steam turbine generator)
- The difference between *plant thermal* and *cycle* efficiencies, e.g., at least *five* percentage points on a net basis

Clearly, while its efficiency on a *gross cycle* basis is quite impressive at 52%, on a *net plant thermal* basis, sCO2 technology does not present a threat to conventional boiler-steam turbine (Rankine cycle) technology. Whether an sCO2 "boiler" of 90% can be designed cost effectively remains to be seen. Even then, the disadvantage associated with the low power density of the sCO2 cycle (nearly one-tenth of that of the steam cycle), i.e., high combustion air flow rate and large air blower power consumption, will always be present. When it comes to the bottom line, the *raison d'être* of an electricity generation plant, namely the ratio of net electric power delivered to the grid to the fuel burned, the sCO2 does not pass muster.

The partial cooling option provides a significant boost in net output, i.e., by about 38 MW or 11.2% with a modest rise in net efficiency by 0.7 percentage points. The improvement can be increased to about 55 MW (16.9%) in output and 1.0 percentage points in net efficiency if the cycle/turbine pressure ratio is increased to 4.57:1 by reducing the exhaust pressure to 51.4 bar (the precompressor PR is 1.4:1). Whether the performance improvement justifies the significant increase in cycle complexity and CAPEX remains an open question until a comprehensive FEED study is carried out. The study presented by Pidaparti et al. [11] claims that the answer is affirmative based on their LCOE analysis.

The next task is to look at the sCO2 closed-cycle gas turbine technology in an application, for which it was originally intended more than half a century ago, i.e., as the nuclear reactor power cycle. In this case, cycle heat addition is accomplished via an intermediate heat exchange loop utilizing an HTF such as molten salt. This loop facilitates the transfer of the heat absorbed from the reactor coolant to the sCO2 working fluid. It should be noted that this heat transfer can be done directly as well. In other words, sCO2 and reactor coolant can exchange heat in a direct configuration. A further modification is that the cycle working fluid, sCO2, *is* the reactor coolant. (For the pros and cons of these options, the reader is advised to consult Chapter 9.) Nevertheless, those variants are not of prime importance to the task at hand, i.e., evaluation of the sCO2 cycle performance.

For this exercise, the intermediate heat exchange loop HTF is molten salt, which enters the heat exchanger at 595°C and heats sCO2 to 580°C. All the other cycle assumptions are the same as in the fired heater example. The flow rate of sCO2 is 2,200 kg/s. The cycle flowsheet is shown in Figure 10.7 and the calculated performance metrics are summarized in Table 10.4. In this application, net plant efficiency is quite respectable, i.e., more than 44% with an open-loop water-cooled heat rejection system. Due to the lower TIT, net output, at about 250 MW net, is lower than that for the fired heater case. Specific power is also low at around 110 kJ/kg (vis-à-vis about 650 kJ/kg for a nuclear steam cycle with saturated steam). In conclusion, this is one application, especially with small modular reactors, where sCO2 is a good candidate as the power cycle.

A third case is calculated with partial cooling modification (see Section 10.1.1). The working fluid flow rate is adjusted to keep the generator output constant at a nominal 300 MWe. (For this application, probably a more appropriate approach is to keep the cycle heat input constant.) Precompressor PR is set to 1.3:1. The efficiency improvement is 0.7 percentage points with a net output increase of 4.5%. At the same cycle heat input, the efficiency improvement stays the same with a net output increase of 1.6% (252 MWe).

It is worth noting that the nuclear power application example presented above is equally valid for a concentrated solar power application. In particular, the molten salt

Table 10.4 Nuclear sCO2 power plant performance

	Case 1	Case 2	Case 3
Cycle Type	Split-Flow Recompression	Split-Flow Recompression	Partial Cooling
Heat Rejection	Air-Cooled	Water-Cooled	Water-Cooled
Turbine Pressure Ratio	3.33:1	3.33:1	4.24:1
sCO2 Flow Rate, kg/s	2,200	2,200	2,050
Flow Split (ψ)	0.69	0.69	0.56
Re-compressor Shaft Power, kW	53,357	53,357	64,119
Pre-compressor Shaft Power, kW			23,913
Net Shaft Output, kW	307,381	307,381	307,312
Generator + Gearbox Efficiency, %	97.4	97.4	97.4
Generator Output, kW	299,438	299,438	299,371
Heat Added to sCO2, kWth	558,606	558,606	574,428
Cycle Gross Efficiency, %	45.2	45.2	45.9
Main Pump/Compressor Motor, kW	46,991	46,991	35,553
Fin-Fan Cooler Motor, kW	13,004	0	0
Circ Pump Motor, kW	5,418	2,290	2,587
Molten Salt Pump, kW	644	644	662
Miscellaneous, kW	299	299	299
Transformer Loss, kW	1,048	1,048	1,048
Total Auxiliary Power, kW	67,404	51,272	40,148
Plant Net Output, kW	232,034	248,166	259,223
Cycle Net Efficiency, %	41.5	44.4	45.1
Specific Power, kJ/kg	105	113	126

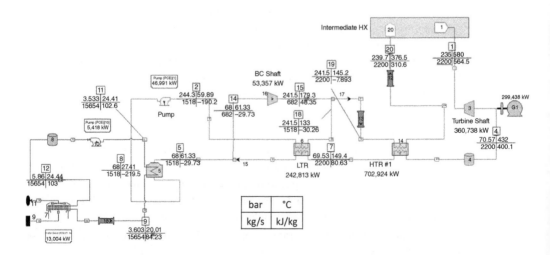

Figure 10.7 Split-flow recompression sCO2 cycle power plant with intermediate heat exchanger loop.

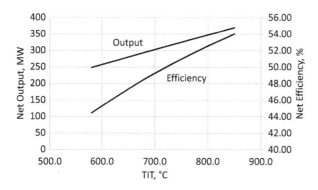

Figure 10.8 sCO2 turbine plant performance as a function of TIT.

HTF can be heated in the receiver of a solar tower (e.g., the *central receiver system* – see Chapter 12). As far as the sCO2 turbine is concerned, the source of heat added to the cycle is immaterial. What counts is (i) its magnitude and (ii) its *quality*, i.e., its temperature. The performance summary in Table 10.4 is for a TIT of 580°C. Net output and efficiency as a function of TIT are plotted in Figure 10.8 (same cycle PR, i.e., 3.33:1 at the turbine). At a TIT of 850°C (1,562°F), net efficiency is 54% (split recompression cycle). Note that the cycle model used in this sample calculation does not account for turbine cooling. Depending on the materials used in turbine casing and flow path (including the turbine rotor), some cooling is likely to be required, with a detrimental impact on cycle efficiency. The other design challenge is associated with material selection and sealing for heat exchanger, piping, and the inlet (throttle) valve (control and stop functions). In essence, temperatures beyond 600°C – especially at pressures pushing 300 bar – are challenging from hardware design, development, and demonstration perspectives. Difficulties encountered in the development of piping, valves, and turbine materials for ultra-supercritical fossil fuel–fired power plant technology provide ample cautionary tales in that regard.

10.2   Operation of the sCO$_2$ Turbine

Supercritical CO$_2$ is a closed-cycle gas turbine. In other words, at the design operating point there is a fixed mass (commonly referred to as the *inventory*) of working fluid circulating in the cycle. The standard way of changing the power output of the gas turbine is by changing the amount of the circulating mass of working fluid. The standard way of doing this, going back to the early Escher–Wyss units, is inventory extraction. What is meant by extraction is to divert the working fluid from the discharge of the compressor to a storage vessel. A simplified closed-cycle gas turbine with recuperation and storage vessel is shown in Figure 10.9.

As shown in Figure 10.9, the gas turbine power output is reduced by extracting the working fluid from the compressor discharge to the storage vessel. Due to the fixed

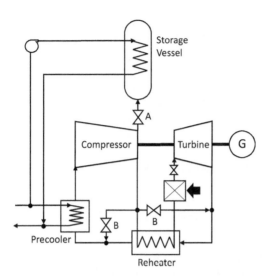

Figure 10.9 Closed-cycle gas turbine with recuperation – simplified diagram.

swallowing capacity of the turbine (S1N area), the cycle pressure is reduced with lower mass flow rate. Since the gas turbine is synchronized to the grid and rotating at a constant speed, the compressor inlet volume flow is constant. In other words, pressure at the compressor inlet (i.e., the lower pressure leg of the cycle) is reduced in proportion. To maintain a constant TIT, the heat addition in the main heater is reduced in direct proportion to the working fluid rate. As far as the simple cycle in Figure 10.9 is concerned, it is easy to see that the gas turbine can be run at part load at essentially the same cycle efficiency.

In a split-flow recompression cycle, the basic principle of part load control via inventory extraction is essentially the same. However, the added complexity of multiple compressors operating at their own optimal rotational speeds, different from that of the turbine rotating at its own, constant rotational speed (connected to the generator through a gearbox), requires compressor inlet guide vane (IGV) control. How that can be done is dependent on the exact nature of the powertrain configuration.

Another method to reduce power output is by closing the throttle valve at the inlet of the turbine. This increases the pressure drop across the valve itself and reduces the pressure drop across the turbine. The power output is thus reduced, but at the expense of lower cycle efficiency due to the exergy destruction across the throttle valve.

Bypass valves at the compressor discharge (labeled B in Figure 10.9) are used to control speed during startup or emergency transients such as turbine trip or load shedding. Due to high cycle pressures and temperatures, turbine casings, pipes, and heat exchanger tubes and shells have thick walls susceptible to thermal stresses during transients such as startup or shutdown. In particular, the turbine rotor, which must be manufactured from austenitic steel for temperatures up to 700°C or higher, cannot be heated at a rate more than 200°C per hour. Another critical component in this regard is the turbine combined stop-control (throttle) valve (similar to the steam turbine SCV upstream of the HP turbine).

For the 10 MW sCO2 power block to be used in a CSP application, the following predictions were made. Assuming 2 MWth incident receiver power, about 50 minutes were estimated to heat up 70 tons of metal (two recuperators plus the solar receiver) from a cold condition to the operating temperature. This could be accelerated by insulating the equipment for 140°F external temperature. The target turbomachinery start time was specified as 15 minutes with an upper limit of 30 minutes. (The intended expander rotor material was Nimonic 105.[2] The rotor was of a single piece (monolithic) construction and manufactured using a 5-axis *Electric Discharge Machining* process.) Numerical simulation results of the key transients of this power block can be found in the paper by Tang et al. [12].

Closed-cycle gas turbine efficiency is highly dependent on the effectiveness of the recuperative heat exchangers and turbomachinery seals. The requisite performance cannot be met by conventional shell-and-tube design. Therefore, a *printed circuit heat exchanger* fabricated out of alloy steels is the preferred technology. These components are highly susceptible to high stresses at block joints and at headers/nozzles that can lead to low cycle fatigue (LCF) problems.

There is not yet any field experience with sCO2 turbines or power plants. The first attempt to design and operate a larger-than-laboratory scale demo/test was the *Advanced Projects Offering Low LCOE Opportunities* (APOLLO, or Apollo) project under the aegis of the US DOE's *SunShot* initiative. The project was a collaboration with General Electric (GE) and the Southwest Research Institute to develop a high-efficiency sCO2 compression system (80% compressor efficiency with variable IGVs and OGVs). The compression system in question comprised two compressors, the main compressor and the recompressor, each rated at 2 MW, of the split recompression sCO2 cycle. The system is intended as a modular power block for CSP applications. The powertrain (27,000 rpm) also comprises a 14 MW four-stage sCO2 expander for net 10 MW (rated) generator output. Turbine inlet conditions were 715°C and 250 bar with about 85 kg/s sCO2 flow rate. The work under the SunShot initiative encompassed six years (2012–2018) with limited success but a wealth of lessons learned for the next step (no pun intended – see below).

The turbomachinery configuration options onsidered in the SunShot program are summarized in Table 10.5. The down-select was based on a trade-off study considering economic and reliability criteria. At the end of the study, the high-speed, geared design was selected (see Figure 10.10). Detailed design analysis of the turboexpander and the recuperator can be found in the project report by Moore [13]. For the testing phase, the test loop was designed and constructed to run at one-tenth of the design flow with a single recuperator and gas fired heater providing the cycle heat input (instead of a solar receiver) – see Table 10.6 for the design data.

During the testing campaign, both the turboexpander and the compressor experienced failures associated with the *dry gas seals* (DGS), which are used to separate the

---

[2] Nimonic 105 is a wrought nickel-cobalt-chromium-base alloy strengthened by additions of molybdenum, aluminum, and titanium. It has been developed for service in temperatures of up to 950°C and combines the high strength of the age-hardening nickel-base alloys with good creep resistance.

Table 10.5 sCO2 powertrain configuration options

IC: Inductively coupled; PM: Permanent magnet

Option	Generator	Compressor	Turbine	RPM
High speed, optimal	A. IC	A. Single stage centrifugal	A. Radial	Optimized for compressor
	B. PM	B. Multi stage pump	B. Axial	
High speed, expander only	A. IC	None	A. Radial	Optimized for expander
	B. PM		B. Axial	
High speed, geared	A. IC	A. Single stage centrifugal	A. Radial	Both expander and compressor run at optimal speed
	B. PM	B. Multi stage pump	B. Axial	
	C. 3,600 rpm			
3,600 rpm integrated	3,600 rpm	Multi stage pump or compressor	Multi stage axial at 3,600 rpm	3,600 rpm
3,600 rpm – expander only	3,600 rpm	None	Multi stage axial at 3,600 rpm	3,600 rpm

Table 10.6 SunShot 1-MW test loop recuperator and gas-fired heater design parameters

	Recuperator Outlet / Heater Inlet	Heater Outlet/ Turbine Inlet
Temperature, °C	470	715
Pressure, bar	251.9	250.9
Mass flow rate of CO$_2$, kg/s	8.410	8.410

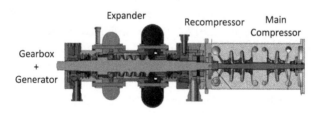

Figure 10.10 SunShot (Apollo) turboexpander and compressors.

oil-lubricated bearings from the working fluid. The failure in the turboexpander was ultimately traced to a vendor design error that led to a thermal runaway condition. Due to the program delay created by the DGS failure (including the RCA and manufacturing of the replacement parts), the SunShot turbine was only able to run for about 40 hours total test time. Even with the lessons learned during the turboexpander test, the Apollo compressor operated for 16 hours at low pressure (20 bar), but only 4 hours total test time at high pressure (85 bar) prior to – again – DGS failure. A detailed RCA was unable to pinpoint the exact root cause.

At the time of writing, a 10 MW, $119 million facility is being constructed on the premises of the *Southwest Research Institute* (SwRI) in San Antonio, TX. The project is undertaken under the US DOE's *Supercritical CO2 Transformational Electric Power* (STEP) program (the US DOE is footing $84 million of the bill). A team led by the Gas Technology Institute, SwRI, and GE has initiated a project to design, construct, commission, and operate the facility. The project goal is to advance the state of the art for high-temperature sCO2 power cycle performance from TRL 3 to TRL 7. The STEP turboexpander is designed to incorporate improvements in the SunShot unit described above, e.g., increased casing and rotor life (100,000 hours vis-à-vis 20,000 hours), shear ring retention rather than bolts, a design for couplings on both shaft ends, and improved aero performance with an increased volute flow area. Special attention is given to the DGS design and thermal management based on the lessons learned from the SunShot.

10.3   The Allam Cycle

The Allam cycle[3] is a semi-closed cycle with oxy-combustion wherein CO_2 constitutes (nominally) 95% of the fluid flow in the combustor (by mass) with the remaining 5% made up of oxygen and fuel. Oxygen for combustion is generated by a cryogenic *air separation unit* (ASU), which is also a critical component of the IGCC system (see Chapter 13). Carbon dioxide generated by the combustion is taken away from the cycle at the CO_2 pump discharge to maintain the cycle mass balance (hence *semi-closed*). The resulting combustion product is roughly 90% CO_2, and the ASU parasitic power consumption is minimized by the lower O_2 requirement. The claimed net LHV efficiency of the Allam cycle is nearly 59% [14–19]. The conceptual cycle with 1,150°C TIT, 300 bara turbine inlet pressure, and a cycle PR of 10:1 is schematically described in Figure 10.11 in its most basic form (i.e., with simple recuperation).

A detailed plant performance analysis based on the Allam cycle is provided by Sifat and Haseli [19]. The authors compare the claimed performance from the 2017 paper by Allam et al. [17] with their own model prediction as presented in Table 10.7. No information is available on the details of the cycle turbine (provided by Toshiba), e.g., number of stages, stage efficiencies, cooling flows. Sifat and Haseli's simple model assumed an uncooled expander with 90% efficiency. Compressor and pump efficiencies were taken as 85% and 75%, respectively. Even with such simple yet optimistic assumptions, the authors could only predict 55% efficiency with the same cycle parameters reported by the original developers. Using the assumptions from the MS thesis of Manso [20], the authors could come up with only 51.8% (lower turbine efficiency).

Scaccabarozzi et al. [21] used an *Aspen Plus*[4] model to investigate the performance of the Allam cycle. Their paper includes detailed information on compressor and cooled

---

[3] Currently referred to as the NET Power cycle after the current owner of the technology, https://netpower.com/.

[4] A process simulation software widely used in the chemical process industry, https://www.aspentech.com/en/products/engineering/aspen-plus.

Table 10.7 Published performance estimates for the Allam cycle

	Allam et al. [17]	Sifat and Haseli [19]	Sifat and Haseli [19], Manso [20]	Scaccabarozzi et al. [21]
Net Output, MWe	303	281.8	401	419.3
Fuel Consumption, MWth	511	511	774	768.3
O$_2$ Consumption, mt/day	3,555	3,555	5,357	5,600
Turbine Flow, kg/s	923	942.1	1,277.5	1,271
TIT, °C	1,158	1,158	1,150	1,154
PR	10:1	10:1	10:1	8.835:1
Turbine Exhaust Temperature, °C	727	786	800	741.2
ASU Power, MWe	56	85	84	90
CO$_2$ Compression, MWe	77	88	72	111.2
Net Thermal Efficiency, %	59	55	51.8	54.6

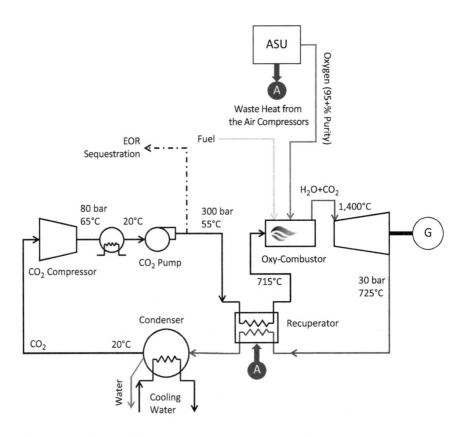

Figure 10.11 Schematic description of the semi-closed oxy-combustion Allam cycle.

sCO2 turbine modeling. An in-depth study of the cooled sCO2 turbine was also published [22]. Their cycle performance with 54.6% net thermal efficiency is included in Table 10.7. In a recent paper, Haseli and Sifat [23] described an optimized Allam cycle

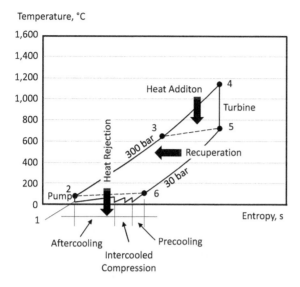

Figure 10.12 Ideal cycle T-s diagram of the Allam cycle.

(also modeled in *Aspen Plus*) with 1,500 K (1,227°C) TIT, 10.9:1 PR, and 59.7% net thermal efficiency. In the paper, a comparison of various Allam cycle study findings is provided, ranging from 54.6% to 59.3% by different authors, e.g., Mitchell et al. [24], Rogalev et al. [25], and Allam et al. [17]. In addition to varying assumptions of component efficiencies and turbine cooling, the key contributors to the disagreement between the findings of different researchers are ASU performance, recuperation scheme, and heat integration of recuperation with ASU heat rejection.

To shed light on the rather wide range of published performance estimates, before diving into the proverbial *weeds* of actual cycle designs, it is instructive to look at the cycle using the first principles. Hence, we start with the ideal cycle diagram shown Figure 10.12. The cycle diagram in the figure is analogous to the air-standard (ideal) Brayton cycle for a conventional gas turbine.

From the T-s diagram in Figure 10.12, one can identify the two key cycle temperatures as follows:

$$\text{METH} = \frac{T_4 - T_3}{\ln\left(\frac{T_4}{T_3}\right)},$$

$$\text{METL} \approx \frac{T_6 - T_1}{\ln\left(\frac{T_6}{T_1}\right)}.$$

For the nominal cycle values of $T_4 = \text{TMAX} = 1{,}150°C$, $T_3 = 715°C$, $T_6 = 80°C$ and $T_1 = \text{TMIN} = 20°C$, METH and METL are found as 1,192 K and 322 K, respectively. The Carnot efficiency is calculated as

$$\text{EFF\_C} = 1 - \text{TMIN}/\text{TMAX} = 79.8\%,$$

Table 10.8 Allam cycle efficiencies from first principles

Carnot Efficiency		79.8%	
Equivalent Carnot Efficiency		73.0% (75.8%)	
Carnot Factor		0.92 (0.95)	
Technology Factor	0.70	0.75	0.80
Actual Cycle Efficiency	51.1% (53.1%)	54.7% (56.9%)	58.4% (60.7%)

whereas the *equivalent* Carnot efficiency turns out to be

$$\text{EFF\_C} = 1 - \text{METL}/\text{METH} = 73\%,$$

which corresponds to a *cycle factor*, CF = 73.0/79.8 = 0.92. What to expect from the actual cycle implemented with real hardware in the field is shown in Table 10.8 as a function of *technology factor*. The numbers in parentheses represent

- the equivalent Carnot efficiency of a "perfect gas CO$_2$" ideal cycle ($\gamma = 1.18$) with perfect recuperation/regeneration, i.e., $T_3 = T_5$ and $T_2 = T_6$, with isothermal heat rejection and isothermal compression, i.e., $T_2 = T_1$, and
- actual cycle efficiencies based on that reference performance

It would be naïve to expect that this first-of-a-kind (FOAK) technology will exceed the maturity level of the advanced combined cycles (represented by a TF of about 0.8 at a TIT of about 1,700°C). Thus, based on first principles and the existing technology state of the art, with 1,150°C TIT and 10:1 cycle PR, a power plant based on the Allam cycle is not likely to record a *cycle efficiency* higher than about 61%. Depending on the ASU performance and other auxiliary loads, actual *net (thermal) efficiency* in the field can be significantly lower. However, ingenious heat integration between the ASU and the power cycle can create surprises as will be discussed below.

For a deeper look into the performance of a power plant operating in an Allam cycle, a heat and mass balance model is developed using THERMOFLEX flowsheet simulation software (see Section 16.2). Basic cycle assumptions are listed below:

- ISO ambient conditions (15°C, 1 atm)
- Circulating cooling water at 20°C
- Inter- and aftercooler, precooler exit temperature 25°C with 10°C coolant temperature rise
- Cycle heat rejection via fin-fan coolers (20 kWe fan power consumption per each MWth of heat rejection)
- Inter- and aftercooler, precooler pressure loss 2%
- Turbine inlet at 1,150°C and 300 bar (nominal), nominal cycle PR = 10:1
- CO$_2$ flow 700 kg/s
- 100% CH$_4$ fuel from "pipeline" at 37.92 bar and 25°C
- Fuel booster compressor 70% polytropic efficiency, 75% adiabacity (to mimic intercooling effectiveness)

- Combustor pressure loss 4%, heat loss 0.3% (of fuel heat input)
- ASU operating at 25°C and 5 bar to generate 99.5%(v) O_2 (0.5%(v) Ar)
- No ASU heat integration (for simplicity)
- Three-stage, intercooled CO_2 compressor with electric motor drive (90% stage polytropic efficiency, 97.3% motor efficiency)
- CO_2 pump with 90% polytropic efficiency (97.3% motor efficiency)
- Cooled turbine, nonchargeable flow 7%, chargeable flow 13% (referenced to the CO_2 flow at the recuperator exit). This is an admittedly *conservative* estimate based on convective cooling of first stage stator vanes and rotor blades (no thermal barrier coating). It is roughly comparable to the cooling flow budget of a vintage E Class gas turbine (nominal 1,300°C TIT).
- Nominal (uncooled) turbine stage polytropic efficiency 90%
- Recuperator (5°C minimum pinch) with pressure losses 4% and 1% on the hot and cold sides, respectively
- O_2 heater (5°C minimum pinch) with pressure losses 4% and 1% on the hot and cold sides, respectively

Note the generous component efficiencies used in the model (i.e., 90% polytropic). Based on information from equipment OEMs, available compressor and pump efficiencies are significantly lower. Pressure losses are assumed to cover piping losses as well. Also, note that the separate O_2 heater is merely a calculation convenience. In an actual implementation, O_2 would be mixed with the "cold" recycled CO_2 stream and heated in the recuperator. Theoretical stoichiometric combustion implies that combustion products have no oxygen in the mixture. However, a certain amount of O_2 is required for flame stability and to avoid excessive flame temperature and dissociation products [26]. Herein, excess O_2 results in 1.11%(v) O_2 in combustion products so that recycled CO_2 after water knock-out in the precooler is 97.8%(v). A simplified cycle heat and mass balance diagram is shown in Figure 10.13.

Key performance parameters are calculated as follows:

- Cycle METH = 1,110 K (837°C)
- Cycle METL = 315 K (42°C)
- Carnot efficiency 79.75% (TMAX = 1,150°C, TMIN = 15°C)
- Equivalent Carnot efficiency 71.63%
- Cycle factor is 0.90

A cycle and plant heat rejection summary is given in Table 10.9. The cycle fuel input, motor and generator power, and net performance summary is provided in Table 10.10.

Clearly, there is nothing wrong with claiming nearly 59% *cycle* efficiency with TF = 0.82 and it is certainly well aligned with the projection made earlier based on fundamental thermodynamic calculations (see Table 10.8). For plant efficiency, however, we must first look at parasitic power consumers, i.e.,

- O_2 compression to 300+ bara at the cycle control volume
- Fuel gas compression to 450+ bara at the cycle control volume

Table 10.9 Allam cycle heat rejection summary

	kWth
Condenser Duty	59,809
Intercooler 1 Duty	23,735
Intercooler 2 Duty	32,160
Aftercooler Duty	124,695
Total Cycle Heat Rejection	251,866
ASU Heat Rejection	46,577
Total Plant Heat Rejection	286,976

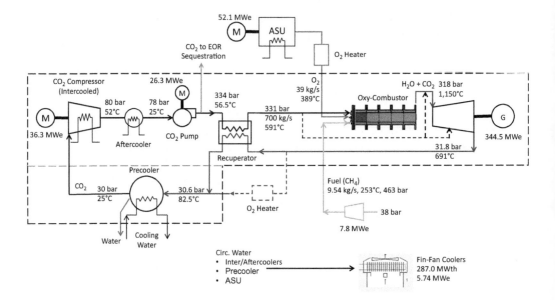

Figure 10.13 Simplified Allam cycle heat and mass balance diagram (from THERMOFLEX).

Accounting for those two power consumers, the performance of the *expanded* cycle is summarized in Table 10.11. The net thermal efficiency dramatically drops by more than five percentage points to 52.1%.

With some caveats, this performance could even be advertised as a marketing number, i.e., with readily available sources of cooling water and oxygen. However, oxygen must be manufactured in a very capital and electric power-intensive plant, namely, the cryogenic ASU. On top of that, in many places in the world, returning circulating cooling water carrying a significant amount of heat at 30°C cannot be dumped into a lake, river, or ocean; not anymore. This will require an electric fan–driven heat rejection system such as a dry cooling tower or bank of fin-fan coolers. Finally, one must account for miscellaneous power users in the plant (0.1% of gross output) and generator step-up transformers in the switchyard (0.35% of gross output).

Table 10.10 Allam cycle output and efficiency

Fuel Burn, kWth	477,248
Generator Output, kWe	344,504
Motor 1 (Compressor), kWe	36,336
Motor 2 (Pump), kWe	26,287
Cycle Net Output, kWe	281,881
Cycle Net Efficiency, %	59.06
Technology Factor	0.82

Table 10.11 *Expanded* Allam cycle output efficiency

Cycle Gross Output, kWe	281,881
O_2 Compressor, kWe	18,029
Fuel Gas Booster Compressor, kWe	7,620
Cycle Net Output, kWe	256,460
Cycle Net Efficiency, %	53.74
Technology Factor	0.75

Table 10.12 Allam cycle *power plant* output and efficiency (numbers in parentheses are with an open-loop (once-through) water-cooled condenser with 15°C cooling water)

Cycle Net Output, kWe	256,460
Main Air Compressor (ASU), kWe	34,328
Fin-Fan Cooler, kWe (Once-through, Open-Loop)	5,740 (1,435)
Miscellaneous (0.1% of Gross), kWe	345
Transformers (0.35%), kWe	1,206
Net Plant Output, kWe	214,842 (219,147)
Net Plant Thermal Efficiency, %	45.02 (45.92)
Technology Factor	0.63 (0.64)

With all this extra parasitic power consumption, the realistic power plant efficiency ends up at 45.0%, i.e., significantly below publicized numbers, as shown in Table 10.12.

Total ASU power consumption, MAC, and O_2 compressors, add up to 52,129 kWe, which corresponds to 1,337 kJ/kg of O_2 based on 39 kg/s O_2 flow (refer to Figure 10.13). The numbers found in the literature range between 900 kJ/kg [25] and 1,354 [24] to 1,391 kJ/kg [22,27]. In a recent paper mentioned earlier, Haseli and Sifat [23] came up with an optimized number of 59,250 kJ/kmol of O_2 per mol of fuel (from Figure 6 in Ref. [23]). This is about 40% higher than the number herein, which comes to about 42,770 kJ/kmol of O_2 per mol of fuel, for specific ASU power of about 1,850 kJ/kg of O_2. The reason for the much higher number is the non-intercooled MAC and a "booster compressor" in the ASU (producing liquid oxygen or LOX) adopted by the authors for maximum heat integration with the power block.

The performance of a power plant based on the Allam cycle is strongly dependent on three factors: (i) recuperation effectiveness, (ii) parasitic power consumed by the ASU, and (iii) heat integration of the of the ASU and the power block. As can be seen in Figure 10.13, simple recuperation is rather ineffective, i.e., the sCO2 exit temperature, 591°C, is 100°C lower than the turbine exhaust gas temperature, 691°C. (Heating oxygen with turbine exhaust gas is clearly not good for cycle efficiency.) One factor in this result is the absence of integration with the ASU.

To investigate the impact of a complex recuperator including heat integration with the ASU, the model is updated as follows:

- Advanced ASU producing *liquid oxygen* (LOX) at 18°C and 340 bar (1,350 kJ/kg of O_2)
- Heat transfer from compressed air (from the MAC in the ASU) at 226°C to recycled O_2
- Turbine coolant at 223°C (14% of turbine inlet flow)
- Recycled CO_2 heated to 689°C with turbine exhaust at 710°C

Net performance with once-through (open loop) water cooling is found to be 235.5 MWe and 53.6% (7.7 percentage points better than in Table 10.12). This corresponds to a TF of 0.73.

Currently available "off-the-shelf" ASU technology can cost-effectively produce oxygen with 95% purity at the requisite quantities (about 3,000 tpd in the example above), which is adequate for oxy-combustion of coal in a boiler or ATR/POX (see Section 8.2.4). A semi-closed supercritical CO_2 cycle with oxy-combustion and CO_2 removal requires high-purity O_2, ideally 99.5% but no lower than 98%. Otherwise, the low purity of recycled CO_2 negatively impacts cycle performance (increased compression and pumping duty) and requires a costly purification unit for the separated CO_2 stream. (In this example, with 99.5% pure O_2, recycled CO_2 purity is 97.8%.) High-purity O_2 production imposes significantly higher parasitic power and CAPEX on the ASU [28].

In conclusion, the Allam cycle is rather unlikely to be a cost-effective competitor to a gas turbine combined cycle power plant with advanced class gas turbines with post-combustion carbon capture. The ease of CO_2 recovery at high pressure from the semi-closed power cycle is certainly a strong advantage of the Allam technology. Nevertheless, unless a "free" oxygen source is available, parasitic power consumed by the ASU along with the need to compress the natural gas fuel to more than 300 bar, render the *bottom-line* net plant efficiency uncompetitive. To reach the best possible efficiency, i.e., 53.6% herein or 55–59% in cited references (mostly on paper), extensive integration between the ASU and the power block as well as a highly complex recuperator design (most likely comprising a multitude of compact heat exchangers) is required. The current TRL of the technology is estimated as 6.[5] For an in-depth analysis of the oxy-combustion sCO2 cycle and its design challenges, the reader is referred to the recent paper by Gülen et al. [29].

---

[5] A press release dated November 16, 2021, stated that the Allam cycle turbine had delivered electricity onto the ERCOT grid from its 50 MWth test facility in La Porte, Texas.

For a recent "key knowledge deliverable" (KKD) report on a pre-FEED study done by by 8 Rivers Capital and its subcontractors for the Department for Business, Energy and Industrial Strategy (BEIS) of the UK Government under the Carbon Capture and Storage Innovation program, the reader is pointed to the relevant government website.[6] Three performance cases are included in the report:

- Base case: 900°C turbine inlet temperature with unintegrated ASU (279.4 MWe and 50.9% net LHV; estimated auxiliary power consumption is 107.6 MWe, for ISO conditions)
- Alternative case: 925°C turbine inlet temperature with unintegrated ASU (303.3 MWe and 55.1% net LHV)
- Optimized alternative case: 925°C turbine inlet temperature with unintegrated ASU balancing CAPEX and efficiency (296 MWe and 53.8% net LHV)

10.4 Other Oxy-Combustion sCO$_2$ Cycles

It is worth nothing that the Allam cycle is neither the only oxy-combustion cycle out there nor is it the only oxy-combustion sCO2 cycle. For a summary of oxy-combustion technology, the reader is pointed to the handbook chapter written by the author [30]. For a complete review of oxy-combustion cycles, see the IEAGHG report authored by Mancuso et al. [28]. For a condensed version of the report findings, including technical and economic assessment of different technologies, refer to the paper by Ferrari et al. [31]. The authors concluded that the Allam cycle offered the best performance and LCOE (at least, on the proverbial *paper*).

Matiant and Allam cycles are the two sCO2 cycles. The *Matiant cycle* is essentially an intercooled, recuperated reheat cycle with sCO2 as the working fluid and O$_2$ as the oxidizer [32]. Not surprisingly, as one hopes at this point, when the ASU power consumption is accounted for, the cycle does not look attractive from an efficiency perspective. A combined cycle version, the *CC-Matiant cycle*, eliminates intercooling and uses a steam bottoming cycle with the recuperator limited to the hottest (exhaust) and coolest (compressor discharge) working fluid streams. The net plant efficiency is limited to 47–49% and does not seem to justify the requisite high-temperature component development effort (1,300°C turbine inlet at 300 bars).

In other oxy-combustion cycles, the moderator or diluent in the combustor is H$_2$O. In one proposed variant [33–34], natural gas fuel and oxygen are supplied to the oxy-combustor operating at about 80 bar (compared with 300 bar in the Allam cycle). Water is injected to maintain a flame temperature of about 1,700°C (around 3,100°F) and deliver diluted combustion gas (roughly 85–90% H$_2$O and 10–15% CO$_2$ – therefore it is called "water cycle") to the *high-pressure turbine* (HPT) at 500–600°C (930–1,110°F). Thus, the HPT is essentially a steam turbine with a PR subject to optimization.

---

[6] https://www.gov.uk/government/publications/carbon-capture-usage-and-storage-ccus-innovation-8-rivers-key-knowledge-deliverables (last accessed December 10, 2022)

Cycle performance is enhanced by reheat combustion before the *intermediate-pressure turbine* (IPT). Typically, the IPT can be selected from the existing gas turbine technology and the inlet temperature can be as high as 1,200°C (about 2,200°F) or even higher.

There are other oxy-combustion cycles similar to the one described above, e.g., the Graz cycle and the *partial oxidation gas turbine* (POGT). The reader can read about these cycles including their pros and cons and relevant bibliography in Refs. [28,30,31]. The author's verdict is the same for all of them. They are at very low TRL (see Section 16.4) with complex, FOAK equipment and are highly unlikely to stand a chance against an advanced GTCC with post-combustion capture in CAPEX or performance.

10.5   References

1. Frutschi, H. U. 2005. *Closed-Cycle Gas Turbines: Operating Experience and Future Potential*, New York: ASME Press.
2. Lee, J. C., Campbell, Jr., J. and Wright, D. E. 1980. Closed-cycle gas turbine working fuels, ASME Paper 80-GT-135, Gas Turbine Conference & Products Show, March 10–13, New Orleans, LA.
3. Gülen, S. C. 2016. Supercritical CO$_2$ – What is it good for?, *Gas Turbine World*, September/October, 26–34.
4. Conboy, T. M. , Wright, S. A., Pasch, J., et al. 2012. Performance characteristics of an operating supercritical CO$_2$ Brayton cycle, *Journal of Engineering for Gas Turbines and Power*, 134, 111703.
5. Fleming, D., Conboy, T. M., Rochau, G., Holschuh, T. and Fuller, R. 2012. Scaling considerations for a multi-megawatt class supercritical CO$_2$ Brayton cycle and path forward for commercialization, GT2012–68484, ASME Turbo Expo, June 11–15, Copenhagen, Denmark.
6. Held, T. J. 2014. Initial test results of a megawatt class supercritical CO$_2$ heat engine, *The 4th International Symposium – Supercritical CO2 Power Cycles,* September 9–10, Pittsburgh, PA.
7. Hejzlar, P., Dostál, V., Driscoll, M. J., et al. 2006. Assessment of gas cooled fast reactor with indirect supercritical CO$_2$ cycle, *Nuclear Engineering and Technology*, 38(2), 109–118.
8. Johnson, G. A., McDowell, M. W., O'Connor, G. M., et al. 2012. Supercritical CO$_2$ cycle development at Pratt & Whitney Rocketdyne, *GT2012–70105, ASME Turbo Expo, June 11–15*, Copenhagen, Denmark.
9. Utamura, M. 2010. Thermodynamic analysis of part-flow cycle supercritical CO$_2$ gas turbines, *Journal of Engineering for Gas Turbines and Power*, 132, 111701-1.
10. Neises, T. and Turchi, C. 2019. Supercritical carbon dioxide power cycle design and configuration optimization to minimize levelized cost of energy of molten salt power towers operating at 650°C, *Solar Energy*, 181, 27–36.
11. Pidaparti, S. R., Weiland, N. T. and White, C. W. 2020. A performance and economic comparison of partial cooling and recompression sCO$_2$ cycles for coal-fueled power

generation, *The 7th International Supercritical CO2 Power Cycles Symposium March 31 – April 2*, San Antonio, TX.

12. Tang, C.J., McClung, A., Hofer, D. and Huang, M. 2019. Transient modeling of 10 MW supercritical CO_2 Brayton cycles using NPSS, *GT2019–91443, ASME Turbo Expo, June 17–21*, Phoenix, AZ.

13. Moore, J. 2018. Development of a high-efficiency hot gas turbo-expander and low-cost heat exchangers for optimized CSP supercritical CO_2 operation, *DE-EE0005804, SwRI Project*, Southwest Research Institute, San Antonio, TX, https://doi.org/10.2172/1560368.

14. Isles, J. 2014. Gearing up for a new supercritical CO_2 power cycle system, *Gas Turbine World*, November/December, 14–18.

15. Allam, R. J., Palmer, M. R., Brown Jr, G. W., et al. 2012. High efficiency and low cost of electricity generation from fossil fuels while eliminating atmospheric emissions, including carbon dioxide, *Energy Procedia*, 37, 1135–1149.

16. Allam, R. J., Fetvedt, J. E., Forrest, B. A. and Freed, D. A. 2014. The oxy-fuel, supercritical CO2 Allam cycle, *Proceedings of the ASME Turbo Expo 2014: Turbine Technical Conference and Exposition. Volume 3B: Oil and Gas Applications; Organic Rankine Cycle Power Systems; Supercritical CO_2 Power Cycles; Wind Energy. Düsseldorf, Germany. June 16–20*.

17. Allam, R. J., Martin, S., Forrest, B., et al., 2017. Demonstration of the Allam cycle: An update on the development status of a high efficiency supercritical carbon dioxide power process employing full carbon capture, *Energy Procedia*, 114, 5948–5966.

18. Fernandes, D., Wang, S., Xu, Q., Buss, R. and Chen, D. 2019. Process and carbon footprint analyses of the Allam cycle power plant integrated with an air separation unit, *Clean Technologies*, 1, 325–340, doi:10.3390/cleantechnol1010022.

19. Sifat, N. S. and Haseli, Y. 2018. Thermodynamic modeling of Allam cycle, *Proceedings of the ASME 2018 International Mechanical Engineering Congress and Exposition. Volume 6A: Energy, November 9–15*, Pittsburgh, PA.

20. Manso, R. L. 2013. *CO2 Capture in Power Plants Using the Oxy-Combustion Principle*, MS Thesis, Norwegian University of Science and Technology, Norway.

21. Scaccabarozzi, R., Gatti, M. and Martelli, E. 2016. Thermodynamic analysis and numerical optimization of the NET Power oxy-combustion cycle. *Applied Energy*, 178, 505–526.

22. Scaccabarozzi, R., Martelli, E., Pini, M., et al. 2022. A code for the preliminary design of cooled supercritical CO_2 turbines and application to the Allam cycle, *Journal of Engineering for Gas Turbines and Power*, 144, 031012.

23. Haseli, Y. and Sifat, N.S. 2021. Performance modeling of Allam cycle integrated with a cryogenic air separation process, *Computers and Chemical Engineering*, 148, 107263.

24. Mitchell, C., Avagyan, V., Chalmers, H. and Lucquiaud, M. 2019. An initial assessment of the value of Allam Cycle power plants with liquid oxygen storage in future GB electricity system. *International Journal of Greenhouse Gas Control*, 87, 1–18.

25. Rogalev, A., Grigoriev, E., Kindra, V. and Rogalev, N. 2019. Thermodynamic optimization and equipment development for a high efficient fossil fuel power plant with zero emissions. *Journal of Cleaner Products*, 236, 117592.

26. Liu, C. Y., Chen, G., Sipöcz, N., Assadi, M. and Bai, X. S. 2012. Characteristics of oxy-fuel combustion in gas turbines, *Applied Energy*, 89(1), 387–394.

27. Mancuso, L., Ferrari. P. Chiesa, N., Martelli, E. and Romano, M. 2015. Oxy-combustion turbine power plants. International Energy Agency (IEAGHG) Report No. 2015/5.

28. Allam, R. J. 2009. Improved oxygen production technologies, *Energy Procedia* 1, 461–470.
29. Gulen, S. C., Taher, M. and Lyddon, L. G. 2022. Oxy-fuel combustion supercritical CO$_2$ power cycle - A critical look, *GPPS-TC-2022-0134, GPPS Chania22, September 12–14,* Chania, Greece.
30. Gülen, S. C. 2019. Advanced Fossil Fuel Power Systems. In D. Y. Goswami and F. Kreith (eds.), *Energy Conversion*, 2nd ed. Boca Raton, FL: CRC Press.
31. Ferrari, N., Mancuso, L., Davison, J., et al. 2017. Oxy-turbine for power plant with CO$_2$ capture, *Energy Procedia*, 114, 471–480.
32. Mathieu, P. and Nihart, R. 1998. Zero emission MATIANT cycle, *Proceedings of the ASME 1998 International Gas Turbine and Aeroengine Congress and Exhibition. Volume 3: Coal, Biomass and Alternative Fuels; Combustion and Fuels; Oil and Gas Applications; Cycle Innovations*, June 2–5, Stockholm, Sweden.
33. Anderson, R. F., Viteri, R., Hollis, A., et al. 2010. Oxy-fuel gas turbine, gas generator and reheat combustor technology development and demonstration, *ASME Paper GT2010–23001, ASME Turbo Expo, June 14–18*, Glasgow, Scotland.
34. Anderson, R. E., MacAdam, S., Viteri, F., et al. 2008. Adapting gas turbines to zero-emission oxy-fuel power plants, *Proceedings of the ASME Turbo Expo, GT2008–51377*, Berlin, Germany.

11 Carbon Capture

Carbon dioxide (CO_2) constitutes the largest fraction of *greenhouse gases* (GHG), which are widely believed to be a major contributor to climate change. As such, significant research and development effort has been dedicated to reducing or eliminating emissions of CO_2 into the atmosphere. The combustion of fossil fuels, especially coal, in electric power plants is a significant source of CO_2. As shown in Figure 11.1, natural gas–fired gas turbine combined cycle (GTCC) is superior to all other technologies and fuels, even integrated gasification combined cycle (IGCCs), in terms of specific CO_2 emissions. Simply replacing coal-fired power plants by gas–fired gas turbine power plants can reduce CO_2 emissions significantly. (Based on actual generation and emission data from the US EIA, the reduction is more than 60%.) This is simply a result of the chemical composition of the fuel (see Table 11.1).

The bar chart data in Figure 11.1, i.e., CO_2 in the flue gas in kg/MWh, is calculated from known efficiency (heat rate) using the equations:

$$\text{Gas: } e_{CO2} = 1000 \cdot \frac{HR}{LHV} \cdot \frac{MW_{CO_2}}{MW_f}, \tag{11.1}$$

$$\text{Other: } e_{CO2} = 1000 \cdot \frac{HR}{LHV} \cdot c \cdot \frac{MW_{CO_2}}{MW_C}, \tag{11.2}$$

where HR is the plant heat rate in kJ/kWh (HR = $3{,}600/\eta$), LHV is the lower heating value of the fuel in kJ/kg, c is the fraction of carbon in coal or oil per ultimate analysis, and MW refers to the molecular weights of fuel gas (~16 kg/kmol), CO_2 (44 kg/kmol), and carbon (12 kg/kmol).

In fossil fuel–fired power plants, there are three methods to eliminate or reduce CO_2 emissions:

- Post-combustion capture (eliminate CO_2 from the flue gas)
- Pre-combustion capture (eliminate CO_2 from the fuel gas)
- Oxyfuel combustion with CO_2 recycling or chemical looping

For a brief coverage of these technologies, refer to the handbook chapter by the author (Ref. [30] in Chapter 10). Oxyfuel combustion was touched upon in detail in Chapter 10 (i.e., the Allam cycle in Section 10.3). Herein the focus is on post-combustion CO_2 capture from the heat recovery steam generator (HRSG) stack gas in a GTCC power plant. The reason for that is simple: GTCC with advanced class gas turbines and post-combustion capture represents the most cost-effective technology

Table 11.1 CO_2 from fossil fuel combustion (per unit of thermal energy)

Fuel	CO_2, t/TJ
Lignite	101.2
Subbituminous Coal	96.1
Coking Coal	94.6
Residual Fuel Oil	77.4
Diesel Fuel	74.1
Natural Gas	56.1

Source: EU Guidelines for the monitoring and reporting of greenhouse gas emissions (2004), after IPPC (1996).

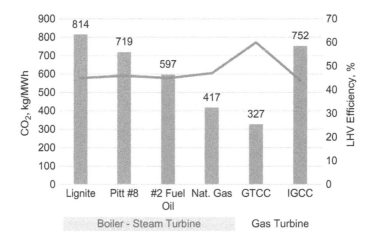

Figure 11.1 CO_2 emissions from different fuels and technologies (GTCC is fired with natural gas, IGCC is fired Pittsburgh #8 bituminous coal).

for carbon-free electric power generation from fossil fuels. This is clearly illustrated in the US DOE prepared chart in Figure 11.2, which also suggests that IGCC (see Chapter 13), even with a "transformative" hydrogen-fired gas turbine (still on paper), or the oxyfuel Allam cycle stand little chance of becoming competitive. A natural gas-fired Allam cycle, while it seems to give the GTCC a run for its money, raises the obvious question of the wisdom of making an investment into a *first-of-a-kind* (FOAK) technology when one can do equally well or even better with a decades old mature technology armed with an equally proven technology (i.e., generic amine-based chemical absorption) to scrub CO_2 from its flue gas. As already discussed in Section 10.3 with detailed calculations and analysis, the Allam cycle performance claims are overoptimistic to begin with. The efficiency shown in the DOE chart in Figure 11.2, about 48% HHV (53.2% LHV), is significantly lower than reported in marketing-oriented articles and papers from the technology developers. This perform-ance is in line with the best possible performance in Section 10.3, i.e., 53.6% net LHV. Still, it is much higher than the more realistic value (e.g., see Section 10.3), of

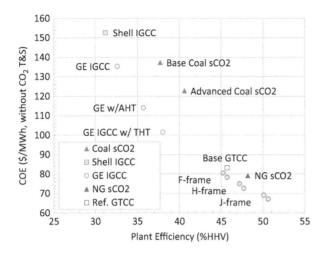

Figure 11.2 Comparison of different technologies for carbon-free power generation from coal and natural gas.[1] (GE and Shell refer to the gasifier technologies; AHT: advanced hydrogen turbine, THT: transformative hydrogen turbine, NG: natural gas, coal used in calculations is Illinois #8. Supercritical CO_2 (sCO2) cases use Shell gasifier technology and are based on the Allam cycle.)

about 45% net LHV, accounting diligently for parasitic power consumers such as the air separation unit (ASU) and fuel booster compressors (no integration with the ASU).

A closer look at the GTCC cost and performance is provided in Figure 11.3.[2] Each single percentage point improvement in GTCC efficiency is worth a 5 kg/MWh reduction in CO_2 emissions. This is equivalent to 22,500 tons of CO_2 for a 750 MW power plant running for 6,000 hours per year. Based on a fuel economy of about 22 mpg and 11,500 miles of annual driving, a typical passenger vehicle emits about 4.6 tons of CO_2 per year. Thus, each percentage point improvement in GTCC efficiency is equivalent to removing 5,000 cars from the roads. The best performance in the chart is for a GTCC where the gas turbine, referred to as a *rotating detonation engine* (RDE), is equipped with a *rotating detonation combustor*, which is a practical approximation for the theoretical constant volume combustion. As it will be shown via rigorous thermodynamic analysis in Chapter 14 on *constant volume combustion* (a theoretical proxy for *pressure-gain combustion*), GTCC performance with RDE in Figure 11.3 is not credible by a long shot (at least two percentage points overblown). In any event, the efficiency of an RDE GTCC at ISO base load is cited as 66.4% net LHV for a 1,041 MW (net) power plant in a presentation by the Aerojet Rocketdyne.[3]

[1] From a presentation at the ASME, IGTI 2019 Turbo Expo – Future Power Systems Session, by Richard Dennis, Technology Manager, Advanced Turbines and sCO2 Power Cycles Programs, US DOE NETL, Phoenix, AZ, June 18, 2019.
[2] The data is from the US DOE report *Current and Future Technologies for Natural Gas Combined Cycle (NGCC) Power Plants*, DOE/NETL-341/061013.
[3] *Rotating Detonation Combustion for Gas Turbines – Modeling and System Synthesis to Exceed 65% Efficiency Goal*, under the US DOE contract DE-FE0023983 (The findings are freely available on the NETL website).

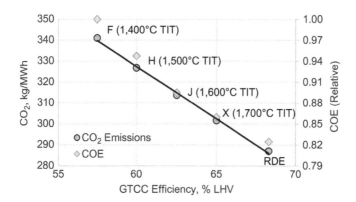

Figure 11.3 GTCC efficiency impact on CO_2 emissions and cost of electricity (COE). (TIT: Turbine inlet temperature; RDE: Rotating detonation engine.)

There are many similar studies done by other private and public organizations as well as academic institutions. One of them is a study done by a consulting engineering firm for the Department for *Business, Energy, and Industrial Strategy* (BEIS), a branch of the government of the United Kingdom (UK).[4] The report looked at different options listed in Table 11.2, including eight existing (current) and two "next generation" (Cases 6 and 7 in the table) power plant technologies with carbon capture and three fuels, i.e., natural gas, coal, and biomass. The study findings are summarized in Table 11.3 (technical performance) and Table 11.4, which summarizes the levelized cost of electricity (LCOE) in relative terms to GTCC with post-combustion capture as the basis (100%). For details of the technological, economic, and financial assumptions used in the study, the reader is advised to consult the cited report (available online). Some of those are listed below for better interpretation of the results:

- GTCC technology is based on GE's 9HA.01 (2014–15 vintage)
- IGCC technology is based on GE's F Class syngas turbine
- Post-combustion capture is based on Shell's proprietary CANSOLV technology
- Pre-combust capture (IGCC) is based on Selexol technology
- See Section 10.3 for Case 6 (the Allam Cycle)
- In Case 7, fuel cell plus GTCC hybrid, CO_2 is captured in the fuel cell itself
- Listed efficiency is the average LHV value over 25 years of operation with assumed output and heat rate degradation
- LCOE should be interpreted as the average electricity price that would be needed by a project to break even, i.e., NPV = 0

The findings summarized in Table 11.4 are pretty much as one would expect and in support of the chart in Figure 11.2. The study is done for the UK-specific economic

---

[4] *Assessing the Cost Reduction Potential and Competitiveness of Novel (Next Generation) UK Carbon Capture Technology – Benchmarking State-of-the-Art and Next Generation Technologies,* Document: 13333-8820-RP-01, Rev. 4A, October 8, 2018.

Table 11.2 Power plant technologies with carbon capture

(NG: natural gas, CC: carbon capture, SCPC: supercritical pulverized coal, POSTCCC: post-combustion CC, PRECCC: pre-combustion CC, ATR: autothermal reforming, CFB: circulating fluidized bed, OXY: oxy-combustion, MCFC: molten carbonate fuel cell)

Case	Technology
1	NG-fired GTCC with POSTCCC
2	NG-fired GTCC with ATR and PRECCC
3	Coal-fired SCPC with POSTCCC
4	Coal-fired SCPC with OXY and CC
5	Coal IGCC with PRECCC
6	Allam Cycle (Natural Gas)
7	NG-fired GTCC with MCFC and CC
8	Biomass CFB with POSTCCC
9	Biomass CFB with OXY and CC
10	Biomass IGCC with PRECCC

Table 11.3 Performance of technologies listed in Table 11.2

Case	Fuel Type	Net MWe	Net Efficiency, %	Technology	Capture Rate, %	CO_2 Footprint, kg/MWh
1	NG	1064.6	52.0	POST	90.8	34.3
2	NG	817.9	40.7	PRE	90.4	45.8
3	COAL	814.2	34.7	POST	90.0	94.6
4	COAL	832.6	35.5	OXY	89.2	100.2
5	COAL	799.8	33.5	PRE	90.3	90.4
6	NG	848.4	52.3	OXY	90.0	37.1
7	NG	1508.6	56.6	MCFC	92.1	27.1
8	BIOM	396.2	30.6	POST	90.0	146.5
9	BIOM	402.1	31.1	OXY	89.9	146.2
10	BIOM	356.1	32.1	PRE	90.8	101.9

Table 11.4 Ranking of the technologies listed in Table 11.2 in terms of increasing LCOE (BIOM: Biomass)

Case	Fuel Type	CAPEX, %	Fuel Cost, %	OPEX, %	Emissions Price, %	Storage and Transport, %	Total LCOE, %
1 (BASE)	NG	100	100	100	100	100	100
7	NG	115	92	140	79	93	101
6	NG	156	99	128	107	99	115
3	COAL	221	59	192	259	241	133
4	COAL	237	57	204	276	233	137
2	NG	179	128	169	131	127	143
5	COAL	343	60	306	259	249	173
8	BIOM	329	165	286	403	374	243
9	BIOM	375	163	317	400	369	255
10	BIOM	474	196	424	290	296	292

and financial criteria, e.g., fuel and carbon prices. In countries with different criteria and location-specific differences such as technical (grid related) and environmental regulations, some variation in the findings can be expected. Nevertheless, on a relative basis, the general order of the ranking is unlikely to vary appreciably.

Clearly, natural gas–fired GTCC with post-combustion capture (PCC) is hard to beat. In fact, using a generic amine such as monoethanolamine (MEA) for PCC, with lower CAPEX and OPEX, this advantage is likely to be even more pronounced. Biomass suffers from high fuel prices and low net output due to the limited commercial experience at higher boiler capacities. Coal is unable to compete with natural gas due to its high CAPEX and OPEX and, even after capture, its carbon footprint is much higher. Its low fuel price advantage is not as pronounced as it once was and with the added burden of carbon capture, it is not enough to save coal.

11.1   Post-combustion Capture

To date, post-combustion CO_2 removal from the stack gases via deployment of aqueous amine-based absorber-stripper (chemical absorption) technology is the only commercially available option, which is applicable to new units as well as to the retrofitting of existing plants. The stack gas of a modern GTCC power plant with advanced F, H, or J Class units contains about 4% CO_2 by volume at near-atmospheric pressure (about 4.5%(v) on a dry basis). Low flue gas pressure and density result in large volume flows requiring large piping, ducts, and equipment, which are reflected in the plant footprint and total installed cost. The only commercially available absorbents active enough for the recovery of dilute CO_2 at very low partial pressures are aqueous solutions of *alkanolamines* such as monoethanolamine (MEA), diethanolamine (DEA), methyl-diethanolamine (MDEA), and the newly developed sterically hindered amines (e.g., piperazine).

As shown in Figure 11.4, the PCC system consists of two main components:

- an *absorber* in which the CO_2 is removed, and
- a *regenerator (stripper)* in which the CO_2 is released in a concentrated form and the solvent is recovered.

Prior to the CO_2 removal, the flue gas (at around 90°C at the HRSG stack for the most efficient GTCC power plants) is typically cooled to about 45–50°C and then treated to reduce particulates that cause operational problems and other impurities, which would otherwise cause costly loss of the solvent (e.g., in a direct contact cooler or "quench tower"). The amine solvent absorbs the CO_2 (together with traces of NOx) by chemical reaction to form a loosely bound compound. A booster fan (blower) is necessary to overcome the pressure loss in the capture plant and is a significant (parasitic) power consumer.

The largest penalty imposed on the power plant output by the PCC system is due to the large amount of heat required to regenerate the solvent. The temperature level for regeneration is normally around 120°C. This heat is typically supplied by steam

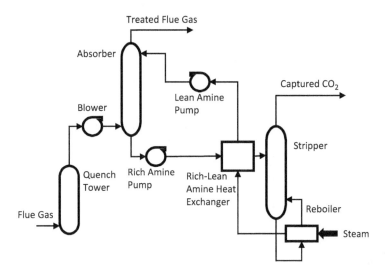

Figure 11.4 Highly simplified schematic diagram of CO_2 capture from the power plant flue gas via aqueous amine-based absorption.

extracted from the bottoming cycle and reduces steam turbine power output and, consequently, the net efficiency of the GTCC significantly. As is the case for all other carbon capture technologies, electrical power is consumed to compress the captured CO_2 for transportation to the storage site and injection into the storage reservoir.

Technologies for gas sweetening and syngas purification using alkanolamines have been extensively utilized in the *chemical process industry* (CPI) over the past century. The emergence of the chemical absorption technology utilizing solvents can be traced back to a 1930 US patent granted to R. R. Bottoms.[5] The technology was applied to make CO_2 used for enhanced oil recovery (EOR) in the 1970s [1]. Capturing and storing CO_2 for climate change mitigation was first proposed in 1977. For a brief history and timeline of CCs, the reader is referred to the paper by Brandtl et al. [2].

Despite the long experience base of the underlying technology, large-scale recovery of CO_2 from the flue gas of a fossil fuel burning power plant poses several serious challenges. Most important of these (for a GTCC), e.g., low CO_2 partial pressure and high regeneration energy, have already been mentioned. In addition, oxygen in the flue gas (about 12% by volume at the HRSG stack) can cause corrosion and solvent degradation. (Due to the absence of many impurities, which are amply present in coal-fired power plant flue gases, e.g., SOx (negligible), soot, fly ash, and mercury, the only significant degrading agent to worry about in GTCC flue gas is oxygen.) While inhibitors have been reasonably effective in mitigating these effects, the need for continuous removal of unavoidable solution contaminants adds to the operating costs.

---

[5] R. R. Bottoms, United States Patent No. 1783901, Process for separating acid gases (1930).

Table 11.5 Comparison of efficiencies and costs for post-combustion CO_2 capture from natural gas–fired power plants [1]

	Post-Combustion Capture			
Base year	2000		2002	
Study	IEAGHG		SINTEF	
	Original	With CO_2 Capture	Original	With CO_2 Capture
Power plan type	NGCC	NGCC	NGCC	NGCC
Plant size. MW el	790	663	400	338
Net efficiency (LHV), %	56	47	5Q	49
CO_2 emissions, kg/MWh el	370	61	363	60
Captured CO_2, kg/MWh el		380	–	370
Avoided CO_2, kg/MWh el[1]		309	–	303
Specific investment, EUR/kWel	410	790	625	1,515
O&M costs, EUR/MWh el	2	4	2.5	6
NO_x cleaning, EUR/MVh el[2]	–	–	1	1.5

[1] Avoided CO_2 emissions = (CO_2 per MWh, original) – (CO_2 per MWh, with CO_2 capture)
[2] SINTEF assumes NO_x cleaning from 20 down to 5 ppmv (15% O_2, dry)
  SINTEF assumes compression of captured CO_2 to 100 bar, IEA assumes 110 bar

Table 11.5 shows performance and cost impact of a post-combustion absorption system with MEA [3]. According to the source of the table, performance (efficiency and output) and cost data was calculated and estimated based on its wide use in CPI applications, albeit at much smaller scales. The original (base) case is a state of the art GTCC power plant (note the vintage, circa 2000). The construction of a new plant based on existing technology is assumed. The difference in plant size chosen in the studies made by IEA GHG[6] [4] and SINTEF[7] [5] and inflation (at least to some extent) should be responsible for the differences in the specific investment numbers. (The former obviously uses a $2 \times 2 \times 1$ GTCC as a basis whereas the latter uses a $1 \times 1 \times 1$ configuration.) Nevertheless, the economies of scale between the two, $410/625 = 0.656$, is too optimistic. Based on the published budget prices [6], 0.9 seems to be a more appropriate factor.

The price/cost data presented in Table 11.5 is first converted from 2000 and 2002 to 2015 currency (in EUR). The marginal cost of adding the carbon capture block to the power block is calculated in EUR and then converted to USD using an average 2015 EUR-USD conversion rate of 1.11. After doing this, the SINTEF data, which includes CO_2 compression to 100 bar, translates into about additional $1,080 per kilowatt (net with capture) for the carbon capture block (extra $365 million divided by 338 MWe). The optimism of this number will be readily apparent below.

[6] International Energy Agency (IEA) Greenhouse Gas R&D Programme
[7] SINTEF (Norwegian: Stiftelsen for industriell og teknisk forskning), headquartered in Trondheim, Norway, is the largest independent research organization in Scandinavia.

A useful yardstick for evaluating the cost of capture is the *annualized* cost in dollars per captured tonnes of CO_2. Using the conversion assumptions above, the SINTEF data in Table 11.5 is evaluated as

- $67 CAPEX (5,000 hours per annum and capital charge factor of 12%) and
- $13.8 OPEX for
- $80.8 total per tonne of capture CO_2.

It is reasonable to assume that the "real" project cost/price should be expected to be significantly higher than what is predicted in studies available in the open literature similar to those presented in Table 11.5. The only way to gauge the "true" cost of building an industrial plant is to have a full-blown front-end engineering design (FEED) study and account for all the financial bells and whistles that go into funding the project under consideration.

Two FEED studies undertaken by an EPC contractor in 2009 (for Gassnova in Norway [7]) and 2021 (for the US DOE [8]) came up with higher numbers. The first study was an 85.3% capture retrofit (MEA-based chemical absorption technology) to a 420 MW natural gas–fired GTCC power plant in Norway. The power plant was in $1\times1\times1$ configuration with a Siemens F Class gas turbine. The GTCC power plant in the second study was in $2\times2\times1$ configuration with the same gas turbine as in the first study but with fired HRSGs (719 MW fired, 594 MW unfired). The retrofit was 85% capture (with 35% MEA) of a slipstream (unfired operation) equivalent to the flue gas of the smaller plant in the first study.

In the 2009 study, the capture block retrofit CAPEX was estimated as NOK 2.24 billion, which translated into USD 472.5 million (2016 dollars) with the currency conversion rate of 1 USD = 5.3 NOK and the inflation factor of 1.12. In the 2021 study, the estimate was USD 477 million. Using a capital charge factor of 11.2% (determined in the 2021 study), the CAPEX per tonne of captured CO_2 is tabulated as a function of annual operating hours in Table 11.6 for the two FEED studies and SINTEF data in Table 11.5.

In the 2009 study, the annual OPEX was estimated as NOK 114.5 million or USD 24 million (2016 dollars). The same number in the 2021 study was USD 19.4 million. The OPEX per tonne of captured CO_2 is tabulated as a function of annual operating hours in Table 11.7. The total capture cost is provided in Table 11.8.

Table 11.6 Annualized CAPEX (per tonne of CO_2) for two FEED studies and SINTEF data in Table 11.5 (in 2020 US dollars)

Annual Op. Hours	2009 Study, $	2021 Study, $	SINTEF, $
4,000	112	104	85
5,000	90	83	68
6,000	75	69	57
7,750	58	70	44

Table 11.7 OPEX (per tonne of CO_2) for two FEED studies and SINTEF data in Table 11.5 (in 2020 US dollars)

Annual Op. Hours	2009 Study, $	2021 Study, $	SINTEF, $
4,000	51	39	19
5,000	41	31	15
6,000	34	26	13
7,750	26	20	10

Table 11.8 Total capture cost ($/tonne of CO_2) for two FEED studies and SINTEF data in Table 11.5 (in 2020 US dollars)

Annual Op. Hours	2009 Study, $	2021 Study, $	SINTEF, $
4,000	164	143	104
5,000	131	115	83
6,000	109	95	70
7,750	84	74	54

The FEED studies discussed above can be downloaded from the internet. They contain the most reliable CAPEX and OPEX data publicly available at present. Clearly, on a per tonne of captured CO_2 basis, the annualized cost is a function of the utilization of the power plant in question, i.e., its capacity factor. At the time of writing, on average, capacity factors of the GTCC power plants in the USA hover around 60%. Thus, using 5,000 hours per annum as the basis, PCC from GTCC flue gas is in the $110–120 range (per tonne of captured CO_2). For 6,000 hours per annum, the range is around $95–100. In general, as evidenced by the comparison in the tables above, studies published by research organizations and academia underestimate the CAPEX and OPEX significantly. It should also be pointed out that the numbers cited above are based on 85% capture with a generic amine-based chemical absorption process (i.e., "open art"). For more aggressive capture targets, i.e., 90% or 95%, with proprietary amines and unique process schemes, those costs can be somewhat higher. Unfortunately, such data/information is highly proprietary and not available to the public.

In terms of CAPEX, the SINTEF report estimate of $365 million or $1,080/kW (divided by 338 MWe) is $110 million or about 25% lower than the number determined by a FEED study. Thus, a more likely CAPEX number is $475/365 \times 1,080 = \$1,400/kW$. As the two examples discussed in Section 11.2 will make amply clear, even with diligent FEED and detail design efforts in hand, unforeseen changes in market conditions and subpar technical performance can render even the best planned project infeasible in a hurry.

As illustrated by the data presented above, PCC comes at a steep price, both in terms of parasitic power consumption (i.e., lower thermal efficiency) and capital cost, which is made worse on a per kilowatt basis due to reduced net power output.

(One should also add the increased operating and maintenance costs to the mix.) Thus, not surprisingly, the current research efforts are based on capture systems with reduced impact on plant thermal efficiency (mostly steam *stolen* from the GTCC bottoming cycle or the Rankine steam cycle), including

- Advanced amines
- Other solvents such as NH_3 (*ammonia*, the lowest form of amine)
- Adsorption
- Membrane separation
- Mechanical separation ("Supersonic CCS")
- Cryogenic separation
- Biomimetic[8] and microalgae systems

For an in-depth discussion of these technologies, the reader can consult Ref. [30] in Chapter 10.

For the application of PCC to a GTCC, the biggest impediment to its viability is the low CO_2 content (i.e., partial pressure) of the flue gas (that is, HRSG stack gas), i.e., of the order of 4%(v). The translation of this into chemical reaction terms for solvent-based absorption process to strip CO_2 from the flue gas is that the "driving force" for the reaction is small. The result is large and costly equipment, i.e., tall and wide absorber and stripper columns to achieve meaningful capture (90% or above). Methods have been proposed to rectify this *shortcoming* of the flue gas (the term is used from the capture perspective, otherwise, low CO_2 content is unquestionably a strong positive) and are described in several patents, e.g., by Gülen [9] and Børseth and Fleischer [10]. In either case, the flue gas is fully or partially sent to another gas turbine for additional combustion and power generation to increase flue gas CO_2 content (and pressure in the second patent). Requisite modifications to the *re-combustion* gas turbine can only be made by an OEM and may turn out prohibitively complex and expensive.

11.2 Commercial Experience

The *Petra Nova* project was designed to capture and store 1.4 million mt of CO_2 per year. At a flue gas stream equivalent to 240 MWe, this was the largest post-combustion CO_2 capture project installed on an existing coal fueled power plant (and the only one in the USA). (Note that only a portion of the 610 MWe power plant flue gas, a "slip stream," is sent to the CCS plant.) It is built around an advanced amine-based carbon capture process and reuse of the captured CO_2 for *EOR* at West

---

[8] Biomimetic refers to synthetic methods that mimic biochemical processes. One biomimetic approach to CCS is based on accelerating the rate of carbon capture by the engineering of an enzyme called carbonic anhydrase, which is combined with a conventional MEA.

Ranch oil field in Vanderbilt, TX (82 miles away). (The technology was deployed earlier in a 25 MWe equivalent pilot plant in a coal-fired power station in Alabama [11].) The capture system, which cost nearly $1 billion (about $4,200 per kW) to build (including a $190 million grant from the US DOE under the *Clean Coal Initiative* and a $250 million loan from two Japanese export credit agencies, the Japan Bank for International Cooperation and Nippon Export and Investment Insurance) began operations in January 2017. While the project was conceived when oil prices were at the $100 per barrel mark, at the time of commissioning the price was down to about $50 per barrel. Ultimately, on May 1, 2020, the plant operator, NRG Energy, shut down the Petra Nova capture plant. Even though the Covid-19 pandemic was cited as the reason, the writing was pretty much on the wall well before that. Apart from the low oil prices, the capture plant was beset by chronic mechanical problems and routinely missed its targets. Since January 2017, according to a technical report provided by the operator to the DOE in March 2020, the plant suffered outages for a total of 367 days. Issues with the carbon capture block accounted for more than a quarter of the outage days, followed by problems with the plant's dedicated natural gas power unit, which supplied the steam to the stripper for solvent regeneration. It captured 3.8 million *short tons* (st, 1 st = 0.907185 mt) of CO_2 during its first three years, nearly 20% shy of the 4.6 million st developers had expected.

Another full-scale PCC initiative resulted in the second-largest commercial-scale CCS project, a 115 MW coal-fired power plant in Saskatchewan, Canada. An old unit of the *Boundary Dam Power Station* was retrofitted with a proprietary amine-based carbon capture process technology for 90% capture (Shell CANSOLV). The reported capital cost was $600 million (Canadian), which translates into $435 million (US) at the time or $3,750/kW (nominal). The captured CO_2 was transported by pipeline to nearby oil fields in southern Saskatchewan to be used for EOR. Other byproducts captured from the project were also planned to be sold. For example, captured SO_2 is converted to sulfuric acid and sold for industrial use. Fly ash would be sold for use in ready-mix concrete, pre-cast structures, and concrete products. This project was one of the ten large-scale[9] PCC projects (all of them power plants) identified by the *Global CCS Institute* in 2014 (and the only one that made it to the execution stage).

The plant went into operation in 2014, i.e., three years before Petra Nova. In late 2015, reports in the media suggested that the project had run into design, operational, and cost problems.[10] In particular, the process was reported to have captured just over 400,000 tonnes of CO_2, well short of the planned one million mt annually. This

---

[9] Large-scale projects are defined as those that involve the capture, transport, and storage of CO_2 at a scale of at least 800,000 mt/year of CO_2 for a coal-fired power plant, or at least 400,000 mt/year of CO_2 for other emissions-intensive industrial facilities (including GTCC). For details refer to the article on pp. 21–23 in the April 2014 issue of the *Chemical Engineering* journal.

[10] http://www.powermag.com/saskpower-admits-to-problems-at-first-full-scale-carbon-capture-project-at-boundary-dam-plant/; last accessed November 10, 2015.

was explained by mechanical problems, which resulted in 40% availability – even though the capture plant, when operational, performed as planned to achieve 90% capture [12]. Such teething problems are to be expected in the initial field-deployment of a FOAK technology. Nevertheless, it is instructive in the sense that, even when all the major components are mature and field-proven (after all, absorber-stripper systems are mainstays of CPI plants all over the world), their deployment in a different application is subject to many unforeseen pitfalls.

Not being able to sell the contracted amount of CO_2 to the EOR customer, the operator, SaskPower, not only lost on sales revenue but also had to pay penalties. Eventually, the contract was renegotiated to reflect the lost capacity, which made the project economics unpalatable. In July 2018, SaskPower announced that it would not retrofit Units 4 and 5 in Boundary Dam with CCS and that Unit 3 would be shut down in 2024 (the mandated date). No information is publicly available about the operating status of the Boundary Dam Unit 3 carbon capture system at the time of writing (early 2021).

Based on the available information, it is safe to assume that the carbon capture block for the two coal-fired power plants comes with a price tag of around \$4,000 per kilowatt (in 2015 dollars). Let us do a "back of the envelope" type comparison with this number and the number deduced from the data in Table 11.5, \$1,080/kW, for a combined cycle power plant. At a CO_2 emission rate of 363 kg/MWh, the combined cycle capture block cost is $1,080 \times 1,000/363 \sim \$2,975$ per kg/h of CO_2 from the stack. Based on the more reliable FEED study estimates reported in Section 11.1, the CAPEX is more likely to be \$1,400/kW or \$3,850 per kg/h of CO_2 from the stack. Assuming 900 kg/MWh for the coal-fired power plant's CO_2 emission rate, the capture block cost comes to $4,000 \times 1,000/900 \sim \$4,450$ per kg/h of CO_2 from the stack.

Even earlier than the two projects mentioned above, was the 320 MWe natural gas–fired combined cycle power plant in Bellingham, MA, with small-scale PCC. The facility used Fluor's proprietary amine-based solvent, *Econamine*, for the capture of 85–95% of CO_2 in the stack gas at a rate of 320–350 tonnes per day (tpd). The power plant comprised two Westinghouse W501D5 (now Siemens SGT6–3000E) gas turbines and HRSGs that produced low-pressure steam for use in the adjacent capture plant. A single W501D5 gas turbine (107 MWe ISO base load rating) emits about 1,500 tpd of CO_2. Thus, the stack gas stream diverted to the capture plant containing $320/0.85 = 375$ tpd of CO_2 constitutes $375/1,500 = 0.25$ or 25% of the stack gas of only one gas turbine-HRSG train. The captured CO_2 was of food and beverage quality and sent to a nearby soft drink bottling plant (i.e., there was no storage). The capture plant ran from 1991 to 2005 but was closed due to the increase in natural gas prices. It was the only commercial-scale CO_2 capture facility in the world operating on gas turbine exhaust gas containing 3%(v) CO_2 and 14%(v) O_2. The power plant is still in operation as a peaker unit.

In the remainder of the chapter, the focus will be based on generic amine-based chemical absorption technology for the capture of CO_2 from the HRSG stack gas. However, before jumping into that discussion, a few words on the selection of this technology are in order.

11.2.1   A Case for Post-Combustion Carbon Capture

According to a recent article in POWER magazine,[11] nearly half of almost $3 billion spent by the US DOE since 2010 to develop advanced fossil energy technologies was dedicated to nine CCS demonstration projects – but only three were active at the end of 2017, and only one was at a power plant (the Petra Nova plant discussed above).

In a report prepared for the Senate Energy and Natural Resources, the Government Accountability Office (GAO) says that the DOE provided $2.66 billion in funding or obligations to 794 fossil energy research and development (R&D) projects. Of the total $2.66 billion it spent since 2010 (the year that the DOE's current data management system came into use), about $1.12 billion was provided – in amounts ranging from $13 million to $284 million – to nine "later-stage, large demonstration projects that were to assess the readiness for commercial viability of CCS technologies." Six of these projects used coal, and the other three used methane, ethanol, and pet coke. Industry paid an additional $610 million in cost-share for these projects, the GAO found.

However, as stated in the article and the GAO report, out of these nine projects,

- The DOE withdrew its support for four projects (costing the DOE $475 million)
- Two other projects were withdrawn by the recipient (costing the DOE $30 million)

The factors leading to the demise of these six projects ranged from a lack of technical progress to changes in the relative prices of coal and natural gas that made the projects economically unviable. Of the remaining three, one was the Petra Nova (discussed earlier) and it was terminated for good in 2020. Under the light of this rather bleak experience base, one might reasonably raise the question: "Why bother?"

Indeed, why is ink wasted, in this book or elsewhere, on post-combustion carbon capture? Here is why. First, the technology can be readily applied to the most advanced and proven fossil fuel–fired technology out there, i.e., GTCC with advanced class gas turbines. With ISO base load rating efficiencies well above 60% net LHV and already more than 60% lower CO_2 emissions *without* capture vis-à-vis coal-fired power plants (the best of them nonetheless) as discussed in the beginning of the chapter, GTCC *with* capture is more efficient than even the most advanced, ultra-supercritical coal-fired power plant technology *without* capture. While there is a valid argument to be made about the size and cost of the capture block due to the low CO_2 content of the HRSG stack gas (about 4% by volume at most), compensated by the low CAPEX of the power block, this renders the LCOE of the GTCC with CCS the lowest among the competing technologies (e.g., see Figure 11.2).

"Good point," one might remark and then add "but, see, the capture technology does not work; look at the failed projects." Prima facie, this seems to be a valid argument, but it is deeply flawed. Both, Petra Nova and Boundary Dam were based on "novel" solvents with the claim to minimize the parasitic steam consumption and thus maximize the thermal efficiency of the coal-fired host power plant. These solvents require specialized

---

[11] Sonal Patel, *DOE Sank Billions of Fossil Energy R&D Dollars in CCS Projects. Most Failed*, October 9, 2018.

system design and additional equipment or control systems quite different from the conventional acid gas scrubbing technology with generic amines such as MEA (monoethanolamine), i.e., the combination widely used in CPI that can be considered a mature and proven technology. This prevents the engineers and EPC contractors from developing "reference plant" designs that can be standardized in manner similar to the current state of the art in three-pressure, reheat (3PRH) combined cycle bottoming steam cycle design. As a result, at the time of writing, PCC is at a *technology readiness level* (TRL) of 9 but it is far from a *commercial readiness index* (CRI) of 6 (i.e., a "bankable" asset[12]). At present, PCC has a CRI of 2, i.e., it is at the commercial trial stage.[13]

Whatever the chosen technology and solvent combination is, the best route for covering the gap between CRI 2 and 6 is most likely the "open art" approach, which has two components:

1. *Open technology*, which means that the operators of a PCC plant are in full control of the technology used in the plant rather than purchasing a "black box" unit.
2. *Open access*, which means that open-technology PCC projects use a generic, nonproprietary solvent as the design solvent.

The open art approach can do this by leading to construction of a relatively small fleet of full-scale plants (e.g., five to ten 400–750 MW GTCC plants) in successful operation for several years. Learnings from the construction and operation of similar plants can be expected to lead to myriad innovations to improve efficiency and reduce CAPEX and OPEX. This can open the door to wider scale deployment as a bankable asset at CRI of 6.

11.2.2   Solvent Selection

Monoethanolamine (MEA), typically at 30%(w) concentration, is the most widely used solvent in chemical absorption. It is a member of the family of alkanolamines, amines that contain an amine group and one or more hydroxyl groups, favored by their increased solubility and reduced volatility vis-à-vis other amines. MEA is characterized by its high chemical reactivity with CO_2 and low cost. Its disadvantages are high energy consumption during solvent regeneration (supplied by steam extracted from the power block steam cycle), corrosiveness (mitigated by use of inhibitors and alloy materials), and degradation in the presence of O_2 (most significant for GTCC applications). Tertiary amines such as MDEA have a lower energy requirement for regeneration but suffer from slower reaction kinetics, resulting in higher overall solvent flow and larger equipment sizes. (Typically, for MEA, the solvent mass flow rate is less than half of that of MDEA and DEA.) For an in-depth comparison of amine solvents,

---

[12] For the CRI definition, see Section 16.5. A mature technology can be considered as a "bankable" grade asset class with known standards and performance expectations. Investment decisions are not driven by market and technology risks. Drivers are proponent capability, pricing, and other typical market forces.
[13] At this stage, the technology is deployed in small-scale, first-of-a-kind projects funded by equity and government project support. Commercial proposition is backed by evidence of verifiable data typically not in the public domain.

refer to Lang et al. [13]. One solution is using a mixture of amines, e.g., MDEA with piperazine at 35%(w) concentration to speed up the reaction kinetics while maintaining the regeneration energy advantage. Another option is to use a catalyst, e.g., 20% (w) MEA with 2%(w) catalyst. See the LLNL report by Jones [14] for more details (US DOE Contract DE-AC52–07NA27344).

An interesting alternative solvent is *ammonia* (NH_3), which has a lower energy requirement for solvent regeneration and higher absorption capacity vis-à-vis MEA, DEA, and MDEA. However, due to its high volatility, prevention of the ammonia slip into the flue gas stream after absorption is a major technical issue and environmental risk. Consequently, there is significant R&D effort underway into finding the "Goldilocks" solvent with high CO_2 absorption, low energy consumption during production and regeneration, but at low cost, and low environmental impact.

In terms of the environmental impact, one concern mentioned in the literature is the direct and indirect CO_2 emissions during the production of MEA, which is produced via an exothermic reaction (i.e., no catalyst is required) of aqueous ammonia and *ethylene oxide* (EO). Byproducts are DEA or *triethanolamine* (TEA) depending on the NH_3/EO ratio. Ammonia is produced from steam-methane reforming, which typically generates 1.15–1.4 kg of CO_2 per kg of NH_3 produced (this does not include the CO_2 in combustion products). The physical and chemical CO_2 absorption capacity of MEA is a function of temperature, pressure, presence of additional gases, and the aqueous MEA concentration. For 30%(w) concentration, a reasonable estimate is 0.4 kg of CO_2 per kg of MEA [15]. To produce 3 kg of MEA, approximately 1 kg of NH_3 is required, i.e., per the reaction formula

$$NH_3 + C_2H_4O \rightarrow H_2NC_2H_4OH.$$

Thus, producing 1 kg of MEA can generate (indirectly) as much as $0.28 \times 1.4 = 0.4$ kg of CO_2. Thus, the CO_2 balance of MEA production-usage cycle is not unfavorable, i.e., roughly 0.5 kg CO_2 generated during MEA solvent production for each kg of CO_2 captured from the flue gas of a fossil-fired power plant. Furthermore,

- CO_2 from ammonia production is highly concentrated and therefore readily and inexpensively compressed and sequestered. Additionally, most of the CO_2 produced is combined with a side stream of ammonia to produce urea, a salable product.
- More importantly, MEA captures many times its weight in CO_2 over its approximately one-year life in the system. As an example, if net capture per cycle is 0.25 kg/kg (pessimistic) and cycle time is 1 hour, then 0.5 kg of CO_2 is offset in only 2 hours. That leaves 8,758 hours for positive capture balance. Even after throwing in losses, leaks, process imperfections, etc. into the mix, the math is rather strongly in favor of MEA.

In any event, a similar accounting should be made for any amine-based solvent to have the bigger picture in sight. For a detailed look into this aspect of using amines in CCS, refer to the paper by Luis [16].

Another environmental concern associated with amine-based CO_2 capture via chemical absorption is the emission of MEA or other amines to the atmosphere with the clean flue gas via liquid carryover from the acid wash section of the absorber column. While amines themselves are not likely to be of toxicological concern, their degradation products, e.g., *nitrosamines* and *nitramines*, which are carcinogenic compounds, present a significant health hazard. Evidence of the formation of these compounds in carbon capture applications have been discovered [16]. This requires the installation of high-efficiency mist eliminators at the top of the absorber column to increase the capture of 2 μm droplets to 99.99%.

Whether to go with a generic solvent/technology or a proprietary one can be answered by examining the findings of a recent World Bank funded prefeasibility study [17]. In particular, the goal of the study in question was to assess and recommend the most appropriate commercially available PCC technology for a 250 MW GTCC power plant, Poza Rica in the State of Veracruz in Mexico, and to develop a conceptual design of a capture pilot plant. Six advanced solvent-based absorption PCC technology developers and licensors participated in the study.

The technology screening, evaluation, and comparison of the six proprietary processes were based on the responses received to a technology survey questionnaire based on the study design basis. A generic 30% MEA-based PCC process design was selected to serve as a benchmark for comparing the performance of the six PCC technologies participated in the study, assessing their claimed improvement, and substituting for missing data needed for the analysis. The incremental PCC *cost of electricity* (COE) for the six proprietary technologies and the benchmark technology is summarized in Table 11.9. (Note that the numbers in the table correspond to a capture cost of $102 ± $7 per tonne of captured CO_2 – assuming a 11.2% capital charge factor and 5,000 operating hours per annum.)

An impartial examination of the data in Table 11.9 would lead one to conclude that, within the level of data accuracy for a study of this type, none of the top proprietary PCC technologies stands out among the rest. Furthermore, all of them are amine-based technologies that operate along the same basic principles as a generic MEA plant. Not surprisingly, the study recommendation was that the capture plant retrofit to the 250 MW Poza Rica GTCC should be designed for generic MEA, but with additional design features to grant it the flexibility to allow for the testing and validation of other amine-based technologies. The study sponsor, the World Bank, agreed with the suggested approach.

The last part of the recommendation is significant. An open art PCC designed for generic equipment and amine solvent does *not* preclude the use of advanced solvents. (Naturally, the proprietary nature of the solvent becomes moot in this scenario.) Granted, such a PCC will not be the optimum platform to draw upon the full benefits of a particular advanced solvent. However, prior experience gained in test sites within the *International Test Center Network*[14] such as *Technology Centre Mongstad*

---

[14] The ITCN (https://itcn-global.org/about-the-itcn/) is a global coalition of R&D facilities that aims to accelerate the CCUS technologies to CRI 6 by sharing knowledge of construction and operation of large test facilities to establish a level playing field for technology vendors to reduce costs and risks (technical, environmental, and financial). The ITCN was co-founded by the US DOE National Carbon Capture Center (NCCC) and Technology Center Mongstad.

Table 11.9 Incremental PCC cost of electricity for selected technology licensors

(Note that Shell CANSOLV® was used in Boundary Dam; MHI technology was used in Petro Nova.)

Incremental Costs to Poza Rica NGCC without CO$_2$ Capture[1]	Estimated Post-Combustion CO$_2$ Capture Costs						
	Generic 30% MEA PCC Design	Alstom	BASF/ Linde	Fluor	HTC Purenergy	MHI	Shell CanSolv
CAPEX Estimate, $MM US USGC$_2$							
PCC Plant + CO$_2$ Compression[2]	181.4	234.7	187.7	181.9	194.5	178.8	194.9
Flue Gas Blower	14.2	14.2	14.2	14.2	14.2	14.2	14.2
Poza Rica Plant Modifications	32.8	32.4	30.4	31.9	29.1	30.9	30.4
TOTAL	228.4	281.4	232.3	228.0	237.8	223.9	239.5
O&M Estimate, $MM US							
Variable Costs[3]	7.6	7.6	7.6	7.5	7.3	7.5	7.5
Fixed Costs	11.0	13.3	11.1	10.9	11.4	10.8	11.6
TOTAL	18.5	21.0	18.7	18.5	18.7	18.3	19.1
Estimated Cost of Electricity (COE), $/MWh[4]	37.6	41.4	35.3	36.5	36.2	35.1	36.0
Ranking based on COE	*N/A*	*6*	*2*	*5*	*4*	*1*	*3*

[1] Values presented here are Nexant's interpretation of the data provided by the PCC licensors.
[2] All figures except Nexant's "Generic 30% MEA Design" are based on vendor-provided data, which are considered proprietary.
[3] Major component is the amine replacement costs, which are considered proprietary.
[4] Incremental to estimated existing Poza Rica NGCC COE of $40.69/MWh.

showed that a wide range of proprietary solvents can be tested in the same facility with limited modifications to the equipment and controls. The information and experience accumulated via testing of different solvents, e.g., long-term solvent degradation characteristics and atmospheric emissions, when shared with the interested parties, can enable rational evaluation of potential projects and risk minimization. This will in turn result in competitive EPC bids and selection that will lower construction costs. (A "black box" approach by the capture technology licensors prohibits competitive bidding and limits the choices available to project developer.)

11.3   CO$_2$ Utilization

Once the CO$_2$ is captured, before or after combustion, one must do something with it. There are two possibilities: *storage* (i.e., sequestration) or *usage* (utilization), i.e., to do something with the captured CO$_2$. For the latter, by far the most common example is *EOR*. Nevertheless, there are other possibilities such as using CO$_2$ as a feedstock for production of, say, urea or methanol. That part of the CO$_2$ "value chain" is beyond the scope of the present treatise. The reader is referred to a recent paper by Jarvis and Samsatli for a comprehensive coverage of available technologies, including CAPEX, OPEX, electricity consumption, TRL, and product price [18]. Possible pathways can be summarized as follows:

- Storage
  - Depleted oil and gas fields
  - Deep saline formations
  - Aquifers
  - Mineral storage
  - Reforestation or afforestation
- Direct use
  - Solvent
  - Working fluid, e.g., EOR
  - Heat transfer
- Conversion (chemical, biological, electrochemical, photochemical) utilizing "green" resources, e.g., wind, solar, geothermal, hydro, and reacting with water, hydrogen, and other chemicals:
  - Feedstock (biodegradable polymers, urea, isocyanates, carbamates, carboxylates, lactones, inorganic and inorganic carbonates)
  - Energy vector (renewable fuels, syngas, methane, formic acid, methanol, and DME)

Capturing CO$_2$ is one thing, what to do with it is yet another. The volume of CO$_2$ that must be captured to stop the increase of this GHG in the atmosphere and even reverse the trend is a task of gigantic proportions. Let us look at the numbers.

Based on preliminary analysis, the global average atmospheric carbon dioxide in 2020 was 412.5 parts per million in volume (ppmv). On June 7, 2021, the *Scripps Institution of Oceanography* and the *National Oceanic and Atmospheric Administration* announced that the atmospheric CO$_2$ (carbon dioxide) concentration had reached its highest level since accurate measurements began 63 years ago. The monthly average CO$_2$ concentration for 2021 is 419 ppmv. Assuming the molecular weight of the atmospheric air as 29 kg/kmol, on a mass basis, the CO$_2$ concentration is $419 \times 44/29 = 636$ ppmw *(parts per million by weight)*. Taking the total mass of atmospheric air as 5,150,000 Gt, the mass of atmospheric of CO$_2$ is found as $(636/10^6) \times (5.15 \times 10^6) = 3{,}274$ Gt.

From the US EIA data, in 2019, CO$_2$ emissions from electricity generation (excluding cogen plants) utilizing natural gas was 0.56 Gt vis-à-vis 0.95 Gt from coal-fired generation. According to the EIA, in 2020, CO$_2$ emissions from the power sector declined by 3.3% (or 450 Mt), the largest relative and absolute fall on record (probably boosted by Covid-19). Thus, using the 2019 total for 2021 is a reasonably good assumption (assuming some recovery from the Covid-19 debacle), i.e., 1.51 Gt total

CO_2 emissions from natural gas and coal-fired electricity generation, which corresponds to $1.51/(5.15 \times 10^6) \times 10^6 = 0.294$ ppmw or $0.294 \times 29/44 = 0.194$ ppmv.

Let us do a quick *Gedanken* experiment. Let us assume that, somehow, the entire gas- and coal-fired power plant fleet is retrofitted overnight with PCC. The amount of CO_2 that we must dispose of is 1.51 Gt, or 1,510 million metric tons. According to a 2010 US DOE NETL report, EOR operations account for 9 million metric tons of CO_2, equivalent to about 80% of the industrial use of CO_2, every year.[15] On top of that, at 88% of the total, EOR represents by far the largest industrial use of CO_2. Beverage carbonation, food industry, fabricated metal products, and other uses account for the rest (12%). Roughly speaking, less than 15% of the CO_2 used in EOR is pulled from anthropogenic sources like natural gas processing and hydrocarbon conversions. The rest, i.e., over 85%, comes from terrestrial sources, i.e., a few big natural CO_2 reservoirs under the surface of the earth. When CO_2 is injected underground for EOR, most of it, around 90 to 95%, stays there, trapped in the geologic formations where the oil was once trapped. Thus, undoubtedly, EOR represents a good means of captured CO_2 disposal, if not the ideal or most preferable.[16]

Today, let us assume that, annually, 10 million metric tons of CO_2 is needed for EOR. Even if we increase it ten-fold, we are looking at 100 million metric tons. The CO_2 that we are looking to get rid of is 1,500 million metric tons and it amounts to barely 0.2 ppmv. Clearly, we must find other uses and "holes in the ground" to put the captured CO_2 into. And even if it could be done, it is barely a drop in the proverbial bucket. For an even more dramatic illustration of the disconnect between what must be done and what can be done, wait until the Epilogue (Chapter 15).

11.4   System Description

Major components of a two-column chemical absorption system are shown in the simplified diagram in Figure 11.4 at the beginning of the present chapter. Some details are left out of the diagram (e.g., absorber sections, bleeds, the reclaimer) for simplicity. They are going to be included in the process description below.

The absorber and stripper columns are made from stainless steel. Typically, they are fitted with stainless steel structured packing to achieve a low-pressure drop across the vessel internals. Stainless steel is the material most suitable to equipment and piping handling amine flows to avoid corrosion problems associated with the high amine concentration and to avoid onerous inhibitor programs.

In general, plate-and-frame heat exchangers are selected for most plant heat transfer services. Heat exchangers of this type have a smaller plot space requirement for a given area than other types of heat exchangers. Also, they can have much lower

---

[15] Carbon dioxide enhanced oil recovery – Untapped domestic energy supply and long-term carbon storage solution, US DOE NETL, March 2010.

[16] One would, of course, justifiably object to utilizing captured CO_2 for the extraction of even more hydrocarbon fuel, the culprit for CO_2 emissions in the first place.

approach temperatures than other configurations and, therefore, increase the amount of heat exchanged between streams and increase the plant's energy efficiency.

11.4.1 Flue Gas Duct

In many theoretical treatises, the connection between the HRSG stack and the capture block is a simple, straight line implying a *trivial* system. Nothing could be further from the truth. Especially in *retrofit* cases, where the capture block is an *add-on* to an existing power plant, design and construction of the flue gas duct connecting the two plants is one of the most crucial tasks undertaken in the FEED study.

In a typical GTCC, the carbon steel stack of the HRSG exhausts the gas turbine flue gas, cooled to around 80–100°C (depending on the gas turbine technology, bottoming cycle, and presence of duct firing), to the atmosphere. The stack is equipped with a motor-operated stack damper, which is closed during overnight or weekend shutdowns to bottle up the HRSG to maintain the drums and the tube banks in a warm condition to allow rapid restart of the power plant.

Whether it is a proverbial "blank sheet" design or a retrofit feature, the addition of the CO_2 capture block requires the modification of the standard HRSG stack design (in essence, a vertical cylinder) to allow different operating scenarios, i.e., no capture (all flue gas flow is through the main vertical cylindrical portion), part capture (i.e., some of the flue gas flow is sent to the capture block through a dedicated ductwork), or full capture (i.e., all flue gas flow is sent to the capture block). This modification comprises an exterior support structure constructed from structural steel around the stack, which provides (i) lateral bracing of the main stack and (ii) the breeching attached to the main stack near its base. In a new design, the HRSG foundation is designed to accommodate the support structure. Otherwise, new foundations must be built.

A generic and mainly conceptual flue gas duct arrangement is shown in Figure 11.5. The actual length and routing of the duct is highly variable, based on the project's requirements, e.g., *green-field* or retrofit (i.e., *brown-field*), site conditions and land availability, flue gas conditions across the operating regime of the power plant, and other financial and economic criteria. The goal is to optimize the duct dimensions (cross-sectional area and length), support structure, construction materials, flue gas cooling, and booster fan location to strike a good balance between performance (i.e., pressure loss in the duct and *booster fan* power consumption) and installed cost. As shown in Figure 11.5, there is a *fogger* shown (for flue gas cooling) downstream of the blower in addition to the *direct contact cooler* (DCC), which is shown as an integral part of the absorber (ABS) column. In most cases, there will be a choice between the two and, typically, the fogger system in the flue gas duct is the less expensive option. For extremely clean natural gas fuel used in the power block and in the presence of a *selective catalytic reduction* (SCR) system in the HRSG (for the removal of NOx and CO), this is acceptable from an amine degradation perspective as well. Otherwise, one must go with the DCC option. Another option is to place the booster fan (the blower) downstream of the DCC or the fogger (as shown in Figure 11.5).

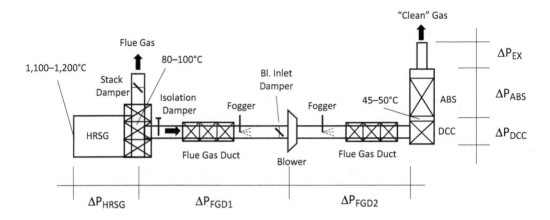

Figure 11.5 Conceptual, simplified depiction of the flue gas duct and associated equipment connecting the power and capture plants.

The blower creates a negative pressure in the duct, causing the flue gas to be pulled into the capture block. It is driven by a *variable speed drive* (VFD) and, typically, is also equipped with an inlet (suction) damper, which enables accurate control of the flue gas flow and pressure with essentially no flue gas exiting the HRSG stack during operation with capture. Precise pressure control in the duct is imperative to avoid any negative impact on the power plant operation and to prevent air ingress into the capture plant through the stack opening. Whether there are two 50% units, or one 100% unit, is a function of cost, performance, and RAM study in the FEED process.

As shown in Figure 11.5, the design boundary conditions are the flue gas temperatures at the HRSG exit and the absorption column inlet and the ambient pressure. (The *stack effect* for the HRSG main stack and the absorber column's discharge chimney can be assumed to be baked into ΔP_{FGD1} and ΔP_{EX}, respectively). The key design goal is to minimize the blower power consumption, which, for a given isentropic efficiency, is a function of flue gas inlet temperature and blower head. The latter is the total of the pressure losses identified in Figure 11.5, i.e.,

$$H_{FAN} = \Delta P_{FGD1} + \Delta P_{FGD2} + \Delta P_{DCC} + \Delta P_{ABS} + \Delta P_{EX}.$$

Sample calculations are carried out for a generic F Class gas turbine with an exhaust gas flow rate of 680 kg/s. At the HRSG exit, the flue gas temperature is assumed to be 93°C. The results are summarized in Table 11.10.

The takeaways from Table 11.10 are as follows:

- Without upstream flue gas cooling, the increase in blower power consumption is about 1 kJ/kg (same head).
- Each 10 mbar of additional head is worth about 1.3 kJ/kg in blower power consumption.
- Each percentage point in blower isentropic efficiency is worth about 0.22 kJ/kg in power consumption.

Table 11.10 Sample flue gas booster fan (blower) performance calculations (flue gas flow 680 kg/s)

Blower Efficiency, %	Blower Head, mbar	Fogger Effectiveness	Blower T_{in}, °C	Blower, kWe	Blower, kJ/kg
80	84	95%	50	6,967	10.2
80	84	85%	54	7,045	10.4
80	84	N/A	93	7,684	11.3
80	127	N/A	93	11,451	16.8
75	127	N/A	93	12,212	18.0

Clearly, minimizing the blower head via careful flue gas duct design provides the bigger (proverbial) "bang for the buck." Typical values for pressure losses are

- 30–40 mbar for the absorber
- 5–10 mbar for the DCC
- 10 mbar or higher for each duct section
- A few millibars for the fogger nozzle grid

In order for the clean flue gas (at 45–50°C) leaving the absorber column to have enough buoyancy to be ejected to the atmosphere via the absorber chimney, it may be necessary to heat it to a temperature of 70–80°C. There are several options to accomplish this. Exergetically, heat integration is the best route, i.e., combining flue gas cooling and clean gas heating via an intermediary coolant (e.g., a water-glycol mixture) loop. However, this may not be feasible from a cost, size, or footprint perspective. Such considerations can only be done on a case-by-case basis during the FEED study.

The flue gas duct is typically of rectangular cross section. Its dimensions (i.e., the cross-sectional area) are determined by the flue gas flow rate. The typical construction material is carbon steel. If there is a flue gas cooler utilizing water injection and evaporation, stainless steel is used from the fogging system grid all the way to the absorber. The ductwork is supported by bridge steel to permit offsite fabrication and installation at the site by crane. This also permits the individual duct sections to be removed for future heavy haul access as required. The duct is fully insulated for thermal and noise protection.

Another possible flue gas duct construction material is fiberglass-reinforced plastics (FRP). FRP has been used in the wet flue gas desulfurization (FGD) systems of coal-fired power plants due to its cost effectiveness and reliability going back almost four decades. In that particular application, FGD equipment made from FRP is about half the cost of nickel alloy-clad steel. FRP has also been used successfully in stack liners, ductwork, absorber vessels, and limestone slurry piping. The trade-off between FRP and stainless steel is a function of temperature resistance of the former and its cost (CAPEX *and* OPEX). (Ducts made from FRP typically have a circular cross section.)

Two guillotine-type isolation dampers in series are provided at the HRSG stack to allow positive shutoff of the flue gas. These are strictly open-and-closed devices with a seal air system to provide positive pressure between the dampers. The reason for that is to prevent the flue gas from the HRSG (in operation) from entering the ductwork during duct maintenance. Platforms are provided for access to instrumentation and to service these isolation dampers.

For precise control of the flue gas flow during the power plant operation, with or without capture, the HRSG stack damper, the isolation dampers, and the blower inlet damper must be operated in a seamlessly integrated manner. At startup, the HRSG stack damper is open, permitting the flue gas to exit the stack at the top. Once the capture block becomes available, the flue gas blower is started while the HRSG stack damper is slowly closed. (This requires that the stack damper is modified with the addition of a pneumatic controller outfitted with a local air reservoir to permit the damper to be modulated from full open to full closed at the desired rate.) This process is controlled by the distributed control system (DCS) in coordination with the blower VFD and inlet dampers. The goal is zero flue gas flow from the HRSG stack during operation with capture. Nevertheless, some small amount of stack damper leakage can be expected. However, for most practical purposes, unless the dampers or the blower malfunction, emissions from the HRSG stack during operation with the capture block in action are essentially zero.

11.4.2   Flue Gas Cooling

There are three methods of cooling the flue gas from the HRSG:

- Indirect cooling using water-cooled fin-tube coils
- Evaporative cooling using direct contact water spray or fogging
- Direct-contact cooling using a packed bed with a side-water cooler

The first option is typically quite expensive. In essence, the cross-sectional area of the cooler must be the same as that of the duct. If seawater cooling is used inside the tubes, the cold water may cause condensation of water vapor in the flue gas on the fin-tube walls. This requires the use of stainless steel fin-tube material. Furthermore, the flue gas entering the absorber column must be saturated with water. Otherwise, it can dry the rich amine solvent before it is sent to the scrubber. This cooling system, being dry, would not provide the requisite water for saturating the flue gas.

The second option, i.e., evaporative cooling by a spraying or fogging system, can reduce the flue gas temperature and saturate the flue gas with water. This is a low-cost option because water spraying does not require special equipment and provides a very efficient means of cooling via water evaporation. However, it cannot reduce the flue gas to below its dew point. Typically, the HRSG flue gas has a dew point of about 45–50°C, which sets the flue gas temperature entering the absorber. This temperature is adequate and provides a reasonable balance between capital and operating costs.

The third option is to use a packed bed, direct-contact cooling section, either in a separate vessel (quench tower) or as an integral packed bed in the absorber bottom section. The packed bed can be sized to cool the flue gas to as low as 25°C if a sufficiently cold outside water source is available (e.g., seawater). If this is the case, this option may be the preferred one.

Starting from the lower left corner of the system diagram in Figure 11.4, flue gas enters the quench tower, which is the *DCC*, a vertical packed column, where the flue gas is cooled down and saturated with water. The objective is to increase absorption efficiency as well as to reduce water loss and keep the water balance in the absorption process.

In the DCC, cooling water is circulated through the packed bed of the column in a closed loop. Circulating cooling water is then cooled in a plate-and-frame type heat exchanger. In general, the flue gas from the HRSG is between 80 and 100°C. In modern three-pressure reheat (3PRH) systems, the lower end of the range should be expected. Heat transfer from the flue gas to the cooling water takes place via direct contact in the DCC packing in counter-flow (i.e., water flows downward and flue gas flows upward). The cooling water temperature at the DCC inlet is 25–30°C and warm water leaves the column at 45–50°C. Water vapor that condenses from the flue gas is purged from the closed circulating cooling water loop and sent to a water tank. It is treated for ammonia and reused as process water within the capture block. Approximately 50% of this treated water is used as amine make-up water. The remainder is used as make-up for the water wash sections and other purposes.

11.4.3 CO_2 Absorption

The absorber is a vertical packed column comprising three sections, i.e., one absorption section and two wash sections (see Figure 11.6). Each section is a packed bed with unidirectional flow of vapor from the bottom section to the top section via chimney trays. *Lean amine* from the stripper is distributed into the absorber column above the packing material. Flue gas enters the column from the bottom of the absorption section where a design provision is made to minimize gas entrainment into the rich amine solvent. The lean amine (liquid) comes into contact with the upcoming flue gas in a counter-current flow arrangement. Carbon dioxide diffuses into the liquid and chemically binds to the amine in the liquid and is removed from the flue gas stream. The remaining gas passes through the chimney tray above the absorption section and enters into the absorber off-gas washing sections. The solution loaded with CO_2 is called *rich amine*, which is collected at the bottom of the absorber and pumped to the amine regeneration side via a rich side amine filter.

What is the purpose of the wash sections in the absorber? The exothermic heat of reaction from the CO_2 absorption increases the overall absorber section temperature thus resulting in solvent being evaporated. Entrainment is also expected due to handling a huge volumetric flow rate of flue gas. The wash sections at the top of the absorber are utilized to cool the off-gas and recover entrained solvent as well as condense water vapors to maintain a plant water balance. The main objective is to minimize significant amine losses due to carryover and evaporation and to eliminate the pollutant emissions. Two packing sections are provided to achieve this goal and meet the off-gas specification. In the first wash section, a large flow rate of wash water is circulated over the packed bed in a closed loop. Reflux water from the reflux drum is used for make-up. In this section, hot off-gas is cooled down to approximately 35–40°C, thereby condensing the majority of amines and water at the off-gas

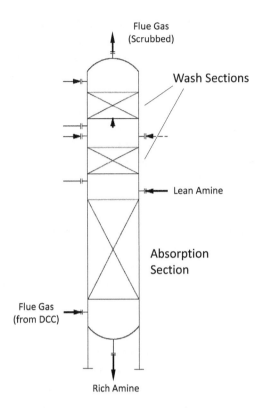

Flue Gas
(Scrubbed)

Wash Sections

Lean Amine

Absorption
Section

Flue Gas
(from DCC)

Rich Amine

Figure 11.6 Simplified schematic diagram of the absorber column.

saturation point. Saturated off-gas is then passed in to the second wash section located above the first wash section via a chimney tray.

The second wash section is required to meet environmental regulations, especially pertaining to ammonia emissions, which is one of the amine degradation products. In the second wash, a large quantity of wash water is required to reduce the ammonia concentration in the off-gas to less than 5 ppm. Due to environmental as well as economical constraints restricting the use of such a large quantity of fresh water in a once-through system, this water is continuously treated and reused for washing. The wash water is circulated in a closed loop within the system via a circ pump and cooled in the wash water cooler. As the wash water trickles down the packing bed, it absorbs the ammonia and is collected in the chimney tray located between wash section 1 and wash section 2. Wash water loaded with ammonia is purged continuously (about 2–3%) and treated in the water treatment plant for reuse.

11.4.4   CO$_2$ Desorption

A portion of the rich amine downstream of the filter passes through the lean/rich heat exchanger, where this portion of rich amine is heated by the hot lean amine from the

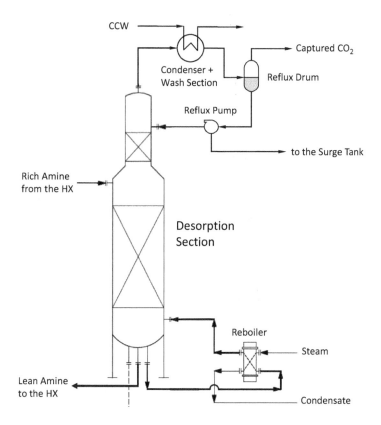

Figure 11.7 Simplified schematic diagram of the stripper column.

stripper. The other portion of rich amine after filter is mixed with wash section bleed and then preheated in stripper overhead heat exchanger, amine preheater, by the hot gas stream leaving the stripper top. The heated rich amines are introduced into the stripper column from the top (see Figure 11.7). It then falls through the packed section of the stripper, where it is counter-currently heated with steam generated in the stripper reboiler. As the temperature of the rich amine increases, the CO_2 is *desorbed* from the solution.

Plate-and-frame thermosyphon or kettle-type reboilers can be used to provide the heat to release the CO_2 from the solution. Liquid is diverted into the stripper reboiler via a segregated sump compartment. Excess liquid overflows from the reboiler sump compartment to the lean amine sump compartment for return to the absorber. The heating medium supplied to the stripper reboiler is saturated steam extracted from the bottoming steam cycle (3–4 bara).

The level in the bottom of the stripper is controlled by balancing the flow of rich amine feed to the stripper with the lean amine fed to the CO_2 absorber in a three-element arrangement. The vapor stream (mainly H_2O and CO_2 with a small amount of solvent) leaves the top of the stripper and is fed to the overhead stripper condenser, cooling the stream to approximately 50°C. The condensed solvent and water are separated from the gas stream in the stripper reflux drum. From the stripper reflux drum, approximately 10% of the liquid is pumped back to the column via the reflux pump, while the balance flows to the process water surge tank. The gas stream leaving the stripper reflux drum is approximately 97% CO_2, with the balance of the composition being predominantly water vapor.

11.4.5   Amine Reclamation and Storage

As mentioned at the beginning of the chapter, during normal operation, the amine solvent is subject to thermal degradation and undergoes chemical reactions producing undesirable compounds. In particular, chemical reactions with O_2 and SO_2 and other flue gas contaminants in the flue gas can produce acids, which combine with the amine to produce *heat stable salts* (HSS). The amine bound in HSS does not absorb CO_2 and reduces the absorber performance.

On an annual basis, approximately 0.5–1% of the lean amine mass flow from the lean amine pump discharge is fed to an *amine reclaimer*. In the amine reclaimer, the amine is released from the HSS by heating it with intermediate pressure steam (typical conditions are 7 bara and 180°C) and through a neutralization reaction with *soda ash* (Na_2SO_3). Water is added to the reclaimer to improve the amine recovery and reduce the amount of amine degradation that occurs in the reclaiming process. Much of the amine and water from the lean amine solution is vaporized and returned to the stripper. The residual sludge (a very viscous liquid when hot with usable heating value) is removed for proper disposal.

From the discussion above it follows that the lost amine should be made up. Fresh, concentrated amine (typically, 85–15% in amine-water by weight) is delivered by truck and stored in the amine storage tank. To maintain the solution at system design concentration i.e., at 30–35% amine by weight, fresh amine is made up to the system as required using a metering pump. A dedicated storage tank is used to store the lean amine removed from the system based on the absorber tower level control as required. During a plant shutdown, this tank is also used to store the amine inventory. This tank is sized to accommodate the amine solution from all the plant's equipment and piping. The amine is returned to the system using the lean amine solvent fill pump.

For shutdown, the amine circulation continues without the flue gas flow to ensure that the solution is lean throughout all the equipment. The equipment is pumped out to a minimum level directly, or through other vessels, to the lean amine solvent storage tank. The remaining solution is drained by gravity though the amine waste piping to the amine waste sump. From the sump, the amine solution is pumped through filters to the amine solvent storage tank.

11.4.6 CO_2 Treatment and Compression

The CO_2 leaving the stripper is saturated with water. During the intercooled compression of captured CO_2, much of the water carried over from the stripper is knocked out in the intercoolers. Typically, for pipeline transmission the final moisture content is specified at very low levels, e.g., less than 50 ppmw (by weight). To satisfy such a stringent requirement, the CO_2 stream must be dried using desiccant dryers. A typical desiccant train contains vessels, valving, instrumentation, and other associated equipment. The two vessels are operated in parallel, i.e., one in regeneration mode and one in operation. The location of the dryers is dependent on the operating pressure, typically, 450–480 psig for triethylene glycol (TEG) dehydration units. If the inlet pressure is atmospheric, four to five compression stages may be needed to get to that pressure level. The regeneration method is extremely efficient because the dryer vessel is not depressurized and virtually no CO_2 is vented or recycled. The desiccant dryers can remove more than 90% of the water coming in with the gas flow to achieve very low CO_2 moisture, e.g., less than 50 ppmw.

The selection of the CO_2 compressor is primarily a function of the overall pressure ratio, flow rate, and the operation envelope. The suction pressure is around atmospheric, i.e., one bar. The discharge pressure is a function of the end use for the captured CO_2 and the length of the pipeline transporting the CO_2 from the power plant to the location of the said end use. Therefore, there is a wide range of possibilities, e.g., from the low 90s (in bara) to as high as 150 bara for pipeline applications. The discharge pressure for sequestration applications is usually above 200 bar. Such high pressures necessitate the use of multiple compression *sections* with intercoolers in between. Each section is a *multi-stage* (i.e., multiple impellers in a row) centrifugal compressor driven by an electric motor through an integral gearbox (to ensure optimal rotational speed for each compressor). (Per ASME PTC-10, a *section* is defined as one or more *stages* having the same mass flow rate without external heat transfer other than natural casing heat transfer.) The final compressor design is subject to optimization by the compressor OEM. There are two options:

- In-line (also referred to as "beam style") centrifugal compressor train
- Integrally geared centrifugal compressor (IGC)

Each configuration has its advantages and disadvantages. Large in-line compressor trains comprise multiple casings (usually two, low-pressure [LP] and high-pressure [HP]), each comprising several stages (impellers). The LP and HP in-line compressors have side streams. They can be designed in a "back-to-back" arrangement. They may operate at constant speed, i.e., on a common shaft, or at different speeds via speed-changing gearboxes. There is intercooling between the casings to reduce compression power and to keep compressed gas temperature at acceptable limits. The casings are packaged either in barrel (for very high pressures) or horizontally split style. They have been in wide use in natural gas compression, petrochemical refineries, and LNG stations.

IGCs are compact designs with a single "bull gear" connected to the drive shaft and one or more pinion shafts, each driving a different compressor stage. This enables the stages to operate at optimal design speeds for good overall efficiency while the driver runs at constant speed. Their architecture enables the incorporation of *variable inlet guide vanes* and *variable diffuser vanes* into each stage for good off-design performance. They have a smaller footprint and allow for intercooling between each stage.

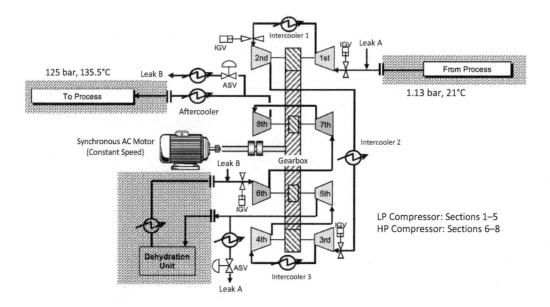

125 bar, 135.5°C

Leak B

To Process

Aftercooler

ASV

IGV

2nd   1st

Intercooler 1

IGV   Leak A

From Process

1.13 bar, 21°C

8th   7th

Intercooler 2

Synchronous AC Motor
(Constant Speed)

Leak B   Gearbox

6th   5th

LP Compressor: Sections 1–5
HP Compressor: Sections 6–8

IGV

4th   3rd

IGV

Dehydration
Unit

ASV

Intercooler 3

Leak A

Figure 11.8 Schematic flow diagram of an integrally geared CO_2 compressor. Rotational speed of sections 1 and 2 is 7,800 rpm; sections 3 and 4 rotate at 11,200 rpm, sections 5 and 6 at 15,000 rpm, and sections 7 and 8 at 18,000 rpm. Synchronous ac motor speed is 1,000 rpm. Sections 1–4 are equipped with labyrinth seals, sections 5–8 with dry gas seals.

Their disadvantages used to be cost and complexity associated with increased number of bearings and shaft seals. However, since their introduction in the 1950s, they can be deemed to be at a stage of design maturity to minimize this in sight of their better efficiency. IGCs have been used as HP CO_2 compressors at pressures up to 200 bar in urea processes.

As an example, see the schematic diagram of an integrally geared centrifugal CO_2 compressor with eight compression stages, three intercoolers, one aftercooler, and one dehydration unit in Figure 11.8.

As shown in Figure 11.8, the compressor is divided into HP and LP sections. For operational flexibility, sections 1, 2, 3, and 6 (HP inlet) are equipped with inlet guide vanes (IGVs). Rotational speeds cited in the figure caption are for a specific design by one OEM. However, for different designs from different OEMs, these numbers do not show a big variation. After all, they are determined by the fundamental aerothermo-dynamics governing the centrifugal compression process and the selected working fluid. The front-end selection or design procedure is governed by the nondimensional parameters, i.e., specific speed and specific diameter, which are functions of compressor speed (rpm), inlet volume flow (m³/s), and polytropic head (m). (The latter is a measure of the compressor pressure ratio and power consumption.) The underlying theory can be found in any undergraduate textbook on fluid mechanics. For practical applications, the selection process is neatly summarized in the widely used Balje chart (see the paper by Khan [19] for a concise description).

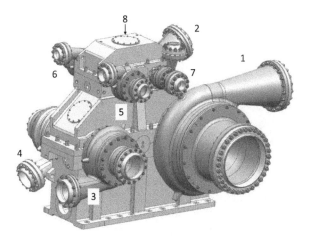

Figure 11.9 Integrally geared, eight-stage centrifugal CO_2 compressor.

The actual compressor plus gearbox package, schematically depicted in Figure 11.8, would be similar to that shown in Figure 11.9. These units are designed and manufactured in accordance with API Standard 617[17] with the OEM's comments and exceptions. Some OEM configurations incorporate two separate LP and HP compressor modules, each driven by a high-speed induction motor. Each compressor comprises two sections with intercoolers between sections of LP and HP compressors also between the LP and HP compressors. Typically, each compressor is driven by its own electric motor. Another option is that a single electric motor is coupled to the LP and HP compressors, each running at a different speed, through two gearboxes. The motor is powered by a *variable frequency drive* (VFD) that permits the motor to be soft started to avoid significant electrical power system inrush loading. Each compressor/motor is expected to run at 100% speed under most operating conditions. A 3D installed depiction of the compressor with its intercoolers and accessories is shown in Figure 11.10.

Thermodynamic performance data for the compressor in Figure 11.8 is presented in the flowsheet diagram in Figure 11.11 (from THERMOFLEX – see Section 16.2). The electric motor power consumption is 14 MWe, which translates to roughly 365 kJ/kg or 16 kJ/kmol of captured CO_2. There is an aftercooler after the compressor discharge to reduce the CO_2 temperature to 35°C prior to entry into the pipeline. Heat rejected from the intercoolers, the dehydration unit and the aftercooler to the circulating cooling water is 22 MWth. This is low-exergy heat (worth about 4 MW in exergy with a "dead state" temperature of 15°C – see Table 11.11), which is very difficult to utilize cost-effectively for additional electric power generation, i.e., less than 1 MWe. The heat picked up by the circulating water is typically rejected to the atmosphere in the plant's cooling tower. Having said that, in some cases, limited integration with

---

[17] Axial and Centrifugal Compressors and Expander-Compressors, 8th ed., September 2014.

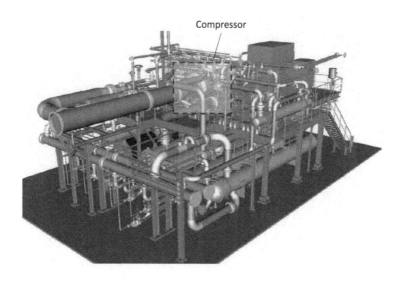

Figure 11.10 CO_2 compressor as installed with the intercoolers.

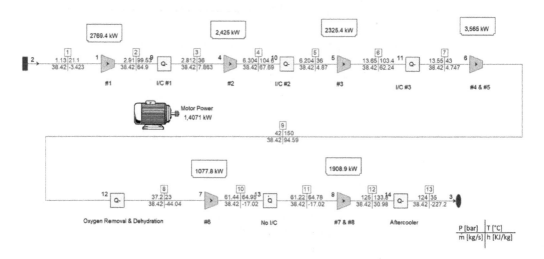

Figure 11.11 Heat and mass balance data of the intercooled, eight-stage integrally geared centrifugal CO_2 compressor (from the THERMOFLEX model).

low-level heat users in other parts of the power and capture blocks may be possible. This is subject to a diligent cost-performance trade-off in the FEED phase on a case-by-case basis.

In this context, it should also be pointed out that a conventional, in-line compressor design (most likely with a final supercritical CO_2 "pump"), with proven centrifugal process compressors (i.e., better RAM), offers a better opportunity for heat recovery from the intercoolers. Thus, it may turn out to be a better option in a diligent cost-performance

Table 11.11 Intercooler (IC), dehydrator, and aftercooler (AC) heat rejection – second law analysis.

	T_{in}, °C	T_{Out}, °C	Heat, kWth	T_{Avg}, °C	Carnot Efficiency, %	Exergy, kW
IC1	99.5	36	2,191	66.8	15.2	334
IC2	104.6	36	2,414	69.2	15.8	382
IC3	103.4	43	2,209	72.3	16.6	367
DEHYD	150.0	23	5,326	82.7	19.0	1,014
AC	133.8	35	9,918	82.1	18.9	1,874
Total			22,058			3,969

trade-off study. In other words, it should not be dismissed off-hand in favor of the IGC in the FEED phase.

Off-design performance can be determined from the OEM-supplied performance curves. These curves typically include discharge pressure, polytropic head, and gas power consumption (i.e., excluding electric motor and gearbox losses) as a function the actual inlet volume flow. Gas power curves for the LP and HP compressors in Figure 11.11 are shown in Figure 11.12. Note that, turndown (i.e., lower CO_2 flow rate) is controlled by the IGVs with the discharge pressure kept constant. This method is more advantageous from a power consumption perspective vis-à-vis recirculation.

11.5   Performance

As far as most calculations are concerned, one needs two pieces of information to gauge the performance impact of carbon capture: (1) electric power consumption of the capture block equipment and (2) steam consumption of the stripper reboiler. The latter is used to measure the lost steam turbine generator output in the power block. The two largest contributors to the former are the CO_2 compressor and the fuel gas booster fan. Let us look at the steam consumption of the stripper reboiler first.

In the stripper column, absorbed CO_2 (from the flue gas) in the "rich" amine is released as a concentrated stream by adding heat to reverse the chemical reaction of capture. The resulting "lean" amine at 110–120°C is sent back to the absorber to complete the loop. From an absorption perspective, the colder the amine the better. Thus, the lean amine is cooled in a two-step heat exchange process prior to entry into the absorber. The typical absorber operating temperature is 45°C. Thus, the rich amine (solvent plus absorbed CO_2) exits the absorber column at about 45°C. For the desorption process, a higher temperature is desirable. Heat exchange between the rich and lean amines ensures that warmer stream goes into the stripper and colder one goes into the absorber. (This constitutes the first step of lean amine cooling, which is followed by a second, "trim cooling," step using circulating cooling water.) Consequently, an optimal heat exchanger design sets the stripper bottom temperature and operating pressure. In particular,

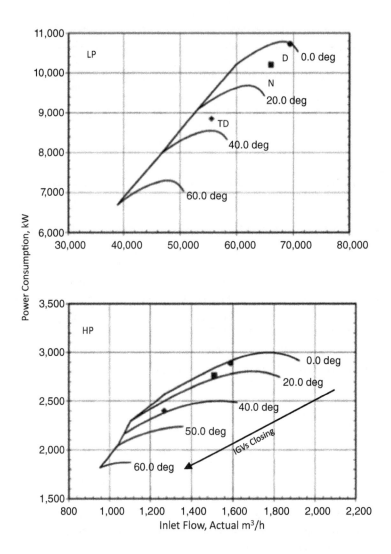

Figure 11.12 Compressor performance curve (D: design, N: normal, TD: turndown).

- Higher stripper operating pressures would be advantageous from CO_2 compression perspective.
- Lower stripper operating pressures would be advantageous from steam extraction (from the bottoming cycle of the GTCC) and "lost" steam turbine generator (STG) output perspective.

A detailed look at this optimization process is provided in a paper by Warudkar et al. [20]. For the most common and proven solvent, 30% MEA, the ideal was found to be around 120°C (~250°F). This corresponds to a pressure of about 1.9 bara in the stripper bottom and sets the steam pressure at 3–3.5 bara (about 50 psia).

The energy required for the absorbent regeneration in the stripper is supplied by a kettle-type reboiler utilizing the steam extracted from the bottoming cycle of the

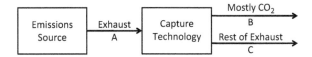

Figure 11.13 Carbon capture process inlet/outlet material streams [20].

power block. The energy requirement of the reboiler has three constituents: (1) the heat of reaction for the desorption of the CO_2, (2) the energy required to generate the stripping vapor (i.e., steam), and (3) the sensible heat required to raise the temperature of the incoming rich amine to the stripper operating temperature. Thus,

$$\dot{Q}_{reb} = \dot{Q}_{rx} + \dot{Q}_{str} + \dot{Q}_{sens}. \tag{11.3}$$

This energy is supplied by the condensing steam (from the power block) in the reboiler, i.e.,

$$\dot{Q}_{reb} = \dot{m}_{stm}h_{fg}(P_{stm}), \tag{11.4}$$

where h_{fg} is the latent heat of condensation of steam at pressure, P_{stm}. For proper evaluation of the terms on the right-hand side of Equation (11.3), one needs to model it in a chemical process simulation software such as *Aspen Plus* or *ProMax*. Such software packages are extremely expensive with hefty maintenance fees and steep learning curves. In short, they are only available to those who are employed by private or public R&D organizations, engineering companies, and OEMs. Nevertheless, it is still possible to make reasonably good estimates starting from the first principles using a spreadsheet. By far the best method is based on the *minimum separation work* (MSW) principle, which is described in minute detail with step-by-step examples in the superb monograph by Wilcox [21]. (For anyone who is interested in the subject matter, this book is *the* must-have reference.)

The minimum thermodynamic work required to separate CO_2 from a gas mixture in an isothermal and isobaric process is equal to the negative of the difference in the Gibbs free energy between the final initial states shown in the simplified capture process diagram in Figure 11.13 (after figure 1.10 on p. 23 in Wilcox [21]). Specifically, MSW can be found from

$$MSW = \Delta G_B + \Delta G_C - \Delta G_A, \tag{11.5}$$

$$\begin{aligned} MSW = R_{unv}T_0\{&(n_{B,CO2}\ln y_B + n_{C,CO2}\ln y_C - n_{A,CO2}\ln y_A) \\ &+ (n_B - n_{B,CO2})\ln(1-y_B) + (n_C - n_{C,CO2})\ln(1-y_C) \\ &- (n_A - n_{A,CO2})\ln(1-y_A)\}, \end{aligned} \tag{11.6}$$

where $R_{unv} = 8.314$ J/mol-K, $T = 298.15$ K, and

- n_A, n_B, and n_C are the total number of moles in streams A, B, and C, respectively,
- $n_{A,CO2}$, $n_{B,CO2}$, and $n_{C,CO2}$ are the total number of moles of CO_2 in streams A, B, and C, respectively,
- y_A, y_B, and y_C are the mole fractions of CO_2 in streams A, B, and C, respectively.

Equation (11.6) is equivalent to equation 1.11 of Wilcox [21, p. 24]. While the equation is relatively straightforward, its evaluation with actual data is quite tedious. For convenience, `Min_Sep_Work` below is an Excel VBA function calculating the MSW as given by Equation (11.6).

```vba
Function Min_Sep_Work(sTemp As Single, sMdot As Single,
sCapRate As Single, sPurity As Single, vComp As Variant)
As Single
' 1 'N2
' 2 'O2
' 3 'H2O
' 4 'CO2
' 5 'Ar
' 6 'SO2
' 10 'Air
' 7 'CH4
Dim sGasMolWeight As Single, sMoleFlowRate As Single
' A: Exhaust
' B: Mostly CO2
' C: Rest of Exhaust
Dim n_A_CO2 As Single, n_B_CO2 As Single, n_C_CO2 As
Single
Dim n_A_Others As Single, n_B_Others As Single,
n_C_Others As Single
Dim y_A_CO2 As Single, y_B_CO2 As Single, y_C_CO2 As
Single
Dim y_A_Others As Single, y_B_Others As Single,
y_C_Others As Single
Dim i As Integer
Dim sTemp_K As Single
Dim G_A As Single, G_B As Single, G_C As Single
Const Rgas As Single = 8.314 'kJ/kmol-K

sTemp_K = (sTemp - 32) / 1.8 + 273.15 'temp. in Kelvins

sGasMolWeight = MWgas(vComp, 10) 'lbm/lbmole or kg/kmole
sMoleFlowRate = (sMdot * 0.4536) / sGasMolWeight 'kmole/s

y_A_CO2 = vComp(4)
y_A_Others = 1 - y_A_CO2
n_A_CO2 = y_A_CO2 * sMoleFlowRate
n_A_Others = sMoleFlowRate - n_A_CO2
```

```
(cont.)


n_B_CO2 = sCapRate * n_A_CO2
n_B_Others = n_B_CO2 / sPurity - n_B_CO2
y_B_CO2 = n_B_CO2 / (n_B_CO2 + n_B_Others)
y_B_Others = 1 - y_B_CO2


n_C_CO2 = (1 - sCapRate) * n_A_CO2
n_C_Others = sMoleFlowRate - (n_B_CO2 + n_B_Others) -
n_C_CO2
y_C_CO2 = n_C_CO2 / (n_C_CO2 + n_C_Others)
y_C_Others = 1 - y_C_CO2


G_A = Rgas * sTemp_K * (n_A_CO2 * Log(y_A_CO2) + n_A_Others *
Log(y_A_Others))
G_B = Rgas * sTemp_K * (n_B_CO2 * Log(y_B_CO2) + n_B_Others *
Log(y_B_Others))
G_C = Rgas * sTemp_K * (n_C_CO2 * Log(y_C_CO2) + n_C_Others *
Log(y_C_Others))


Min_Sep_Work = G_B + G_C - G_A 'in kW


End Function
```

Evaluation of MSW as a function of $y_{A,CO2}$ and T shows that MSW increases:

- as the concentration of CO_2 in stream A decreases or
- as the process temperature, T, increases.

This is a distinct disadvantage for carbon capture from the HRSG stack gas in a GTCC power plant with $y_{A,CO2}$ around 0.04 (4% CO_2 by volume). In coal-fired power plants, the boiler stack gas CO_2 concentration is around 0.14 (14% CO_2 by volume).

Note that MSW is the theoretical minimum value for the capture process. In other words, it is analogous to the Carnot equivalent efficiency of an actual heat engine (e.g., a gas turbine) or the exhaust gas exergy of a gas turbine. As such, it is unachievable in practice. To convert it into a realistic estimate, we need a "technology factor." This technology factor is the second law (exergetic) efficiency defined by Wilcox on pp. 26–27 in his book and plotted in Figure 1.12 of Ref. [21, p. 27]. For a combined cycle power plant, the said technology factor can be taken as 0.22 so that the *actual separation work* (ASW) is found from

$$ASW = \frac{MSW}{0.22}. \qquad (11.7)$$

To get a feel for the magnitudes involved, consider that for an F Class gas turbine combined cycle rated at 400 MWe net, the following data is available:

- Flue gas flow rate 1,264.1 lb/s (HRSG Stack)
- Flue gas temperature 192.3°F
- Flue gas MW 28.33 lb/lbmol (kg/kmol)
- Flue gas molar composition
  - Nitrogen 74.13%
  - Oxygen 11.711%
  - Water vapor 9.148%
  - Carbon dioxide 4.118%
  - Argon 0.893%

For 90% capture and 99.9% CO_2 purity, using `Min_Sep_Work`, i.e., Equation (11.6), we obtain 8.8 MW for MSW. Using Equation (11.7), ASW is found as 8.8/ 0.22 = 40.5 MW. The auxiliary power consumption for CO_2 compression and conditioning is 17–19 MJ per kmol of captured CO_2. This can be estimated as follows. We start by estimating the isentropic compression work, w_{is}, from

$$w_{is} = N \frac{R_{unv} T_0}{k} (PR^{\frac{k}{N}} - 1), \qquad (11.8)$$

where N is the number of compressor casings with intercooling in between, PR is the overall compressor ratio (typically, 150, i.e., compressing CO_2 from roughly 1 bar to 150 bar), and $k = 1 - 1/\gamma$, with γ being the specific heat ratio of CO_2 (1.289). With $N = 3$, Equation (11.8) returns a value of 15 MJ/kmol, which translates into 17–18 MJ per kmol depending on the isentropic efficiency (i.e., compressor technology).

Continuing with the calculation, the captured CO_2 molar flow rate is

$$(1264.1/28.33) \times 0.04118 \times 0.90/2.2046 = 0.75 \text{ kmol/s},$$

so that total compression power requirement is

$$0.75 \text{ kmol/s} \times 17.5 \text{ MJ/kmol} \approx 13 \text{ MW}.$$

Another 2 MJ/kmol should be allocated to amine recirculation and other capture unit pumps, i.e., the balance of plant (BOP) of the capture block.

Typically, the PCC plant is connected to the power plant via a flue gas duct, whose total length, cross-sectional area, and layout is site specific. A booster fan is required to push the flue gas and compensate for myriad pressure losses in the system. Without a detailed design at hand, an allocation of 10 MJ/kmol for booster fan duty is reasonable. Thus, the total BOP and booster fan power requirement of the capture plant is found as

$$0.75 \text{ kmol/s} \times (10 + 2) \text{ MJ/kmol} = 9 \text{ MW}.$$

Finally, the total capture penalty is estimated as

$$40.5 + 13 + 9 \approx 62.5 \text{ MW},$$

which amounts to 62.5/400 = 15.5% of the plant rated output with no capture. Thus, as a front-end estimate, equipping a GTCC power plant with an amine-based carbon capture system results in a 15–16% reduction in plant net output. Since the fuel consumption of the gas turbines is not affected, this translates into a 15–16% increase in the plant net heat rate or, equivalently, a 15–16% (in relative terms) reduction in net thermal efficiency. In other words, for a net 60% H Class GTCC, a carbon capture retrofit brings down the net thermal efficiency to about 51%.

A key number to be used in amine-based carbon capture calculations is the stripper reboiler energy requirement, which determines the extraction steam flow from the bottoming cycle and, as a result, the "lost" steam turbine power output. Based on the numbers used for this example, the CO_2 flow rate is

$$0.75 \text{ kmol/s} \times 44 \text{ kg/kmol} = 33 \text{ kg/s}.$$

The ASW is calculated as 40.5 MW, which is the lost steam turbine work. The typical conversion efficiency is 35–36%. This means that the net energy input to the reboiler (the left-hand side of Equation (11.3)) is

$$\dot{Q}_{reb} = 40.5/.35 = 115.7 \text{ MWth}.$$

Using the saturated vapor/steam and liquid enthalpies at 3.5 bara to calculate h_{fg}, the requisite steam flow is found from Equation (11.4) as

$$115,700/(2731.5 - 582) = 53.8 \text{ kg/s}.$$

Furthermore, the heat of desorption can be found as

$$115,700 \text{ kWth}/33\text{kg/s} = 3,500 \text{ kJ per kg of captured } CO_2 \text{ (about 1,500 Btu/lb)}.$$

This is indeed about right for 30% (by weight) MEA solvent.

11.5.1   Sample Calculation

In this section, we take a closer look at the impact of PCC on the performance and cost of an advanced class gas turbine combined cycle power plant using Thermoflow, Inc.'s GT PRO software. The gas turbine selected for this exercise is GE's 7HA.02, as represented by model #681 in the software's engine library. This gas turbine is the 60 Hz variant of the HA class product line of the OEM with a nominal 385.5 MWe output at ISO base load with 42.7%. The key design information is as follows:

- Single-shaft GTCC
- Three-pressure, reheat steam bottoming cycle with air cooled condenser
- 168.9 bar – 585°C HP steam
- 29.3 bar – 585°C hot reheat steam
- 3.9 bar – 314°C LP admission steam
- 101.6 mbar condenser pressure

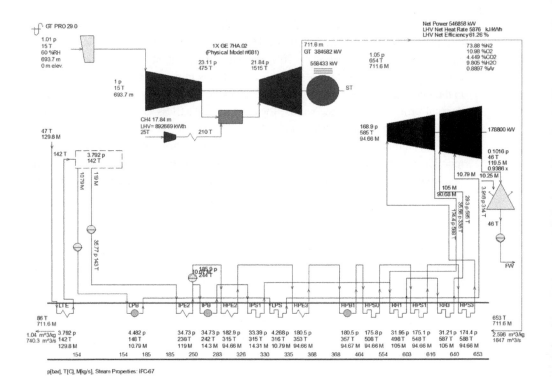

Figure 11.14 Single-shaft GTCC with GE's 7HA.02 gas turbine – heat balance diagram.

The cycle heat and mass balance diagram is provided in Figure 11.14. Thermal performance is calculated as 546.9 MWe at 61.26% net LHV efficiency. (The fuel gas is 100% methane.)

The PCC is also modeled in GTPRO with the following assumptions:

- 95% capture efficiency
- Stripper energy requirement 3,721 kJ/kg
- Stripper reboiler steam at 3.447 bar (saturated) supplied from steam extracted from the IP-LP crossover of the steam turbine
- Booster fan pressure head 124.5 mbar
- DCC gas exit temperature 45°C
- Clean flue gas exit temperature 35°C
- Captured CO_2 compressed to 151.7 bar

A schematic diagram of the capture block is shown in Figure 11.15. The electric power consumption of the PCC block is summarized in Table 11.12. A comparison of GTCC performance with and without PCC is provided in Table 11.13.

In conclusion,

- Capture penalty is 14% in net plant output, 8.6 percentage points in net LHV efficiency (16.3% increase in net LHV heat rate)

Table 11.12 Capture block electric power consumption

Total	**36,088**
CO$_2$ compressor	15,639
Booster fan	10,214
Solvent circulation pumps	1,448
Cooling water pump	1,634
Condensate pump	5
Miscellaneous	464
Fin-fan coolers	6,684

Table 11.13 Performance comparison – 7HA.02 single-shaft GTCC

	With PCC	No PCC
GT shaft output, kW	384,581	384,581
ST shaft output, kW	136,417	178,800
Generator losses, kW	4,628	4,949
Gross output, kW	516,370	558,432
Total auxiliaries and transformer losses	46,096	11,575
Net output, kW	470,274	546,857
Gross efficiency, %	57.85	62.56
Fuel consumption – kWth	892,669	892,669
Net efficiency, %	52.68	61.26
Net heat rate, kJ/kWh	6,833	5,877

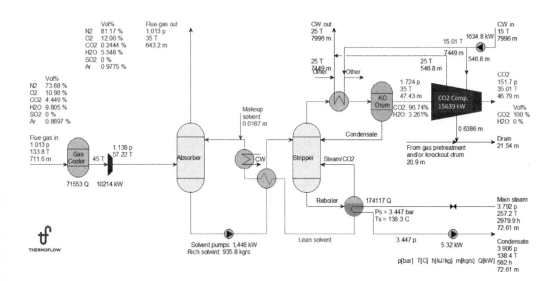

Figure 11.15 PCC with amine-based chemical absorption – schematic diagram.

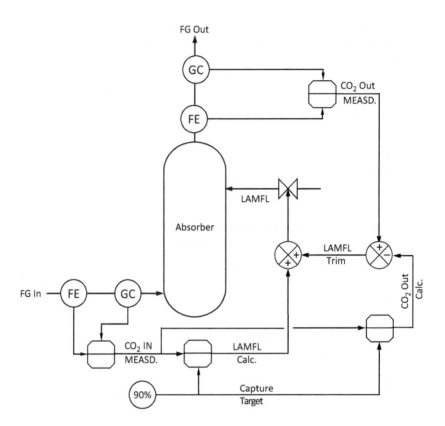

Figure 11.16 Simplified schematic description of the CO_2 capture master control scheme.

- In absolute terms, capture penalty is 76.6 MWe, of which 42.4 MWe is due to lost steam turbine output (about 55% of the total)
- Reduction in ST exhaust steam flow reduces air-cooled condenser (ACC) fan and condensate pump power consumption

GT PRO's PEACE add-in estimated the EPC (overnight construction) cost of the base GTCC as $330 million or $600/kW. With the capture block add-in, the cost increases to $831 million or $1,765/kW.

11.6   Operation

The key operating scenarios are capture block startup and shutdown. The exact nature of control sequences governing these plant transients is strongly a function of the specific technology. Nevertheless, a generic control philosophy for amine-based capture operation can be provided for guidance. The conceptual *master control* scheme to achieve a desired CO_2 removal rate from the flue gas is described in a highly simplified diagram (Figure 11.16). In essence, the *lean amine flow rate* into the absorber is determined from calculations using the incoming flue gas mass flow and CO_2

concentration measurements in a *feed-forward* control scheme. The final control trim of the *flow control valve* (FCV) is based on a *feedback* scheme using a comparison of the CO_2 removal rates, i.e., by calculating the difference between the desired value and the calculated value from the CO_2 inlet and outlet flow (indirect) measurements.

The capture plant startup process is described as follows:

1. Make-up is added to the absorber and stripper bottom levels, water wash chimney tray, condensate blowdown tank, and compressor steam generators as needed.
2. Cooling water circ pumps are started.
3. The auxiliary boiler start sequence is initiated.
4. The LP steam line is warmed via the drain line until the minimum required superheat is reached. Steam is then admitted to the reboiler, and the drain valve is closed.
5. The stripper column and reflux drum are pressurized, and the reflux flow is established.
6. System pumps, e.g., the lean amine, rich amine, and flash drum pumps are started in recirculation mode.
7. Minimum lean/rich amine flow is established. At this point, CO_2 loading in the rich amine flow is minimal due to zero flue gas flow through the absorbers.
8. The wash water pump is started, and wash water flow is established.
9. The CO_2 compressor is started at minimum speed in recirculation mode.
10. The flue gas duct isolation dampers are opened.
11. The flue gas blower is started at minimum speed.
12. The CO_2 master output gradually increases from zero, admitting flue gas to the first absorber.
    i. In the beginning, the blower damper controls flow until it reaches the maximum throttling position, at which point it goes open and the blower speed increases to take more flue gas flow.
    ii. The HRSG stack damper is modulated to maintain steady pressure in the flue gas duct.
    iii. Lean amine flow to the absorber is controlled by the CO_2 controller.
    iv. Steam flow to the stripper reboiler increases as the amine flow increases.
    v. The stripper reflux drum *pressure control valve* (PCV) is modulated to maintain constant pressure in the stripper. As the CO_2 flow increases, the PCV is fully opened and the compressor speed is used to control constant stripper pressure.
13. The CO_2 master output is fixed to a minimum load until the CO_2 product system has been filled and started and the pipeline has been filled and is in service.
    i. As CO_2 begins to fill the surge drum, air vents from motor-operated valves located on the CO_2 product send-out pumps and piping upstream of the CO_2 product control valve.
    ii. Once the system is vented, the motor-operated valves close and a CO_2 product send-out pump starts.
    iii. Flow to the CO_2 pipeline is strictly controlled until the pressure in the pipeline approaches operating pressure.

14. The CO_2 master output begins to increase again.
15. Once the unit has stabilized, the CO_2 master is released for fully automatic control.
16. Flow is established through the carbon filter package after the system is fully stabilized.
17. Semi-lean amine flow to the absorber is controlled in proportion to the rich amine flow.

The basic steps during capture plant shutdown process are as follows:

1. CO_2 master control is deactivated
2. Flue gas flow is slowly reduced to zero
3. LP steam flow to the stripper reboiler is shut down
4. IP steam flow to the reclaimer, if it is in operation, is shut down
5. Flow is circulated until stripper bottoms temperature drops below 65°C
6. Stripper vessel and reflux drum are slowly depressurized
7. Absorber wash water pump is stopped
8. Lean and rich amine pumps are stopped

The following conditions must be met before flue gas is allowed to flow into the absorber columns:

- Minimum lean/rich amine flow is established
- Adequate temperature in the stripper reboiler is reached
- Wash water system is in operation. Note that wash water pump is tripped if
  - Absorber chimney tray level is low (LLL)
  - Discharge flow is low (LL)
- Stripper reflux drum level is normal

Rich/lean amine pumps, reboiler, reclaimer, and stripper reflux system also have their own trip alarms, e.g., low net positive suction head (NPSH) for the pumps.

There are several emergency cases that require *emergency shutdown* (ESD) of the capture unit:

- Manual engagement of the ESD (i.e., operator presses a button)
- Loss of power (to the ESD system or plant wide)
- Instrument air loss
- High (HHH) level in the stripper reflux drum
- High (HHH) or low (LLL) absorber bottom level
- High (HHH) or low (LLL) stripper bottom level
- High (HH) LP steam pressure

Major upset conditions are classified as H/L, HH/LL, or HHH/LLL. For H/L conditions, the DCS issues an alarm code and records it. For HHH/LLL conditions, the ESD system is activated. For HH/LL conditions, in addition to the alarm, the control takes a specific action. For example, the following occurs for the absorber column bottoms:

- At HH level, the solvent return valve is opened to remove excess solvent from the absorber

- At LL level, the solvent make-up valve is opened to add solvent to the system

Similarly, for the stripper column bottoms,

- At HH level, the rich amine FCV position and rich amine pump speed are automatically decreased
- At LL level, the LP steam flow to the reboiler is decreased

For the stripper column pressure, at HH reading, the LP steam flow to the reboiler is decreased. If the LP steam flow to the reboiler is lost, the CO_2 level in the lean amine will increase. This will increase the corrosiveness of the lean amine. Although this is not an ESD trigger in and of itself (no immediate danger), it is prudent to shut the capture plant down if the steam flow cannot be restored in a timely manner.

When the ESD is triggered, the system closes the FCVs automatically. Flue gas and lean amine flow into the absorber as well as rich amine flow to the stripper along with the LP steam flow to the reboiler are isolated. All pumps are shut down (with the exception of the cooling water circ pump). HRSG flue gas dampers are activated to divert the full flue gas flow to the HRSG stack. The absorber column and its associated piping (lean/rich amine and wash water) depressurize to static conditions. The stripper column and the reflux system will remain pressurized or at vacuum, depending on the temperature at the time of ESD, due to the pressure control action of the CO_2 product valve (the gas is not flammable). The pressure is subsequently released via the CO_2 product, recycle valves, or manual vents. The reclaimer also stays pressurized until the stripper is depressurized.

11.6.1   Sample Calculation

The *Department for BEIS* in the UK has commissioned an EPC consultant to investigate potential improvements to the startup and shutdown times of a GTCC with PCC. A report summarizing the findings of the study based on a reference case, 95% capture with 35% MEA from an advanced class GTCC can be found on the internet.[18] The reader can consult the full report for modeling details and findings of the study (including a literature review). Salient parts of the capture plant/block startup are presented below because they are illustrative of the key process dynamics in an amine-based chemical absorption process.

The approach used in the study is a *quasi-static* analysis combining steady-state model results for the GTCC (using Thermoflow, Inc.'s THERMOFLEX software) at different points. These are referred to as *snapshots*, where each snapshot corresponds to an intermediate point in a transient process, e.g., A-to-B (e.g., loading of the gas turbine from full speed, no load [FSNL] to full speed, full load [FSFL]). In other words, while the actual time-dependent process goes through thermodynamically *nonequilibrium* states between the starting state A and end state B, the snapshots

---

[18] https://assets.publishing.service.gov.uk/government/uploads/system/uploads/attachment_data/file/929283/startup-shutdown-times-power-ccus-main-report.pdf (last accessed on May 7, 2021)

Table 11.14 PCC steady-state data (per train)

Lean amine circulation rate, kg/s	1,500
Amine rich/lean loading, mol/mol	0.45/0.25
Reboiler temperature, °C	125
Stripper condensate temperature, °C	50
Reboiler Duty, MWth	336
PCC auxiliary power, MWe	44.9
PCC heat rejection, MWth	537
CO_2 design rate into PCC plant	85
Capture rate, %	95
CO_2 residual emissions, kg/s	4.1

from the steady-state (i.e., time-independent) simulation model are used to represent those nonequilibrium states. For processes with small "relaxation times," i.e., the time required to reach the equilibrium state, this is a reasonably good approximation. This is the case for the gas turbine exhaust gas flow (pressure, temperature, mass flow rate, and composition), which is the input to the capture plant as the HRSG stack gas, from which CO_2 is to be stripped. The exceptions are processes where the relaxation time is long compared to the time scale of the flow process, e.g., those involving components with large metal mass, which take a long time to heat or cool.

The sample GTCC power plant comprises two $1 \times 1 \times 1$ powertrains (single-shaft, generator in the middle with a "triple S" (SSS) clutch) with a Siemens SGT5–9000HL gas turbine (ISO base rating 588 MW). The estimated plant capacity is 1,740 MW gross (at the generator terminals) at site conditions of 9°C, 80% relative humidity, and 1.013 mbar. The bottoming cycle is 3PRH with nominal 277 MW steam turbine output (per train). The HRSG is equipped with an SCR for 90% NOx removal. The heat rejection system is a water-cooled condenser with wet cooling tower.

In addition to the flue gas duct and axial booster fans (90 mbar head), the 95% capture plant comprises the standard equipment described earlier in the chapter. The solvent is 35% MEA. The stripper operates at 2.2 bara and 125°C to regenerate the amine loading from 0.45 mol/mol to 0.25 mol/mol (i.e., 0.25 *lean loading*). See Table 11.14 for key PCC process data. There is a lean amine storage tank for draining the system during shutdown. Captured CO_2 is dehydrated and compressed to 150 bara for export.

During operation, PCC plant heat rejection is handled by the plant mechanical draft (wet) cooling tower. The power plant only heat rejection (per train) is 414 MWth. Due to the LP steam extraction for stripper reboiler (equivalent to 74 MWe lost STG output), this duty is reduced to 136 MWth. Thus, combined power-capture plant heat rejection duty is $537 + 136 = 673$ MWth. The extra electric power consumed by cooling water circ pump and cooling tower fans is about 3 MWe, and the carbon dioxide compressor electric power consumption is 25.1 MWe (about 310 kJ/kg). The total capture penalty is $44.9 + 74 = 118.9$ MWe (13.7% of gross power output

Table 11.15 GTCC startup sequences

Stage		Time, min		Flue gas
		Hot	Cold	kg/s
0	START	0	0	0
1	First fire	5	15	511
2	Ramp to 50%	20	25	681
3	Steam export	25	60	823
4	Full load	30	200	1021

Table 11.16 GTCC shutdown sequences

Stage		Time	Flue gas
		min	kg/s
0	STOP	0	1020
1	Ramp down to 30%	5	547
2	ST shutdown complete	20	547
3	GT load hold at 5%	25	499
4	Breakers open	30	0

per powertrain). Gas turbine fuel consumption is 1,389 MWth LHV (per powertrain) for a gross LHV efficiency of 1,389/1,740 = 62.6%. The net efficiency with plant auxiliary, 23 MWe, is 60.9%. With PCC, the net powertrain output becomes 840 − 23 − 118.9 = 728.1 MWe. (In the report, powertrain output with capture is given as 722.7 MWe. The reason for the discrepancy of 5.4 MWe is not clear.)

The study divides the plant startup into four stages from the pushing of the START button:

1. Gas turbine ignition, roll to FSNL, synchronization and load to 15% load (see Table 11.15)
2. Gas turbine ramp up to 50% load
3. Gas turbine ramp up to 75% load (hot start, 8 hours or less after last shutdown) or hold at 50% load (cold start, 64 hours or more after last shutdown)
4. GTCC startup complete (all steam bypass valves are closed); gas turbine ramp up to full load (hot start) or hold at 50% load (cold start)

The shutdown is also divided into four stages from the pushing of the STOP button:

1. Gas and steam turbine ramped down to 30% load and held there (see Table 11.16)
2. Steam turbine completes its shutdown sequence while the GT load is held
3. Gas turbine ramped down to 5% load and held while the breakers are opened
4. Breakers are opened and the SSS clutch disengages the gas turbine

The amine inventory is based on 30 minutes of process at full circulation rate, thus

$$1{,}500\,\frac{\text{kg}}{\text{s}} \times 30\,\text{min} \times 35\,\text{wt}\% \times \frac{1\,\text{kmol}}{61\,\text{kg}} = 15{,}492\,\text{kmol MEA}.$$

At 0.25 mol/mol lean loading (amine supplied from the storage tank) the quantity of CO_2 dissolved in the lean amine is given by

$$0.25\,\frac{\text{mol}\,CO_2}{\text{mol MEA}} \times 15{,}492\,\text{kmol MEA} = 3{,}873\,\text{kmol}\,CO_2.$$

During the first 30 minutes of operation, fresh lean amine is passed through the absorber from the lean amine storage tank. Rich amine is then returned to the tank. Assuming that the tank is always well mixed, the loading is recalculated at each startup stage based on the total quantity of CO_2 absorbed.

The loading in the amine tank throughout the initial period during the 30-minute rich amine collection period is calculated as a function of:

- Quantity of CO_2 already stored in the lean amine (3,873 kmol), and
- Quantity of CO_2 absorbed from the flue gas as the gas turbine is ramped from minimum load to load at 30 minutes.

Using the basis of 15,492 kmol total MEA inventory, the overall amine loading throughout the startup stages is calculated as follows:

$$\text{Loading at end of Snapshot} = \text{Loading at start of Snapshot}$$
$$+ \frac{\text{kmols }CO_2\text{ absorbed during Snapshot}}{15{,}492\,\text{kmol MEA}},$$

where:

$$\text{Loading at start of Snapshot} = 0.25\,\frac{\text{mol}}{\text{mol}}\ \text{for Snapshot 1}$$

and rises thereafter,

$$\text{mols }CO_2\text{ absorbed during Snapshot} = \frac{\text{mass }CO_2\text{ absorbed during Snapshot}}{44\,\dfrac{\text{kg }CO_2}{\text{kmol }CO_2}},$$

$$\text{mass }CO_2\text{ absorbed during Snapshot 1}$$
$$= (\text{mass flow of }CO_2\text{ from HRSG at GT loading}$$
$$- \text{mass flow of }CO_2\text{ in treated gas to stack}) \times \text{duration of Snapshot,}$$

mass CO_2 absorbed during Snapshot 1 = GT taken as 25%,

$$\text{so } \left(31.6\,\frac{\text{kg}}{\text{s}}CO_2 - 0.192\,\frac{\text{kg}}{\text{s}}CO_2\right) \times 15\,\text{min} = 31.4\,\frac{\text{kg}}{\text{s}} \times 15\,\text{min} = 28{,}267\,\text{kg or }28.27\,\text{t}$$

Table 11.17 Amine loading at each startup stage

Snapshot (Stage)	1	2	3	4	5	6	7
Time after start, min	15	22	25	30	37	48	82
Duration, min	15	7	3	5	7	11	34
Stripper status	No steam	Preheat	Preheat	Preheat	Preheat	Preheat	Preheat
Gas turbine load, %	25	50	75	100	100	100	100
CO_2 from GTCC, kg/s	31.6	50	80	85	85	85	85
CO_2 after PCC, kg/s	0.192	0.42	0.93	3.93	35	44.5	85
CO_2 absorbed in stage, mt	28.27	20.82	14.23	24.32	21	26.73	0
Amine loading, mol/mol	0.29	0.32	0.34	0.38	0.41	0.45	0.45

$$\text{Loading at end of Snapshot } 1 = 0.25 \frac{\text{mol}}{\text{mol}} + \frac{28,267 \text{ kg}}{15,492 \text{ kmol MEA} \times 44 \frac{\text{kg}}{\text{mol}}} = 0.29 \frac{\text{kmol}}{\text{kmol}}.$$

Loading for the rest of the startup stages is calculated stage by stage in a similar method and summarized in Table 11.17. The amine inventory is seen to saturate with CO_2 by the end of snapshot 6 (0.45 mol/mol), indicating that the standard configuration is no longer able to capture CO_2 and continuing emissions from the GT are effectively unabated in CO_2.

The stripper heating time at startup is calculated from the sum of sensible heat input required:

- To heat the amine inventory from ambient (9°C) to the stripper normal operating temperature (125°C)
- To heat the metal mass of the stripper, fittings, reboiler, and interconnecting piping from ambient to the stripper normal operating temperature

There are 30 minutes of amine inventory circulation to be heated within the stripper column (2,700,000 kg inventory at 1,500 kg/s). Assuming an heat capacity of 3.34 kJ/kg-K for lean amine at startup (0.25 mol/mol), the heat requirement is calculated as

$$Q = 2,700,000 \text{ kg} \times 3.34 \frac{\text{kJ}}{\text{kg} \cdot \text{K}} \times (125°C - 9°C) = 1,046,088,000 \text{ kJ, or } 1,047 \text{ GJ.}$$

The stripper column metal mass calculation is based on the column wall thickness calculation with 30% design margin to allow for column dished ends, mass transfer packing, piping, and associated mechanical equipment. Assuming a LP column where vacuum normally dictates the design metal thickness, the calculation is shown below:

$$P_c = 2.2 E_y \left(\frac{t}{D_0}\right)^3,$$

where $P_c = 0.101325$ MPa for vacuum-rated column, $E_y = 193$ GPa for 316L steel, and $D_0 = 10$ m (internal diameter) $+ 2t$, so that

$$\frac{t}{10\text{m} + 2t} = \sqrt[3]{\left(\frac{0.101325}{193000 \times 2.2}\right)} = 0.0062\text{m}.$$

Rearranging gives :

$$t = \frac{0.062\text{m}}{1 - 0.0124} = 0.063\text{m}, \, or \, 63\text{mm}.$$

With an overall column height of 40 m to accommodate the fittings and sump in addition to the packing, the column volume is calculated as follows:

$$\text{Area} = \pi \times \left(R^2 - r^2\right) = \pi \times 5.063^2\text{m}^2 - 5^2\text{m}^2 = 1.99\text{m}^2,$$

$$\text{Volume} = \text{height} \times \text{Area} = 40\text{m} \times 1.99\text{m}^2 = 83\text{m}^3.$$

Assuming a metal heat capacity of 0.5 kJ/kg-K, the thermal mass and startup heat requirement for the stripper column metal is found from

$$Q = 862,000 \text{ kg} \times 0.5 \frac{\text{kJ}}{\text{kg.K}} \times 116\text{K} = 49,996,000 \text{ kJ}, \, or \, 50 \text{ GJ}.$$

Consequently, the total heat input is $1,047 + 50 = 1,097$ GJ, or, rounding up, 1,100 GJ.

The stripper startup calculation requires LP steam extraction from the GTCC bottoming cycle to feed the reboiler. The extraction location is the crossover between the IP and LP turbines. Based on the startup timetable in Table 11.15, the time at which first steam extraction is available can be set to 25 minutes and 60 minutes for hot and cold starts, respectively. A "quick and dirty" estimate is to set the steam requirement to gas turbine load, i.e., for the hot start, 75% steam demand met when the gas turbine is at 75% load (252 MWth) and 100% steam demand met 100% load (336 MWth).

11.7   References

1. Roberts, H. A. 1983. The logistics and economics of a CO2 flood, International Energy Agency Workshop, 25 August, Vienna, Austria.
2. Brandtl, P., Bui, M., Hallett, J. P. and Mac Dowell, N. 2021. Beyond 90% capture: Possible, but at what cost?, *International Journal of Greenhouse Gas Control*, 105, 103239.
3. VGB PowerTech e.V. 2004. CO_2 capture and storage, a VGB Report on the state of the art, Essen, Germany.
4. IEA GHG. 2000. Leading options for the capture of CO_2 emissions at power stations, International Energy Agency, Paris (France), Greenhouse Gas R&D Programme, Report Number PH3/14.
5. Bolland, O. 2002. Gaskraftverk med CO_2-håndtagning. Studie av alternative teknologier, SINTEF Energiforskning AS, Report nr. TR A5693, Trondheim, Norway, 2002. ISBN Nr: 82-594-2358-8 (In Norwegian).

6. Gas Turbine World. 2015. *2014–2015 Handbook, Volume 31*, Fairfield, CT: Pequot Publishing, Inc.

7. Bechtel Overseas Corporation. 2009. CO_2 capture facility at Kårstø, Norway – Front end engineering and design (FEED) study report, 25474-000-30R-G04G-00001, 13 January, Revision 1, Released for Public Distribution on 3 April 2019, 10112936-PB-G-DOC-0005.

8. William, Elliott R. 2021. Front-end engineering design (FEED) study for a carbon capture plant retrofit to a natural gas-fired gas turbine combined cycle power plant, Work Performed Under Agreement by Bechtel National, Inc. for the US DOE National Energy Technology Laboratory (NETL), DE-FE0031848.

9. Gülen, S. C. 2020. Gas Turbine Combined Cycle Optimized for Post-Combustion CO_2 Capture, US Patent 10,641,173.

10. Børseth, K. and Fleischer, H. 2021. Carbon Capture Comprising a Gas Turbine, US Patent Publication 2021/0060478.

11. Iijima, M., Nagayasu, T., Kamijyo, T. and Nakatani, S. 2011. MHI's energy efficient flue gas CO_2 capture technology and large-scale CCS demonstration test at coal-fired power plants in USA, *Mitsubishi Heavy Industries Technical Review*, 48(1), 26–32.

12. Reitenbach, G. 2015. SaskPower admits to problems at first "full-scale" carbon capture project at Boundary Dam plant, *POWER*. Available at www.powermag.com/saskpower-admits-to-problems-at-first-full-scale-carbon-capture-project-at-boundary-dam-plant/

13. Lang P., Denes F. and Hegely L. 2017. Comparison of different amine solvents for the absorption of CO_2, *Chemical Engineering Transactions*, 61, 1105–1110, doi:10.3303/CET1761182.

14. Jones, D. A. 2019. Technoeconomic evaluation of MEA versus mixed amines and a catalyst system for CO_2 removal at near-commercial scale at Duke Energy Gibson 3 pulverised coal plant and Duke Energy Buck natural gas combined cycle (NGCC) plant, Lawrence Livermore National Laboratory, LLNL-TR-758732, https://doi.org/10.2172/1499969.

15. Huertas, J. I., Gomez, M. D., Giraldo, N. and Garzón, J. 2015. CO_2 Absorbing capacity of MEA, *Journal of Chemistry*, 2015, 965015.

16. Luis, P. 2016. Use of monoethanolamine (MEA) for CO_2 capture in a global scenario: Consequences and alternatives, *Desalination*, 380, 93–99.

17. World Bank. 2016. Pre-feasibility study for establishing a carbon capture pilot plant in Mexico, Report no. AUS8579-2, https://www.gob.mx/sener/en/documentos/pre-feasibility-study-for-establishing-a-carbon-capture-pilot-plant-in-mexico?idiom=en.

18. Jarvis, S. M. and Samsatli, S. 2018. Technologies and infrastructures underpinning future CO2 value chains: A comprehensive review and comparative analysis, *Renewable and Sustainable Energy Reviews*, 85, 46–68.

19. Khan, M. O. 1984. Basic practices in compressors selection, *International Compressor Engineering Conference, Paper No. 509*, https://docs.lib.purdue.edu/icec/509.

20. Warudkar, S. S., Cox, K. R., Wong, M. S. and Hirasaki, G. J. 2013. Influence of stripper operating parameters on the performance of amine absorption systems for post-combustion carbon capture: Part I – High pressure strippers, *Interntational Journal of Greenhouse Gas Control*, 16, 342–350.

21. Wilcox, J. 2012. *Carbon Capture*, New York, NY: Springer.

12 Concentrated Solar Power

Like most renewable power generation technologies, *concentrated solar power* (CSP) suffers from intermittency and unpredictability, which render the technology undispatchable. There are means to alleviate this problem to a certain extent such as by using *thermal energy storage* (TES), which adds to plant size and capital cost appreciably (but not necessarily to cost of generation) [1]. An obvious remedy is to supplement the solar thermal power input by burning a fossil fuel, which obviously defeats the original purpose of deploying the renewable technologies [2]. Current industry practice is to dispatch renewable power when it is available and bring in fossil-fired generation (primarily gas turbine-based simple or combined cycle power plants in spinning or non-spinning reserve mode) when renewable power sources are not available (e.g., at night, cloud cover, no wind) but there is still a power demand to meet [3].

Traditional solar-fossil hybrid concepts typically use solar thermal as the supplementary thermal power input. A well-known example is the *integrated solar combined cycle* (ISCC) power plant [4]. In this concept, high or intermediate pressure (HP or IP, respectively) feed water from the heat steam recovery generator (HRSG) is sent to the CSP boiler (e.g., on top of a solar tower) and the steam generated is sent back to the HRSG [5]. Another example is using a solar tower or a parabolic trough system to heat the feed water or supplement the coal-fired boiler steam in a Rankine steam turbine power plant [6]. Yet another example is the *solar hybrid combined cycle* (SHCC), also known as *hybrid solar gas turbine* (HSGT), where gas turbine Brayton cycle heat addition via combustion is augmented by solar thermal power. One can certainly think of SHCC as a special type of ISCC with integration between solar and thermal technologies shifted to the topping cycle, although they are treated as two different systems and studied as such [7]. (Another way to inject low-grade solar thermal power into the topping cycle is via an absorption chiller for GT inlet cooling.) There are also proposals for solar thermal power injection between the Brayton and Rankine cycles [8].

For a thorough understanding of existing CSP technologies, the reader is referred to Kalogirou [9]. A concise recap of solar-augmented thermal power plant technology and pertinent literature can be found in Refs. [10,11]. Since there is an immense body of work on CSP (types of technology, cost, performance, etc.), details pertaining to it will be treated as given. (Needless to say, the same goes for the gas turbine combined cycle [GTCC], which is the other major player in the ISCC.) The focus in this chapter is on the optimal integration of CSP and GTCC via the bottoming cycle of the latter in an ISCC framework, which can be considered as a currently available (if not truly mature) technology. The SHCC and GT exhaust gas reheat variants involve significant technical challenges (e.g., a high pressure and temperature volumetric receiver to heat compressed air for the SHCC) and have a long way to go before becoming of interest to developers, original equipment manufacturers (OEM), and engineering, procurement, and construction (EPC) contractors. As such, they are not considered in this treatise, which is primarily intended for practicing engineers.

This chapter has an abundance of efficiency terms and other solar energy related terms not common to conventional gas and steam turbine power plants. Hence, for user convenience, a nomenclature is provided in Section 12.8.

The original coverage of the subject matter in this chapter was done in British units. To prevent errors and confusion, no attempt has been made to convert them into the SI units.

12.1   Integrated Solar Combined Cycle

The conceptual layout of a typical ISCC power plant is shown in Figure 12.1.

The interaction between the GTCC and the CSP plants can be simplified into two water/steam streams:

1. Boiler feed water from the HRSG to the solar steam generator at pressure and temperature, PFW and TFW, respectively
2. Saturated or superheated steam from the solar steam generator to the HRSG at pressure and temperature, PSTM and TSTM, respectively

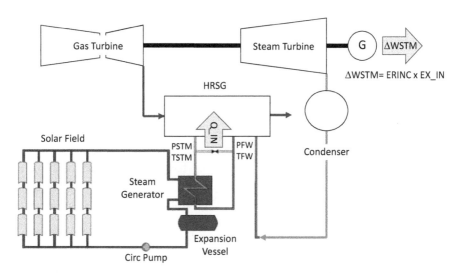

Figure 12.1 Schematic description of a generic ISCC with a GTCC and CSP plant comprising a solar field and solar steam generator. Note that EX_IN denotes the exergy of solar heat input Q_IN and ERINC is the *incremental* Rankine cycle (exergetic) efficiency, which quantifies the fraction of solar exergy converted to additional steam turbine output, ΔWSTM.

Table 12.1 Proven CSP technology comparison. (It should be pointed out that a recent development in LFR technology claims higher steam temperatures up to 500°C [12].)

HTF = heat transfer fluid, MS = molten salt, WF = working fluid.

CSP	Reflector	Absorber	WF	CR	EOPT, %	TREC, °C	Solar-to-Electric Efficiency, %
LFR	Flat	Linear	H_2O	40–50	~65	<300	9–11
PTR	Parabolic	Linear	HTF	70–80	~75	<400	14–16
CRS	Flat	Point	H_2O, MS	300–600	~65	up to ~600	15–22

The state of the art in GTCC technology comprises three-pressure, reheat (3PRH) steam Rankine bottoming cycles (built around multi-casing steam turbines (ST) with high component efficiencies and large-diameter exhaust annuli to exploit low condenser pressures) with advanced F, G, H, and J Class GTs as the topping Brayton cycle. Three CSP technologies are potential candidates for the ISCC:

1. Parabolic trough (PTR)
2. Linear fresnel (LFR)
3. Central receiver (CRS)

(The last one is also referred to as *solar tower*.) From a system perspective, each CSP technology is founded upon a *solar collector system* (SCS) comprising two components: (1) reflector/mirror (parabolic or flat) and (2) receiver/absorber (tubular/linear or point). From a performance perspective, they are characterized by three key parameters:

1. Concentration ratio (CR)
2. Optical efficiency, EOPT, which is a combination of several factors, including the cosine effect, mirror reflectivity, atmospheric attenuation and blocking and shading
3. Receiver temperature, TREC

CSP information pertinent to the subject is summarized in Table 12.1.

PTR with synthetic oil receiver and molten salt energy storage is a proven commercial technology option with several existing CSP plants already in operation, e.g., Andasol 1, a 50 MW plant in Spain that has been operation since 2009, Solano, a two unit 2×140 MW plant with a common molten salt thermal storage system, and Solafrica's 50 MW Bokpoort CSP plant near Upington, South Africa. The last one has been in commercial operation since early 2016.

LFR was developed for direct steam generation in the solar field. There are existing plants in Australia, California, and Spain. The technology is not readily amenable to molten salt energy storage. LFR with molten salt receiver and storage has been investigated but no applications are available.

CRS with molten salt receiver and storage is a proven technology. The first demonstration plant was built in California and operated for two years (see *Solar*

Two in Section 5.1). Since then, many plants have come online. A notable one is the *Ivanpah Solar Electric Generating System* in California that became operational in 2014. The 392 MW facility comprises three solar thermal power plants and use BrightSource Energy's Luz Power Tower 550 (LPT 550) technology, which heats the steam to 550°C directly in the receivers. (There is no storage.)

12.2   ISCC Thermodynamics

The thermodynamic principle governing the optimal combination of CSP and GTCC into ISCC is a direct manifestation of the second law of thermodynamics. Specifically, it is the principle of maximum exergy (see Ref. [6] for a similar approach). The total solar energy input rate from the CSP plant to the GTCC, Q_IN, is a fraction of the *direct normal irradiance* (DNI) that falls on the solar field. It is the net thermal power absorbed by the receiver of the particular SCS, which is utilized for steam generation. The ratio of the net thermal power to incident solar power is the SCS thermal efficiency, ETHM, which is the product of EOPT, receiver absorptance, α, and the receiver efficiency, EREC, i.e.,

$$ETHM = EOPT \cdot EREC \cdot \alpha, \tag{12.1}$$

$$Q\_IN = ETHM \cdot (AAPT \cdot DNI), \tag{12.2}$$

where AAPT is the total *aperture* area where solar radiation enters the collector. (Typical values of α and EREC are ~0.95 each.) In terms of steam generation in the receiver, the additional thermal power input to the bottoming cycle, Q_IN, is given as the product of steam flow, MSTM, and enthalpy change, ΔH, i.e.,

$$Q\_IN = MSTM \cdot \Delta H, \tag{12.3}$$

$$\Delta H = HSTM(PSTM, TSTM) - HFW(PFW, TFW). \tag{12.4}$$

(Note that piping losses between the GTCC and the CSP are ignored in this discussion.) The additional steam turbine output, $\Delta WSTM$, is a fraction of the solar contribution, Q_IN. The theoretical maximum of $\Delta WSTM$ is given by the solar *exergy* contribution, EX_IN, which can be related to Q_IN by

$$EX\_IN = Q\_IN \cdot E\_CNT, \tag{12.5}$$

$$E\_CNT = 1 - \frac{TREF}{METH}, \tag{12.6}$$

which is the maximum possible (theoretical) efficiency with which Q_IN can be converted into useful work, $\Delta WSTM$. In Equation (12.6), TREF, is a *reference temperature* (a logical choice for which is the prevailing ambient temperature) and METH is the *mean-effective heat addition temperature* and given by

$$\text{METH} = \frac{\Delta H}{\Delta S},\tag{12.7}$$

where S denotes the *entropy* and ΔS is given by

$$\Delta S = \text{SSTM}(\text{PSTM}, \text{TSTM}) - \text{SFW}(\text{PFW}, \text{TFW}).\tag{12.8}$$

Combining Equations (12.3), (12.4), and (12.7), one can write

$$\text{METH} = \frac{\text{Q\_IN}}{\text{MSTM} \cdot \Delta S}.\tag{12.9}$$

Practical engineering value of the conversion efficiency, which can be achieved with an economically feasible design, is only a fraction of the Carnot efficiency, E_CNT defined by Equation (12.6), say, EACT, which can be thought of as a second law-based *conversion effectiveness*. Thus, for the contribution of the solar thermal power to the ISCC power output, which is the increase in ST power output due to the steam imported from the CSP plant, one can write

$$\Delta \text{WSTM} = \text{ERINC} \times \text{Q\_IN},\tag{12.10}$$

$$\text{ERINC} = \text{EACT} \times \text{E\_CNT},\tag{12.11}$$

where ERINC is the *incremental* Rankine cycle efficiency. This fundamental equation tells the designer that for the maximum contribution from the solar field, i.e., for *maximum* ΔWSTM, at given site ambient conditions (represented by TREF), METH should be maximized. This requirement is met when two conditions are satisfied:

1. TSTM and PSTM are as high as possible (HP steam is more valuable than LP steam, i.e., higher EACT)
2. TFW is as close to TSTM as possible (in passing, this is the thermodynamic principle underlying feed water heating in fossil steam turbines)

The importance of ERINC for an optimally designed solar-thermal hybrid system such as ISCC is easily appreciated when it is put into the context of the entire solar power formula, i.e.,

$$\Delta \text{WSTM} = \text{ESOL} \times (\text{AAPT} \times \text{DNI}),\tag{12.12}$$

$$\text{ESOL} = \text{ERINC} \times \text{ETHM}.\tag{12.13}$$

The biggest cost and complexity adder in an ISCC is the SCS (in installed terms with all the requisite balance of plant [BOP], this is the *solar field*). Thus, the ultimate goal of the designer is to get the highest possible contribution (in watts of electricity) from each watt of DNI at a given site with the smallest possible solar field, which is proportional to AAPT. (In reality, however, economic feasibility considerations force the designer to compromise on that goal.) For a given ΔWSTM objective, the first part of the goal is equivalent to maximizing ERINC, which is the equivalent of maximizing METH. (It will be shown later in the discussion that this statement, while certainly true for different levels of steam, e.g., HP vs. LP, is not so universally.) This, however,

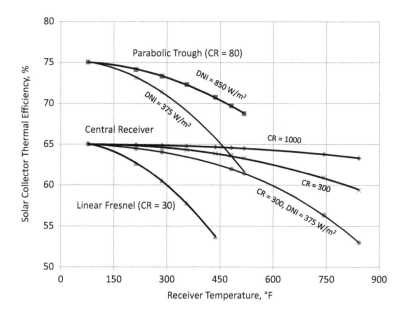

Figure 12.2 Solar collector thermal efficiency for three major CSP technologies as a function of TREC, CR, and DNI.

conflicts with the second part of the goal because maximizing METH means maximizing TREC and leads (counterintuitively in fact) to lower ETHM (e.g., see Kelly et al. [13]). Solar collector system thermal efficiency is given by (ignoring convection and conduction losses; see Ref. [9] for a derivation)

$$ETHM = EOPT - \frac{\sigma \cdot \varepsilon \cdot \left(TREC^4 - TAMB^4\right)}{DNI \cdot CR}, \qquad (12.14)$$

where $\sigma = 5.67 \cdot 10^{-8}$ W/m^2-K^4 is the Stefan–Boltzmann constant and ε is emissivity (unity for a true *black body*). Representative thermal efficiencies for the three different CSP systems are plotted in Figure 12.2 as a function of TREC ($\varepsilon = 0.9$, TAMB $= 25°$C (77°F), DNI $= 850$ W/m^2). The plots clearly illustrate the detrimental impact of lower DNI and higher TREC on thermal efficiency. One important takeaway from Figure 12.2 is the lower susceptibility of the technology with higher CR (i.e., central receiver, CRS) to either detrimental impact. This is important due to the annual variation in DNI (with significant impact on total annual electric generation) and the requirement of high temperatures for high power generation (exergetic) efficiency (with positive impact on solar field size and cost). Considering the conflicting tendencies in ETHM and ERINC with increasing TREC, flatter ETHM (TREC) curvature is clearly an advantage. Note that the efficiency curves in Figure 12.2 should be treated as conceptual approximations useful for front-end optimization considerations similar to the treatment herein. Nevertheless, they are representative of values obtained from more comprehensive analytical treatments and/or field or laboratory measurements (e.g., see Schenk et al. [14] or Kutscher et al. [15]).

12.3   Design Considerations

A key selling point for ISCC is to utilize solar thermal power as extensively and reliably as possible while the sun is shining and not worry about the times when it is not. Without TES, accounting for daily and seasonal variation in DNI, meteorological events such as cloud passage and power block transients, CSP has an annual capacity factor of ~20–25% (about 2,000 hours at full load). To achieve higher capacity factors and make a real impact on the environment, large CSP plants with high *solar multiples* (SM) and TES are required (at nominal costs of $6,000 per kW [16,17] – much higher for a real project). (The solar multiple is the ratio of the actual size of a CSP's solar field to the field size requisite for generating steam to run the steam turbine at rated full load at the design DNI.) ISCC enables the deployment of relatively small CSP blocks (50 to 100 MWth) with commensurately low technology and financial risks. From a power generation perspective, ISCC is equivalent to a supplementary fired GTCC with similar power boost, economic incentives (hot day power augmentation), and limitations (mainly steam turbine throttle and condenser pressure variation between "solar fired/unfired" operation modes). Thus, the design principles are also similar.

To establish representative design rules, a typical 600 MW, $2 \times 2 \times 1$ (two GTs and HRSGs, one ST) GTCC block is considered (see Table 12.2). The system is modeled in Thermoflow, Inc.'s GT PRO/MASTER (design and off-design) heat balance simulation software [18]. Generic external heat addition to the bottoming cycle at high, medium (intermediate), and low pressure HRSG sections (HP, IP, and LP, respectively) is evaluated as shown (conceptually) in Figure 12.3. The selection of an *air-cooled condenser* (ACC) is commensurate with most geographical locations suitable to solar power generation with scarce resources of naturally available cooling water. The selection of the throttle pressure strikes a reasonable balance between steam cycle pressures with and without external heat addition.

As shown in Figure 12.3, feed water from the economizer sections at three different locations with increasing "thermal distance" to the evaporator is sent to the external heat source (e.g., a solar steam generator). Saturated steam is returned to the HRSG at the evaporator exit. The heat added to the HRSG is evaluated by Equation (12.3) and is primarily driven by the flow rate of feed water (steam) sent to the external *boiler*. The results for the HP variant with feed water source A in Figure 12.3 are summarized

Table 12.2 Typical $2 \times 2 \times 1$ GTCC with an advanced F Class GT – used as basis for the technology curves. The ballpark installed cost for this plant is around $600 million (typical US location at the time of writing).

Gas Turbine	Bottoming Cycle	Performance
Advanced Frame 7	3PRH	610 MW – 56.0% (ISO)
200 MW Rating (ISO)	1,800 psig – 1,050°F HP Steam	580 MW – 55.8% (Hot)
	Air-Cooled Condenser	
1,115°F Exhaust (ISO)	2.0 in. Hg @ ISO	
1,125°F Exhaust (Hot)	3.9 in. Hg @ Hot Day	

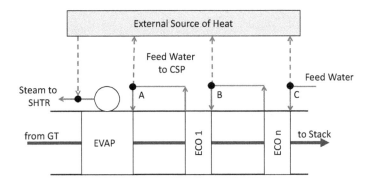

Figure 12.3 Conceptual HRSG section layout (HP, IP, or LP) for external heat addition via additional steam generation.

in Figure 12.4. (Note that this is the case closest to HRSG supplementary (duct) firing in concept as well as numerically.)

In essence, external heat input via HP steam generation increases ST throttle flow and pressure (the direct proportionality between the two for fixed ST inlet capacity is described by the famous formula of Stodola) as well as the condenser pressure via higher LP turbine exhaust flow. The limiting factors are

1. ST casing design pressure
2. condenser pressure (typically 5 in. of Hg is the maximum allowable)

Myriad design solutions are possible but involve significant cost–performance trade-offs, which can only be resolved via a FEED study undertaken by experienced OEM or EPC teams on a case-by-case basis. Examples include lower ST throttle pressure at the design point (e.g., 1,400 or 1,600 psia, sacrificing some *non-augmented* performance), lower condenser pressure (i.e., larger ACC with extra cells [19] or even aggressively water-cooled systems where possible), operational philosophy adjustments (e.g., running the GT at part load when external *boiler* is on – thus, the GTCC is run at a partially or fully *fuel saving* mode [20] rather than at a *power boost* mode), and solutions in the category of "variations on the theme." One example of the latter is admitting the (superheated) IP steam generated in the CSP system at the ST cold reheat (instead of saturated IP steam at the IP evaporator). This will be mentioned again later in the chapter in the context of system selection.

Figure 12.4 clearly demonstrates why there is a limit to power augmentation via external heat input to the bottoming cycle. (Note that the x-axis in Figure 12.4 is often referred to as *solar fraction* (SF) within the solar-thermal hybrid context.) In this example, ~20% extra steam turbine output (equivalent to ~20/3 = 6.5% improvement in GTCC heat rate with constant GT fuel input) is reached at a throttle pressure of $1.3 \times 1,800 = 2,340$ psia with Q_IN equivalent to ~10% of GT heat consumption (i.e., SF ~ 0.1).

While system-specific, optimized designs are expected to show some variation, as will be explored below, and they are unlikely to achieve a boost to the ST generator output of more than ~25%. As an example, Figure 12.4 shows the extension of

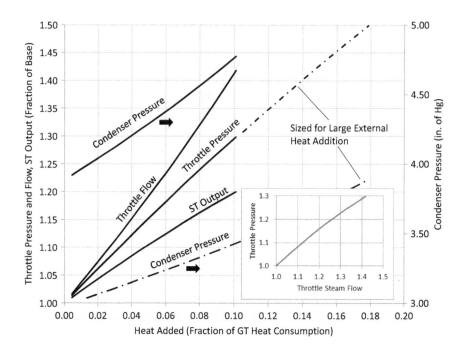

Figure 12.4 External GTCC bottoming cycle heat addition via additional HP steam generation with feed water take-off from A in Figure 12.3.

external heat addition capability by setting the bottoming cycle design (with no external heat addition) as follows:

- 1,450 psia (100 bara) throttle
- superheaters sized for 1,100°F (593°C) steam (controlled to 1,050°F (565°C) with desuperheating)
- 3.0 in. Hg (100 mbar, absolute) condenser pressure (on a hot day)

As a shorthand, this design will be referred to as "large-sized" bottoming cycle from here on.

The capability afforded by larger HRSG and ACC does not come cheap; the GTCC cost is ~5% higher than the typical base cost indicated in Table 12.2 (~$30 million), which should be added to the CSP cost. The key cost/performance driver is the ACC sizing and matching with ST exhaust annulus (i.e., last stage bucket size or number of LP ends).

A frequent optimization issue is the location of the HRSG feed water take-off point. The exergy maximization principle clearly dictates that it should be A in Figure 12.3. However, interesting trade-off opportunities exist – as can be seen in Figure 12.5. (Note that this degree of freedom mostly pertains to the HP option; it is very limited for the IP and essentially nonexistent for the LP.) Below are the main takeaways from the figure:

1. At the same heat input as in point 1, sending HP feed water to the external boiler from the lowest temperature source (C in Figure 12.3, point 2 in Figure 12.5)

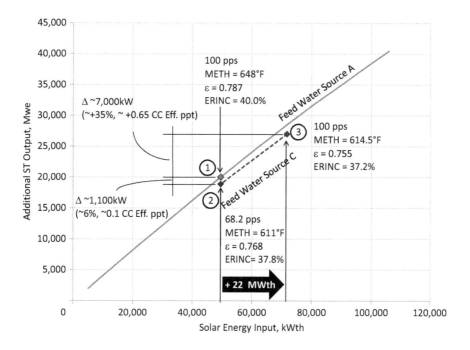

Figure 12.5 Impact of switching feed water take-off location from A to C in Figure 12.3.

reduces HP steam production by 30+% but its detrimental impact is greatly compensated by increased IP and/or LP steam production (because more exhaust gas heat is available for those sections – downstream of the HP section).

2. At the same additional HP steam generation (i.e., point 3 in Figure 12.5), however, Q_IN increases by ~45% (because more heat is required to preheat the colder feed water) with commensurately larger impact on ST output (~7 MW higher) and overall GTCC thermal efficiency (+0.65 percentage points).

This brings up the crucial question at the heart of ISCC optimization: Does the exergy maximization principle fail here? After all, the ultimate objective is maximization of solar thermal power utilization and the prima facie message of Figure 12.5 is quite clear:

Highest Q_IN (e.g., solar thermal input) takes place at lowest EACT.

The answer to the question posed above is, in fact, yes. The objective herein is *not* exergy (incremental Rankine cycle efficiency) maximization but, rather, incremental ST power output (i.e., solar heat input) maximization. There are two analogies to this situation, which should make grasping this very important observation easier.

1. Consider the simple gas turbine Brayton cycle. Cycle efficiency is a strong function of the cycle pressure ratio. With increasing cycle pressure ratio, cycle efficiency increases (e.g., aero-derivative gas turbines) but at the expense of (i) specific power output and (ii) combined cycle efficiency.

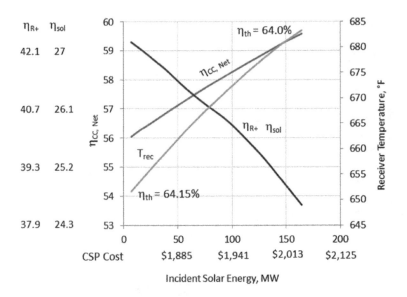

Figure 12.6 Key performance-cost trade-off drivers for ISCC based on HP steam generation in a CRS-based solar plant.

2. In maximizing gas-steam turbine combined cycle efficiency, the bottoming cycle design objective is *not* maximizing the steam cycle efficiency but, rather, minimizing HRSG stack temperature (i.e., maximizing the heat transfer from the gas turbine exhaust gas to water/steam). For a rigorous, mathematical proof of this assertion, the reader is referred to chapter 8 of **GTCCPP**.

A related question is quite obvious: At what cost does the extra ~7 MW ST output come? The increase of ~22 MWth in Q_IN comes at a cost of lower ESOL (by about 1.5 percentage points) and, consequently, higher incident solar radiation (i.e., larger solar field). For a solar thermal efficiency of 64.1%, extra solar input (AAPT×DNI) to compensate for lower ESOL is ~34 MW.

It is difficult to make a case using absolute cost numbers – CSP technology is far from that stage of maturity. Suffice to say that, whatever the actual cost of the particular CSP system might be (for the HP steam generation case, it must be CRS), the additional $/kW for ~7 MW extra ST output by shifting the feed water take-off point from A to C is ~30% higher than the additional $/kW for the base case 1 (as can be verified by numbers in Figure 12.5). Even using the published (probably optimistic) cost data (e.g., see Ref. [16], which suggests $0.5–0.75 million per incident solar MWsol (*solar megawatts*) – excluding the power block) this becomes a losing proposition; the maximum exergy principle stands and the overall picture is summarized in Figure 12.6.

The specific cost ($/kW) for the CSP plant (no TES and no separate power block and associated BOP) in Figure 12.6 is based on a solar multiple (SM) of 1.0 and $0.5 million per one megawatt of incident solar radiation (MWsol) and given by

$$C\_CSP = \left(1{,}835 + QSOL + 5.62 \times 10^{-8} \cdot QSOL^4\right) \cdot SM \cdot CF, \qquad (12.15)$$

where QSOL is the total solar incident radiation in MWsol and CF is a cost factor with reference to the base cost assumed here (i.e., if the "real" cost is, say, $600,000 per MWsol, CF is 600/500 = 1.2). Thus, using Figure 12.6, for example, an ISCC with 100 MWsol solar CRS field (SM of 1.2) at $600,000/MWsol (CF = 1.2) would cost ~$1,940 × 1.2 × 1.2 ~ $2,800 for additional ~26 MW ST output at the design point (a much more palatable number than $6,000+ for typical CSP plants). In comparison, the recently published *budgetary price* for a GTCC in the size range of 500–600 MW is ~$550/kW [21], which can easily go up to $800+/MW when real projects and associated costs are considered.

Boosting the output of a ~600 MW rated GTCC at ~$800/kW by ~25 MW at an additional cost of $2,800 might seem a hopeless proposition. However, one should also consider the worth of efficiency improvement or, equivalently, heat rate reduction that comes with it, which from Figure 12.6 is about 2.40 net percentage points (i.e., [(625/600 − 1) × 56 ~ 2.4] or about 250 Btu/kWh *lower* heat rate). As demonstrated in a paper by Gülen [22], this is a rather tricky endeavor. In terms of $/kW, the capital-equivalent heat rate reduction value can be as high ~$1,500/kW for $5/MMBtu natural gas fuel (e.g., see equation 1 in Ref. [22]). This makes the maximum acceptable increase in overall plant cost about $2,300/kW ($800 + $1,500). Thus, depending on the price of fuel and other projected economic criteria, a strong case *might* exist for ISCC. Due to the uncertainty involved in key economic and financial factors (e.g., capital cost, fuel price, financing, and "bankability" issues) any assertion beyond that would be speculative. (A potential remedy via probabilistic analysis is proposed by Mehos et al. [23].)

The most important takeaway from Figure 12.6 is the absence of a well-defined maximum or minimum point, which renders a traditional optimization conclusion infeasible. Several key facts are unmistakable, e.g., the monotonous decrease in both solar and incremental Rankine cycle efficiencies and the relative insensitivity of solar thermal efficiency. The latter is driven by the high CR of the CRS technology, which is the CSP option most suitable to HP steam generation (see Figure 12.2). The impact on the bottom line is the decrease in the proverbial "bang for the buck," i.e., the increased marginal cost of higher solar megawatts, which renders CSP sizes beyond a certain MWsol level infeasible. The exact cut-off point, to reiterate the earlier comment, is subject to uncertainty and can only be determined via an in-depth study by experienced OEMs and EPCs on a case-by-case basis.

The decrease in ERINC is, however, somewhat puzzling. From Equation (12.11), since METH and, consequently, E_CNT definitely improve with increasing Q_IN, the driver of the trend in Figure 12.6 must be the reduction in the conversion effectiveness, EACT. This can be explained by examining the incremental exergy loss in the HRSG, ST, and condenser as a function of solar fraction (see Figure 12.7). Steam turbine and condenser combined exergy loss is driven primarily by the condenser pressure. For a fixed condenser design (i.e., heat transfer surface area), additional steam generation increases the quantity of steam to be condensed and, at the same condenser size, the condenser pressure increases with an accompanying loss in ST power output. Another

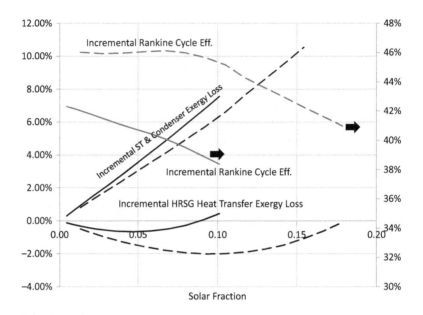

Figure 12.7 Exergy loss in the HRSG, ST, and condenser as a fraction of GT exhaust gas exergy at the HRSG inlet. The dashed curves are for the large-sized bottoming cycle.

notable takeaway from Figure 12.7 is the difference between ERINC for the two different bottoming cycle designs, which is also driven by the same exergy loss mechanisms – especially that of the HRSG as will be explained below.

HRSG heat transfer exergy loss is qualitatively measured by the divergence of GT exhaust gas cooling and water/steam heating lines from each other on a *heat release diagram* (temperature is the *y-axis* and cumulative heat transfer in the HRSG is the *x-axis*) – also known as *Q–T diagram* (see Figure 12.8). The largest divergence is in the evaporators, where water boils at constant pressure and temperature (the HP evaporator is the largest one, followed by the IP and LP evaporators).

Initially, shifting of HP steam production to the solar generators reduces the HRSG HP evaporator duty and evaporator exergy loss (i.e., *shorter* horizontal line on the Q–T diagram). However, eventually, increasing the amount of solar steam forces the total water/steam flow in the HRSG beyond the capability of the *fixed size* cross-flow heat exchangers (vertically arranged tube rows in a horizontal HRSG) to achieve high enough steam temperatures and, thus, leading to higher exergy destruction (i.e., gas and steam temperatures start to diverge again). This is best illustrated by the inset in Figure 12.8 quantifying the reduction in superheater steam exit temperatures. If the HRSG is sized "large" with no external heat addition, the point of diminishing returns (i.e., the minimum in HRSG heat transfer exergy loss in Figure 12.7) takes longer to reach. This is so because of (i) exergy loss recovery potential at low HP pressure (graphically, the HP evaporator line in Figure 12.8 is longer and lower) (ii) the ability of the superheaters to accommodate larger steam flows with a relatively small drop in steam exit temperatures (or none at all).

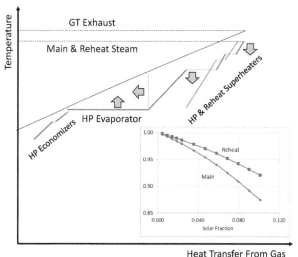

Figure 12.8 HRSG heat release diagram (the inset shows the decrease in main and reheat steam temperatures).

12.4   General ISCC Potential

12.4.1   Power Augmentation Perspective

The discussion so far has considered the best possible scenario of CSP-GTCC integration for power augmentation: solar HP steam generation in a CRS solar system. There are other possibilities, three of which are considered herein to complete the ISCC optimization picture. Two of them are obvious: IP or LP steam generation per the concept in Figure 12.3 (feed water take-off at A – as shown to be the optimal choice). The third one is described conceptually in Figure 12.9. As shown in the figure, the external heat source assumes a portion of reheat superheating duty so that less GT exhaust energy is spent in the HRSG reheat superheater sections. The net effect is increased HP steam production and ST output.

All four options are evaluated, and the results are summarized in Figures 12.10–12.11 and Tables 12.3–12.4. (Also shown in Figures 12.10–12.11 are the impacts of system sizing and running in fuel-saving mode (GT at part load) – these will be discussed in more detail below.) The HP case is the same as presented in the previous section. In IP and LP steam generation cases, the limit is set by increased condenser pressure. The reheat superheating case is limited to diverting 50% of the cold reheat steam to the external heat source. On paper, even more reheat superheating duty can be shifted to the external source (e.g., the solar field). However, one should be cognizant of performance (to prevent excessive cold and hot reheat pipe pressure loss) and cost ramifications (long alloy piping).

The picture emerging from Figures 12.10–12.11 is quite clear. LP steam generation is not a particularly appealing option – it is quite costly as well (see Table 12.4).

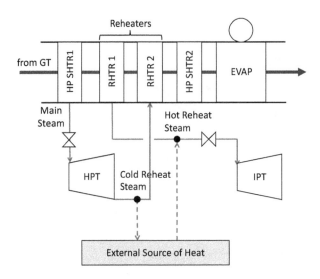

Figure 12.9 Conceptual HRSG-ST layout for external heat addition via reheat superheating.

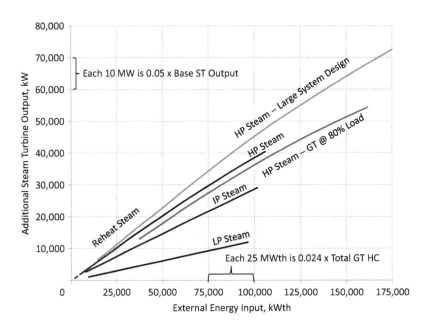

Figure 12.10 Impact of HRSG steam level on additional ST power output via external heat addition. (The plots can be used for generic 3PRH GTCC bottoming cycle studies using the fractional information on x and y-axes.)

Reheat superheating and HP steam generation are essentially at par, but the latter has far more capacity, which is desirable because the ultimate goal is to maximize the (renewable) solar thermal power deployment. IP steam generation is definitely less attractive than HP from a pure performance perspective, but it can utilize the most

Table 12.3 Average performance parameters for four principal external heat addition options (over the range in Figures 12.10–11)

	LP	**IP**	**HP**	**RHT**
T_{in}	307.7	471.8	646.4	824.1
η''	0.430	0.670	0.804	0.725
T_{rec}	327.7	491.8	666.4	844.1
η_{R+}, %	12.5	27.8	40.8	41.7
CSP	LFR	PTR	CRS	CRS
η_{th}, %	58.9	69.4	64.1	63.3
η_{sol}, %	7.3	19.3	26.1	26.4

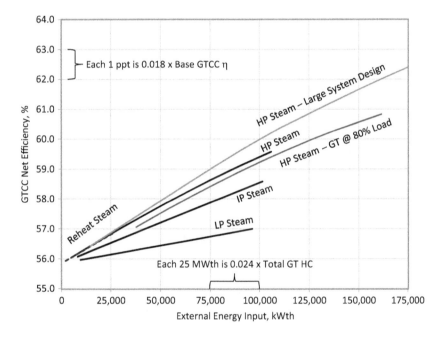

Figure 12.11 Impact of HRSG steam level on GTCC net (thermal) efficiency output via external heat addition (constant GT heat consumption).

proven CSP technology available, i.e., PTR, and thus might be an easier sell to developers. This is especially true if one can also make a cost argument in favor of PTR.

This brings the discussion to the shakiest part of the ISCC optimization problem: the cost of the CSP system. It is definitely worth repeating that while there is no doubt about the governing thermodynamic principles and the performance aspects of the integration problem, the capital cost of a particular CSP system is subject to extreme

Table 12.4 Incremental cost for four principal external heat addition options (for given incident solar radiation in MWsol)

	LP	**IP**	**HP**	**RHT**
CSP	LFR	PTR	CRS	CRS
QSOL	100	100	100	41
SF	0.057	0.067	0.061	0.025
ΔST MWe	7.3	19.5	25.8	10.5
SM/CF	1.5/0.6	1.3/0.95	1.2/1.0	1.2/1.0
$/kW	6,150	3,175	2,325	2,325

speculation. This makes a comparison of different options on a truly apples-to-apples basis very hard. Only a diligent study for a given project can establish a reliable basis. Nevertheless, a few assertions can be made with reasonable certainty (in accordance with the main objective of this discussion).

The first obvious conclusion is that CSP-GTCC integration via LP steam generation is truly suboptimal from a performance as well as cost perspective. Even assuming a significant cost advantage for LFR (CF of 0.6 in Table 12.4), it has limited capacity at nearly twice the cost. (Note the assumed SM of 1.5 for LFR is most likely too optimistic –1.8 would be more realistic and make the cost even higher.)

Integration via HP steam generation is the clear winner over IP in terms of both performance and cost. However, one should consider that the HP case in Table 12.4 requires steam generation at ~2,300 psia (150–160 bar). Receiver design for direct steam generation and operability challenges (daily startup of the system from cold conditions) can make this an infeasible option. The high pressure difficulty can certainly be remedied by large-sizing the bottoming cycle to fit the solar-augmented operation but at a significantly higher cost (see below).

One should also consider the maturity of PTR, which is by far the most dominant CSP technology (e.g., more than 90% of solar power plants in operation with more than 1,800 MW installed capacity going back to the early 1980s), which makes it a more "bankable" technology than CRS. Thus, a very strong case can be made for CSP-GTCC integration via IP steam generation utilizing PTR technology. (Interestingly, IP steam generation utilizing CRS technology was adopted by one OEM in their most recent ISCC offering [23]. The configuration is based on superheated IP steam from a CRS receiver admitted at the cold reheat line.)

Integration via reheat system is an intriguing option, which might be worthy of consideration for a low cost introductory system (especially if CF < 1.0 due to a simpler heat exchange arrangement). It can also be combined with the HP steam generation (since the CRS technology can support main *and* reheat steam superheating) to alleviate the ST design pressure limitation. The final arrangement is subject to optimization based on site-specific criteria.

Sizing the GTCC to accommodate higher solar contribution is obviously an attractive option. It results in higher ESOL (as implied by Figure 12.10) and lower solar $/kW. However, when the added cost of the larger GTCC is factored in, it becomes more expensive. For example, Q_IN can be increased to 200 MWsol (double the value used in Table 12.4) and 54 MWe extra ST output by sizing the GTCC, as discussed earlier, with an extra cost of ~$30 million, which makes the incremental cost ~$2,650/kW (~15% higher). A proper evaluation would require hour-by-hour simulation over the entire year with a given load profile (e.g., see Dersch et al. for a discussion of this [24]).

12.4.2   Fuel Saving Perspective

The evaluation so far has considered the external (i.e., solar) energy contribution in a "power boost" mode. An alternative take on the hybrid system can be made in a "fuel saving" mode. In this mode, the GT is run at part load with reduced fuel consumption (and reduced emissions). The power deficiency resulting from the reduction in the topping *and* bottoming cycles (the latter via reduced exhaust energy) is made up by external heat added to the bottoming cycle. To investigate this mode, sample runs are made by considering an ISCC with HP steam integration and changing the GT load from 100% to 80% (hot day). The results are summarized in Table 12.5. (The impact on system performance for 80% load case can be seen in Figures 12.10–12.11.)

It is possible to save fuel and maintain the base output down to about 90% GT load. Beyond that, the solar thermal power contribution is unable to compensate for the reduction in GTCC power output without exceeding hardware design limits (e.g., steam turbine admission and condenser pressures). Whether the savings in fuel expenditure justify the CSP system cost is dependent on the fuel cost and site-specific annual solar irradiation and ambient conditions. Consider the following (implausibly optimistic) scenario: the plant operates in solar-augmented mode for 3,000 hours a year (~8 hours per day) at the same single-point performance as in Table 12.5. For the 85% GT load case, noting that 99.2 MW ~ 339 MMBtu/h, the annual fuel saving at $12/MMBtu (HHV) with a levelization factor of 1.169 and natural gas fuel HHV/LHV ratio of 1.109 would be given by

$$\text{Fuel saving} = 339 \times (\$12 \times 1.109 \times 1.169) \times 3,000 = \$15.8 \text{ million.}$$

Thus, even ignoring the unrecovered generation loss of 10 MW, the payback period would be about 10 years. It should be pointed out that the key driver in optimal fuel saving mode design is the part load performance of the gas turbine. State of the art gas turbines in general and sequential combustion (reheat) gas turbines, in particular, have much better part load characteristic than the sample unit in Table 12.5. In particular, the key metric is the marginal GT efficiency, which is given by

$$\frac{\Delta \dot{W}_{GT}}{\Delta HC} = EGTM = EGT0 \cdot \left[\frac{L - 1}{L/E - 1} \right], \qquad (12.16)$$

Table 12.5 ISCC via HP steam integration – "fuel saving" mode. The base GTCC design is per Table 12.2 (i.e., the bottoming cycle is not "large-sized").

GT Load	100	95	90	85	80
Δ CC Output, MWe		0	0	−10.0	−31.3
Δ Heat Consumption, MWth		−32.5	−65.6	−99.2	−133.3
HP Pressure, psia	1,800	2,060	2,360	2,515	2,515
HP Temperature, °F	1,050	1,004	911	835	838
Reheat Temperature, °F	1,049	1,025	967	913	918
Condenser Pressure, in. Hg	3.9	4.2	4.7	5.0	5.0
Q_IN, MWth	NA	83.6	185.6	257.2	260.8
SF	NA	0.053	0.122	0.175	0.184
Δ ST Output, MWe	NA	19.4	40.0	50.4	48.3
ERINC, %	NA	36.3	33.7	30.6	28.8
ESOL, %	NA	23.3	21.6	19.6	18.5
Solar Cost, MM$	NA	$50	$110	$155	$155

where L is the gas turbine load as a fraction and E is the normalized efficiency, EGTM/EGT0, at that particular load (i.e., x and y-axis, respectively, of the GT part load efficiency curve). It can be shown that to have a feasible design (i.e., no lost GTCC output) in fuel saving mode, the following inequality should be satisfied

$$\text{ERINC} > \frac{(1 - L/E) \cdot (\text{EGTM} + (1 - \text{EGTM}) \cdot \text{ERSC})}{\text{SF}}, \qquad (12.17)$$

where ERSC is the Rankine steam bottoming cycle efficiency (~30%). For example, taking the 80% GT load case in Table 12.5 (L = 0.8), with E = 0.923 from the GT PRO model (built-in part load curve) one finds from Equation (12.16) that EGTM ~ 0.57, i.e., for a 20% reduction in GT output (~77 MWe for two units), 77/0.57 ~ 135 MWth worth of fuel is saved. Compensating the lost GT output (along with the lost ST output) requires, as per Equation (12.17), an ERINC of ~38% at SF = 0.25, which, per Figure 12.7, does not seem to be possible. Assuming that the GT in question is an advanced unit with E = 0.98 at L = 0.8, EGTM = 0.413, and a 20% reduction in GT output saves 77/0.413 ~ 186 MWth in fuel consumption. Unfortunately, this is not achievable because it would require a SF of ~0.3 but the bottoming cycle hardware is already at its design limit at SF ~ 0.18 (see the last column in Table 12.5). However, using the large-sized bottoming cycle, the goal of constant GTCC output at L = 0.8 can be achieved with SF ~ 0.25 (refer to the dashed line in Figure 12.7).

12.4.3   Thermal Storage Perspective

It is duly recognized that evaluating CSP technologies on a $/kW basis is a rather crude approach. A proper study should be made on an annual electricity generation basis taking into account site-specific load demand as well as ambient and solar DNI characteristics (see, for example, Ref. [24]), which especially impacts ETHM in a significant way for the three CSP technologies under consideration here (e.g., see Figure 12.2). Even then, it is highly unlikely that LP steam generation (i.e., existing LFR technology) will be a viable option. The competition is most likely to be between (i) HP and IP steam generation (the former maybe supplemented by reheat) and (ii) PTR or advanced LFR and CRS technologies (or a "mix-match" approach, e.g., see Ref. [23]).

A related subject in that regard is the inclusion of TES in ISCC. One of the appealing aspects of the ISCC concept is exclusion of TES with its added cost (~$90/kWth [24] but expected to go down to $25/kWth [25]) and complexity (i.e., utilization of molten salt as heat transfer fluid *and* storage medium vis-à-vis direct steam generation). Adding, say, 6 hours of TES capability would increase SM to 1.8–2.0 for PTR and CRS [24] with the concomitant impact on $/kW in Table 12.4 – *excluding* the cost of the TES system. One benefit of a limited TES system can be to reduce steam turbine pressure and temperature transients for better thermal stress control and faster startup via increased gland steam temperatures during overnight shutdown. A recent study indicated up to 10% more daily energy generation [26]. This can be especially crucial for HP steam systems with high steam pressures (well above 2,000 psia) with lost generation capacity due to slow thermal transients.

There are quite a few CSP plants with a parabolic trough collector field and molten salt TES. Some notable examples are the Solana Generation Station in Arizona (280 MW capacity) and the Andasol 1 in Spain (150 MW capacity). As far as *molten salt solar tower* (MSST) technology, the landscape is rather sparse: there are only three known (to the author at least) examples in North America and Europe:

1. **Solar Two**; a 10 MWe demo plant with 3-hour TES in Barstow, CA, which operated between June 1996 and April 1999 to validate MSST technology (see [7] in Chapter 5).
2. **Gemasolar** (formerly Solar Tres) near Seville, Spain; nominal 20 MWe commercial plant with 15-hour TES, which went on-line in May 2011 and is currently operational. This power station uses concepts pioneered in the Solar One and Solar Two demonstration projects.
3. **SolarReserve's Crescent Dunes** (Tonopah, NV); nominal 110 MWe with 10-hour TES. This project began operation in September 2015. It went off-line in October 2016 due to a leak in a molten salt tank and returned to operation in July 2017. Production in the facility ceased in April 2019.

Based on the operational experience of the Solar Two and Gemasolar MSST plants, using the US DOE's TRL scale (see Table 16.1 in Section 16.4), one might reasonably conclude that MSST is at TRL 9. As far as the technology (commercial) maturity, two separate sources [27,28] place generic solar tower (central receiver) CSP approximately at the same stage: at the cusp of commercial deployment (no distinction is made about TES or HTF (i.e., molten salt or water/steam) in either source).

The position of a particular technology in the maturity spectrum is a function of "how much capacity has been installed to date, how much is expected in the near future, and how much momentum the technology currently has for the future [27]." Nevertheless, considering that MSST is about 30% of operating solar tower capacity and Solar Reserve's Crescent Dune is about one-fifth of the solar tower capacity in construction, it is safe to assume that MSST is well within the generic solar tower CSP maturity range provided in Ref. [16].

A key concern in new technology development is the *scalability*, i.e., the Solar Two and Gemasolar plant capacities are only a fraction of that of SolarReserve. In general, scalability of steam and power generation systems should not be a concern. These are extremely mature and proven technologies. To a certain extent, one would think that the same can be said of the two-tank thermal storage system as well. In particular, the BrightSource Ivanpah *Solar Energy Generating System* (SEGS) can be presented as a case of scalability of a single tower/field ST/CR system to 100 MWe (albeit in direct steam generation mode). However, the dismal performance and eventual bankruptcy of the Crescent Dunes project (the plant suffered several design, construction, and technical problems, only achieving about a 20% capacity factor in 2018, resulting in lawsuits and changes of control) once again proved that this would be a misguided expectation.

One can argue (and with good reason) that shell-and-tube heat exchangers, where one fluid is *liquefied* nitrate salt, cannot be lumped that cavalierly with traditional systems. However, based on field experience (e.g., Solar Two) the real issue is to ensure that the salt stays as a liquid (i.e., above its freezing temperature), which is more of a control system design issue rather than size – up to point, though (That point *might* have been reached in the case of one particular CRS project, eSolar, with *kilometers* of molten nitrate salt (cold and hot) piping).

12.5   CRS Technology

It is reasonable to assume that CRS (Solar Tower), preferably with TES, is the most suitable solar thermal technology for ISCC application. Its key components are:

1. Heliostat and solar field to reflect sunlight onto a centralized solar receiver located on a tower. Heliostat subcomponents include a reflector, azimuth and elevation drives, and a support structure and pylon.
2. The solar receiver to transfer solar radiation energy to a heat transfer fluid using the reflected and concentrated sunlight.
3. The steam generator is a heat exchanger located on an elevated platform or at grade level through which the heated molten salt transfers the energy to water, thus producing steam.
4. Steam turbine generator.

5. A two-tank molten salt system – hot salt storage and cold salt storage.
6. The control system adjusts the azimuth and elevation of each reflector to follow the sun during the course of the day, and control solar flux on to the solar receiver to control steam and power production. It is assumed that each heliostat will have its own stand-alone control system for tracking the sun that is linked via a wired or a wireless network.

There are two steam generator design options:

1. molten salt as a heat transfer fluid (HTF) circulating through the low-pressure solar receiver, or
2. *direct steam generation solar receiver* (DSSR) using high pressure and temperature steam as a heat transfer agent.

The solar receiver sits atop of the tower. It is designed to absorb the solar energy that is directed by the heliostats and transfer that energy to a heat transfer fluid, i.e., either water to steam or molten salt. The unit typically has a substantially cylindrical shape consisting of multiple flat panels. The heat transfer panels of the receiver are designed for radiative heat transfer with their outer surface facing the solar field. The outer surface exposed to insolation is typically coated by a black coating to ensure that high solar absorptivity and low thermal emissivity values are maintained. The backs of the panels are insulated to minimize conductive losses. For the cases where molten salt is utilized, the heated salt is directed to the steam generator located below the receiver but above grade.

The key to a good receiver design is to maximize the heat flux from the receiver surface to the HTF without overheating the heat exchanger tubes or the HTF itself. This requires striking a good balance between the tube material and the heat transfer capability of the HTF as well as minimizing the energy losses, e.g., spillage of incoming solar flux, radiation from the receiver surface, and conduction and convection losses to the supporting structure and the surroundings.

The tower structure consists of either a steel tower of structural members, a cast-in-place cylindrical concrete shell, or a tubular steel structure. The steel design should be capable of supporting piping and other equipment. Within the concrete cylinder is a structural steel frame that is also used to support piping and equipment as well as providing access to the solar receiver. This interior framing carries its own gravity load to the foundation and transfers its lateral load to the concrete cylinder. The design details of the tower structure (height, base dimensions, etc.) are supplier dependent, based on the final size of the receiver, solar field arrangement, and constructability considerations.

For good performance, in stand-alone CSP plants steam conditions are 565–580°C at an outlet pressure of 175+ bar at 100% of *maximum continuous rating* (MCR). The exact conditions are project- and supplier-dependent. There is no reheat in stand-alone CSP applications. The higher efficiency resulting from reheat is justified based on the additional capital cost and increased operational complexity, especially for a cycling unit.

The molten salt to steam generator system consists of multiple heat exchangers. These include the superheater, steam drum, evaporator, preheater, and feedwater heater. In all cases except the feedwater heater, the design is with the water or steam within the tubes and the heated molten salt on the shell side of the exchanger. These heat exchangers are typically located within the tower structure at an elevation sufficient to allow molten salt drainage by gravity to either the hot or cold thermal storage tanks. Materials may range from stainless steels for the superheater and reheater to carbon steel for the steam drum, preheater, and feedwater heater. The evaporator is constructed of alloy steel for corrosion purposes.

12.5.1   Falling Particle Solar Central Receivers

Fundamental thermodynamics dictates that the thermal efficiency of a heat engine cycle is strongly dependent on the temperature of the working fluid at the turbine inlet. With steam as the working fluid, the upper limit to turbine inlet temperature (TIT) is 565–580°C (600°C in modern, advanced steam turbines). When using the molten salt as a heat transfer and energy storage medium, a similar limit applies. For advanced sCO2 cycles planned for CSP applications, however, TIT values as high as 700–715°C are envisioned. While, as the power cycle working fluid, sCO2 is eminently capable of handling such high temperatures, the problem of selecting appropriate receiver and storage heat transfer medium remains. One proposed solution is the use of solid particles to directly absorb and store energy from concentrated sunlight. In particular, the *falling particle receiver* concept was originally conceived in the 1980s. The concept consists of a thin sheet of opaque, highly absorptive ceramic particles, e.g., spherical, sintered bauxite particles, with a sheet thickness of 10–60 particle diameters (particle diameter is of the order of 100 μm), which is cascaded through a beam of concentrated solar energy.

From a purely thermodynamic perspective, falling particle CRS and TES systems are identical to conventional systems operating with steam and/or molten salt. The receiver sits atop a solar tower and receives a concentrated solar beam from a surrounding field of heliostats. The proverbial devil is in the design details of the CRS, storage silos, and particle-to-sCO2 heat exchangers. For more information on these aspects of the system design and particle properties the reader is referred to the paper by Siegel et al. [29] and the report from Sandia National Laboratories [30]. In a typical falling particle system, particle flow from the receiver to the hot storage tank ("hot silo") and the heat exchangers (typically, of moving packed bed and plate design) takes place under the action of gravity. The transfer of particles from the heat exchangers to the cold storage tank ("cold silo") and back to the receiver requires a mechanical system with bins, bin feeders, and hoists.

The industry term for the particles in question is "proppant." Proppants, e.g., sand, treated sand, or artificially manufactured ceramic materials, are used in the fracking industry to keep induced hydraulic fractures open. They are used during or after the process of inducing the actual fracture. Proppants are typically added to a base and thus form the fracking fluid. CARBO Ceramics, Ltd. is one of the leading manufacturers of proppants. The properties of their product specifically developed for energy storage applications, CARBOBEAD, can be found in their brochures available for download on the company website. The material is alumina silicate ceramic with a density of about 3,500 kg/m³. Typical particle bed porosity is about 40% so that the density of the heat transfer "fluid" comprising CARBOBEAD particles is about 2,000–2,100 kg/m³ (with a base fluid of air). The average particle diameter (of the HSP variant) is about 400 μm. The bulk-effective specific heat of the falling particle material is a function of temperature and varies between 950 and 1,100 J/kg-K between 200 and 800°C [29]. Bulk-effective thermal conductivity is also a function of temperature and, to a lesser extent, the base fluid, i.e., air or nitrogen. Typical values of thermal conductivity are between 0.3 to 0.6 W/m-K.

12.5.2 Air Receivers

Pressurized volumetric receivers using air as the HTF have been considered for solar gas turbines (see Section 12.7). Another variant is the "open" volumetric receiver, which has the advantages of technical simplicity and ability to achieve high HTF temperatures. While no commercial plants with this receiver type are in operation or under construction, active research and development is in progress – especially in Europe. The reader is referred to the paper by Stadler et al. [31] for in-depth information and an extensive bibliography.

In a CRS using open volumetric technology and air as an HTF, temperatures of up to 900°C are possible. This enables utilization of a high-efficiency power cycle, e.g., supercritical CO_2 recompression cycle with 750°C TIT. While air is a much more suitable candidate as HTF vis-à-vis molten salt, it is not a good thermal energy storage medium. One particular option for TES is utilization of a "packed bed" with high thermal capacitance, as shown in Figure 12.12.

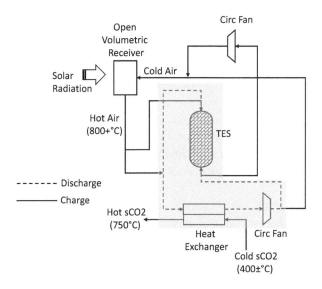

Figure 12.12 Open volumetric air receiver with packed bed TES.

As shown in Figure 12.12, air is heated in the receiver to 800°C or higher. Hot air is used to heat the sCO2 working fluid of the power block to 750°C. The power block can be run during the day (charging) or at night (discharging). During charging, part of the hot air from the receiver is sent to the power block heat exchanger and the remainder is directed to the TES.

There are several types of solid, bed-packed media for TES, including volcanic rock and ceramic bricks. During charging, hot air flows into the TES container (in the case of volcanic rock, a concrete enclosure insulated with aerated concrete block and other materials) and creates a *thermocline* toward the cool end as the rock volume is heated. Cooled air from the cold end of the TES container is moved back to the receiver by the action of circulation (circ) fans. During the discharge mode, the receiver and those circ fans are isolated by the control system. Hot air from the hot end of the TES container is sent to the heat exchanger and cold air from the heat exchanger is circulated back to the cold end. Selection of the storage medium is a question of balancing cost, availability, and performance (i.e., pressure drop of air as it flows from the hot end of the TES block to the cold end).

The primary challenge in the heat exchanger design is the large pressure difference between the sCO2 at nearly 300 bar and air at nearly atmospheric pressure. In a typical design, air flows on the shell side and sCO2 flows through the tubes (or plates) of the heat exchanger. The end result is very high tube wall stress during operation. A related problem is the avoidance of thermal stresses during transients such as startup, shutdown, and load ramps.

12.6   Operational Aspects

The DSSR for HP steam generation is designed with thick-walled components subject to thermal stresses and can negatively impact the life of the unit considering the cycling nature of operation. There is limited experience in the industry for boilers designed for similar conditions. The receiver, with its panels open to the environment, is subjected to more severe cycling vis-à-vis the HRSG of a GTCC plant. This is because the HRSG heat transfer sections are enclosed within an insulated casing, which allows better heat retention during shutdowns. Thus, a daily cycling HRSG is exposed to a large number of hot startups. In contrast, the pressure parts of the solar receiver are subject to larger cooling during nightly shutdowns, and, as a result, the equipment is operated in a predominantly warm (or even cold) startup mode of operation.

The molten salt solar receiver, on the other hand, is subject to much lower operating pressures due to the very low vapor pressure of molten salt at high temperatures. As a result, the minimum wall thickness of the pressure parts is substantially smaller vis-à-vis those in a DSSR. Consequently, this type of receiver is less prone to the fatigue damage due to daily cycling. In any event, an adequate evaluation of the cyclic (fatigue and creep fatigue) life of the DSSR and molten salt solar receiver is of the utmost importance.

Solar receiver panels are designed to withstand high heat fluxes. Under these conditions, an accurate detection of the tube wall temperature is vital. The adequacy of specific instrumentation, although manageable from an availability point of view, may impact the reliability of the installation. One remedy is that the heat flux measurement as well as the tube wall temperature monitoring is based on the readings provided by the infrared (IR) cameras or, possibly, dedicated heat flux monitors installed on the heat transfer surface. To ensure that the heat transfer surface operates within allowable limits, the accuracy of such devices should be as high as possible.

There are certain concerns associated with the accuracy of the IR cameras operating in the desert environment typical for many solar thermal power installations (after all, that is where most sunshine is available). The attenuation associated with the site location, distance from the receiver to the IR camera, angle of aim associated with the elevation of the receiver, clarity of the air, etc. could impact the margin of error of the measurement equipment significantly. Heat flux monitors are typically installed on the heat transfer surface of the receiver and are thus subjected to the elements. This makes them prone to installation errors and difficulties with regular maintenance, which would impact the accuracy of the instrumentation. Thus, IR cameras are to preferred but the accuracy of their readings should be validated by the vendor by performing tests in the environment simulating the conditions typical of the particular site.

During operation, a temperature differential exists along the perimeter of a single receiver tube, located in the heat transfer panel of the solar receiver, because the front part of the tube crown sees high temperature and heat flux, whereas the back of the tube is located in the shadow. The resulting temperature delta generates thermal stresses and create a fatigue mechanism that should be evaluated during the design stage. Obviously, this issue is not unique to CSP because it occurs in conventional boiler furnaces firing all kinds of fuels and HRSGs. Nevertheless, flow distribution flaws in both the molten salt solar receiver as well as DSSR heat exchangers should be carefully analyzed during the design phase.

In steam generator superheaters, a combination of high outlet steam temperature with maximum superheater heat flux generated by the solar field can result in tube wall temperatures exceeding the allowable oxidation limits for widely used materials such as T11, T22, and T91 for the exit panels of the superheater. Thus, the exit panels of the superheater are typically fabricated from high temperature resistant alloys such as Super 304H, Inconel 617, HR6W, or HR3C, which have already been approved by the ASME Code. There is limited experience regarding systems designed for high pressure and temperature and operating in daily cycling mode for CRS steam generator applications. The same criterion is also applicable to molten salt steam generator that will also go through the same cycling operation, with one added complication – the presence of any chlorides in the salt that may result in corrosion. This should be mitigated by specifying a technical grade salt with minimum chloride content.

The routing of the interconnecting piping between the steam drum and evaporator outlet headers can result in a complex, asymmetrical piping network. This can create a concern that associated variances in the pressure drop would impact the characteristics of individual circulation circuits. As a result, the flow through the evaporator may

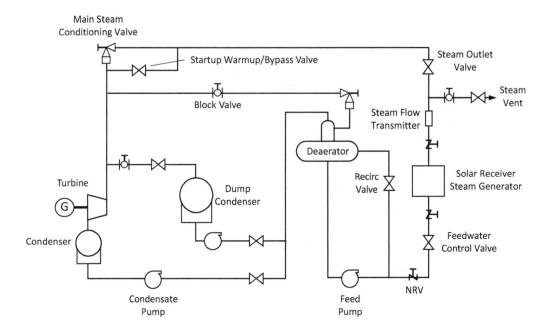

Figure 12.13 CSP steam cycle with a central solar receiver steam generator.

become unstable and pulsations, common to the forced circulation applications, may take place. These flow pulsations may become especially detrimental for such high heat flux application for the CSP tower steam generator. Consequently, designing a symmetrical layout of the evaporator piping, to the extent possible, is imperative to ensure that every sub-circuit of the main circuit would have similar pressure drop.

A process flow diagram of a typical CSP steam system is shown in Figure 12.13. Superheated steam is generated in a *solar receiver steam generator* (SSG) and utilized in a steam turbine for electric power generation. Control of the system is based on three parameters: steam flow, pressure, and temperature. Of these, steam flow rate is the most important (responsive) one. A change in the feedwater flow into the SSG has an immediate impact due to the relatively small volume of fluid and gas within the unit and the absence of a steam drum.

Steam flow rate is controlled by the feedwater pump. The feedwater pump is a single speed centrifugal pump with a variable frequency drive to enable operation in a given pressure range (e.g., 45–85 barg). If there are multiple SSGs, the feedwater control valve regulates the flow entering into each SSG. In the case of an emergency, the SSG can be taken off-line by closing the feedwater control valve.

Steam pressure is the most stable of the operating parameters in the system. During the warmup phase, steam generation pressure is established by using the back pressure valve located at the exit of the SSG. Once the desired pressure is established, the steam outlet valve is partially opened to warm and pressurize the main steam line to the desired level.

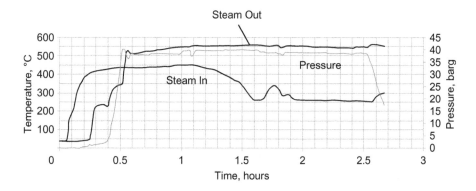

Figure 12.14 Typical run of a CSP plant with LFR field (evaporator) and central receiver superheater.

The superheated steam temperature at the SSG exit can only be altered by changing the flow rate through the SSG. As the feedwater flow rate increases, the temperature of steam is reduced. Similarly, the steam temperature can be increased by reducing the feedwater flow rate. This is a rather primitive form of control and is not very responsive. It is limited by the SSG materials as well as the maximum and minimum flow rates that can be practically achieved through the unit. There is a desuperheating station upstream of the steam turbine to control the steam admission temperature if necessary.

When the steam turbine generator is off-line or if it trips, steam is directed into the dump condenser. A steam pressure reduction can be achieved by venting the steam into the vent stack. This can be used during startup in lieu of the steam bypass. It is a wasteful method and not recommended unless required in an emergency (e.g., a steam turbine trip) or other operational need.

In certain designs, the solar receiver acts as a superheater. Steam can be generated, for example, in a linear Fresnel receiver field (LFR). In that case, feedwater from the feed pump is sent to the LFR field. The output from the LFR field is sent to a separator vessel, the liquid from which is recirculated back to the LFR field. Steam from the separator is sent to the SSG for superheating. If, at any point during normal operation, the flowrate from the separator reduces to a defined minimum (dictated by the steam turbine minimum flow requirement), the LFR bypass flow control valve will allow some saturated water from the separator to bypass the LFR field to be mixed with the steam leaving the separator and entering the SSG. This will allow the minimum flowrate to be maintained and prevent the steam turbine from tripping. Steam temperatures and pressures in a typical run of a CSP plant with LFR field (evaporator) and central receiver superheater are shown in Figure 12.14.

12.6.1 Best Practices and Lessons Learned

For in-depth information on operability concerns, the reader is referred to a recent report by the National Renewable Energy Laboratory (NREL) of the US Department of Energy (DOE) [32].[1] The report covers the best practices and lessons learned from the engineering, construction, commissioning, operations, and maintenance of existing CSP parabolic trough and power tower systems. The report is primarily based on the feedback from the CSP stakeholders in interviews (close to 50 of them). The coverage was extensive, with participants representing nearly 80% of CSP plants operating worldwide. Detailed information obtained in these interviews is included in the report. Furthermore, the authors synthesized the raw data, especially those of broader significance, into best practice recommendations. Two of their findings are related to the concerns expressed above:

• Performance models that accurately model the operation of the plant accounting for transient behavior of the plant, include startups, shutdown, intermittent clouds, and operational transitions are sine qua non. The authors emphasize that these models should be provided by independent third parties that are transparent, have been independently validated, and are accepted by the financial community.
• Plants and equipment must be designed for the transient behavior that they will see. Due to the nature of solar energy, CSP plants are fully expected to cycle multiple times a day. Consequently, plant designers need to understand allowable equipment temperature gradients and make sure that the design of the plant prevents those gradients from being exceeded.

The reader is strongly encouraged to study this comprehensive report (269 pages) for a thorough understanding of the problems encountered in the field, especially in the steam generation and thermal storage systems.

12.7   Other Integration Possibilities

One possibility for integrating solar energy with a gas turbine combined cycle power plant is through the gas turbine itself. Probably the easiest way is to use solar energy for performance fuel gas heating. This will be a relatively easy undertaking and can even be done on a retrofit basis. The conventional method is to utilize the IP feed water extracted from the HRSG in a shell-and-tube heat exchanger to heat the natural gas fuel. The penalty associated with this method is the small drop in steam turbine output via reduced IP steam flow. The benefit in fuel saving, of course, more than makes up for this so that performance fuel gas heating is a standard feature in modern GTCC power plants.

The more difficult option is to replace the combustor of the gas turbine partially or completely with a solar receiver. The resulting system is referred to as a *solarized* or *solar-hybrid* gas turbine. Conceptually, this can be done in two ways. The most obvious and, arguably, the easiest way to extract air from the compressor discharge and reroute it through the solar receiver. This can possibly be done as a retrofit to a

---

[1] NREL is a national laboratory of the US Department of Energy Office of Energy Efficiency & Renewable Energy and is operated by the Alliance for Sustainable Energy, LLC.

recuperated gas turbine such as Solar's Mercury 50 or a vintage E Class machine with a silo combustor. Easier said than done but the only other option is to design a gas turbine from scratch.

Even after solving the problems associated with the architecture of the gas turbine, there are significant design challenges associated with combustor stability and controls. Solar irradiation variations are going to result in significant fluctuations in compressed air temperature in the inlet of the combustor. These fluctuations will cause controls problems relating to combustor sleeve and transition piece cooling, prevention of autoignition, premix ratios, humming, and other instabilities. On top of this, one should also account for the thermal inertia of the solar receiver part of the overall system, which is liable to adversely impact the dynamic response of the gas turbine load ramps and sudden transients such as trips.

There have been several attempts to design and developed pressurized solar receivers for solarized gas turbine applications. One example is the pressurized volumetric solar receiver (REFOS, an acronym derived from the German for *receiver for solar-fossil power plants*) developed within the purview of the European SOLGATE project. In this system, pressurized air is heated in a volumetric absorber placed into a pressurized vessel with a parabolic quartz window for solar radiation incidence. The technology was tested in a 250 kW gas turbine at over 800°C in the early 2000s. Plans were in place to develop ceramic volumetric absorbers to operate at 1,200°C. For a review of pressurized air receivers, including REFOS, the reader is referred to Lubkoll et al. [33]. At the time of writing (2021), there is little expectation that this technology will ever reach a stage where its cost and performance shortcomings (excessive pressure drop, durability of the quartz window, thermal stresses in the combustor, etc.) – also noted in Ref. [33] – are anywhere near to resolution.

Most of the problems weighing down the technology are not that different from those that prevented closed-cycle gas turbines from gaining a commercial foothold. In essence, the external heat addition equipment and piping have to operate at high temperatures *and* pressures while their construction and layout have to be such to minimize the pressure loss of air between compressor discharge and turbine (or combustor) inlet. In that sense, the transition of the concept from paper to the field can be more manageable for a microturbine, e.g., see Kalathakis et al. [34].

There are several other hybridization schemes. In addition to the basic compressed air preheating scheme above, solar energy input can be accomplished by steam injection into the gas turbine combustor. This is a variation on the conventional STIG technology in that the injected steam is generated in a concentrated solar system. There are two options: direct steam generation in either a parabolic trough field or in a solar tower (central receiver). In the case of the solar tower, if the gas turbine is small enough, it can be (at least conceptually) situated at the top of the tower along with the receiver. This would eliminate an expensive piping arrangement to transfer steam from the receiver at the top of the solar tower to the gas turbine on the ground. Solar STIG is essentially a hot day power augmentation scheme (provided that, of course, there is no cloud cover). Steam is generated to augment the gas turbine power output when the sun shines, i.e., an occurrence likely to coincide with a hot summer day. In

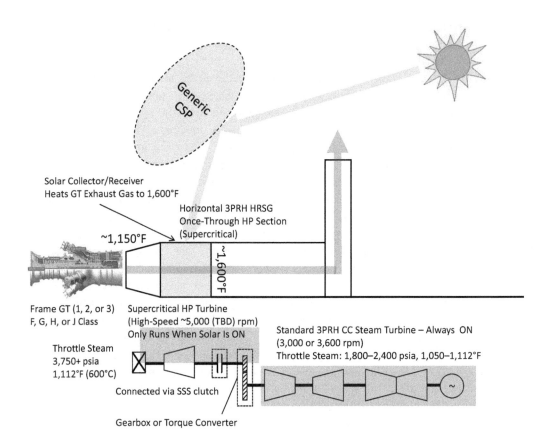

Solar Collector/Receiver
Heats GT Exhaust Gas to 1,600°F

Horizontal 3PRH HRSG
Once-Through HP Section
(Supercritical)

~1,150°F

~1,600°F

Frame GT (1, 2, or 3)
F, G, H, or J Class

Supercritical HP Turbine
(High-Speed ~5,000 (TBD) rpm)
Only Runs When Solar Is ON

Throttle Steam
3,750+ psia
1,112°F (600°C)

Standard 3PRH CC Steam Turbine – Always ON
(3,000 or 3,600 rpm)
Throttle Steam: 1,800–2,400 psia, 1,050–1,112°F

Connected via SSS clutch

Gearbox or Torque Converter

Figure 12.15 Solar "fired" HRSG with a supercritical steam turbine.

the cold winter period or when there is a cloud cover, there is no solar irradiation and so there is no steam generation and no power augmentation. Obviously, combining solar STIG with TES would alleviate this problem but it is highly unlikely that the resulting system would be remotely cost effective.

For numerical evaluation of these hybridization schemes, including different shaft configurations for simple air preheating (e.g., recuperated single-shaft, twin-shaft, and two-spool, three-shaft, intercooled and recuperated), the reader is pointed to another paper by Kalathakis et al. [35]. Another version of STIG covered in the cited paper is what the authors refer to "hybrid" STIG, which is similar in spirit to the ISCC. In this system, there is a single-pressure, simple HRSG utilizing the energy of the gas turbine exhaust gas to produce the injection steam. The HRSG steam generator is assisted by a CSP plant, e.g., a CRS, in exactly the same way as the ISCC. Thus, whether there is solar steam generation or not, there is always steam for injection into the gas turbine combustor. In other words, the system is a hybrid of conventional and solar steam generation.

Finally, another, admittedly somewhat outlandish (maybe), solar hybrid scheme was proposed by the author in a US patent application [36]. The patent describes a novel ISCC concept (see Figure 12.15) comprising two unique components:

- Solar receiver integrated into the exhaust gas path between the gas turbine and the HRSG inlet
- Separate supercritical high-pressure steam turbine (SC-HPST) connected to the main steam turbine (MST) via a gearbox and synchro-self-shifting (SSS) clutch

In combination, these two modifications to the basic GTCC accomplish two objectives:

- Introduction of solar energy into the GTCC bottoming cycle at the *exergetically* most advantageous point
- Utilization of added solar energy for steam production in the HRSG and steam expansion in the ST for useful work generation without being limited by the swallowing capacity of the MST

While the concept eliminates the complicated HRSG modification and solar steam system with intricate piping and controls, its success is ultimately dependent on the successful design, development, and construction of a large volumetric solar receiver where 1,500–2,000 pps exhaust gas from an advanced F, G, H, or J Class gas turbine is heated by concentrated solar rays from 1,150–1,200°F up to 1,600°F. Considering the difficulties associated with a much smaller receiver (e.g., the REFOS discussed above), envisioning a contraption of such a challenging scope could be called foolhardy.

Nevertheless, the performance carrot is rather large. Specifically, with a base GTCC of 56% efficiency, on a hot day (85°F, 50% relative humidity), 64.5% net LHV efficiency is possible in solar firing mode with the proposed scheme shown in Figure 12.15, with only a 1,450°F gas temperature at the exit of the solar receiver. The incremental performance is based on

- Solar thermal input 21% of GT heat consumption
- SC-HPST throttle conditions, 4,250 psia, and 1,111°F

This result indicates that when utilizing a state of the art H or J Class gas turbine with 60% base efficiency in GTCC mode, 68.5% is possible in solar firing mode on a hot day. Similarly, with a 1,600°F gas temperature at the exit of the solar receiver, with an H or J Class GTCC, 70% efficiency is possible in solar firing mode. Detailed calculation results can be found in Ref. [36].

12.8   Nomenclature

12.8.1   Key Parameters

CF = Cost factor (Equation (12.15))
CR = Concentration ratio
DNI = Direct normal irradiation (in MWsol)
EACT = Second law conversion effectiveness
EGTM = Marginal gas turbine efficiency

EOPT = Solar field optical efficiency
EREC = Receiver efficiency
ERINC = Incremental Rankine bottoming cycle efficiency
ESOL = Solar efficiency
ETHM = Solar collector thermal efficiency
EX_IN = Rate of solar thermal exergy input to ISCC bottoming cycle, kW or MW
H = Feed water or steam enthalpy (same subscripts as temperature), Btu/lb
METH = Mean-effective (average) temperature of solar thermal power addition to ISCC bottoming cycle, °F
MSTM = Steam mass flow rate, lb/s
P = Feed water or steam pressure, psia
Q_IN = Rate of solar thermal energy input to ISCC bottoming cycle, kW or MW
S = Feed water or steam entropy, Btu/lb-R
SF = Solar fraction
SM = Solar multiple
TAMB = Ambient temperature, °F
TFW = Feed water temperature, °F
TREC = Receiver temperature, °F
TSTM = Steam temperature, °F
ΔWSTM = Additional ST output resulting from solar thermal power input to ISCC bottoming cycle, kW or MW

12.9   References

1. Glatzmaier, G. 2011. Summary report for concentrating solar power thermal storage workshop, *NREL/TP-5500-52134*, Golden, CO: National Renewable Energy Laboratory.
2. Turchi, C. S., Ma, Z. and Erbes, M. 2011. Gas turbine solar parabolic trough hybrid designs, GT2011–45184, ASME Turbo Expo, June 6–10, Vancouver, Canada.
3. Cox, J. 2010. Implications of intermittency, *Modern Power Systems*, pp. 22–23.
4. Armistead, T. F. 2010. Integrating solar, conventional energy resources, *Combined Cycle Journal*, 2Q, pp. 106–111.
5. Ugolini, D., Zachary, J. and Park, H. 2009. Options for hybrid solar and conventional fossil plants, *Bechtel Technology Journal*, 2(1), 133–143.
6. Camporeale, S. M., Fortunato, B. and Saponaro, A. 2011. Repowering of a Rankine cycle power plant by means of concentrating solar collectors, GT2011–45736, ASME Turbo Expo, June 6–10, Vancouver, Canada.
7. Barigozzi, G., Franchini, G., Perdichizzi, A. and Ravelli, A. 2013. Simulation of solarized combined cycles: Comparison between hybrid GT and ISCC plants, GT2013–95843, ASME Turbo Expo, June 3–7, San Antonio, TX.
8. Horn, M., Führing, H. and Rheinlander, J. 2004. Economic analysis of integrated solar combined cycle power plants: A sample case: The economic feasibility of an ISCC power plant in Egypt, *Energy*, 29, 935–945.
9. Kalogirou, S. A. 2009. *Solar Energy Engineering: Process and Systems*, Burlington, MA: Academic Press.

10. Pihl, E. and Johnsson, F. 2012. Concentrating solar power hybrids – Technologies and european retrofit potential, *ISES Eurosun*, September 17–21, Rejika, Croatia.

11. Petrov, M. P., Salomon Popa, M. and Fransson, T. H. 2012. Solar augmentation of conventional steam plants: From system studies to reality, *World Renewable Energy Forum (WREF)*, May 13–17, Denver, CO.

12. Selig, M. and Mertins, M. 2010. From saturated to superheated direct solar steam generation – Technical challenges and economic benefits, *SolarPaces*, September 21–24, Perpignan, France.

13. Kelly, B., Herrmann, U. and Hale, M. J. 2001. Optimization studies for integrated solar combined cycle systems, *Proceedings of Solar Forum*, April 21–25, Washington, DC.

14. Schenk, H., Hirsch, T., Feldhoff, J. F. and Wittmann, M. 2012. Energetic comparison of linear fresnel and parabolic trough collector systems, *ES2012–91109, ASME 6th International Conference on Energy Sustainability*, July 23–26, San Diego, CA.

15. Kutscher, C., Burkholder, F. and Stynes, K. 2010. Generation of a parabolic trough collector efficiency curve from separate measurements of outdoor optical efficiency and indoor receiver heat loss, *NREL/CP-5500-49304*, Golden, CO: National Renewable Energy Laboratory.

16. CSP Today. 2013. CSP solar tower report: Cost, performance and key trends.

17. CSP Today. 2012. CSP parabolic trough report: Cost, performance and key trends.

18. Thermoflow, Inc. GT PRO, GT MASTER Version 23.0, Southborough, MA.

19. Bohtz, C., Gokarn, S., Conte and E., 2013. Integrated solar combined cycles (ISCC) to meet renewable targets and reduce CO_2 emissions, *PowerGen Europe*, 4–6 June, Vienna, Austria.

20. Ojo, C. O., Pont, D., Conte, E. and Carroni, R. 2012. Performance evaluation of an integrated solar combined cycle, *GT2012–68134, ASME Turbo Expo*, June 11–15, Copenhagen, Denmark.

21. Gas Turbine World. 2013. *Handbook*, Volume 30, Fairfield, CT: Pequot Publishing, Inc,.

22. Gülen, S. C. 2013. What is the worth of 1 Btu/kWh of heat rate?, *POWER*, June, 61–63.

23. Mehos, M. S., Finch, N. S., Ho, C. K., Turchi, C. S. and Wagner, M. J. 2012. Probabilistic analysis of power tower designs and technology improvement opportunities to meet sunshine goals, *SolarPaces*, September 11–14, Marrakech, Morocco.

24. Dersch, J., Geyer, M., Herrmann, U., et al. 2004. Trough integration into power plants – A study on the performance and economy of integrated solar combined cycle systems, *Energy*, 29, 947–959.

25. Glatzmaier, G. 2011. Developing a cost model and methodology to estimate capital costs for thermal energy storage, *NREL/TP-5500-53066, National Renewable Energy Laboratory*, Boulder, CO.

26. Spelling J, Jöcker M and Martin A. 2011. Thermal modeling of a solar steam turbine with a focus on start-up time reduction, *Journal of Engineering for Gas Turbines and Power*, 134 (1), #013001.

27. Emerging Energy Research, LLC. 2011. US solar power: Markets and strategies: 2011–2025 (Market Study Excerpt, June).

28. Bechtel. 1999. Topical report on the lessons learned, project history, and operating experience, Solar Two Project, Daggett, CA, Bechtel Document 30C-R-013, February 15 (unpublished).

29. Siegel, N. P., Gross, M. D. and Coury, R. 2015. The development of direct absorption and storage media for falling particle solar central receivers, *Journal of Solar Energy Engineering*, 137, 041003.

30. Carlson, M. D., Albrecht, K. J., Ho, C. K., Laubscher, H. F. and Alvarez, F. 2020. High-temperature particle heat exchanger for sCO2 power cycles, Sandia Report, SAND2020–14357.
31. Stadler, H., Tiddens, A., Schwarzbözl, P., et al. 2017. Improved performance of open volumetric receivers by employing an external air return system, *Solar Energy*, 155, 1157–1164.
32. Mehos, M., Price, H., Cable, R. et al. 2020. Concentrating solar power best practices study, *Technical Report, NREL/TP-5500-75763, National Renewable Energy Laboratory*.
33. Lubkoll, M., von Backström, T. W. and Kröger, D. G. 2014. Survey on pressurised air receiver development, *SASEC 2014, 2nd Southern African Solar Energy* Conference, 27–29 January, Port Elizabeth, South Africa.
34. Kalathakis, C., Aretakis, N. and Mathioudakis, K. 2018. Solar hybrid micro gas turbine based on turbocharger, *Applied System Innovation*, 1, 27.
35. Kalathakis, C., Aretakis, N., Roumeliotis, I., Alexiou, A. and Mathioudakis, K. 2017. Assessment of solar gas turbine hybridization schemes, *Journal of Engineering for Gas Turbines and Power*, 139(6), 061701.
36. Gülen, S. C. 2015. Solar fired combined cycle with supercritical turbine, US Patent 20150128558 A1, Published on May 14.

13   Coal Redux

13.1   Integrated Gasification Combined Cycle

Once touted as a "savior" of coal, at the time of writing (late 2020), especially after the debacle in Kemper IGCC, this particular "clean coal" technology is unlikely to make a comeback. Nevertheless, for the sake of completeness, it deserves a brief look. One reason for that is that *gasification* is the commercially available technology for syngas production from all forms of hydrocarbon feedstock, i.e., not only coal, as well as biomass. (Of course, coal gasification is the earliest use of the technology and goes back to the 1800s when it was used to make "town gas" for local cooking, heating, and lighting.)

Gasification is the substoichiometric reaction of coal with oxygen and steam under high pressure and temperature to form a gaseous product consisting primarily of carbon monoxide (CO) and hydrogen (H_2). The gasification process typically takes place at temperatures of 1,200–1,400°C (2,200–2,600°F) and pressures of 30–60 bar (435–870 psi). The resulting gas contains CO, H_2, CH_4, and other fuel constituents, which can be referred to as the desirable products (i.e., suitable to be used as GT fuels) in addition to undesirable products such as CO_2, H_2S, NH_3, particulate matter, and neutral products such as N_2 and H_2O vapor. For detailed coverage of the fundamental physical and chemical principles and gasification technologies, see Probstein and Hicks [1], Higman and van der Burgt [2] and Rezaiyan and Cheremisinoff [3].

When the destination of the gaseous product of gasification is a gas turbine combustor, the entire system is referred to as *integrated gasification combined cycle* or IGCC (because the power block is generally a gas turbine combined cycle). For a concise but well-rounded review of IGCC technology with different gasifiers, the reader is referred to the book by Khartchenko [4]. A simplified method for IGCC performance estimation can be found in Gülen and Driscoll [5] and in chapter 18 in **GTCCPP**.

The IGCC power plant has been the most promising clean coal technology for the last four decades. Several large-scale commercial IGCC power plants have been in operation for many years to the point that the technology can safely be considered field proven. For a brief but informative discussion of these plants, refer to the article by Barnes [6]. Based on the experience gained from those plants, their cost and staggering complexity, resulting in extremely high CAPEX and OPEX, prohibited the technology from becoming a commercial success. (Indeed, the IGCC plants in

Europe, North America, and Japan have accumulated tens of thousands of hours of operation with a range of coals and other feedstocks.) Under the light of recent developments, i.e., availability of shale gas at very low prices, its inherent "cleanliness" vis-à-vis coal in terms of emissions of criteria pollutants and CO_2 and the amenability of the GTCC power plant to post-combustion carbon capture, have pretty much put the last proverbial nail on coal's proverbial coffin even in an IGCC configuration. (There might still be a future for the coal-fired IGCC in China and, maybe, in India.) For a brief coverage of IGCC technology and the prognosis regarding its future, refer to the gasification section in the handbook chapter by the author [7].

13.2   IGCC Complexity

The hopelessness of a cost-effective and satisfactory solution of the complexity problem can be easily seen from the highly simplified conceptual schematic of a generic IGCC power plant shownin Figure 13.1. In essence, the IGCC is a GTCC power plant with a complex fuel skid, which converts a solid fuel such as coal or biomass (or a liquid fuel such as heavy oil) into a gaseous fuel, syngas, or *substitute natural gas* (SNG). As such, it comprises three separate plants (sometimes referred to as "blocks"):

1. Power block (gas and steam turbine combined cycle power plant)
2. Oxygen plant (air separation unit or ASU) – generates O_2 used in the gasifier[1]
3. Process block (gasification and cleanup)

Furthermore, there is a high degree of interconnection between these three plants or blocks. In particular:

1. Syngas fuel from the gasification block to the power block
2. Diluent N_2 from the ASU to the gas turbine combustor
3. Air from the gas turbine to the ASU (to reduce the main air compressor load)
4. Feedwater from the heat recovery steam generator (HRSG) to the gasifier
5. Steam from the gasifier to the HRSG
6. Miscellaneous feedwater and/or steam exchange between the blocks, e.g.,
   a. Syngas moisturization
   b. Water-gas shift reaction
   c. Syngas cooling

These interactions are illustrated in a generic block diagram in Figure 13.2.

Each of the three individual plants or blocks depicted in Figure 13.1, on their own, represents mature technology with tens or even hundreds of thousands of hours commercial operation under its belt. Cryogenic plants for manufacturing oxygen

---

[1] Some gasifier technologies use air instead of O_2, but these are generally not considered for IGCC (except the technology developed by MHI in Japan).

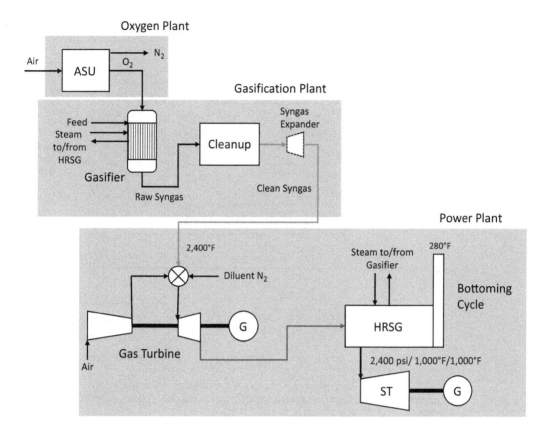

Figure 13.1 Simplified schematic of a generic IGCC power plant.

and acid gas scrubbing units are mainstays of the chemical process industry (CPI). Similarly, gas and steam turbine combined cycle power plants have nearly five decades of design, operation, and EPC experience in electric power generation across the world. The same is true of gasification technology. In a 2014 paper by Higman [8], the number of gasification projects in 2014, according to the *GTC Worldwide Gasification Database*, was stated as 862 projects, consisting of 2,378 gasifiers (excluding spares), of which 272 projects with 686 gasifiers were active commercial operating projects. The output of operating gasifiers was 116.6 MWth with 82.8 and 109.2 MWth in the construction and planning phases, respectively. Chemical production is the most important application of gasification, followed by fuel production (gaseous and then liquid), with power generation a distant fourth. Coal is the dominant feedstock in capacity and actual production. Petroleum, petcoke, and biomass constitute just a fraction of coal. In terms of gasification technology, GE-Texaco and Shell (both now owned by Air Products) occupy the first two spots, followed by Lurgi.

In conclusion, the ingredients are proven, mature technologies but the final mix, the IGCC, is far from being so even though almost four decades went into its development. At first glance, this may be a vexing problem for the engineers and developers.

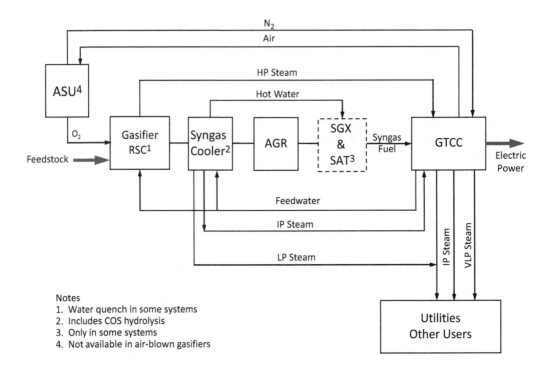

Figure 13.2 IGCC block diagram with key process streams (RSC: Radiant Syngas cooler, AGR: acid gas removal, SGX: Syngas expander, SAT: Syngas saturator (moisturizer), ASU: air separation unit, VLP: very low pressure)

However, this outcome is not surprising at all and can be explained by an analogy. Consider a diesel truck engine. As a prime mover, it is probably the most mature technology that one can think of. With regular (scheduled) maintenance and proper usage following the OEM's guidelines, a diesel engine can serve its owner for hundreds of thousands of miles without a single glitch. Now, consider that one includes a mini fractional distillation tower in the same truck, which uses the crude oil as the feedstock. Diesel fuel is used in the truck engine. Other products are either stored or burned in the furnace. Exhaust gas is used in the furnace as combustion air (possibly with supplementary air blown in). This crazy contraption would look something like that shown in Figure 13.3 (components not drawn to scale). The futility of this scheme is obvious even to a layman. While there is no reason that it shouldn't work (on paper, that is), its complexity, size, weight, cost and, before all of them, control system design difficulty for an application that requires operation over a wide range of torques and speeds would render it impossible. Yet, when one really thinks about it, what is being aspired to in an IGCC power plant is not too different from this exercise in futility.

Conceptual considerations similar to those above along with field experience make it abundantly clear that the only way to make the IGCC a technology with high reliability and availability is to minimize the interaction between the main process and

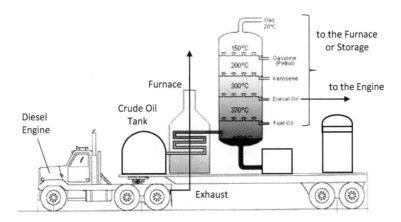

Figure 13.3 Diesel engine truck with diesel fuel production as an analogy to the IGCC power plant.

power blocks (as shown in Figure 13.2) as much as possible. Even then, whether the resultant product will be cost-effective or not remains to be seen. It can only be determined after a full-blown FEED study. We will therefore focus on the control and operability features of the generic IGCC technology as depicted in Figures 13.1 and 13.2.

13.2.1   A Historical Perspective

As it turns out, the "crazy" example of a diesel truck with its own gasifier and fuel generation plant in Figure 13.3 is not that crazy after all. During World War II, almost every motorized vehicle in continental Europe was converted to use firewood. They were known as "wood gas cars" or "producer gas cars." In Germany alone, around 500,000 wood gas cars were in operation by the end of the war[2] (see Figure 13.4). Apparently, even today, trucks running on wood gasifiers are in use in North Korea.[3]

The development of wood gasifiers can be tracked to the early 1900s. The process is very simple. Wood chips are burned incompletely in a fire box, producing wood gas, solid ash, and soot. Solid byproducts are removed periodically from the gasifier. The wood gas is filtered for tars and soot/ash particles, cooled, and directed to an internal combustion engine as fuel. Gas composition and heating value is a function of the gasification process and input stream properties, i.e., oxidizer (air, oxygen, or steam) and feedstock moisture. For more information, the reader is pointed to the article cited in footnote 2.

---

[2] From *Low Tech Magazine*, https://www.lowtechmagazine.com/2010/01/wood-gas-cars.html; accessed November 21, 2021.

[3] From *The Atlantic*, https://www.theatlantic.com/photo/2013/01/a-look-inside-north-korea/100432/, accessed November 21, 2021.

Figure 13.4 German wood gas car (Bundesarchive, Bild 183-V00670A /CC-BY-SA 3.0).

13.3   Operability Considerations

First, let us start with the basics. The product of the IGCC power plant is electric power supplied to the grid. This power is generated by the prime mover generators of the GTCC power plant, which is burning syngas fuel in the gas turbine combustors and, in some cases, in the HRSG duct burners. (In gasifiers with quench cooling, there might be a small contribution from the syngas expander.) Consequently, the master controller of the GTCC power plant generates the syngas and combustor diluent (usually N_2 from the ASU) flow demand based on the power generation set-point.

The syngas demand signal (a feed-forward signal) generated by the GTCC controller is the set-point for the gasifier master controller, which then generates the feedstock and O_2 demand to generate the requisite syngas flow. The O_2 demand, of course, becomes the set-point of the ASU master controllers. These three master controllers continuously interact with each other so that syngas production and consumption is matched to the level required for the desired electric power output.

Let us now consider the details within this very simple process. GTCC syngas flow indirectly determines the O_2 demand from the ASU, which then determines the requisite airflow rate. This, however, sets the N_2 rate of production automatically. If this rate of production is equal to or less than the diluent N_2 requirement of the gas turbine combustors at the power output, there is no problem. Otherwise, this generates feedback to the GTCC master controller so that the power output is pulled back until syngas, O_2, and N_2 flows reach an equilibrium. Obviously, this should not be a problem during operation at or near the design point because, as the name itself suggests, the entire system is designed for that performance point. It is, of course, quite possible, in theory and in practice, to replace N_2 with steam as the diluent, either fully, as a backup, or as a co-diluent. This, however, comes at the expense of reduced thermal efficiency (because steam is "robbed" from the steam turbine) and increased system complexity and cost (additional piping, valves, and controllers). Field experience from the existing IGCC power plants so far has shown that minimizing system and control complexity is vital for the feasibility of IGCC power plants.

At off-design operation conditions, especially at loads lower than 100%, and important plant transients such as startup, load ramps (up or down), and shutdown, this control system interaction can become very complicated because of the operating envelope limits of the gasifier and the AGR. Especially during load down ramps and shutdown, the process should be controlled at a precisely determined rate to prevent gasifier upsets and syngas flaring. Although there is no change in temperature or pressure inside the gasifier vessel at gasifier loads between 50% and 100%, care must be taken to ensure that the control ramps do not disturb the oxygen/carbon (O/C) ratio during the transients. Even a small deviation in the O/C ratio can create a substantial change in the operating temperature. The same complexity arises during unplanned transients such as gasifier, ASU, or prime mover trips. In the event of a gas turbine trip, for example, the combined cycle output of a $2 \times 2 \times 1$ GTCC power plant drops to 50% and the gasifier load is reduced at the fastest possible rate (4% per minute as stated by Pisacane et al. [9]). In that case, excess syngas generated in the gasifier must be flared.

As discussed in some detail in Barnes [5], a high degree of interaction, especially between the gas turbines and the ASU, was the main source of operability problems encountered during startup, commissioning, and the early years of operation in commercial IGCC plants (e.g., Puertellano in Spain). Lengthy startup times and long plant stabilization periods, as long as five days, led several operators to the conclusion that a startup air compressor was needed for the ASU. (Note that IGCC gas turbines have to be started with natural gas before switching to syngas and it takes time for them to get ready before they can send air to the ASU to reduce the main air compressor load.)

Thus, prima facie, for high availability and reliability, either to eliminate the ASU altogether by using an air-blown gasifier (e.g., Nakoso IGCC in Japan) or to have a stand-alone ASU (as in Tampa Polk IGCC) should be seriously considered. For the differences between the O_2-blown and air-blown technologies (e.g., equipment size and CAPEX), the reader can consult the references cited at the beginning of the chapter. Reduced integration is likely to result in lower efficiency, but, so far, operational experience has shown that trying to squeeze the last decimal point in efficiency from a highly complicated system is probably not a good approach to plant design with high RAM. Whatever gains/savings are predicted on paper during the design phase are almost guaranteed to be squandered during operation.

Another source of problems encountered during design and operation of the IGCC power plants, with their staggering complexity, is the expectation that these plants are to be cycled just like a conventional natural gas–fired GTCC power plant. As mentioned above, the ASU presents itself as a bottleneck for startup times and load ramp rates for the IGCC power plant. A proposed remedy for this is integrating liquid oxygen and liquid nitrogen storage into the control system for oxygen supply to the gasification and separate startup air compressor. This, of course, will add significantly to the installed cost of an already expensive technology.

Gasifier vessels are massive structures not readily amenable to fast transients. Cold starts for a membrane wall gasifier typically take about 2 hours. For refractory lined gasifiers, heating up the refractory requires 24–30 hours. Maintaining the refractory in a hot standby condition during plant down times can reduce this time to 2 hours. For single-burner gasifiers, 50% turndown is possible, but the system must be specifically designed for that requirement. For multiple-burner gasifiers, a lower turndown ratio is possible.

Gasifiers used in CPI and refineries are not designed for large turndowns or fast transients. (This is not surprising because there is really no need for that type of performance in those plants.) As far as those used in IGCC plants are concerned, there is scant data available. Certainly, the flexibility characteristics of a given plant are expected to be a function of the gasifier technology, oxidant type, syngas cooling, and cleanup methods. Pisacane et al. cite 2% per minute [9]. In another article, ramp rates over the range of 50%–100% load are stated to fall in the range of 3–3.5% per minute [10]. In the same article, the AGR unit is cited as another bottleneck because at ramp rates higher than 2–3%, there is a potential risk of losing the syngas quality. Another item requiring attention is the change of temperature profile in the water-gas shift reactor bed during rapid turn-down. Occasional occurrences are not of undue concern; however, if such ramping is performed on a regular basis the impact on the reaction catalyst can be detrimental. Higman et al. suggest the use of an *isothermal* reactor instead of an *adiabatic* reactor with intercooling as a potential solution [10].

The difference in load ramp from 50% to 100% between a natural gas–fired GTCC and IGCC is depicted in the chart in Figure 13.5 [10]. One technology cited to enhance the ramp rate of the IGCC was the "natural gas boost," which refers to the short-term addition of natural gas (or other storable fuel such as LPG or methanol) to the gas turbine fuel supply to improve the overall IGCC ramping performance beyond the limitation of syngas ramping [10].

A plant unloading event from 100% load to 50% in a $2 \times 2 \times 1$ IGCC with HRSG duct firing is shown in Figure 13.6 [9]. In this case, the plant load is reduced at 2% per minute.

13.3.1   A Real-World Example

There is nothing like seeing something in the proverbial "flesh" and in operation with your own eyes. Anybody with passable mechanical skills, a big backyard, some spare cash, and ample spare time (for reading technical references, tinkering, assembling or reassembling parts, and experimenting) can build a home-made gasifier and learn everything about the system (and more). There are videos online

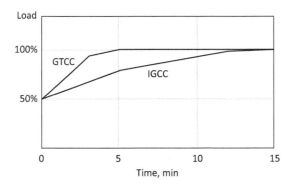

Figure 13.5 Load ramp characteristics of natural gas–fired GTCC and IGCC.

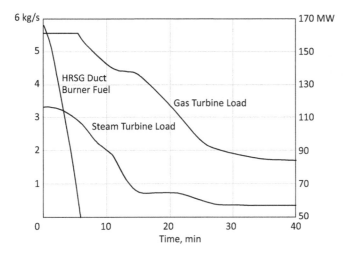

Figure 13.6 Load ramp (down) characteristics of an IGCC with fired HRSG.

describing such endeavors. One that is particularly liked by the author is "Using Wood to Fuel a Generator! (How to Build a Wood Gasifier w/Demonstration)" by a *YouTube* poster[4] (see Figure 13.7). I leave it to the reader to watch the video (or others if they find them) and learn. This may be a first (maybe even a *sacrilege?*) in a technical treatise. However, I guarantee that it is much more instructive than reading hundreds of pages full of impenetrable formulae and charts – including mine in the preceding paragraphs.

---

[4] See the link https://www.youtube.com/watch?v=AyTqo4mCUUY for the YouTube video. You can also check out the poster's own website https://www.instructables.com/How-a-Wood-Gasifier-Works-and-How-I-Built-Mine-Out/ for more details on the construction of his wood gasifier. Both were last accessed on November 21, 2021.

Figure 13.7 Do-it-yourself wood gasifier.

13.4   Coal to Gas

The exercise in futility touched upon in Section 13.1 also points the way to the only viable solution to burning coal (or a similar feedstock) "cleanly" for utility-scale electric power generation in a gas turbine power plant. In an analogy to a large refinery producing different grades of fuel from crude oil for distribution to the points of end use, a large gasification plant can be run at base load, around the clock to produce *substitute* (or *synthetic*) *natural gas* (SNG) and/or hydrogen (with carbon capture and sequestration), which is then distributed nationwide via the existing pipeline network. Ideally, such plants are located in locations near large coal reserves with easy access to the pipeline network. Breaking the connection with the end user, i.e., the GTCC, and base load operating regime without load ramps and minimal start-stop cycles ensures that such a plant will operate with high reliability and availability.

This process is demonstrated in the *North Dakota Gasification Plant* in Beulah, ND, where approximately 18,000 tons per day of lignite is converted to 160 million standard cubic feet of methane. The calculated thermal efficiency is 61.9% for conversion of the heating value of lignite to the heating value of the methane produced. The $2.1 billion plant began operating in 1984. Lignite is gasified in 14 Lurgi Mark IV gasifiers. The raw gas is cooled after it exits the gasifier, removing tar, oils, phenols, ammonia, and water via condensation from the gas stream. These products are then purified, transported, or stored for later use as fuel for steam generation. After cooling, the gas is further treated to remove impurities.

Following cleanup, the syngas gas is sent to a methanation unit where CO and most of the remaining CO_2 are reacted over a nickel catalyst with free hydrogen to form methane (CH_4). The product gas, i.e., SNG, is then further cooled, dried, and compressed. The compressed gas with a heating value of 975 Btu/ft^3 (about 36 MJ/m^3) leaves the facility for sale. The product joins the *Northern Border Pipeline*, which supplies natural gas to the eastern United States. Incidentally, this project also captures some of the CO_2 from the process and sells it to Canada for EOR. A simplified process diagram is shown in Figure 13.8. To achieve the correct stoichiometry in the methanation unit, some of the CO_2 in the raw syngas is removed in the AGR (Rectisol®) unit. To meet the SNG specification, a final Rectisol unit can be used for additional CO_2 removal. If the methanation unit is replaced by a *pressure swing adsorption* (PSA) unit, the end product is hydrogen. In that case, the gasifier can be replaced by a POX reactor. The AGR can be a Selexol™ unit.

Another technology for hydrogen production is hydromethanation, also known as *direct gasification*, in which methane is produced through the combination of steam, coal, and a catalyst in a fluidized bed reactor. The advantages of hydromethanation are twofold: (i) no ASU is required and (ii) gasification and the water-gas shift reaction are combined in a single unit. The disadvantage of the process is the difficulty of separating the catalyst from ash/slag and the loss of reactivity of the catalyst. For thermodynamic details of these processes, the reader is referred to the paper by Chen et al. [11].

To get an idea about the magnitudes of product yields, let us look at gray or blue H_2 from gasification of coal or another solid or liquid hydrocarbon feedstock. Unfortunately, in this case, feedstock composition and key processes are not amenable to a quick analysis using only first principles. Nevertheless, using some reliable numbers from an example in Higman and van der Burgt [2], a pretty good idea can be obtained. The example is a hydrogen plant defined as follows:

- Feedstock is visbreaker residue (20,420 kg/h)[5]
- The feedstock and oxygen are sent to a POX reactor (the flow ratio is roughly 1:1)
- After the raw gas shift, a Selexol™ process is used to separate H_2S and CO_2 (42,988 kg/h) from the syngas
- Clean gas is sent to a PSA block to generate hydrogen (4,464 kg/h)

Hydrogen produced by this process can be used in a GTCC to generate electric power. Assuming a net 60% LHV efficiency, the specific output is calculated as 72 MJe/kg. For a fair power accounting, one should subtract from this the number of ASU (to produce the oxygen used in the POX reactor and Claus unit) and the gasifier power consumption. Let us assume that the net specific output thus drops to 60 MJe/kg. This translates into net CO_2 emission of 578 kg/MWh.

---

[5] A visbreaker is a thermal (i.e., non-catalytic) oil refinery process unit, in which the quantity of residual oil produced in the distillation of crude oil is reduced and the yield of more valuable middle distillates (e.g., heating oil and diesel) are increased. The name refers to the reduction of the viscosity of the feedstock (residual oil). Residues from the visbreaking process include tar and coke. They are used for different purposes like roofing and the manufacture of dry cells.

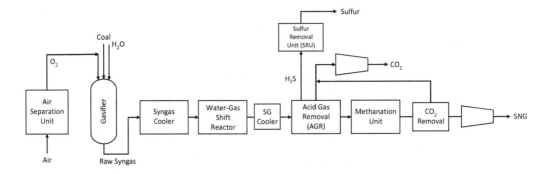

Figure 13.8 Block diagram of a coal to SNG plant.

The feedstock for this gasification process, visbreaker residue, is pretty nasty stuff. Its typical composition is about 85%(w) carbon (C), 10%(w) hydrogen (H), and 4% sulfur (S) with vanadium, nickel, iron, sodium, and calcium in the mix. Its density is 1,100 kg/m^3 and the LHV is about 39 MJ/kg (from table 4.10 on p. 62 in Ref. [2]). Under normal circumstances, this residual heavy oil is not suitable as a gas turbine fuel. In the old days, such ash-bearing fuels were utilized for marine, refinery, and oil-field applications. Significant pre-treatment (e.g., washing) and use of additives (inhibitors) were required to prevent turbine hot gas path deterioration. The reader is referred to the GE white paper by Kaufman for details [12]. A further difficulty presents itself in the HRSG with exhaust gas containing high levels of sulfur. There is only one GTCC power plant in the world designed to burn refinery bottoms for power and steam cogeneration: *Kalaeloa Cogeneration Plant* in Hawaii, USA.[6] Needless to say, its performance was not stellar, and the operators had a tough time with extreme fouling of HRSG tube banks due to sulfuric acid formation.

Consider the most optimistic assumptions, i.e., a GTCC based on an E or (maybe) F Class gas turbine with

- fuel preparation with washing and additives,
- diffusion flame combustors with steam injection, and
- a two-pressure, no-reheat HRSG with high stack temperature (to prevent sulfuric acid condensation on economizer tubes).

A quick calculation, using Thermoflow, Inc.'s GT PRO heat and mass balance simulation software, came up with the following results:

- 7E.03 gas turbine (nominal TIT 1,134°C)
- Two-pressure, no-reheat bottoming steam cycle (100 bara, 521°C main steam)
- Steam turbine heat rejection to water-cooled condenser (48 mbar vacuum) with a mechanical cooling tower

---

[6] See the Combined Cycle Journal article for detailed information on this particular power plant, https://www.ccj-online.com/2q-2009/kalaeloa-cogeneration-plant-continual-improvement-defines-black-oil-fired-combined-cycle/ (last accessed on November 22, 2021).

- 161°C stack temperature
- 132,414 kWe net with 25,320 kg/h fuel flow (visbreaker residue)
- 18.8 MJe/kg specific power
- CO_2 in stack gas 79,124 kg/h (597.6 kg/MWh)
- SO_2 in stack gas 2,024 kg/h (15.3 kg/MWh)

Improvements can certainly be made but, clearly, at around 20 MJe/kg specific power with 600 kg/MWh stack CO_2 emission (not to mention a lot of SOx), utilizing ash-bearing fuels such as crudes, residuals, and mixtures thereof directly for electric power generation is not a good option. As mentioned, so far there has been only one such plant that went into commercial operation and the field experience has been quite painful.

If we return to the gasifier example above,

- if one uses the 20,420 kg/h of the gasifier feedstock directly as fuel in a GTCC power plant, the net electricity generation would be (20,420/25,320) × (132,414/1,000) = 106.8 MWe with 598 kg/MWh CO_2 and 15.3 kg/MWh SOx.
- net GTCC output with 100% H_2 fired GTCC (60% net LHV efficiency), using the gray H_2 generated by gasification, would be (4,464/3,600) × 60 = 74.4 MWe with 578 kg/MWh CO_2 but pretty much no SOx.

The first option is not realistic. However, the viability of the second one, with the addition of CC to reduce CO_2 emission to, say, 57.8 kg/MWh (90% capture), i.e., "Blue H_2," depends on the $/MW CAPEX (when everything is thrown in, i.e., from the raw feedstock to the final megawatt-hours of electricity delivered to the grid), OPEX, and $/MWh LCOE.

In conclusion, if the CAPEX and OPEX numbers can be made to work out, gasification of a "not green" feedstock such as coal, petcoke, refinery residues (or biomass for that matter), with carbon capture and storage to produce blue H_2 can be a viable way to ensure clean, sustainable electric power generation.

13.5 Magnetohydrodynamics

About 50 years ago coal was declared ("unquestionably" at that) as the fossil fuel of the future. Back then, although nearly 90% of all fossil fuel reserves were in the form of coal, oil and natural gas together accounted for close to 80% of the US energy consumption. However, at the time, it was thought that the country faced the danger of running out of natural gas. What prevented coal from taking over was the high sulfur content (*flue gas desulfurization* systems were in their infancy) and environmental damage from mining, which were both big concerns preventing coal-fired plants from wider deployment. Half a century later, at the time of writing this book, "the fossil fuel of the future" is in its death throes and natural gas (especially after the fracking revolution) is the new king of the fossil fuel realm.

In its heyday, the two technologies that could bring coal to eminence so that the ample coal reserves in the country could be utilized cleanly and safely for power generation were gasification, i.e., making syngas or SNG from coal, and coal-fired *magnetohydrodynamics* (MHD) generator technology. Of the two technologies, gasification, especially in the form of IGCC, made a valiant attempt at technical and commercial maturity but failed (see Sections 13.1 and 13.2). MHD, on the other hand, despite its attractiveness from efficiency and emissions perspectives, did not make it beyond the R&D stage. In the USA, extensive work was conducted by the DOE in collaboration with industry from 1987 to 1994 in a *proof of concept* (POC) program, which included a 50 MWth coal-fired MHD generator operating in Butte, Montana, and component tests for a coal-fired MHD bottoming (steam) cycle at the University of Tennessee.

It is probably premature to declare MHD a dead-and-gone technology. At least, the US DOE's NETL is still looking at the technology for improved efficiency for fossil-fuel power plants and reducing the costs of implementing carbon capture (by combining MHD with oxy-combustion).[7] Thus, a brief look is provided herein at MHD and its "Oxy-MHD" variant, which can rejuvenate this dormant technology for utilizing an otherwise "untouchable" fuel by burning it cleanly and capturing the CO_2 from the flue gas.

In a gas turbine, the thermal energy of hot gas (i.e., combustion products) is first transformed into kinetic energy in the stator and then imparted on the rotor (blades attached to a disc) resulting into shaft rotation. In the ac generator, the rotational (kinetic) energy of the shaft is transformed into electrical energy. In the MHD generator, the kinetic energy of the hot (ionized) gas is converted directly to electric energy as it expands in a channel formed by the magnetic poles (the magnetic field). The output of the generator is *direct* current (dc) and is converted to *alternating* current (ac) in an inverter and supplied to the grid.

The operating principle of MHD is elegant and simple. An ionized (i.e., electrically conducting) gas is produced at high pressure by the combustion of a fossil fuel, e.g., coal, natural gas, or any hydrocarbon feedstock. The fast-moving, ionized gas flows through a strong magnetic field. Oppositely charged particles are deflected away from each other in the field as described by the *Lorentz equation* and collected on electrodes. The current density (force) in A/m^2 through the electrodes connected to an external load is given by the Lorentz equation,

$$J = \frac{i}{A} = \sigma(E - uB),$$

where i is the current through the electrodes (in amperes, A) with a total area of A in m^2, σ is the particle *conductivity*, which is the inverse of *resistivity* ρ in Ωm, u is the particle velocity in m/s, B is the strength of the magnetic field in T (in a direction perpendicular to the particle velocity),[8] and E is the external (load) voltage drop per

---

[7] From an NETL article dated June 17, 2020 on https://www.netl.doe.gov/node/9790 (last accessed on December 27, 2021).

[8] Tesla, T, is the unit of magnetic induction or magnetic flux density in SI units and is equal to one weber (Wb) per square meter. One Wb is equal to one J/A or one $\Omega \cdot C$ (C is one coulomb) or one V·s.

unit length of interelectrode gap width in V/m (perpendicular to the B-u plane). Defining a load factor $0 \leq K \leq 1$, such that

$$K \equiv \frac{E}{uB},$$

where the denominator quantifies the voltage generated in the MHD, the current density becomes

$$J = \sigma u B (1 - K).$$

Electrical power delivered to the load per unit volume in W/m^3 is defined by

$$P = -J \cdot E,$$

so that

$$P = \sigma u^2 B^2 K (1 - K).$$

The conversion efficiency for maximum power density can be shown to correspond to $K = 0.5$, i.e.

$$P_{max} = \frac{1}{4} \sigma u^2 B^2.$$

Thus, to maximize MHD power density, high gas velocities (reaching supersonic levels via expansion through a *Laval nozzle* and diverging channel) and strong magnetic fields are required. (Ionized gas conductivity is relatively low compared to the product of the other two.) For example, for $u = 1{,}000$ m/s, $B = 3.5$ T, and $\rho = 0.05$ Ωm, P_{max} can be calculated as

$$P_{max} = 0.25 \times (1/0.05) \times (1{,}000 \times 3.5)^2 / 10^6 = 61.25 \text{ MWe}.$$

The maximum possible conversion efficiency has been difficult to achieve in the pilot plants and test facilities. There are several loss mechanisms, e.g., *Ohmic* (resistive) *heating* power loss, *Hall effect* loss,[9] end loss, and friction and heat transfer losses. The conversion efficiencies achieved so far have been around 20%, corresponding to a thermal efficiency of about 30%, which made MHD uncompetitive vis-à-vis conventional (steam Rankine cycle) thermal power plants.

The reader can consult the wide literature on MHD for the historical development of the technology, types of MHD generators, open and closed cycle systems for power generation, and other aspects of the technology in more detail. For a "crash course," the best source that the author can recommend is the short section (pp. 200–208) in Khartchenko [4]. For a compact (only 20 pages) but reasonably comprehensive review, refer to chapter 4 in the *IEA Clean Coal Center* report by Zhu [13]. The

---

[9] This loss is due to the fact that the current flow between the electrodes is not in the same direction as the field. The deviation can be as high as 80°, increasing with low pressures (desirable for increasing σ), high magnetic fields, and high electron mobilities. The resulting axial component of current flow (which should ideally be zero) leads to inefficiency.

Achilles' heel of the technology in terms of actual hardware was the long-term high-temperature component reliability and durability required for commercial readiness (hot, ionized gas is highly corrosive and erosive, which limited parts life for ducting, nozzles, and valves). Materials needed for durable insulators, electrodes, and heat exchangers (e.g., an air preheater, which is requisite for good efficiency) were unavailable or exorbitantly expensive. Other technical problems were electrode plasma arcing and the complex power transfer process from plasma to the electrical grid. The efficiency problem mentioned above can be solved by making use of the hot gas leaving the MHD generator to make steam in a boiler and expand it in a steam turbine. In such a binary MHD-ST power cycle, efficiencies of up to 60% were shown to be possible (at least on paper) but only with significant technology development in the areas of superconducting magnets (for strong magnetic fields with B up to 6 T), materials, and plasma fluid dynamics (high gas velocities). The best possible efficiency, with hot gas from the MHD generator entering the steam boiler at about 1,900°C, is around 45%, i.e., about the same as advanced *ultra-supercritical pulverized coal* power plants.

In the Oxy-MHD system, the fuel is natural gas or syngas generated by coal gasification. The oxidant is O_2 generated in an *ASU*. Thus, combustion products are CO_2 and H_2O, enabling easy separation of CO_2 for sequestration (similar to the Allam cycle in Section 10.3). Power required to drive the ASU is provided by the MHD generator. One advantage of 100% O_2 as oxidant is the elimination of the air (atmospheric or enriched with O_2) preheater to achieve the high temperatures requisite for electric conductivity of the gas (close to 3,000°C, which requires air preheating to well above 1,000°C). Hot gas from the MHD generator is expanded in a turbine (no heat recovery boiler is necessary) for additional power generation. The turbine exhaust is passed through a condenser to knock H_2O out of the gas. Wet CO_2 is dehydrated, compressed, and sent to the site of sequestration or usage. The 500 MW system envisioned by the US DOE is looking at 3,000°C (5,432°F) gas temperature in the MHD and 1,000°C (1,832°F) gas at the turbine inlet for 52.5% efficiency (the eventual development target is 1,760°C and 69% efficiency). Enabling technologies available today (which were not available back in the early 1990s, when the POC program was folded after spending more than $200 million) are as follows:

- High-temperature superconducting magnets
- Advanced alloys and ceramics
- Improved computational plasma fluid dynamics
- Computer systems aid in design and power conditioning
- Control technologies to enhance fault protection and mitigate arcing

For a detailed review of the MHD technology in general and the Oxy-MHD technology in particular, the reader is referred to a report by Hustad et al. [14].

The Oxy-MHD concept is amenable to utilization of green or blue H_2 as the fuel. The only combustion product is steam. For adequate conductivity, as in conventional MHD with coal or natural gas as fuel, steam can be seeded with an easily ionizable salt such as potassium carbonate (K_2CO_3). Even if the significant design challenges mentioned above can be solved cost-effectively, Oxy-MHD with green H_2 is unlikely to be a competitive alternative to GTCC with green H_2 or natural gas–fired GTCC with PCC. However, Oxy-MHD can be a very attractive technology with coal-based blue H_2.

13.6   Coal in RICE

Coal-to-liquids (CTL) refers to the process of converting coal to fuels such as diesel or gasoline. While there are three distinct routes to CTL, commercially and technologically, the most mature one is *indirect liquefaction*. The two-step process involves first gasification of coal (or another solid feedstock) and then conversion of the resulting syngas to a liquid. The resultant liquid fuel is known as synthetic fuel or *synfuel*. Synfuels are used as an alternative to oil and can be used to make petroleum and diesel fuels, as well as synthetic waxes, lubricants, chemical feedstocks, and alternative liquid fuels such as methanol and dimethyl ether. The best-known synthesis process is the *Fischer–Tropsch* synthesis widely used by Germans during World War II. The reader can consult the extensive literature on the subject to learn more on CTL, including the two other routes, *pyrolysis* (for liquefaction of shale oil and tar sands) and *direct liquefaction* (hydrogenation). In this author's opinion by far the best source is the book by Probstein and Hicks [1]. Herein, the focus will be on a less-known variation of turning coal into a *slurry* for combustion in internal combustion engines.

Advanced natural gas–fired reciprocating internal combustion engines (colloquially known as *diesel engines*) with close to 50% electric efficiency (LHV basis) represent the most efficient technology for burning fossil fuels. Diesel engines rated up to 20 MWe are widely available for electric power generation, especially with cogeneration, across a wide load range, i.e., from peaking and/or distributed power to utility-scale applications with multiple engines. The modular nature of such power plants makes them highly amenable to cyclic duty. In certain geographic locations with limited access to LNG and smaller grids (e.g., island nations), they can be used for base load duty firing diesel or number 2 fuel oil as well.

Interestingly, the first fuel in Rudolf Diesel's mind for his internal combustion engine was coal dust (in the 1890s). Not surprisingly, though, he quickly switched to petroleum-based liquid fuels, which were easier and safer for his engine. Since then, coal-fueled diesel engine research and development has continued in a sporadic fashion in industrialized countries such as Germany, Japan, the US, and Australia. Coal slurry fuel was successfully injected, ignited, and burned in test engines as well as commercial ones with good stack emissions in studies going back to the late 1960s (especially with locomotive applications in mind). Furthermore, coal-fired diesel engine technology, commonly referred to as *direct injection carbon engine* (DICE), can retain the efficiency advantage of its natural gas–fired brethren to a large degree. This has been verified by rigorous calculations, test-bench experimental studies, and field experience obtained in engines fired with *Orimulsion*.

The efficiency of the engine-only (simple cycle) configuration can be further enhanced via turbocompounding and the addition of a bottoming cycle.

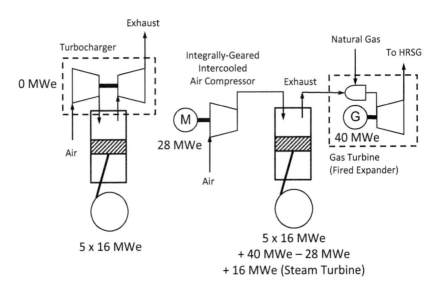

Figure 13.9 Turbocharging versus turbocompounding.

Turbo*compounding* is different from conventional turbo*charging* in that the exhaust gas expander (turbine) in the former, in addition to driving the charge air compressor, contributes to the engine shaft output (see Refs. [23–24] in Chapter 2). In turbocharging, the compressor and expander constitute a self-balanced shaft with zero net power output. This distinction is illustrated in Figure 13.9. The numbers in the figure correspond to a modular power plant design with five engines (16 MWe each) and a hot gas expander with 1,400°F inlet. The charge air is compressed to about 75 psia whereas exhaust gas pressure at the turbocharger expander inlet is about 60 psia.

A further enhancement of the turbocompound DICE combined cycle is via *reheat combustion* between the engine exhaust and expander inlet (see Figure 13.9). This further increases the expander output as well as exhaust gas energy for a higher bottoming cycle contribution. The underlying thermodynamics is graphically illustrated in the *temperature-entropy* (T-s) diagram in Figure 13.10. In the figure, ideal internal combustion engine and gas turbine processes are represented by air-standard *Atkinson* and *Brayton* cycles, respectively. For the same "mechanical compression" pressure ratio (between state points 2 and 1) and cycle maximum temperature (state points 3 and 3A), constant volume heat addition (2–3A) of the Atkinson cycle results in higher efficiency via a higher overall cycle pressure ratio (i.e., higher mean-effective heat addition temperature and lower mean-effective heat rejection temperature). Combining the two cycles in a reheat configuration (the "hybrid" cycle) leads to a better approximation of the ultimate ideal, i.e., the *Carnot* cycle {1-2C-3-4C-1}. The added benefit of the hybrid (turbocompound plus reheat) cycle is higher bottoming cycle potential for an even better approximation of the Carnot limit (i.e., triangular area {1-4-4C-1} vis-à-vis {1-4A-4D-1}).

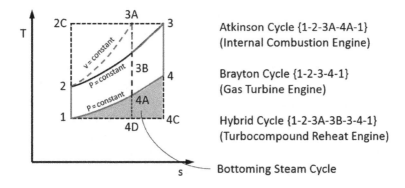

Figure 13.10 Temperature-entropy diagrams of ideal, air-standard RICE and gas turbine cycles.

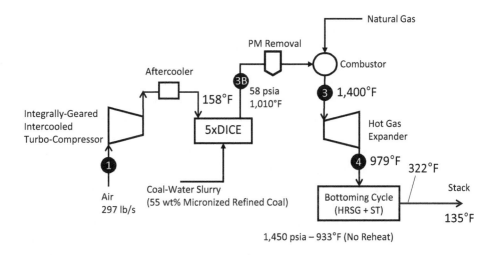

Figure 13.11 DICE-gas turbine compound reheat combined cycle.

A simplified system diagram of the *DICE-gas turbine compound reheat combined cycle* (DICE-GT CRCC) is shown in Figure 13.11. Coal feedstock is low-sulfur, subbituminous *Powder River Basin* coal (less than 1% by weight sulfur), which is *micronized* and physically *beneficiated* to an ash content of 2.2% by weight on a dry basis. The coal-water slurry burned in DICE is 55% *micronized refined coal* (MRC) and 45% water by weight with an LHV of 14,513 kJ/kg. There are five DICEs in the plant, each rated at 15.7 MWe and 42.5% (net LHV) efficiency. Natural gas is burned in the reheat combustor (18% of total plant fuel energy input). The net plant output is 106 MWe at a net LHV efficiency of 45.6%.

It is highly unlikely that burning coal without carbon capture is a feasible proposition in the foreseeable future – at least not in Europe and North America. One additional advantage of the DICE combined cycle with reheat combustion is the increased CO_2 content of the stack gas (9.5% by volume vis-à-vis about 4% for conventional gas turbine combined cycles), which makes it amenable to post-combustion CO_2 capture with the amine-based chemical absorption-desorption process, which is the most mature and readily available capture technology. Rigorous system modeling showed that the generation of enough steam in the HRSG to satisfy the demand of the desorption tower (the *stripper*) reboiler (a kettle-type evaporator) requires supplementary (duct) firing. However, the oxygen content of the DICE exhaust gas (about 10% by volume) is insufficient to sustain reheat combustion and duct firing without supplementary air. This adds cost and complexity to the system. Consequently, to minimize the requisite engineering development cost and simplify the system, the introductory version of the DICE combined cycle with post-combustion carbon capture does not include a reheat combustor. With a carbon capture add-on, the DICE-GT CRCC output drops to about 83 MWe with net LHV efficiency of 31.2% (30.4% in higher heating value or HHV). On an overnight (EPC) basis, CAPEX is estimated at north of $5,000/kW with PCC.

To a large extent, DICE-GT CRCC is based on mature, off-the-shelf technology with some modifications.[10] (One exception is the reheat combustor, manufactured to spec by a qualified supplier, which, however, is not included in the introductory offering.) The stock diesel engine is a medium speed (500 rpm), large bore (460 mm) heavy fuel oil–fired V18 engine available from manufacturers such as Wärtsila or MAN. The turbocharger module of the engine is removed to make it ready for turbo-compounding (see Figure 13.9). (Charge air to all engines is supplied by the main air compressor – see Figure 13.11.) For firing coal-water slurry fuel, retrofit consider-ations include atomizer nozzle wear, piston ring jamming, abrasive wear, ignition delay, and exhaust valve seat wear. Also important are fuel system issues such as blockage, fuel stability, and corrosion.

Known solutions to atomizer nozzle wear include diamond compact or silicon carbide nozzles and the use of lower speed engines (with increased time for combus-tion giving a higher tolerance to poor atomization). Good atomization (MRC slurry is like house paint in thickness), low speed, and large bore cylinders with large clear-ances are key to the prevention of piston ring jamming. Abrasive wear can be reduced by plasma-spray carbide coatings. Ignition delay (due to the slow burning characteris-tics of the MRC slurry) and knocking problems can be rectified by pilot injection of diesel fuel to ensure reliable ignition, particularly at low load.

The bottoming cycle is very simple with a one-pressure (no-reheat) HRSG and single-casing, condensing steam turbine with high- and low-pressure casings (HP and LP, respectively). During operation with capture, the LP turbine is idle and separated from the powertrain via a "triple-S" (SSS) clutch. Exhaust steam from the HP turbine at 60 psia is sent to the stripper reboiler. Particulate removal is accomplished by a *third stage separator* commonly used in *fluid catalytic cracking* (FCC) applications for the same purpose to satisfy the requirements of the hot gas expander (also common in FCC applications). The HRSG contains an SCR/CO catalyst module to scrub NOx

---

[10] This is by no means a guarantee of the proverbial overnight success; see Sections 16.5 and 16.6.

and CO from the stack gas. Coal beneficiation can remove the nonorganic sulfur from MRC so that SOx in the stack gas can be removed in a direct contact cooler upstream of the capture block.

DICE-GT CRCC can burn coal efficiently (comparable to ultra-supercritical boiler-steam turbine technology available at gigawatt scale) and cleanly at 100 MWe scale without a need for exorbitant investment in R&D. Past efforts came to naught not due to insurmountable technical difficulties but rather due to the wide availability of cheap oil and, later, natural gas for transportation and electric power generation applications. Another factor is the cost of manufacturing MRC and storing the MRC slurry in large quantities to facilitate experimental runs of long durations (e.g., 6,000 hours of demonstration run intended in the US DOE's *Clean Coal Technology* program). In the face of rising public opposition to fossil fuels in general and coal in particular, large engine manufacturers in the developed world are (understandably) leery of embarking on coal-fired diesel development. Still, the reality is that coal will continue to be a major energy source in other parts of the world for the foreseeable future. This may translate into a lucrative market for DICE-based power plants (especially with large, centralized MRC slurry fuel production facilities akin to oil refineries).

13.7   References

1. Probstein, R. F. and Hicks, R. E. 2006. *Synthetic Fuels*, Mineola, NY: Dover Publications, Inc.
2. Higman, C. and van der Burgt, M. 2008. *Gasification* (2nd ed.), Oxford, UK: Gulf Professional Publishing.
3. Rezaiyan, J. and Cheremisinoff, N. P. 2005. *Gasification Technologies – A Primer for Engineers and Scientists*, Boca Raton, FL: CRC Press, Taylor & Francis Group.
4. Khartchenko, N. V. 1998. *Advanced Energy Systems*, Washington, DC: Taylor & Francis.
5. Gülen, S. C. and Driscoll, A. V. 2013. Simple parametric model for quick assessment of IGCC performance, *Journal of Engineering Gas Turbines and Power*, 135, #010802.
6. Barnes, I. 2013. Recent operating experience and improvement of commercial IGCC, CCC/222, IEA Clean Coal Centre.
7. Gülen, S. C. 2019. Advanced Fossil Fuel Power Systems. In D. Y. Goswami and F. Kreith (eds.), *Energy Conversion*, 2nd ed. Boca Raton, FL: CRC Press.
8. Higman, C. 2014. State of the Gasification Industry: Worldwide Gasification Database 2014 Update, Gasification Technologies Conference, October 29, Washington, DC.
9. Pisacane, F., Domenichini, R. and Fadabini, L. 1998. Dynamic modeling of the ISAB energy IGCC complex, Gasification Technologies Conference, San Franciso, CA. Available at https://api.semanticscholar.org/CorpusID:16656180
10. Higman, C., Marasigan, J., Todd, D., Kubek, D. and Sporensen, J. 2016. Increasing the flexibility of IGCC, *Modern Power Systems*, May.
11. Chen, L., Nolan, R. and Avadhany, S. 2009. Thermodynamic analysis of coal to synthetic natural gas process. Available at https://www.researchgate.net/publication/228441605_Thermodynamic_Analysis_of_Coal_to_Synthetic_Natural_Gas_Process.

12. Kaufman, E. 2003. Considerations when burning ash-bearing fuels in heavy-duty gas turbines, GE Power Generation, GER-3764A.
13. Zhu, Q. 2015. High-efficiency power generation – review of alternative systems, Report CCC/247, London, UK: IEA Clean Coal Centre.
14. Hustad, C. W., Coleman, D. L. and Mikus, T. 2009. Technology overview for integration of an MHD topping cycle with the CES oxyfuel combustor, Final Report, Houston, TX: CO2-Global, LLC.

14 A Technology Leap?

The second law of thermodynamics unambiguously asserts that the maximum efficiency of a heat engine operating in a thermodynamic cycle cannot exceed the efficiency of a Carnot cycle operating between the same hot and cold temperature reservoirs. This is a direct translation of one of the two famous corollaries of the second law, i.e., the Kelvin–Planck statement. In this most fundamental sense, all practical heat engine cycles are attempts to approximate the ideal Carnot cycle within boundaries imposed by existing design factors, e.g., available materials, mechanical considerations, size, and cost. Examples are internal combustion engine cycles such as Otto, Diesel, Atkinson, and Brayton and external combustion engine cycles such as Ericsson, Stirling, and Rankine. All the aforementioned cycles are idealized "closed" cycles with a pure working fluid that describe the major operating principles of actual engines, e.g., car and truck engines burning liquid fuels such as gasoline and diesel oil, gas turbines burning a wide variety of gaseous and liquid fuels, and steam turbines.

The Brayton cycle with *constant-pressure heat addition* (CPHA) describes the operation of a gas turbine and, in its combined cycle (CC) configuration, it represents the most efficient means of fossil-fuel burning power generation that is currently available. An in-depth evaluation of the thermodynamic potential of the state of the art gas turbine simple and combined cycle power plants can be found in **GTFEPP**. For a quick look at the technology, refer to Section 3.1 in this book. At the time of writing (2020–2021), J Class technology with 1,600°C turbine inlet temperature (TIT) (nominal) and a cycle pressure ratio (PR) of about 25:1, is rated at 64% net LHV efficiency (ISO base load rating). With 1,700°C TIT, 65% net LHV efficiency is in sight. However, further improvement is limited by the availability of advanced materials, Dry-Low-NOx (DLN) combustion technologies, mechanical design considerations, and turbine cooling technologies, which are significant hurdles that need to be overcome in an economically viable manner. In all likelihood, the next goal post of 70% net LHV efficiency requires a change in the thermodynamic cycle of the gas turbine as a possible ploy to enable a transformative (rather than incremental or marginal) increase in power plant efficiency.

It is a well-known fact that by far the most important Brayton cycle deficiency that precludes approximating Carnot cycle performance is the combustion or heat addition irreversibility [1]. Amann et al. demonstrated that, if no limit is imposed on cycle PR, the air-standard internal combustion engine cycle with four key processes (compression, heat addition, expansion, and heat rejection) is the constant-volume heat addition

(Atkinson) cycle [2]. The *constant-volume heat addition* (CVHA) or "explosion combustion" is the main feature of the classic reciprocating (piston-cylinder) engine cycle (i.e., the Otto cycle). First proposed by Holzwarth, whom Stodola [3] credited for "having built the first economically practical gas turbine," the CVHA gas turbine cycle idea is not new. Holzwarth's turbine, in its latest installed version in Mannheim, Germany (1920), with a valve-controlled and water-cooled combustion chamber and operating in an expansion-scavenging-precompression-explosion cycle ultimately proved too complicated and expensive to be a viable product.

One device that can be used to achieve approximately constant-volume combustion is a *pulse combustor* [4]. Also known as *pressure-gain* or *pulsejet* combustion, the idea goes back nearly half a century [5]. Modeling and experimental data have been reported for basic aero-valve pulse combustors at near-atmospheric pressures [6]. Gains of about 4% of compressor exit pressure have been demonstrated at temperature ratios comparable to those of modern gas turbines [7]. Modest potential in terms of gas turbine efficiency improvement and significant difficulties in designing viable systems incorporating this highly unsteady process into the gas turbine structure make pressure-gain combustion utilizing valved chambers an unlikely path to follow. The internal combustion *wave rotor* is another technology for achieving pressure-gain or constant-volume combustion [8]. The concept has been successfully used as a pressure wave supercharger for diesel engines. Recently, interest in the wave rotor to achieve pressure rise and confined combustion simultaneously (with deflagration or detonation modes of combustion) has picked up due to its potential to significantly improve gas turbine efficiency. Akbari et al. [9] can be consulted for an exhaustive review of wave rotor technology and its applications.

Another idea that has also been around for more than half a century is the intermittent or *pulsed detonation combustion* (PDC). For detailed coverage of the theory and concepts underlying PDC and engines using PDC, referred to as *pulse detonation engines* (PDE), the reader should consult Bussing and Pappas [10]. The bulk of the existing literature on the subject is centered on the PDE as an aircraft jet engine. Kailasanath [11] provides a concise history of the application of detonation combustion to aircraft propulsion. In this chapter, the advantage of PDC in gas turbine simple and combined cycle performance will be discussed utilizing fundamental thermodynamic cycle analysis.

14.1   Air-Standard Cycles

Air-standard cycle is a theoretical construction to facilitate simple qualitative analysis of complex power generation cycles. Three major assumptions characterize a "cold" air-standard cycle:

1. a pure working fluid (air) modeled as a calorically perfect gas
2. external heat addition
3. internally reversible processes

Table 14.1 Comparison of four air-standard power cycles

(SSSF: steady-state steady-flow, USUF: uniform-state uniform-flow)

	Brayton	**Atkinson/Humphrey**	**Otto**	**Reynst–Gülen**
Process 1 $(1 \rightarrow 2)$	Isentropic Compression (SSSF) $s_1 = s_2$	Isentropic Compression (USUF/ SSSF) $s_1 = s_2$	Isentropic Compression (USUF) $s_1 = s_2$	Isentropic Compression (SSSF) $s_1 = s_2$
Process 2 $(2 \rightarrow 3)$	Constant-Pressure Heat Addition (SSSF) $p_3 = p_2$	Constant-Volume Heat Addition (USUF) $v_3 = v_2$	Constant-Volume Heat Addition (USNF) $v_3 = v_2$	Constant-Volume Heat Addition (SSSF) $v_3 = v_2$
Process 3 $(3 \rightarrow 4)$	Isentropic Expansion (SSSF) $s_3 = s_4$	Isentropic Expansion (USUF/SSSF) $s_3 = s_4$	Isentropic Expansion (USUF) $s_3 = s_4$	Isentropic Expansion (SSSF) $s_3 = s_4$
Process 4 $(4 \rightarrow 1)$	Constant-Pressure Heat Rejection (SSSF) $p_4 = p_1$	Constant-Pressure Heat Rejection (USUF/SSSF) $p_4 = p_1$	Constant-Volume Heat Rejection (USUF) $v_4 = v_1$	Constant-Pressure Heat Rejection (SSSF) $p_4 = p_1$
Heat In, q_{in} $(2 \rightarrow 3)$	$c_p \cdot (T_3 - T_2)$	$c_v \cdot (T_3 - T_2)$	$c_v \cdot (T_3 - T_2)$	$c_p \cdot (T_3 - T_2)$
Applies To	Gas Turbine with Deflagration Combustion	Holzwarth (Explosion) Gas Turbine with Precompression	Spark-Ignition (Reciprocating) Internal Combustion Engine	Gas Turbine with Detonation Combustion

An implicit assumption, in particular pertaining to turbogenerators such as gas turbines, is that an air-standard cycle comprises four processes connecting four state points, which are representative of actual machine locations (e.g., compressor exit, turbine inlet). Air-standard cycles are typically represented on a temperature-entropy (T-s) diagram. This convention goes back to the Carnot cycle, which incorporates four basic ideal processes: isentropic compression and expansion and constant temperature heat addition and heat rejection, which comprise a rectangle on the T-s diagram.

All major air-standard cycles are attempts to approximate the Carnot cycle as closely as possible. Most (but not all) are idealized descriptions of real production machines, e.g., Otto and Diesel cycles for automotive engines, or the Brayton cycle for aircraft jet engines. A summary and comparison of four air-standard cycles, which are of particular significance for the subject at hand, are provided in Table 14.1. First three are well-known variants that can be found in any standard textbook. The fourth, as far as known to the author, is explicitly defined and developed for the first time in his 2010 paper [12] and will be discussed in detail in the following paragraphs.

14.1.1   Brayton Cycle

The CPHA process in the Brayton cycle shown in Figure 14.1 is described by the *modified* Gibbs equation as follows:

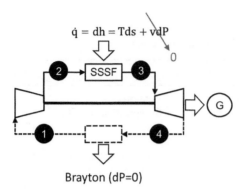

Brayton (dP=0)

Figure 14.1 Brayton cycle (heat addition, dP = 0); cycle heat input is represented by the modified Tds equation.

$$dh = T \cdot ds + v \cdot dP = T \cdot ds. \tag{14.1}$$

Integrating both sides of Equation (14.1) between states 2 and 3, one gets

$$h_3 - h_2 = \int_2^3 T \cdot ds = \mathrm{METH} \int_2^3 ds = \mathrm{METH}\,(s_3 - s_2) = q_{\mathrm{in}}, \tag{14.2}$$

where METH is the Brayton cycle's *mean-effective heat addition temperature*, whose value is between T_2 and T_3, and which is the temperature at which the *Carnot-equivalent* cycle would receive heat from a *high-temperature reservoir* during isothermal and isobaric heat addition. Similarly, for the constant-pressure heat rejection process, one gets

$$h_4 - h_1 = \int_1^4 T \cdot ds = \mathrm{METL} \cdot \int_1^4 ds = \mathrm{METL} \cdot (s_4 - s_1) = q_{\mathrm{out}}, \tag{14.3}$$

where METL is the Brayton cycle's *mean-effective heat rejection temperature*, whose value is between T_1 and T_4, and which is the temperature at which the Carnot-equivalent cycle would transfer heat to a *low-temperature reservoir* during isothermal and isobaric heat rejection. Using Equations (14.2) and (14.3), and noting that $s_1 = s_2$ and $s_3 = s_4$ for isentropic compression and expansion processes, one arrives at

$$\eta = 1 - \frac{\mathrm{METL}}{\mathrm{METH}}. \tag{14.4}$$

Note that for a perfect gas, the mean-effective temperature for any isobaric process between two states, inlet and outlet, can be written as[1]

---

[1] This description is for an open (flow) system or control volume. The same relationship is obtained for a constant volume process in a closed system with given initial and final states.

$$\text{MET} = \frac{T_{\text{in}} - T_{\text{out}}}{\ln(T_{\text{in}}/T_{\text{out}})}. \tag{14.5}$$

The mean-effective heat addition and rejection temperatures can easily be calculated for any given power generation cycle from the fundamental thermodynamic relationship, Equation (14.1).[2] Using the well-known isentropic p-T formula, the Brayton cycle efficiency given by Equation (14.4) can be shown to be a function of overall cycle PR only:

$$\eta_B = 1 - \frac{1}{PR^k}, \tag{14.6}$$

where $k = 1 - 1/\gamma$. The corresponding combined cycle efficiency is

$$\eta_{BCC} = 1 - \frac{T_1}{\text{METH}} = 1 - \ln\left(\frac{T_3}{T_2}\right)\frac{T_1}{(T_3 - T_2)}, \tag{14.7a}$$

$$\eta_{BCC} = 1 - \ln\left(\frac{\tau}{PR^k}\right)\frac{1}{(\tau - PR^k)}, \tag{14.7b}$$

where $\tau = T_3/T_1$. Unlike the simple cycle, Brayton combined cycle efficiency is a function of cycle PR and cycle maximum temperature, T_3. The ultimate theoretical efficiency is the efficiency of the Carnot cycle operating between hot and cold temperature reservoirs at T_3 and T_1, respectively. Based on that, one can evaluate a "Carnot Factor," which is a measure of the "goodness" of the heat engine cycle in question, Thus, for the Brayton combined cycle, the Carnot factor is given by

$$CF_{BCC} = \frac{\tau - \ln\left(\frac{\tau}{PR^k}\right)\frac{\tau}{(\tau-PR^k)}}{\tau - 1}, \tag{14.8a}$$

$$CF_{BCC} = \frac{\tau - \theta\frac{\ln(\theta)}{(\theta-1)}}{\tau - 1}, \quad \theta = \frac{T_3}{T_2} = \frac{\tau}{PR^k}. \tag{14.8b}$$

For a J Class gas turbine technology, the parameters defined above are summarized in Table 14.2. Also included in the table is the nondimensional cycle input given by

$$\beta = \frac{q}{c_p T_1} = \tau - PR^k. $$

14.1.2   Atkinson Cycle

This is an air-standard cycle that describes a piston engine similar to the Otto engine. As such, the working fluid (air in the piston-cylinder assembly) constitutes a *closed* system. However, under the name of *Humphrey*, the similar (but *not* the same) collection of four processes is used for a gas turbine with CVHA (i.e., a modified Brayton cycle) as a surrogate for a PDE. While PDEs so far exist on paper only, the

---

[2] For an open or flow system (i.e., a control volume.) For a closed system the appropriate T-ds equation is the Gibbs equation, du = T·ds − P·dv.

Table 14.2 Air-standard Brayton combined cycle performance with J Class gas turbine parameters

(TIT $= 1,600°$C, PR $= 24:1$)

T_3, K	1873.15
T_1, K	288.15
Carnot Efficiency, %	84.6
PR	24
τ	6.50
METH, K	1,202
k	0.2857
PR^k	2.48
θ	2.62
β	4.02
Brayton SC Efficiency, %	59.7
Brayton CC Efficiency, %	76.0
BCC Carnot Factor	0.90

Humphrey cycle can be considered as the air-standard cycle for Holzwarth's explosion turbine [3], which combined a centrifugal compressor and velocity-compounded turbine (each an *open* system) with an *explosion chamber* (i.e., a closed system) in between (see Figure 14.2).

Using the first law for a closed system, the heat addition process in the Atkinson cycle is given as:

$$du = dq_{in} = c_v \cdot dT. \tag{14.9}$$

All other processes can be described in a manner identical to the Brayton cycle. Using the ideal gas equation of state for the beginning and end of the heat addition process, $2 \rightarrow 3$ and noting that $v_3 = v_2$, one can verify that

$$\frac{P_3}{P_2} = \frac{T_3}{T_2} = \theta.$$

Since the temperature rises as a result of heat addition, during a constant-volume process the pressure will also rise in proportion to the rise in temperature. Combining the pressure ratios for the isentropic compression and expansion processes, and noting that the heat rejection is at constant pressure, i.e., $P_4 = P_1$, the following relationships are obtained:

$$\frac{P_3}{P_2} = \frac{PR}{PR'} = PR'' = \theta, \tag{14.10}$$

$$PR = PR' \cdot PR'' = \theta \cdot PR', \tag{14.11}$$

where PR' and PR'' are pressure ratios of isentropic precompression and CVHA processes, respectively. Using isentropic p-T relationships and going through some algebra, it can be shown that

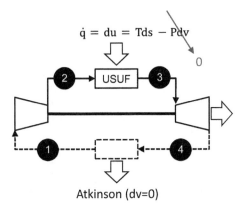

$$\dot{q} = du = Tds - Pdv$$

Atkinson (dv=0)

Figure 14.2 Atkinson (Humphrey) cycle (heat addition, dv $= 0$); cycle heat input is represented by the Gibbs equation.

$$\eta_A = 1 - \frac{\gamma}{\tau_2}\left\{\frac{\theta^{1/\gamma} - 1}{\theta - 1}\right\}, \tag{14.12}$$

where $\tau = PR'^k$ is the nondimensional temperature (via division by T_1) at the cycle state indicated by the subscript. The equivalent of Equation (14.2) can be written for the Atkinson cycle as

$$u_3 - u_2 = \text{METH} \cdot (s_3 - s_2) = q_{in},$$

so that METH for the CVHA process for a perfect gas has the same algebraic expression, i.e.,

$$\text{METH} = \frac{T_3 - T_2}{\ln(T_3/T_2)} = T_1 PR'^k \frac{\theta - 1}{\ln(\theta)}, \tag{14.13}$$

$$\text{METH} = T_1 PR'^k \frac{PR'' - 1}{\ln(PR'')}. \tag{14.14}$$

Equations (14.13) and (14.14) highlight a significant difference between air-standard Brayton and Atkinson cycles. In the former, there is only one degree of freedom, i.e., cycle PR, which sets the cycle efficiency. In the Atkinson cycle, there are several degrees of freedom but only two of them can be set independently. In particular,

- Given PR$'$ and T_3, PT$''$ and PR are determined from Equation (14.10) or (14.11).
- Given PR$'$ and PR$''$ (or PR), T_3 is determined from Equation (14.10) or (14.11).

Consequently, Atkinson combined cycle efficiency can be written in two different ways, i.e.,

$$\eta_{ACC} = 1 - \frac{\ln(\theta)}{PR'^k(\theta - 1)} = 1 - \frac{\ln(\theta)}{\tau_2(\theta - 1)}, \tag{14.15}$$

Table 14.3 Air-standard Atkinson (Humphrey) combined cycle performance with J Class gas turbine parameters

(TIT = 1,600°C, PR = 24:1)

T_3, °C	1,600	1,600	1,835	1,935
Carnot Efficiency, %	84.6	84.6	86.3	87.0
PR′	6.225	12	6.225	12
τ_3	6.50	6.50	7.32	7.66
$\tau_2 = PR'^k$	1.69	2.03	1.69	2.03
$\theta = PR''$	3.86	3.20	4.34	3.77
PR	24.00	38.35	27.01	45.21
METH	1,028	1,108	1,105	1,223
β	3.44	3.19	4.02	4.02
Atkinson SC Efficiency, %	52.8	59.5	53.9	60.7
Atkinson CC Efficiency, %	72.0	74.0	73.9	76.4
Carnot Factor	0.85	0.87	0.86	0.88

$$\eta_{ACC} = 1 - \frac{\ln(PR'')}{PR'^k(PR'' - 1)} = 1 - \frac{\ln(PR'')}{\tau_2(PR'' - 1)}. \tag{14.16}$$

Similarly, for the Carnot factor of the Atkinson combined cycle, one can write that

$$CF_{ACC} = \frac{\tau_3 - \frac{\theta \ln(\theta)}{(\theta - 1)}}{\tau_3 - 1}, \theta = \frac{\tau_3}{PR'^k}, \tag{14.17}$$

$$CF_{ACC} = \frac{\theta PR'^k - \frac{PR'' \ln(PR'')}{(PR'' - 1)}}{\theta PR'^k - 1}, \theta = PR''. \tag{14.18}$$

Repeating the same exercise as before, the air-standard Atkinson (Humphrey) combined cycle is calculated using cycle parameters for a J Class gas turbine. The results are summarized in Table 14.3.

Takeaways from Table 14.3 are summarized below (Atkinson and Brayton combined cycles are referred to as ACC and BCC, respectively):

- For J Class TIT (T_3) and PR, ACC efficiency is lower than BCC efficiency. This is a direct result of the lower METH of the former cycle (1,028 K vis-à-vis 1,202 K).
- Furthermore, ACC *heat input* is lower than that for BCC (i.e., $\beta = 3.44$ vis-à-vis 4.02). This is so because a significant portion of the cycle *energy input* is in the form of *mechanical work* increasing the cycle pressure.
- Higher cycle PR via higher precompression PR′, i.e., 12:1, at the same T_3 increases the ACC efficiency but at the expense of cycle heat input (i.e., $\beta = 3.19$ vis-à-vis 3.44). An increase in METH comes via increased mechanical work, i.e., overall cycle PR, i.e., 38.4:1. Carrying this approach further is futile because the cycle PR becomes too high for a feasible mechanical design.
- For the same cycle heat input, i.e., $\beta = 4.02$, one must increase T_3 and PR′, which results in extremely high T_3 and PR, i.e., 1,935°C and 45.2:1, respectively, pointing to infeasible designs vis-à-vis BCC with state of the art design parameters.

Here, we pause and think. The Atkinson (Humphrey) cycle, for *the same overall cycle PR*, is inferior to the Brayton cycle in terms of simple and combined cycle thermal efficiency. In other words, a gas turbine with CVHA is *not* advantageous. The reason for this unexpected finding is simple. The Atkinson (Humphrey) cycle is not the correct ideal air-standard proxy for the CVHA gas turbine. This will be demonstrated in the next section using rigorous thermodynamic analysis.

14.1.3   Air-Standard Cycle for CVHA

A turbomachine (e.g., a gas turbine) is a *steady-flow* device, which can be naturally described by an air-standard cycle comprising four steady-flow (also known as *control volume*) processes. The resulting cycle, while conceptually identical to the Atkinson cycle on a T-s diagram, is significantly different in performance. This will be elaborated upon below using fundamental thermodynamic considerations and numerical examples. The cycle will be referred to as *Reynst–Gülen* (R–G) cycle in recognition of F. H. Reynst, who conceptually defined a gas turbine cycle with *pressure-gain combustor* (not necessarily at constant-volume) and numerically evaluated the performance [5]. The conceptual system diagram is shown in Figure 14.3.

The T-s diagram representations of Brayton and R–G cycles are shown in Figure 14.4, based on the J Class gas turbine cycle parameters listed in the caption. The cycles are chosen such that the *energy* (i.e., *enthalpy*) addition and overall cycle PR values for both are identical. Since the working fluid for both cycles is air as a perfect gas, this is equivalent to the relationship

$$T_3 - T_2 = T_{3'} - T_{2'} \qquad (14.19)$$

Each cycle is characterized by two parameters:

1. Cycle total PR, which is the same for isentropic compression and expansion of the Brayton cycle but not for the R-G cycle, and
2. Cycle maximum temperature, T_3 or $T_{3'}$, which is at the end of the heat addition and the start of isentropic expansion.

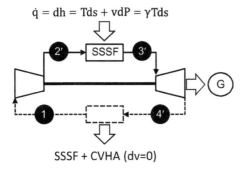

Figure 14.3 Ideal proxy for the CVHA gas turbine cycle.

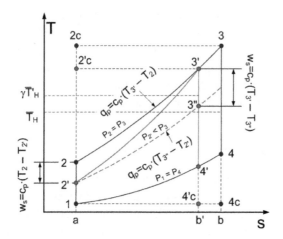

Figure 14.4 Brayton (1-2-3-4-1) and R–G (1-2'-3'-4'-1) cycles on T-s diagram. State 1 is ISO (1.014 bara, 15°C), PR = 24:1, $T_3 = 1{,}600$°C; $\gamma = 1.40$, R = 0.068577 Btu/lb-R, $c_p = 0.24$ Btu/lb-R.

The significant difference in the R–G cycle is the pressure rise during the CVHA process. In other words, the energy added to the working fluid comprises two parts:

1. heat, i.e., temperature increase, and
2. mechanical work, i.e., pressure increase.

The CVHA process in the R–G cycle is described by the modified Gibbs equation, Equation (14.1). Both terms on the right-hand side (RHS) of Equation (14.1) are retained because $dP \neq 0$. The first term in the RHS of the equation describes the temperature rise of the working fluid via heat transfer at constant pressure and the second term represents the temperature rise via the isentropic work done on the working fluid, which increases the temperature *and* pressure simultaneously. Dividing and multiplying the second term on the RHS with ds, Equation (14.1) is rewritten below as

$$dh = (T + v \cdot dP/ds) \cdot ds. \tag{14.20}$$

The exact differential dP is given by

$$dP = (\partial P/\partial s)_v \cdot ds + (\partial P/\partial v)_s \cdot dv. \tag{14.21}$$

For the constant-volume process, $dv = 0$ so that one ends up with

$$dP/ds = (\partial P/\partial s)_v = P/c_v. \tag{14.22}$$

Combining Equations (14.20)–(14.22) and using the ideal gas equation of state gives

$$dh = T(1 + R/c_v)ds = T \gamma \, ds. \tag{14.23}$$

Integration between the states $2'$ and $3'$ gives

$$h_{3'} - h_{2'} = \gamma \int_{2'}^{3'} T \cdot ds = \gamma \, METH' \int_{2'}^{3'} T \cdot ds = \gamma \, METH'(s_{3'} - s_{2'}) = q_{in}, \tag{14.24}$$

$$\gamma \, METH' = \frac{h_{3'} - h_{2'}}{s_{3'} - s_{2'}} = \frac{c_p \cdot (T_{3'} - T_{2'})}{c_v \cdot \ln(T_{3'}/T_{2'})} = \gamma \cdot \frac{(T_{3'} - T_{2'})}{\ln(T_{3'}/T_{2'})}, \tag{14.25}$$

$$METH' = \frac{(T_{3'} - T_{2'})}{\ln(T_{3'}/T_{2''})}. \tag{14.26}$$

Equation (14.25) gives the mean-effective temperature for the *entire* CVHA process between states 2' and 3'. Equation (14.26) gives the mean-effective temperature of the hypothetical heat transfer process at constant pressure between states 2' and 3'. For the two cycles shown in Figure 14.4, the heat addition, q_{in}, was identical for the constant-pressure and constant-volume processes so that combining Equations (14.2) and (14.25) gives

$$\frac{\gamma \, METH'}{METH} = \frac{s_3 - s_2}{s_{3'} - s_{2'}}. \tag{14.27}$$

As can be visually gathered from Figure 14.4, the entropy change in the CVHA process is smaller than the entropy change in the CPHA process. Thus, the RHS of Equation (14.26) is greater than unity, which provides the key to the principal advantage of the R–G cycle with CVHA over the Brayton cycle with CPHA with the same heat input and overall cycle PR:

> **The R–G cycle has a higher mean-effective heat addition temperature than the Brayton cycle, i.e.,**
> $$\gamma \, METH' > METH.$$

In an R–G cycle, energy added to the cycle during the constant-volume flow process between states 2' and 3' manifests itself in two distinct forms:

1. **Heat addition** at constant pressure, Tds, state 2' to state 3'' *or* state 2 to state 3' in Figure 14.4
2. **Work addition** at constant entropy, vdP, state 3'' to 3' *or* state 2' to state 2 in Figure 14.4

Consequently, one can write that

$$w_s = c_p(T_2 - T_{2'}) = c_p(T_{3'} - T_{3''}), \tag{14.28}$$

$$\tilde{w} = \frac{w_s}{c_p T_1} = \tau_2 - \tau_{2'} = \tau_{3'} - \tau_{3''}, \tag{14.29}$$

$$q_p = c_p(T_{3'} - T_2) = c_p(T_{3''} - T_{2'}), \tag{14.30}$$

$$\tilde{q} = \frac{q_p}{c_p T_1} = \tau_{3'} - \tau_2 = \tau_{3''} - \tau_{2'}. \tag{14.31}$$

For the R–G cycle, going through similar algebra as before, it can be shown that the thermal efficiency is given by (note that $\theta = \tau_{3'}/\tau_{2'}$)

$$\eta_{R-G} = 1 - \frac{1}{\tau_{2'}} \left(\frac{\theta^{1/\gamma} - 1}{\theta - 1} \right). \tag{14.32}$$

Note that the thermal efficiency of the Atkinson (Humphrey) cycle is given by

$$\eta_A = 1 - \frac{\gamma}{\tau_{2'}} \left(\frac{\theta^{1/\gamma} - 1}{\theta - 1} \right). \tag{14.33}$$

Thus, we can write a generic formula for heat engines with CVHA as follows

$$\eta_{CVHA} = 1 - \frac{\alpha}{\tau_{2'}} \left(\frac{\theta^{1/\gamma} - 1}{\theta - 1} \right). \tag{14.34}$$

In particular,

- For $\alpha = 1$, Equation (14.34) gives the efficiency of the R–G cycle, which is the true ideal (air-standard) proxy for a gas turbine with CVHA.
- For $\alpha = \gamma$, Equation (14.34) gives the efficiency of the Atkinson (Humphrey) cycle, which is the ideal (air-standard) proxy for a heat engine similar to Holzwarth's gas turbine.

Another variation of Equation (14.34) can be written in terms of major cycle parameters, i.e., precompression ratio, PR', and cycle maximum temperature, $\tau_{3'}$:

$$\eta_{CVHA} = 1 - \frac{\alpha}{PR'^k} \left[\frac{\left(\frac{\tau_{3'}}{PR'^k} \right)^{1/\gamma} - 1}{\frac{\tau_{3'}}{PR'^k} - 1} \right]. \tag{14.35}$$

After some algebraic manipulation, Equation (14.35) can be recast so that the independent parameters on the RHS of the equation are the precompression ratio, PR', and the nondimensional cycle energy input, i.e.,

$$\eta_{CVHA} = 1 - \frac{1}{\tilde{q}} \times \left[\left(\frac{\alpha \tilde{q}}{PR'^k} + 1 \right)^{\frac{1}{\gamma}} - 1 \right]. \tag{14.36}$$

In either case, pressure rise during constant-volume heat (energy) addition, PR'' and, thus, overall cycle pressure ratio, PR, falls out from fundamental thermodynamics, i.e., Equation (14.10) or (14.11).

To conceptualize a gas turbine operating in an R–G cycle, consider the drawing in Figure 14.5, in which the CVHA process is represented by a motor-compressor assembly as described by Reynst [5]. The motor, which drives the compressor, is a heat engine and sends its heat into the compressed air instead of rejecting it to the surroundings [5]. The constant-volume, pressure-gain combustor of the hypothetical turbine converts about 18% of the fuel chemical energy to mechanical work. (In the alternate representation per Figure 14.5, i.e., working fluid heated from state 2′ to state 3″ and compressed to state 3′, also about 18% of the fuel chemical energy is converted to mechanical work.)

Using the cycle parameters listed in the caption of Figure 14.4, the R–G cycle performance is calculated for the same cycle PR and energy (heat) input, $\tilde{q}$, as the air-standard Brayton cycle for a J Class gas turbine, i.e., GE's 7HA.02, whose performance is taken from the *Gas Turbine World 2020 Handbook*. The TIT (T_3) is estimated from a heat and mass analysis (see **GTFEPG**, chapter 7) as 1,676°C for PR = 23.1:1. The results are summarized in Table 14.4. For comparison, the corresponding Brayton and Atkinson (Humphrey) cycle parameters are also included in the table. For the Atkinson cycle, two cases are evaluated,

- Same T_3 as the Brayton cycle and PR′ = PR of the Brayton cycle (Case A)
- Same heat input as the Brayton cycle and PR′ = PR of the Brayton cycle (Case B)

The R–G cycle is evaluated for the same cycle heat input and overall PR as the Brayton cycle.

The state of the art J Class combined cycle efficiency is around 64% (ISO net base load rating). For the specific example used herein, GE's 7HA.02, rating efficiency is 63.4–63.6% (for 1×1×1 and 2×2×1 CC configuration). This corresponds to a "technology factor" of TF = 63.5/76.5 = 0.83. If a gas turbine with a pressure-gain combustor can be developed to the same level of technology, its ISO base load efficiency would be 0.83 × 79.2 = 65.7%. This is a gain of almost 2.5 percentage points at a TIT of

$$\text{TIT} = T_{3'} = \tau_{3'} \times T_1 = 6.02 \times 288.15 = 1{,}735 \text{ K} = 1{,}462°\text{C}.$$

In other words, the stated efficiency advantage, vis-à-vis a gas turbine with a (nominal) 1,600°C TIT, is achieved at an F Class TIT. As a final note, in Figure 11.3 the ultimate gas turbine combined cycle (GTCC) performance is shown as 68.3% net LHV (ISO base load) with a rotating detonating engine (RDE). However, in a presentation by Aerojet Rocketdyne, the efficiency of the RDE GTCC was calculated as 66.4% via rigorous modeling.[3] The futility of using the Atkinson (Humphrey) cycle as an ideal proxy for a gas turbine with pressure-gain combustion is obvious from the numbers in Table 14.4. At the same cycle maximum temperature as the base Brayton cycle, the Atkinson cycle reaches parity in CC efficiency at an overall cycle PR of 63.7:1. (The SC efficiency is better because of high cycle PR, which leads to a lower mean-effective cycle heat

---

[3] https://netl.doe.gov/sites/default/files/2020-04/NETL%20charts%20for%202016%20UTSR%20symposium%20r4.pdf (last accessed on March 26, 2021).

Table 14.4 R–G cycle key performance parameters compared to Brayton and Atkinson (Humphrey) cycles

	Brayton (7HA.02 – GTW 2020)	Atkinson Case A	Atkinson Case B	R–G
T_3, K	1,949	1,949	2,446	1,735
PR′	NA	23.1	23.1	6.57
PR″	NA	2.76	3.46	3.52
PR	23.1	63.7	80.0	23.1
τ_3	6.77	6.77	8.49	6.50
$\tilde{q}$	4.31	3.08	4.31	4.31
$\tau_{3'}$	NA	NA	NA	6.02
$\tilde{W}$	NA	NA	NA	0.74
$\tau_{3''}$	NA	NA	NA	5.28
$\tau_{2'}$	NA	NA	NA	1.71
PR′^{k}	2.45	2.45	2.45	1.71
θ	2.76	2.76	3.46	3.52
METH′	NA	NA	NA	987
γ METH′	NA	NA	NA	1,382
METH	1,225	1,225	1,401	NA
Ideal SC Efficiency, %	59.2	65.4	66.9	66.2
Ideal CC Efficiency, %	76.5	76.5	79.4	79.2
Carnot Efficiency, %	85.2	85.2	88.2	83.4
Carnot Factor (CC)	0.90	0.90	0.90	0.95
Actual CC Efficiency, %	63.5	63.5	65.9	65.7
Technology Factor	0.830			

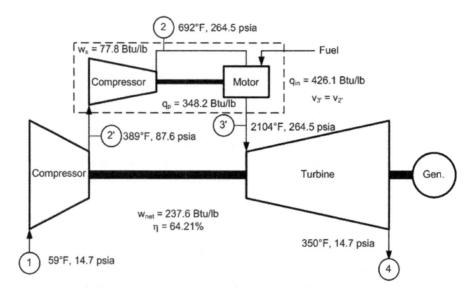

Figure 14.5 Hypothetical Reynst–Gülen cycle GT with constant-volume combustor replaced by a motor-compressor assembly as per Ref. [5]. The state points correspond to those shown in Figure 14.4.

rejection temperature.) This, of course, is a meaningless comparison because, leaving aside the impracticality of such a high cycle PR, the Brayton cycle, at the same 63.7:1 pressure ratio, would be superior to the Atkinson cycle.

14.2   Detonation Combustion

Fundamental thermodynamics via air-standard cycle analysis establishes the superiority of the R–G cycle vis-à-vis the Brayton and Atkinson cycles. The difficulty is in designing (conceptually as well as physically) a steady-flow device to accomplish combustion with temperature *and* pressure rises. Unsteady flow examples approximating the ideal constant-volume combustion are the Holzwarth turbine, the *Pescara* turbine, which is a combined gas turbine-diesel engine machine as described by Reynst, and pulsating combustor engines [5]. The only possibility for an adiabatic and steady-flow process with pressure rise is a *supersonic* flow with a standing *shock* wave (idealized as a *discontinuity* in the flow field.) Similarly, the only possibility for a steady-flow process with pressure rise *and* heat addition is a supersonic flow with a standing *detonation* wave. Therefore, from a conceptual perspective, one option to realize the R–G cycle is a modified Brayton cycle in which the heat addition process between states $2'$ and $3'$ (e.g., see Figure 14.4) is via *detonation combustion* (DC), which will be referred to as DCHA cycle.

Detonation is a "rapid and violent form of combustion" that differs from other modes (e.g., flames) in that the main energy transfer mechanism is mass flow in a strong compression wave, i.e., a shock wave, with negligible contribution from other mechanisms (e.g., heat conduction in flames [13–17].) In other words, detonation is a "composite" wave that has two parts: An ordinary shock wave, which raises the temperature and pressure of a mixture of reactants, followed by a thicker reaction zone, in which the chemical reaction (ideally but not necessarily) goes to completion [17].[4] Strictly speaking, the DCHA cycle is not an ideal air-standard cycle, in that the irreversible shock-driven heat addition process cannot be depicted on an equilibrium T-s diagram.[5] Furthermore, DC is not a true constant-volume combustion process. For sufficiently high PR across the *Chapman–Jouget* (C–J) detonation wave, the specific volume ratio is approximately $\gamma/(1 + \gamma)$, where γ is the specific heat ratio of the combustion products downstream of the wave. For a typical γ of 1.3, this gives about 0.6, which is close to an ideal CVHA process (and $q_{in}/c_v \Delta T$ is about 0.85).

Superficially, from a purely numerical perspective (and incorrectly from a theoretical standpoint), the numbers cited above suggest that the DCHA can be considered as an approximation of the CVHA. The correct thermodynamic argument that DCHA is

---

[4] The most common mode of detonation that is of interest for practical applications and the current study is the *Chapman–Jouget* (C–J) detonation. Please refer to the works cited in the text.

[5] In the literature, the DCHA cycle is commonly shown on a *T-s* diagram along with the other cycles via an artifice: The leading shock wave part of the detonation process is depicted as a dashed line combining the states upstream and downstream of the shock wave [18–19].

an apt *qualitative* surrogate for the CVHA process of an R–G cycle (*caution: not the* Atkinson cycle) stems from the fact that C–J detonation is a composite process comprising an *adiabatic* shock (pressure and temperature rise with no reaction) and a heat addition region with reaction accompanied by pressure and temperature relaxation. Furthermore, for the *frozen* shock the entropy change can be shown to be exactly zero so that it is a very fitting proxy for the process $2' \rightarrow 2$ in Figure 14.4 (i.e., the "compressor" in Figure 14.5.)

Closed-form algebraic solutions of the C–J detonation problem for a perfect gas with constant γ can be found in many published sources, e.g., see Thompson [17]. Previous researchers used these equations to determine the efficiency of the DCHA cycle (e.g., see Ref. [18,19]). Herein, only the final result is presented for the sake of completeness:

$$\eta_{DCHA} = 1 - \frac{1}{\tilde{q}}\left(\frac{1}{\Psi} - 1\right), \tag{14.37}$$

$$PR_{C-J} = \frac{1 + \gamma Ma^2}{1 + \gamma}, \tag{14.38}$$

$$\Psi = \frac{Ma^2}{PR_{C-J}^{1+1/\gamma}}, \tag{14.39}$$

where Ma is the C–J *Mach* number given by the following formula as a function of nondimensional heat addition, Θ:

$$Ma = \sqrt{\Theta + 1} + \sqrt{\Theta}, \tag{14.40}$$

$$\Theta = \frac{\gamma^2 - 1}{2\gamma}\tilde{q}. \tag{14.41}$$

Equations (14.37)–(14.41) are based on the *approximate* solution of the shock and detonation adiabats for the C–J condition where reactants (*upstream* of the detonation wave) and products (*downstream* of the detonation wave) are represented by a single-γ perfect gas model. For the detailed development of the equations, the reader is referred to Thompson [17]. The overall entropy rise for the entire C–J detonation, however, is about 15–20% higher than that for the CVHA process $2' \rightarrow 3'$ of the R–G cycle at the same q_{in} and PR', so that DCHA will always be less efficient than its ideal counterpart, the R–G cycle:

$$\frac{\Delta s_{DCHA}}{\Delta s_{R-G}} = \gamma \frac{\ln(1/\Psi)}{\ln(PR_{R-G})} > 1. \tag{14.42}$$

14.2.1   Pulse Detonation Combustion

It is not possible to design a combustor with steady supersonic flow (Ma of about 3–4) and a standing detonation wave in a land-based power generation turbine. One possibility is to create detonation waves at a high frequency (say tens of times per second) inside a semi-closed channel (tube) utilizing a suitable ignition system. The inherent unsteadiness of the practical detonation combustion led to the concept of an intermittent or *pulsed detonation combustion* or *combustor* (PDC), which has been seriously investigated for aircraft propulsion systems for more than half a century. In its most basic configuration, the resulting aircraft gas turbine engine, widely known as a *pulse detonation engine* (PDE), comprises a semi-closed multi-tubular combustion chamber in which detonations are created at a high frequency, e.g., 80–100 times a second for practical flight units [10]. While instantaneous values of static pressure and temperature of the air-fuel mixture and combustion products can reach extremely high values, the *quasi steady-state* engine performance is determined by a mass or time-average of PDC exhaust flow [20].

For a long time, the standard approach to PDC modeling consisted of either using the Atkinson cycle (under the name of Humphrey) as a proxy or the DCHA cycle directly [19]. Alternate models, e.g., *Fickett–Jacobs* or FJ cycle, have been proposed to investigate the performance entitlement of detonation cycle engines [18]. The bulk of the existing work utilizes ideal models with closed-form algebraic solutions such as Equations (14.37)–(14.39) to describe the detonation process. Recently, realistic calculations via CFD methods have been distilled into simple curve-fit formulas and used in system models [20,21]. In any event, a well-established combustion model for the PDC is not available at the present time. Quite frequently, the non-ideal DCHA cycle, used as a measure of the performance of the PDE, has been "shown" to be more efficient than the ideal Atkinson and Brayton cycles [19].[6] This is patently false and stems from two key assumptions; strictly speaking, neither assumption is incorrect, but both are fundamentally erroneous. The first is the comparison of the proverbial apples with oranges, i.e., comparing cycles on a constant precompression PR basis (instead of the more meaningful overall cycle PR basis.) The second, more serious, erroneous assumption is the substitution of *C–J detonation* (a flow process) for an *explosion* in a closed system. While the two terms, i.e., detonation and explosion, have been frequently used interchangeably they are entirely different physical processes.[7] A meaningful comparison of gas turbine cycles requires two prerequisites:

1. All cycle processes are flow processes (i.e., each cycle component is an open system).
2. Cycles are compared on the basis of same overall PR and heat input (or maximum cycle temperature).

The fallacy of not paying attention to these distinctions has been demonstrated quantitatively in Section 14.1.3.

---

[6] It is important to recognize that from a purely thermodynamic cycle perspective, the monikers *pulse* or *pulsed* have no relevance. They are only relevant within the context of the actual aircraft engine or industrial GT, where the combustion is via detonation. In that sense, PDC and DCHA are interchangeable with the implicit understanding that the former pertains to the actual device, whose ideal performance is described by the latter.

[7] For a discussion of this distinction, please refer to Oswatitsch [23].

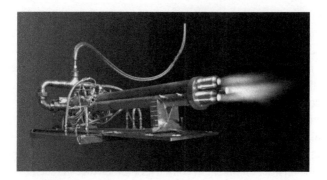

Figure 14.6 General Electric Global Research Center's three-tube detonation rig for PDE investigation (Courtesy: General Electric).

14.2.2   A Brief History of Detonation Engines

The idea of using intermittent/pulsed detonation combustion in aircraft gas turbine engines goes back to the 1950s. (Apparently, German scientists considered it as early as 1940 [11].) In World War II, German engineers briefly toyed with the idea of a gas turbine jet engine with constant-volume combustion (e.g., see Chapter 11 in Kollmann et al. [22] for an example cycle from that time and its thermodynamic analysis.) Since then, the focus has almost exclusively been on aircraft propulsion with emphasis on military applications. Decades-long research and development culminated in the first flight of a PDE powered aircraft in 2008 (detonation/pulse frequency of 80 Hz). The experimental PDE, built for the US Air Force Research Laboratory (AFRL), had four detonation tubes firing at 20 Hz (the shock waves were at about Mach 5). The engine produced a peak thrust of about 200 lbs to power a small aircraft (Burt Rutan's Long-EZ) to just over 100 knots at less than 100 ft of altitude. While it is difficult to see that the public will board airliners powered by PDEs anytime soon (if ever), PDEs recently piqued the interest of major OEMs for land-based power generation applications. As an example, General Electric's (GE's) 3-tube PDE research rig can be seen in a movie on the *GE Global Research* internet site (Figure 14.6).

GE, in cooperation with NASA, also tested a multi-tube PDC and turbine hybrid system (Figure 14.7). The PDC tubes were arranged in a circumferential fashion (like a Gatling gun) "firing" into a single-stage axial turbine (1,000 hp nominal rating at 25,000 rpm). According to GE, the system was operated at frequencies of up to 30 Hz (per tube) in different firing patterns using stoichiometric ethylene-air mixtures to achieve 750 hp at 22,000 rpm.

Government agencies (NASA, AFRL, DARPA), OEMs (GE, Pratt & Whitney, and possibly others outside the USA) and universities, in most cases in a collaborative effort, have done extensive research on PDEs and other variants of pressure-gain combustion (e.g., resonant pulsed combustion, wave rotor, rotating detonation wave) for gas turbine applications. As stated earlier, by far the largest focus is on aircraft propulsion with scattered interest in other applications. At one point, a DARPA

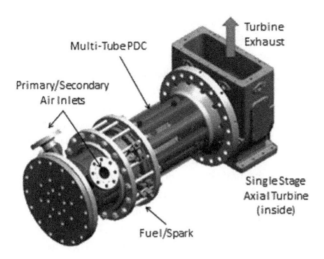

Figure 14.7 GE-NASA multi-tube pulse detonation combustion turbine rig (from Baptista et al. [24]; reprinted by permission of the American Institute of Aeronautics and Astronautics, Inc.).

project (named *Vulcan*, top secret, needless to say), for example, had an objective to develop a constant-volume combustion gas turbine engine for marine propulsion to be deployed in US Navy ships to help the Navy achieve its goal of 20% fuel burn reduction.

Although PDC and PDE concepts were actively investigated for aircraft and marine propulsion as well as land-based power generation applications, they did not lead to a viable product. One obvious impediment was the intermittency of the process (e.g., 30 cycles per second in the GE-NASA turbine rig described above). The practical maximum is limited by about 100 Hz, which is probably not enough to approximate continuous flow requisite for steady and stable aircraft propulsion [25]. Other problems included complex valving (weight and control/timing difficulty) along with the length of tube required to achieve deflagration-to-detonation transition (DDT), i.e., increased size and weight with reduced reliability. Ultimately, PDC research has been dropped in favor of another variant of detonation combustion, i.e., *rotating detonating combustion* (RDC) and RDE. It should be noted that size and weight problems associated with PDC/PDE were not show-stoppers for land-based applications, e.g., as combustors for industrial gas turbines.

Nevertheless, an RDE with a continuous valveless injection of reactants is mechanically and aerodynamically deemed to be superior to the PDE with complex valving because it reduces weight, complexity, and unsteadiness in downstream components. An in-depth review of RDE, including a detailed description of the concept with graphics, tables, selected qualitative and quantitative results, and extensive bibliography, can be found in Shaw et al. [25]. In essence, the RDE is based on an annular combustion chamber (imagine two concentric cylinders with space between them), into which fuel and oxidizer (can be as separate streams or one premixed stream) are fed through a series of orifices. The detonation wave, once initiated tangentially, propagates around the chamber, consuming the reactants (fuel-air mixture) and generating a high-pressure zone of burnt products behind it. In that sense, the underlying principle is similar to the detonation wave propagating in a cylindrical tube. In the RDC, the wave propagates across the annulus in a circular path.

The advantage of this scheme vis-à-vis PDC is two-fold: (i) as long as a fresh fuel-air mixture is supplied (note that no purge is required), the detonation waves are self-sustaining (there are no valves, opening and closing continuously) and (ii) the operational frequency can be as high as 10 kHz, i.e., 10,000 Hz. Inside the chamber (i.e., the annulus) there are contact surfaces, an oblique shock wave, and expansion waves that enhance the ejection of the burnt product out of the combustor almost axially to produce thrust. This process is practically impossible to describe in mere words. There are a variety of CFD simulations as well as video recordings of experiments available on the internet in the form of colorful motion pictures of the wave motion and interaction inside the RDC annulus. The interested reader can easily locate those by a simple Google search. Suffice to say, to an untrained eye, even a three-dimensional visualization of the wave motion and interaction inside the annulus is of not much help to clearly discern the transition from circular wave to axial chamber/annulus exit of burnt products. At the time of writing, active R&D with a focus on aircraft and rocket propulsion on RDE is underway in different countries and in government (mostly defense-related), industry and academic institutions. Said R&D activities include experimental, numerical, and analytical work to bring this highly complex technology to a level amenable to application in a feasible end product.

14.2.3   Simplified Solution of Detonation Problem for GT Performance Calculations

The detonation problem requires the solution of the *jump* equations describing the two *adiabats*, i.e., the leading "frozen" shock with no chemical reaction and the *detonation adiabat* representing all possible end states for the combustion process. It can be shown that the only possible solution to detonations in a semi-closed pipe or tube of length L where the detonation is initiated at the closed end ($x = 0$) and propagates toward the open end ($x = L$), is a C–J detonation [15]. Even a brief description of the C–J detonations is beyond the scope of the current section. The reader who is interested in the full theory of this fascinating subject of the gas-dynamics should consult the classic works written by the established authorities in the field, e.g., Refs. [13–17].

The DC calculations are based on the five-stage PDC cycle described in Ref. [21]. Specifically,

1. Compressed air from the GT compressor is mixed with fuel.
2. The combustible mixture (i.e., reactants) fills the detonation chamber, which is a semi-closed tube.
3. The detonation starts (the method is not specified) instantaneously at the closed end of the tube, which is filled with a (premixed) fuel-air mixture.

4. The detonation wave is a C–J wave, which is the only possible solution (e.g., see Ref. [15]). The C–J shock wave, idealized as a pressure and temperature discontinuity, propagates toward the open end of the tube. Downstream of the C–J wave, the fluid comprises the complete combustion products. A rarefaction wave follows the C–J wave and satisfies the boundary condition of zero velocity at the closed end of the tube, $x = 0$.
5. The pressure of the combustion products drops across the rarefaction wave and the C–J wave exits from the open end of the tube, at $x = L$, where L is the length of the tube. An expansion (rarefaction) wave starts at the open end and moves to the closed end of the tube. A further drop in the pressure takes place across the expansion wave. The burned combustion products evacuate the tube, which is now ready for the next cycle to start with purging and then filling with fresh fuel-air mixture.

The entire cycle is repeated and results in a "periodic high-pressure zone near the closed end of the tube" [11]. In a gas turbine, the fresh air for the detonation process and the purge air are supplied from the discharge of the compressor and the combustion products exiting the tube expand through the turbine section to generate useful shaft work.

The following data is known (or assumed):

1. Fuel mass flow rate, temperature, and composition (i.e., natural gas as 100% CH_4)
2. Air composition, pressure, temperature, and mass flow rate (same as the base GT case) at the exit of the compressor
3. Purge flow rate is assumed to be 20% of the total of air mass flow rate that enters the reaction and the fuel flow

Pertinent assumptions implicit in the calculations are as follows:

- The flow in the PDC chamber, which is a circular tube with constant cross-sectional area, is one-dimensional, inviscid, and adiabatic.
- Reactants (i.e., fuel and air) and combustion products are ideal gases with properties per a JANAF model [26]. In other words, they are *not* assumed to be calorically perfect.
- Valve(s) that control the cycle processes open and close instantaneously with no losses.
- Pressure losses in the PDC are ignored.
- Static pressure and temperature inside the detonation chamber are assumed to be the same as the total values at the compressor exit.
- The fuel-air mixture at the beginning of the detonation is assumed to be at the static pressure of the compressed air.
- Total pressure and temperature at the PDC exit are calculated using stagnation formulas for state 4.
- The state of the working fluid at the turbine stage-1 nozzle inlet is determined by mixing with purge air.
- Mixing with purge air is assumed not to change the total pressure.

The calculation sequence is as follows:

1. Calculate composition, total enthalpy, and temperature of the unreacted air (from the compressor discharge) and fuel gas mixture (state 1).
2. Calculate the state of the combustion products downstream of the C–J wave following the theory outlined in Ref. [17], state 2. The hydrodynamic solution of the shock problem involves three jump conditions (written for a reference frame that is moving with the shock front) and the ideal gas equation of state. The solution is iteratively obtained in Excel using the SOLVER add-in to determine the four unknowns from the four equations.[8]
3. Use the PR relationship for the centered (isentropic) rarefaction fan (from Eq. 3-24b in Ref. [14]) to calculate the properties of the combustion products downstream of the rarefaction wave part of the C–J detonation, state 3.
4. Use the isentropic formula and the overall heat and mass balance to calculate the total properties of the state downstream of the expansion wave, state 4. The centered expansion wave starting from the open end of the tube and propagating toward the closed end carries with it the "information" that the detonation wave has exited the tube with the ensuing drop in pressure. In laboratory experiments, this final expansion (or blow-down) empties the detonation tube and the exiting combustion products end up with the reservoir pressure. Herein, the final pressure at the exit of the PDC, labeled as state 4, is determined from the isentropic p-T relationship and the overall PDC heat balance. This method is different from the time-averaged PR for the detonation combustion, which is the ratio of the time-averaged total pressure at the PDC exit to the time-averaged total pressure at the PDC inlet.
5. Calculate the final state, i.e., state 5, after mixing with the purge air (pressure losses are neglected). Calculate the final temperature before the entry to the turbine section by mixing the combustion products with the purge air and calculating the enthalpy via simple energy balance.

A *conceptual* representation of states 1–4 is shown in Figure 14.8, which qualitatively depicts two pressure transducer signals recorded at two different PDC tube locations in experiments. The numbers are representative for a C–J detonation propagating into a fuel (100% CH_4)-air mixture (fuel-air-ratio of ~0.04) at $Ma = 3.72$.

Calculations are done for a range of precompression PRs, ranging from 2 to 8, and fuel-air ratios (FAR), ranging from 0.009 to 0.035. The results are cast into transfer functions to be used in gas turbine performance calculations.

$$TR = \frac{175 \, (FAR/\tau_2) + 1}{FAR^{5.5174}}, \tag{14.43}$$

---

[8] The equations from Ref. [17] are: Rankine–Hugoniot adiabat, Eq. 7–23, and Rayleigh line, Eq. 7–18, C–J condition ($Ma = 1$ downstream of the detonation wave, Eq. 7–72) and Mach number definition upstream of the detonation wave.

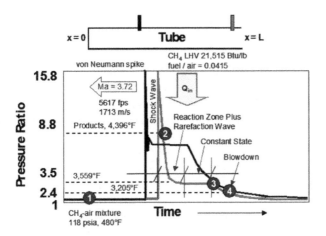

Figure 14.8 Conceptual C–J detonation states (state 5 is not shown).

$$FAR = 1.8211 \left(\frac{1 - TR}{TR - 175/\tau_2} \right)^{1.1106}, \qquad (14.44)$$

$$PR = 0.5289\ TR + 0.3751, \qquad (14.45a)$$

or using the R-G cycle terminology

$$PR'' = 0.5289\ \theta + 0.3751 \qquad (14.45b)$$

In the equations above, TR is the nondimensional PDC tube exit temperature T/T_{mix}, which, in the cycle notation of the states,[9] is T_3/T_2 or, in the detonation wave notation in Figure 14.8, T_5/T_1. (In other words, TR is the nondimensional temperature rise across the PDC tube.) For the calculation of the reactants' temperature, the following relationship is used:

$$T_{mix} = \frac{c_{p,air}\ T_2 + FAR\ c_{p,fuel}\ T_{fuel}}{c_{p,air} + FAR\ c_{p,fuel}}. \qquad (14.46)$$

Although not needed for the PDC modeling, the following transfer function is provided as a comparison to earlier published relationships between PR and enthalpy ratio HR:

$$PR = 0.2472 \left(\gamma_2 \sqrt{\tilde{q}}\ HR^{1-1/\gamma_2} \right) + 0.7196. \qquad (14.47)$$

The formulas are valid for 100% CH_4 gaseous fuel and a TR between 1.5 and 5.0. The PDC performance of Equation (14.45) is comparable to that by Nalim [8] for a "perfect constant-volume combustor" (CVC):

---

[9] Strictly speaking, air-fuel reactants mixture temperature, which is not present on a T-s diagram for the air-standard cycle.

$$PR = TR^{\gamma/\gamma-1} \left\{ 1 + \gamma \left(TR - 1 \right) \right\}^{1/1-\gamma}. \tag{14.48}$$

For the same temperature rise, TR, Equation (14.45) predicts about 25% higher pressure rise PR than Equation (14.48) for heat addition $\tilde{q}$ between 3.0 and 4.0. Whether this performance can be realized or not remains to be seen.

As a final note, combining the ideal gas equation of state with Equation (14.45), one obtains the specific volume ratio across the detonation combustor (inclusive of the expansion) as

$$VR = \frac{1 - \frac{0.3751}{PR}}{0.5289} = 1.8907 - \frac{0.7092}{PR}. \tag{14.49}$$

If the pressure gain in the detonation combustor is equivalent to, say, PR = 2:1, from Equation (14.49), VR is found as 1.5361 (instead of the ideal constant-volume process with VR = 1). (In passing, PR = 1.26:1 for VR = 1.) In this case, a modification of the R–G cycle is necessary as follows. Equation (14.22) now becomes

$$dP/ds = (\partial P/\partial s)_v + (\partial P/\partial s)_v \frac{dv}{ds} = \frac{P}{c_v} - \left(\frac{a}{v}\right)^2 \frac{dv}{ds}, \tag{14.50}$$

where a is the speed of sound, which, for an ideal gas, is given by

$$a = \sqrt{\gamma RT},$$

$$dP/ds = (\partial P/\partial s)_v + (\partial P/\partial s)_v \frac{dv}{ds} = \frac{P}{c_v} - \frac{\gamma RT}{v^2} \frac{dv}{ds}. \tag{14.51}$$

Combining Equation (14.51) with Equation (14.20), using the ideal gas equation of state and the definition of specific heat ratio, one obtains

$$dh = \left(T + \frac{RT}{c_v} - \frac{\gamma RT}{v} \frac{dv}{ds} \right) ds$$

$$= \left(1 - \frac{R}{v} \frac{dv}{ds} \right) \gamma T ds$$

$$= \left(1 - R \frac{d}{ds} \left(\ln v \right) \right) \gamma T ds. \tag{14.52}$$

Integrating Equation (14.52) gives

$$q_{in} = h_{3'} - h_{2'} = \gamma \, METH'(s_{3'} - s_{2'}) - \bar{a}^2 \ln\left(\frac{v_{3'}}{v_{2'}}\right),$$

Where $\bar{a}^2$ is the square of the "mean-effective" speed of sound, which can be expressed as

$$\bar{a}^2 = \gamma \, R \cdot METH',$$

so that, with substitution, one arrives at

$$\frac{h_{3'} - h_{2'}}{(s_{3'} - s_{2'})} = \gamma\,\text{METH}' - \frac{\gamma\,\text{RMETH}'\,\ln\left(\frac{v_{3'}}{v_{2'}}\right)}{c_v\,\ln\left(\frac{T_{3'}}{T_{2'}}\right) + R\,\ln\left(\frac{v_{3'}}{v_{2'}}\right)}$$

$$= \gamma\,\text{METH}\left[1 - \frac{1}{\frac{\ln\theta}{(\gamma-1)\,\ln\left(\frac{v_{3'}}{v_{2'}}\right)} + 1}\right]. \tag{14.53}$$

According to Equation (14.53), the mean-effective heat addition temperature of an ideal (constant-volume) R–G cycle is reduced by the factor

$$\delta = \frac{1}{\frac{\ln\theta}{(\gamma-1)\,\ln\left(\frac{v_{3'}}{v_{2'}}\right)} + 1}. \tag{14.54}$$

For the pulse detonation process given by Equation (14.49), Equation (14.54) becomes

$$\delta = \frac{1}{\frac{\ln\theta}{(\gamma-1)\,\ln\left(1.8907 - \frac{0.7092}{\text{PR}}\right)} + 1}. \tag{14.55}$$

14.3   Reality Check

Mechanical problems associated with the PDC (structural integrity of turbine blades, vibration, acoustics, high stress, and fatigue) has been a big impediment in the development of aircraft engines utilizing the technology. As discussed in Section 14.2.2, the rotating or continuous detonation combustion (RDC) concept is a variant, which has the potential of overcoming most of those problems. Either PDC or RDC is readily applicable to land-based gas turbines for power generation.

Dynamic pressure exchangers or "wave rotors" have also been investigated for pressure gain in combustion. A *wave rotor* (WR) is a rotating, cellular drum between the inlet outlet manifolds [8]. The flows inside the manifolds are steady, which is a significant advantage over the PDC with strongly unsteady, pulsating flow. Pressure gain is generated inside the rotor cells via shock waves generated by stoppage of the outlet flow and compression of the inlet (charge) flow via rotation between the manifolds. While the concept sounds too esoteric at first glance, it should be noted that the *Comprex* supercharger developed by the Brown Boveri Co. to be used in diesel engines was essentially a WR.

The valveless (also called *aero-valved*) *pulse combustor* is another combustion device that generates only a modest pressure gain. It is a variant of the valved pulse combustor used in the German V-1 *Buzz Bomb* in World War II [27]. The mechanical "flapper" valve is replaced by an inlet pipe, which acts as an acoustically modulated aerodynamic valve (i.e., no moving parts).

From a land-based, electric generation perspective the key engineering design criterion is the relationship between pressure and temperature ratios, PR and TR,

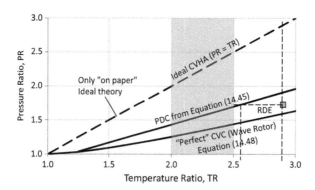

Figure 14.9 Pressure and temperature ratios for PDC and CVC.

respectively, in the pressure-gain combustion (PGC) and the requisite fuel-air ratio (FAR). From that perspective, the PDC offers the largest efficiency gain potential but is beset by many design challenges, primarily the problem of integration between the unsteady outlet gas flow and the steady turbine flow. For large heavy-duty industrial gas turbines (HDGT), the relevant TR range is 2.0–2.5. The pressure rise potential of PDC and WR (using the "perfect" CVC relationship, Equation (14.48)) are shown in Figure 14.9. (Pulse combustion PR is too low to be of much significance, i.e., less than 1.10:1.) The result is in good agreement with formulae developed by different researchers [28].

The present state of the art in land-based, heavy-duty industrial gas turbine technology is the J Class with nominal (introductory, ca. 2013) 1,600°C TIT and PR of 23: to 24:1. (At the time of writing, 2020–21, actual TIT values are closer to 1,700°C.) The largest J Class HDGTs are 50 Hz machines rated well above 500 Mwe generator output and more than 43% thermal efficiency [29]. Smaller 60 Hz machines are rated at 350 Mwe or higher with similar efficiency. On a combined cycle basis, J Class HDGTs have been pushing to a 64% thermal efficiency rating. As the basis of the real cycle analysis, a 60 Hz J Class HDGT with 700 kg/s (nominal) airflow, PR of 23:1, and TIT of ~1,660°C (estimated from a heat and mass balance analysis) is chosen. Thermoflow Inc.'s THERMOFLEX software (see Section 16.2) is used to model the gas turbine with appropriate chargeable and nonchargeable cooling flows (see Figure 14.10).

The base HDGT in Figure 14.10 is modified by replacing the combustor with a PGC and adding a booster compressor (BC) as shown in Figure 14.11. The BC ensures that chargeable and nonchargeable cooling flows for the turbine stage 1 are delivered from the compressor discharge, which is now at a lower pressure. This is so because part of the compression duty is taken over by the *quasi* constant-volume combustion in the PGC. Thus, hot gas pressure at turbine stage 1 is in fact higher than that at the compressor exit.

The pulsed detonation combustion/combustor (PDC) is assumed to be the particular PGC process. The PDC is modeled in THERMOFLEX as a dummy

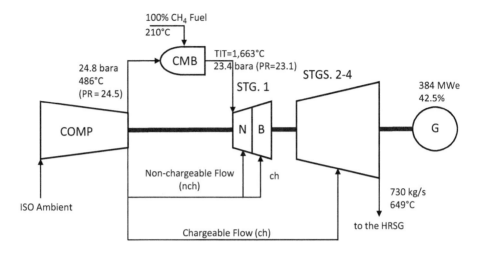

Figure 14.10 J Class gas turbine model and performance (overall, cooled stage 1 isentropic efficiency is 80.4%).

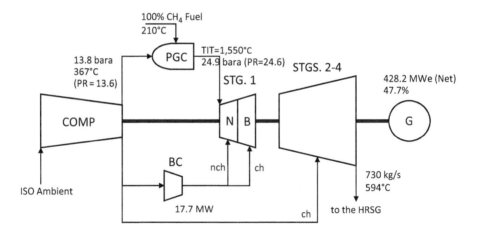

Figure 14.11 Gas turbine with pressure-gain combustor (overall, cooled stage 1 isentropic efficiency is 77.2%).

compressor-combustor combination controlled by a script (to implement temperature and pressure rise via Equation (14.45) along with the fuel consumption via a heat balance). A pressure-temperature changer accounts for pressure and purge-air-dilution loss (5% and ~100°C, respectively) between the PDC and the turbine inlet plus the diffuser (see Figure 14.12). The unfavorable impact of the pulsating gas flow, somewhat "smoothed out" in the transition piece between the PDC and stage 1 nozzle, is quantified by a three-percentage point debit to the stage 1 isentropic efficiency.

To facilitate a correct comparison with the Brayton cycle GTCC benchmark, two cases of GTCC performance with PDC-GT are evaluated:

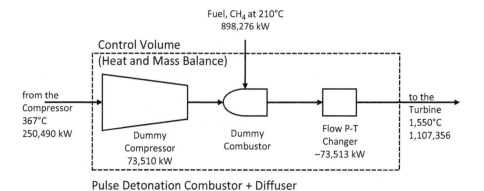

Figure 14.12 Implementation of the PDC in the THERMOFLEX model.

- Same heat input and cycle PR as the base Brayton cycle GTCC (case A)
- Same TIT as the base Brayton cycle GTCC, i.e., about 1,650C (case B)

Gas turbine performance is calculated rigorously in THERMOFEX as described above. Bottoming cycle performance is estimated using the second law (exergy) approach, which is described in detail in **GTFEPG**. The reader can also refer to a paper by Gülen for a concise description and formulae (Ref. [45] in Chapter 2). This simple but highly accurate method enables comparison the of gas turbine technologies on a proverbial "apples to apples" basis without getting bogged down in the details of the bottoming steam (Rankine) cycle modeling. This is neither an empty boast nor a trivial point. Rather, deliberately (in the case of unscrupulous marketing people) or inadvertently (in the case of inexperienced and untrained researchers and engineers), this is where many grossly inaccurate and, in some cases downright ludicrous, performance claims are made in the archival and industry publications. For an in-depth look at this point and some egregious examples, the reader should consult a recent paper by Gülen ([12] in Chapter 2).

Key assumptions used in the performance estimates are as follows. Plant auxiliary power consumption is assumed to be 1.6% of plant gross output. (This is a bare bones minimum for rating purposes.) Fuel compression is not included (i.e., pipeline pressure is assumed to be high enough). BC load is debited to the gas turbine generator output. Selected results are listed in Table 14.5. Bottoming cycle technology is set to the BC exergetic efficiency of the Brayton GTCC, which is representative of the most advanced three-pressure, reheat (3PRH) bottoming cycle and steam turbine generator and low back pressure set by a once-through water-cooled condenser heat rejection system. Adjustment to the base value is a function of the gas turbine exhaust gas temperature. The takeaways from the performance data presented in Table 14.5 are as follows:

- If it can be done right, a gas turbine equipped with a pressure-gain combustor, with a performance as modeled by the pulse detonation combustion process encapsulated by Equations (14.43)–(14.49), one can expect a thermal efficiency gain of two to three percentage points (combined cycle basis).

Table 14.5 Performance comparison of Brayton and PDC-GT GTCC power plants

	BASE BRAYTON	PDC-GT CASE A	PDC-GT CASE B
Gas Turbine Net Electric Output, kWe	383,937	428,215	462,594
Fuel Consumption, kWth (LHV)	903,962	898,276	964,261
Gas Turbine Efficiency, %	42.5	47.7	48.0
PR''	NA	1.81	1.89
θ	NA	2.71	2.86
τ_2	NA	2.22	2.22
FAR	0.0324	0.0338	0.0397
Cycle Pressure Ratio (PR)	23.1	24.6	25.7
Turbine Inlet Temperature (T_3), °C	1,663	1,550	1,650
Exhaust Temperature, °C	648.9	594.2	627.6
Exhaust Flow, kg/s	730	729.9	731.2
Exhaust Gas Exergy, kW	248,482	216,481	236,222
Bottoming Cycle Exergetic Efficiency, %	78.3	77.3	78.0
Steam Turbine Generator, kWe	198,365	170,562	187,774
CC Gross Output, kWe	582,302	598,777	650,368
CC Net Output, kWe	572,985	589,197	639,963
CC Net Efficiency, %	63.4	65.6	66.4

- At the low end, this can be achieved at a gas turbine cycle (overall) PR comparable to the present J Class technology but with roughly 100°C lower TIT.
- At the high end, this can be achieved at the same TIT and comparable cycle PR as the present J Class technology.
- No change in bottoming cycle technology (i.e., 3PRH Rankine steam cycle) is required.

Achieving this, presently "on paper," performance requires significant investment in hardware development. This requires attention to the "usual suspects" encountered in DLN combustor design, e.g., combustion stability, acoustic vibrations, NOx and CO emissions, and others in addition to the new challenges presented by this intermittent firing technology. In this author's opinion, size, weight, and similar considerations, which rendered PDC not very appealing to aircraft propulsion, are not valid for land-based gas turbine applications. The critical item is the firing frequency, which is limited by about 100 Hz for PDC. To determine whether this is going to be sufficient for a quasi-continuous combustion product (i.e., hot gas) flow into the first stage of the gas turbine requires the building of a full scale prototype and extensive testing.

14.4   Operability Considerations

As far as operability goes, the challenges encountered by a GTCC with PDC-GT are similar to those discussed within the context of conventional GTCC. The bottoming cycle is the same and steam turbine stress control considerations should be fully applicable. However, the unique combustion process introduces additional stability considerations into the mix when starting, shutting, and load ramping the gas turbine. The key question is how to reliably initiate and sustain detonations over a range of operational conditions with natural gas fuel that can show some composition variation. This is so because the detonation process is very sensitive to stoichiometry, particle/droplet size (liquid fuels), local degree of mixing, etc. In particular, detonation chamber purging, and refilling must be reliably repeated on very short time scales to ensure against the premature ignition of fresh propellant charges. Furthermore, the turbine downstream is typically designed with a steady flow of gas. The pulsing nature of gas flow from the detonation chamber(s) can be detrimental not only to component efficiency but also to component integrity due to vibrations and high cycle fatigue.

There is a large body of literature on pulse PDEs for aircraft propulsion and the operability challenges unique to that application mode. To the best knowledge of this author, there has not been a design and operational study specifically aimed at land-based applications for shaft power generation. The only PDE, so far, was flown on a modified *Scaled Composites Long-EZ* aircraft in January 2008, in the Mojave desert. The aircraft achieved a speed of over 120 mph and 60–100 feet altitude, which produced greater than 200 pounds of thrust. A jet assist takeoff was used to minimize takeoff roll and provide more runway margin but was subsequently shut down when the PDE provided plenty of thrust for flight – for 10 seconds at an operating frequency of 80 Hz. The engine was developed and manufactured by in-house by members of the *AFRL Propulsion Directorate's Turbine Engine Division*, Combustion Branch and its on-site contractor, *Integrated Silicon Solution Inc.* (ISSI).

Fuel and oxidizer (i.e., reactants) selection affects PDE cycle performance due to the large variation in detonation velocities, compression ratios, and temperatures produced by various types of fuels. Gaseous reactants are advantageous due to their lower detonation energy requirements. Just like in a conventional DLN combustor, liquid fuels can be used with atomization prior to injection. Nevertheless, they require more power from a direct ignition system or a longer deflagration-to-detonation transition section.

Clearly, there is a lot of development work that needs to be done to ensure that a valved (e.g., with tubes) or valveless (e.g., rotating detonation) PDC is going to be ready for prime time. Even then, there still remains the issue how to start a PDC-GT from turning gear speed to full speed, no load (FSNL) and then to full speed, full load (FSFL). An added complication in this respect is the presence of the BC requisite for delivery of hot gas path cooling air into the turbine section. Present day DLN combustors operate in different stages (e.g., diffusion, piloted premix, premix) in a tightly controlled manner subject to a lot of field "tuning." That technology state of the art has been achieved over a time period covering more than four decades with hundreds of thousands of hours of field and lab test-bench operating experience.

There is also the issue of NOx emissions in gas turbines with PDC. There is an expectation that short residence times at high flame temperatures in detonation combustors can actually be favorable to low NOx emissions. The literature on this aspect of detonation combustion is pretty limited. The reader is referred to a recent paper on this subject [30]. Ultimately, EGR can be utilized to mitigate NOx emissions. However, this would mean adding complexity to a system already more complex than a conventional gas turbine. In the end, emissions control difficulties can be the ultimate Achilles' heel for PDC-GT for successful deployment in utility scale power generation applications.

14.5   References

1. Horlock, J., Young, J., and Giampaola, M. 2000. Exergy analysis of modern fossil-fuel power plants, *Journal of Engineering for Gas Turbines and Power*, 122, 1–7.
2. Amann, C. A. 2005. Applying thermodynamics in search of superior engine efficiency, *Journal of Engineering for Gas Turbines and Power*, 127, 670–675.
3. Stodola, A. 1927. *Steam & Gas Turbines*, Authorized Translation From The 6th German Edition by L. C. Löwenstein, Volume 2, pp. 1237, New York, NY: McGraw-Hill Book Company Inc.
4. Narayanaswami, L. and Richards, G. A. 1996. Pressure gain combustion: Part I – Model development, *Journal of Engineering for Gas Turbines and Power*, 118, 461–468.
5. Reynst, F. H. 1961. *Pulsating Combustion – The Collected Works of F. H. Reynst*, M. W. Thring (ed.), Oxford, UK: Pergamon Press.
6. Richards, G. A. and Gemmen, R. S. 1996. Pressure gain combustion: Part II – Experimental and model results, *Journal of Engineering for Gas Turbines and Power*, 118, 469–473.
7. Paxson, D. E. and Dougherty, K. E. 2005. Ejector enhanced pulsejet based pressure gain combustors, AIAA-2005-4216, NASA/TM-2005-213854, 41st Joint Propulsion Conference and Exhibit, Tucson, AZ.
8. Nalim, M. R. 1999. Assessment of combustion modes for internal combustion wave rotors, *Journal of Engineering for Gas Turbines and Power*, 121, 265–271.
9. Akbari, P., Nalim, R. and Mueller, N. 2006. A review of wave rotor technology and its applications, *Journal of Engineering for Gas Turbines and Power*, 128, 717–735.
10. Bussing, T. R. A. and Pappas, G. 1996. Pulse detonation engine theory and concepts, *Progress in Aeronautics and Astronautics*, 165, 421–472.
11. Kailasanath, K. 2000. Review of propulsion applications of detonation waves, *AIAA Journal*, 38(9), 1698–1708.
12. Gülen, S. C. 2010. Gas turbine with constant volume heat addition, ASME Paper ESDA2010–24817, ASME 10th Biennial Conference on Engineering Systems Design and Analysis, July 12–14, Istanbul, Turkey.
13. Fickett, W. and Davis, W. C. 1979. *Detonation*, Berkeley, LA: University of California Press.
14. Liepmann, H. W. and Roshko, A. 1957. *Elements of Gas Dynamics*, New York, NY: John Wiley & Sons.
15. Landau, L. D. and Lifshitz, E. M. 1987. *Fluid Mechanics*, 2nd ed., Oxford, UK: Pergamon Press.
16. Strehlow, R. 1984. *Combustion Fundamentals*, New York, NY: McGraw-Hill.
17. Thompson, P. A. 1988. *Compressible Fluid Dynamics*, New York, NY: McGraw-Hill.

18. Wintenberger, E. and Shepherd, J. E. 2006. Thermodynamic cycle analysis for propagating detonations, *Journal of Propulsion and Power*, 22(3), 694–698.

19. Heiser, W. H. and Pratt, D. T. 2002. Thermodynamic cycle analysis of pulse detonation engines, *Journal of Propulsion and Power*, 18(1), 68–76.

20. Paxson, D. E. 2004. Performance evaluation method for ideal airbreathing pulse detonation engines, *Journal of Propulsion and Power*, 20(5), 945–947.

21. Goldmeer, J., Tangirala, V. and Dean, A. 2008. System-level performance estimation of a pulse detonation based hybrid engine, *Journal of Engineering for Gas Turbines and Power*, 130, 011201-1-8.

22. Kollman, K., Douglas, C. E. and Gülen, S. C. 2021. *Turbo/Supercharger Compressors and Turbines for Aircraft Propulsion in WWII: Theory, History and Practice – Guidance from the Past for Modern Engineers and Students*, New York, NY: ASME.

23. Oswatitsch, K. 1976. *Grundlagen der Gasdynamik*, Vienna, Austria: Springer Verlag, pp. 142–143.

24. Baptista, M., Rasheed, A., Badding, B., Velagandula, O. and Dean, A. J. 2006. Mechanical response in a multi-tube pulsed detonation combustor-turbine hybrid system, *AIAA 2006-1234, 44th AIAA Aerospace Sciences Meeting and Exhibit, January 9–12*, Reno, NV.

25. Shaw, I. J., Kildare, J. A. C., Evans, M. J., et al. 2019. A Theoretical Review of Rotating Detonation Engines. In S. Rao (ed.) *Direct Numerical Simulations – An Introduction and Applications*, London, UK: IntechOpen.

26. Stull, D. R. and Prophet, H. 1971. *JANAF Thermodynamic Tables*, 2nd ed., NSRDS-NBS 37, National Bureau of Standards.

27. Gülen, S. C. 2013. Constant volume combustion: The ultimate gas turbine cycle, *Gas Turbine World*, 43(6), 20–27.

28. Tangirala, V. E., Rasheed, A. and Dean, A. J. 2007. Performance of a pulse detonation combustor-based hybrid engine, *GT2007–28056, ASME Turbo Expo – Power for Land, Sea & Air, June 14–18*, Montreal, Canada.

29. Gas Turbine World. 2020. *Handbook*, Vol. 35, Fairfield, CT: Pequot Publishing.

30. Ferguson, D., O'Meara, B., Roy, A., et al. 2020. Experimental measurements of NOx emissions in a rotating detonation engine, AIAA SciTech Forum, January 6–10, Florida, FL.

15 Epilogue

Wer alles verteidigt, verteidigt nichts.[1]

Friedrich der Große

To paraphrase the great king of Prussia, it is impossible to be all encompassing in covering a broad subject, especially in a book of a few hundred pages. In this book, an attempt is made to cover the role of a particular prime mover, which represents the pinnacle of human technology in the beginning of the third decade of the twenty-first century, i.e., the *gas turbine*, in the current climate of *energy transition* (*die Energiewende*) and the number one goal underlying it, *decarbonization*. (After all, steam is a gas, too, albeit much denser than air and combustion products (gas), and, thus, an integral part of the coverage herein.) In doing this, the focus has been on the following areas with select quantitative examples:

- *Mature* (TRL 9) technologies forming the backbone of electric power generation at the time of writing, i.e., simple and combined cycle gas turbine power plants in the light of
    - Technology factor, i.e., a physics based (specifically, the second law of thermodynamics) measure of technological maturity
    - Operability with acceptable RAM
- *Emerging* (TRL 6 or lower) technologies in energy storage and power generation, which are quite often hyped beyond what they can realistically promise, with the help of the technology factor concept supported by detailed heat and mass balances or first-principles calculations and the operability aspect
- *Other* technologies that have been around for decades, e.g., IGCC, CSP, and nuclear, which, regardless of their TRL 9 status, either failed to make it to the proverbial prime time, mainly due to commercial reasons, or have been unfairly maligned (especially, nuclear) even though they can make a substantial contribution to *decarbonization*
- Hydrogen as an *energy vector* (because it is hailed as the "knight in shining armor" to save the world from a climate catastrophe)
- Pre- or post-combustion carbon capture (as a facilitator of a second lease at life, so to speak, for the fossil-fuel technologies in the era of *decarbonization*)

---

[1] He who defends everything, defends nothing.

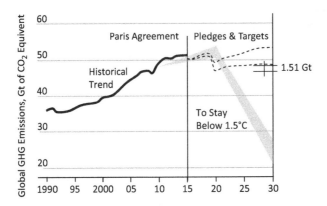

Figure 15.1 Global GHG emissions in gigatons of CO_2 equivalent.

Since the book's focus is on technologies directly or indirectly related to gas turbines and power plants utilizing prime movers and similar types of turbomachinery, technologies such as *direct air carbon capture* (DACC), touted as an enabling factor for the *Paris Agreement* objectives (see Figure 15.1), are not covered. Considering the extremely low concentration of CO_2 in atmospheric air (about 0.4% by volume), one would be inclined to reject DACC off the cuff as a *very* long shot. Indeed, a careful analysis shows that this is likely to be the case.[2] Furthermore, zero carbon renewable technologies such as *wind* and *solar PV*, although they are the leading actors (so far) in the *energy transition*, are also not discussed at any length because they do not comprise any "burning and/or turning" components. The reader should consult treatises focusing exclusively on those technologies.

The Paris Agreement is a legally binding international treaty on climate change with the goal of limiting global warming to well below 2°C, preferably to 1.5°C, compared to pre-industrial levels. It was adopted by close to 200 parties at COP 21 in Paris, on 12 December 2015 and entered into force on 4 November 2016.[3] The level of greenhouse gas (GHG) emissions expressed in terms of Gt of CO_2 equivalent to reach that goal by midcentury is shown in the chart in Figure 15.1. The chart also shows how dramatically the global community falls short of that objective. The gap is somewhere between 23 and 27 Gt as of COP 26 held in Glasgow, Scotland, in 2021.

According to the Paris Agreement, countries had to submit their plans for climate action known as *nationally determined contributions* (NDCs), in which countries communicate the actions they will take to reduce their GHG emissions to reach the agreement coals. Countries also communicate in the NDCs the actions they will take to build resilience to adapt to the impacts of rising temperatures. Alas, updated NDCs

[2] Sudipta Chatterjee and Kuo-Wei Huang. 2020. Unrealistic energy and materials requirement for direct air capture in deep mitigation pathways, *Nature* Communications, 11, 3287.

[3] The Conference of the Parties attended by the countries that signed the United Nations Framework Convention on Climate Change (UNFCCC) – a treaty that came into force in 1994. The 26th COP (COP 26) summit was held in Glasgow from 1 to 12 November 2021, a year later than planned due to delays caused by the COVID pandemic.

in COP 26 made a dent of only 4 Gt in the aforementioned gap. Sectoral initiatives such as the global methane pledge (to reduce human-caused methane emissions by 30% between 2020 and 2030), electrical vehicles, afforestation, and a global coal-to-clean power transition add another reduction of 2.2 Gt (equivalent to the emissions of Germany, Japan, and the UK *combined*) so that the gap reduces to 17–20 Gt.

In Section 11.3, it was noted that CO_2 emissions from natural gas- and coal-fired electricity generation in the USA amounted to about 1,500 million metric tons (1.51 *gigatons*). This number was shown to translate to an atmospheric concentration of 0.2 ppmv, whereas, at present, the CO_2 concentration in the atmosphere is pushing 420 ppmv and the Paris Agreement goal is 350 ppmv. Even eliminating the entire US emissions at a stroke of a pen, somehow, the gap would barely budge.

The author would be the first to admit that the science behind the climate change tied to anthropogenic GHG emissions and the doomsday scenarios are not as precise and unassailable as, say, the solution of a quadratic equation. There is always a possibility, albeit very low, that decades will pass and nothing catastrophic will happen. On the other hand, the alternative has a finality that does not leave room for corrective action. In that regard, the reader is encouraged to read Steven Koonin's recent book *Unsettled* (cited in Chapter 1).

If the thinking is that we do indeed have decades to let the market dynamics work their magic, there is no case to be made. However, the smart assumption, in this author's opinion, of course, would be that (i) we do not have decades and (ii) the problem cannot be left to time and markets. In other words, we do not have time anymore to waste with dilettantism to find "magic wands" to find a wholesale solution. We must stop making excuses for not attacking the problem of manmade climate change. There must be a concerted global attack against growing CO_2 emissions. It must happen now, and all must participate for success.

There are four major options for generating utility-scale electric power.[4] Ignoring the small players (e.g., fuel oil–fired), they are as follows (carbon-free[5] ones are in *italics*):

1. Fossil fuel–fired
   a. Coal
   b. Natural Gas (NG)
   c. Hydrogen (if generated from coal gasification or steam-methane reforming [SMR])
2. *Renewable*
   a. *Hydro*
   b. *Wind*
   c. *Solar*
   d. *Geothermal*

---

[4] It is important to recognize the electric power is only one *relatively minor* part of the *global* GHG emissions problem. In the USA, electricity barely accounts for 30% of GHG emissions whereas industry and transportation account for 50% of it.

[5] Carbon-free during operation only; not "cradle to grave".

3. *Nuclear*
4. *Hydrogen* (from electrolysis using renewable or nuclear power)

Hydrogen is the *odd duck* in this picture. It is not a naturally available energy resource; it is an "energy carrier" (see Chapter 8), i.e., it cannot be mined or extracted. It can only be produced at the expense of large energy consumption. If the energy requisite for its manufacture is green, then H_2 is a green fuel – its combustion produces only water (H_2O). It can also be produced chemically from fossil fuels (e.g., SMR widely used in the chemical process industry for hydrogen production) – with or without carbon capture, i.e., *blue* or *gray* H_2.

The significance of hydrogen comes from its *dual* nature:

- Carbon-free combustion in gas turbines (the most efficient electric power generation technology)
- Energy storage medium (similar to any solid, liquid, or gaseous fuel)

To be carbon-free, fossil fuel–fired technologies require carbon capture accompanied by either (i) putting it to use (e.g., enhanced oil recovery [EOR] or (ii) putting it deep into the ground (i.e., sequestration), hence the acronym CCUS.

There are two opportunities for capturing carbon:[6]

1. Pre-combustion (clean the fuel gas, e.g., clean the gasification product syngas and separate H_2)
2. Post-combustion (clean the stack gas)

A comparison of coal-fired steam power plants and natural gas–fired combined cycle power plants makes it abundantly clear that even the most efficient advanced ultra-supercritical (USC) steam power plant (with the best quality coal) cannot even get close to let alone match the CO_2 emission performance of vintage gas turbine combined cycle (GTCC) technology.[7]

Prima facie, coal has pretty much only one shot left at playing a role in a carbon-free generation portfolio: CCUS. The post-combustion variant is a dead end since modern GTCC plants with advanced class gas turbines can be equipped with post-combustion capture as well – with superior efficiency and less cost (they do not need extra clean-up technologies for criteria pollutants). This leaves gasification and pre-combustion capture as the only viable alternative for coal to play a role in a sustainable power portfolio.

The *energy transition* is based on *sustainable* electric power generation. The moniker "sustainable" can be translated as *net zero carbon* power (i) with no emissions of other toxic pollutants and (ii) via judicious use of natural resources of the earth. There are two constraints to be satisfied. The first is the attainment of this goal

---

[6] One might add "oxyfuel combustion with supercritical CO_2 as the combustion diluent" (i.e., the Allam cycle – still in demo phase) to the list – see Section 10.3.

[7] There is a side effect, though, i.e., natural gas – mostly methane – leaks into the atmosphere during production and pipeline transportation and that methane is much more potent – albeit with shorter life – than CO_2 as a GHG.

with no sacrifice in acceptable living standards for *all* inhabitants of the earth (i.e., not just for a lucky few). The second constraint is time; the goal should be attained with utmost urgency in the shortest possible time. (See the superb book by the late David MacKay on this subject, which was cited in Chapter 1.)

The question to answer is: how do we attain this goal? First, let us recap what we know to be facts and already covered in this book at length:

1. Superiority of gas-fired technology to coal-fired technology in thermal efficiency (by almost 20 percentage points at state of the art ratings; by more than 12 percentage points at average field performance in the USA), i.e.,
2. Coal as a direct fuel is out of the question except
3. Carbon-free conversion of coal or biomass to H_2 with CCUS
4. Carbon-free nature of nuclear power
5. Rapid advances made in solar and wind in the last two decades
6. Intermittency (dispatchability and capacity) problems of both can be alleviated by energy storage
7. Hydrogen's dual nature as an energy storage medium and carbon-free combustion fuel
8. Biofuels for net zero carbon power generation.

It looks like all the ingredients to attain the goal of sustainable electric power generation are available. However, there are significant hurdles to a wide-scale deployment of all the available tools to attain the set goal. These hurdles can be (and are) enumerated in a long list of "lack of. . ." items, which can be found by a quick search on the internet. The result is tedious foot dragging by global parties (governments, private sector, academia, non-profits, and other agencies, organizations, etc.) without doing anything of real consequence. This is not to say that absolutely nothing is done. On the contrary; as an example, consider the progress made by the US power sector. According to the US Energy Information Administration (EIA), between 2005 and 2019, total the US electricity generation increased by almost 2% while related CO_2 emissions fell by 33%. According to the EIA, during that period, fossil-fuel electricity generation declined by about 11%, and non-carbon electricity generation rose by 35%. (To boot, natural gas overtook coal in the US generating capacity.) Alas, this achievement, while certainly welcome, was not the sole result of a targeted effort (excluding tax subsidies, state, and local mandates) but merely a windfall from a *perfect storm* of cheap shale gas, retirement of obsolete and costly-to-run coal-fired power plants, huge capex and opex advantage of GTCC over the new-built replacement of the same, and the rapid deployment (somewhat surprisingly) of solar and wind resources (still less than 10% of the total US generation portfolio). (In real terms, i.e., to get an idea about the magnitude of the *dent* it made in atmospheric CO_2, you can use the calculations made above as a yardstick.)

One thing is certain: Constantly searching for the proverbial silver bullets is not a remedy. This would be acceptable in "normal" times with a "business as usual" state of mind. Currently, the requisite state of mind should be that we are in a "state of war" against potentially destructive climate change. Without the urgency imposed by such a state of mind, where the stakes are the sustainability of the civilization as we know it, no meaningful action should be expected. What does a *state of war* mentality entail? The best analogy to the problem at hand (i.e., sustainable electric power generation) is design and production of "weapons systems." Four priorities in any system, in the battlefield, or in an industrial plant, are

1. maximum *availability* (i.e., being ready to fire)
2. maximum *reliability* (i.e., firing bullets whenever the trigger is pulled without jamming)
3. maximum *maintainability* (i.e., sturdy, quick to disassemble, clean, and reassemble in field conditions)
4. maximum *effectiveness* (i.e., kill as many enemy combatants as possible without getting killed)

Note the use of the term "effectiveness" above in lieu of "efficiency." This is a conscious choice because

- *effectiveness* is the degree to which something is successful in producing a desired result whereas
- *efficiency* is doing it in the best possible manner with the least waste of time and effort.

There is one and only one way to achieve maximum effectiveness with maximum RAM in fighting climate change:[8] *rational planning under determined leadership to come up with design specs requiring*

1. Proven technology (no "wonder weapons")
2. Low-cost manufacturability in large numbers ("big arsenal")
3. Ease of shipment and constructability ("fast deployment")
4. Simple for reliability, availability, and maintainability (in the field)
5. Sophisticated enough to do the job effectively (no more, no less)

Let us look at the current situation in the technology landscape. Mature and field-proven (i.e., well beyond TRL 9) electric power generation technologies for a fully carbon-free portfolio are readily available, e.g.,

- Coal gasification
- Hydrogen production
- Gas cleanup/scrubbing
- Energy storage (to improve capacity and dispatchability of solar and wind)

None of these technologies can be a *panacea* on their own. They can be deployed simultaneously and in large numbers with a well-coordinated plan (i) to ensure that they complement each other optimally, i.e., the strength of one ameliorates the

---

[8] The epitome of *maximum effectiveness with maximum RAM* in the battlefield is the AK-47 "Kalashnikov" assault rifle. (See *The Gun* by C. J. Chivers, Simon & Schuster, 2011.)

weakness of the other, and (ii) to maximize their delivery, constructability, operability, and maintainability. In particular:

1. Nuclear for baseload power (i.e., 24/7 operation at 100% load)
2. Renewables for baseload power (i.e., operation at full capacity whenever the sun shines and the wind blows)
3. Large-scale (several hundred megawatts) and long duration (8–12 hours) energy storage via
   a. H_2 production by electrolysis with excess renewable/nuclear power as the energy source (i.e., eliminate curtailment)
   b. Pumped hydro (already used)
   c. Compressed-air energy storage, CAES[9] (two commercial installations worldwide, one in the USA)
4. Battery storage (small-to-medium scale, short duration)
5. Coal gasification and synthetic natural gas (SNG) and H_2 production
6. Hydrogen-fired GTCC for baseload power or load following with H_2 generation from
   a. Electrolysis ("green hydrogen")
   b. Coal or oil gasification with pre-combustion CCUS
   c. SMR with pre-combustion CCUS
7. Hydrogen-fired simple cycle gas turbine or recips for peak load and load following. These units can be supported by batteries to provide a "spinning reserve" while in standstill.

Alas, there is no vision, no urgency, and no coherent plan to combine the best (i.e., most proven) of them into an "effective system" to generate affordable power with minimal carbon emissions, which can be deployed all over the world *in large numbers today*. The missing link is a coherent policy and legislative/regulatory structure to let the proverbial horses loose, i.e., to stimulate "gargantuan scale" investment in the right technologies.

One symptom of the lack of vision is almost funny. While real achievements in CO_2 and other emissions are few and far in between, there are perennial announcements of "wonder weapon" technologies that are touted as *the* solution to the carbon emissions problem in one fell swoop in, say, ten years. Then they quietly go away (sometimes they come back repeatedly, always ten years in the future), and new ones take their place. One big reason is, of course, that scientists in academia and industry *follow the money*, writing proposals and publishing reams of paper on the subjects that granting agencies and equipment manufacturers favor. *Concepts du jour* are frequently hyped up by the media stoked by the marketing departments of the latter. Hydrogen is the latest; before that it was CCUS, which is still around, before that there was IGCC – now in a coma, before that it was fuel cells – they absolutely refuse to die. The list is too long. The show goes on and on while the dire warnings reach a level of crescendo.

---

[9] Can be combined with H_2 combustion (blended with NG or 100%)

This is why a global effort, supported by all countries in a broad consensus, and coordinated by global agencies in a modern day, i.e., a global "war against GHG emissions" is a must. Similar to the war effort in the 1940s, the populace must be mobilized to join in the effort as dictated by the program. This is actually well recognized by many parties thinking on the subject (in fact, the term is frequently used in the media). The crux of the matter is how to finance this new war effort. The argument herein, as will be stated below, is that *there is no lack of funds*; there are ample funds and there are ways to find ample funds.[10] However, the distribution and allocation problems must be solved.

In 1941–45, General Motors (GM), Ford, GE, Boeing, and many other US industry giants designed and manufactured tens of thousands of tanks, airplanes, trucks, etc. at a scale never seen before – not only for the US Armed Forces but for the Allied powers as well ("lend-lease"). The US paid for the gargantuan effort by raised taxes and increased borrowing ("war bonds"). In the post-war climate of capital controls, the debt was eventually "inflated away" – halved in just ten years by 1955. This focused effort was complemented by the sacrifices made by people at home – not only in terms of foregoing luxuries (for example, no new automobiles were built in the US during the war) but also in terms of shortages in basic goods.

In a similar vein, the government of today in cooperation with major industrial companies (i.e., large manufacturers like GE, GM, Boeing, etc., chemical companies like DuPont, energy companies like Shell, ExxonMobil, BP, and large EPC contractors) and leading scientific organizations (universities, research labs, etc.) should launch a multi-pronged program including (but not limited to):

1. Accelerated deployment of large as well as small modular nuclear power plants with proven fission technology[11] and
2. Continued deployment of renewable resources in synchronicity with
3. Building of large-scale energy storage facilities ("green" hydrogen, pumped hydro and CAES – as well as battery storage in a judicious manner where feasible)
4. Centralized Coal-to-H_2, Coal-to-SNG, and Gas-to-H_2 plants (comprising CCUS, storage and transportation) combined with
5. Extended or new-built pipeline networks to transport SNG and H_2 to GTCC power plants and other end users.

Technology deployment should be planned and implemented in a way to optimally fit the existing power grid, pipeline network, and coal and natural gas mines and fields as well as the retirement (or retrofit) schedule of coal-fired power plants. In that respect, it

---

[10] In this context, the reader is referred to (i) Piketty, T., *Capital in the Twenty-First Century*, Harvard University Press, 2014 and (ii) Raworth, K., *Doughnut Economics*, Chelsea Green Publishing Co., 2017.

[11] On November 9, 2020, around 8 am, Germany and France each generated roughly 60 GWe of electric power. In Germany, about 26 GWe was from coal and 10 MWe from natural gas. In contrast, in France, more than 40 GWe was from nuclear with no coal and about 5 GWe from natural gas. Not surprisingly, the CO_2 emissions of Germany (with more than 100 GWe installed wind and PV capacity) were nearly eight times that of France. Clearly, by shutting down her nuclear power plants in a hurry, Germany did a huge favor to herself and earth(!).

is imperative to ensure that the "cure does not kill the patient." In the 1970s, when nuclear power was in its heyday, it had an unintended negative impact on fossil-fuel consumption. Instead of reducing coal-fired generation (as expected), expansion of nuclear power actually led to *increased* strip-mining of coal to supply power for the gaseous diffusion plants, where U^{235} isotope was enriched for nuclear fuel. In particular, TVA thermal power stations burning strip-mined coal provided the power to the nuclear fuel production plants. Another well-known example is the increased usage of fuel-efficient cars leading to higher overall gasoline consumption because people encouraged by higher gas mileage become less cost-conscious and drive more. In a similar vein, materials (mining and processing) and manufacturing facilities to manufacture exponentially increasing numbers of wind turbines, solar panels, and requisite infrastructure (as *green* advocates hope) can (and most likely will) lead to a jump in electric power consumption that could only be met by increased use of fossil fuels.[12]

For a speedy route to commercial readiness level, CRI of 6 (see Section 16.5), an "open art, open access" approach to technology should be adopted. In other words, the best available technology should be deployed in a "cookie cutter" manner (the industry term is "reference plant") with resulting duplication and learning helping to (i) bring construction costs down and (ii) increase operational RAM. One example is amine-based chemical absorption technologies for CO_2 capture. The benefit to the technology owners will be from the increased volume of sales. The main shortcoming of the technology, i.e., power output/efficiency penalty can be made up by a concerted effort to (i) curb power consumption and (ii) improve energy efficiency at the user end. This is one case where the polio vaccine example should be heeded.[13]

The effort should be globally coordinated and undertaken by all governments along the same basic principles. One country making big strides on its own while others are continuing the same "growth at all costs" path will be counter-productive. All the investment and improvement achieved, say, in the OECD realm can be easily negated by coal-burning power plants in a few developing countries. In World War II, the US was able to be the savior of the free world ("The Arsenal of Democracy") on its own. In this war, the US cannot do it alone. No single country can. All countries in the world, rich and poor, should pull their own weight. This is probably the hurdle most

---

[12] Material requirements for the 300GW capacity of offshore wind needed to replace 60% of UK fossil fuel energy use in 2019 included 32 million tonnes of steel and 150 million tonnes of concrete (from a white paper by G. Kalghatgi, *Scoping Net Zero*, The Global Warming Policy Foundation, Briefing 55, 2021).

[13] Jonas Salk did not patent his invention, the polio vaccine. Whether he did that on principle (as he famously claimed, "can you patent the sun?") or – at least partly – due to practical reasons, he forfeited billions of dollars in licensing royalties. His not doing so kept the vaccine prices low, increased the speed of introduction and production volume (by several manufacturers) and thus saved many lives worldwide and fast.

It would be wrong to translate this unique example into a universal "truth." In most cases, basic economic principles suggest that, without adequate incentives, it is unrealistic to expect individual or institutional inventors to invest substantial resources into finding solutions for myriad problems afflicting the society out of purely altruistic reasons. Then again, when it is time to declare war on a "plague" of enormous proportions, such considerations migrate from reasonably justifiable into unacceptably selfish. Resources should be pooled and applied with no regard to personal gain.

difficult to overcome in this scenario (almost a "fatal flaw" if one is honest). The only way to accomplish this is for one country to assume the leadership and work with other countries and supranational entities to come up with a serious "plan of action" with concrete action items and mechanisms to enforce compliance.

Finally, as already noted above, electric power generation is only one piece of the puzzle. All industrial sectors should be targeted simultaneously. Otherwise, gains made in one power sector will be negated by failures in others. For instance, an examination of data from different industry sectors in Europe between 2013 and 2019 shows a 30% reduction in carbon in the power sector vis-à-vis 27% increase in the aviation sector.

The knee-jerk reaction to this obvious course of action is two-fold: First and foremost, the ever-present "spoilsport," i.e., who will pay for all this? The second one is the argument that developing and/or poor countries have a right to grow just like the rich and developed countries. Leave aside the fact that, when the global ship sinks in the middle of the ocean, steerage passengers will drown along with the first-class passengers and the crew. Unlimited growth at all costs is a fallacy and cannot be sustained. The reader is referred to a landmark book on this subject for a lucid and facts-based dissertation on this subject.[14] Once a "sheet of music" is drawn up, all players must play to it; if need be, under threat of "severe penalties."

The technology *portfolio* to fight climate change tied to global warming effectively, with a reasonable chance (but no guarantee) of turning the tide, exists today. There is no need for new technology inventions or innovations. Alas, a market-based economic system with private finance capital looking for short-term high returns (risk-free or very low risk, of course) cannot and will not be the source of requisite investment into large-scale deployment of the said portfolio. Growing evidence [1] suggests that waiting for the requisite technologies to be bankable, dependent on myriad factors including government regulatory action, is not a feasible path. (Of course, there are those who vehemently deny that.) Similarly, public funding is impossible in the current political climate. (In addition to the wrong priorities, the recent election campaign in the USA made it amply clear that people's irrational fear of socialism will poison the well from the get-go.) While policy wonks saw no problem or harm in providing *quantitative easing* (QE) to banks and financial companies (they are *too big to be left to fail* after all), who caused the crash of 2008 in the first place and spent the QE windfall on the same unproductive money-from-money-from-money derivative shenanigans, when it comes to a similar QE to be spent on real economy, they become rather tight-fisted.

The money is there,[15] both in public and private coffers. There is ample research showing that billions and billions of dollars stashed by a select few in tax havens all

---

[14] Meadows, D., Randers, J. and Meadows, D., *Limits to Growth – The 30-Year Update*, Chelsea Green Publishing Co., 2004.

[15] According to a Congressional Budget Office (CBO) report published in October 2007, the US wars in Iraq and Afghanistan could cost taxpayers a total of $2.4 trillion (or $2,400 billion) by 2017 including interest (from Wikipedia). Current US CO_2 emissions from fossil fuel–fired electricity generation is 1.51 Gt. It is hard to gauge the real cost at this stage but, with the learning curve and economies of scale effects in mind, this magnitude of money would be more than enough to retrofit all coal–fired power plants

over the over world, if properly taxed, could pay for many of the ills suffered by the less fortunate denizens of the world. French economist Piketty wrote a critical tome (widely celebrated as well as critiqued) on this subject, i.e., when making money from money overtakes real growth, the result is concentration and unequal distribution of wealth leading to social and economic instability (e.g., see Piketty's book cited in footnote 10). Given the way the existing political system in much of the world works, of course, Piketty's remedy of progressive tax on wealth is an abject impossibility. (A very clear-eyed examination of the need to revisit the economic market orthodoxy and solve the "distribution problem" can be found in the recent book by Raworth, cited in footnote 10.) However, apart from the increasingly obvious failure of the neoliberal market-based economic system to help the growing "precariat" of the world,[16] the sudden and vicious blow on the global economy by the Covid-19 pandemic in 2020 made it abundantly clear that "business as usual" is over and a "state of war" must be declared. Not only on the pandemic but also on worldwide poverty, economic malaise, and unfair distribution of wealth. The war on climate change can and should be an integral part of this larger war and it can be a remedy with increased (and meaningful) employment driven by the gargantuan scale of requisite engineering, manufacturing, production, and construction activities.

The white elephant in the room when discussing all those wonderful technologies for sustainable power is the "carrying capacity" of the earth. A common marketing gimmick encountered in presentations of leading industry participants in trade conferences is that their products are aiming to increase the "life standards" of the poor of the world. A truly noble goal indeed. Alas, with a world population of 7.8 billion today, which is projected to reach 9.9 billion by 2050, this is a dangerously vague notion. The average power consumption per capita is about 0.3 kW worldwide; 1.4 kW in the US, 0.5 kW in China, 0.14 kW in India, and 0.75 kW in France and Germany. With a 2% annual economic growth assumption, this corresponds to an extra 3,130 GW of electricity consumption. It would be nice to imagine that all those gigawatts would be solar or wind (currently, the installed global capacity of both is around 1,200 GW). Alas, this is highly unlikely to be the case.

It should be noted that population and atmospheric CO_2 concentration go hand in hand. This is illustrated in Figure 15.2, where the data from the 60-year period 1958–2019 is plotted along with the 2050 projection if the current trend continues. There are several regions identified in the chart:

- **Impossible**: The clock cannot be turned back (although proponents of DACC would disagree)
- **Target** of a "state of war" declaration on both growths (i.e., CO_2 growth is arrested, and population growth is minimized)

---

with carbon capture. This is not to say that this is "physically" possible or even necessary (refer to the first sample calculation above to gauge the impact of this "magic event" on the current situation). The point is that the pecuniary problem that we are facing collectively (or that we are being made to believe that we are facing) is not a problem of "availability of funds"; it is a problem of "distribution of funds."

[16] See Standing, G., *The Precariat: The New Dangerous Class*, Bloomsbury Academic, 2011.

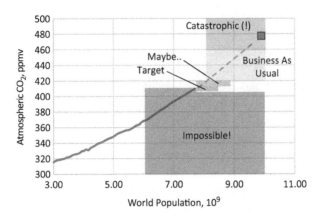

Figure 15.2 World population and atmospheric CO_2 concentration (1959–2019).

- **Maybe**: Best outcome to hope for
- **Business as usual**
- **Catastrophic (!)**: Listen to the climate deniers

The huge disparity in living standards between rich and poor mentioned above introduces a confounding factor into the sustainability efforts. Managing population and economic growth simultaneously while alleviating that disparity is the most challenging task facing us. In other words, how to bring the conspicuous consumption of the leisure classes of the richest countries in the world down to a level so that simultaneous lifting of the poor masses abroad from abject poverty and misery to a decent lifestyle will not negate the efforts to fight GHG emissions and global warming. As an example of the magnitude of the problem, consider that reducing carbon emissions from the US electric power sector to zero overnight would be rendered to naught rather quickly by nearly 2.5 billion people burning wood to cook food (not to mention deforestation and deaths from toxic fumes). Thus, providing those folks with net zero carbon electricity to do their cooking would be a worthy goal indeed but this would not be feasible while we in the US consume nearly five times the world average per capita power (more than 10 times that of India!). Finding a middle ground between overconsumption in some parts of the world and inadequate consumption in others, while very difficult under the best of circumstances, is a sine qua non. Making the populace buy into this mode of thinking is a vital component of "mobilization" (and it is very likely to fail without draconian measures).

In short, any "war on GHG emissions" should simultaneously tackle the complex "growth and distribution problem" in a meaningful way – certain "sacrifices" will have to be made. Otherwise, hairsplitting about how much H_2 to burn in dry-low-NOx combustors and similar engineering problems is just an exercise in futility.

16 Odds and Ends

16.1 Acronyms

A technical subject without acronyms would be like soda pop without gas. Like any field of specialized expertise, gas turbine technical literature is full of acronyms. All acronyms used in this book are defined where they first appear in the narrative. Nevertheless, some acronyms appear so frequently (in this book and in trade/academic literature) that they are defined in a central location anyway for easy reference. (Note that some acronyms, e.g., PR and CDT, are also used as variables in certain formulae.) In fact, 90% of the acronyms used in the text are industry standards (HRSG, DLN, BOP, etc.) and, in all likelihood, most readers would require minimal need for a full-blown glossary.

Acronyms used in the book (and some *not* used in the book but commonly encountered in trade and academic literature) are listed below. Note that some of the terms are really not acronyms per se but rather capitalized and "shortened" parameters such as TFIRE (firing temperature) or TAMB (ambient temperature). To make life easier for the reader unfamiliar with "industry-speak" (whose first reaction would be to go to the acronyms list when coming across an all-capitals term), they are also included herein. (Note that Chapter 12 has its of nomenclature section due to the large number variables and acronyms used in the text. They are not repeated here.)

1PNRH:	One-pressure with no reheat
1PRH:	One-pressure with reheat
2-D or 2D:	Two-dimensional
2PNRH(T):	Two-pressure with no reheat
2PRH(T):	Two-pressure with reheat (steam cycle type)
3-D or 3D:	Three-dimensional
3PNRH(T):	Three-pressure with no reheat
3PRH(T):	Three-pressure with reheat
ABB:	Asea Brown-Boveri
AC or ac:	Alternating current
ACC:	Air-cooled condenser
AGR:	Acid gas recovery (unit) – Selexol™ or Rectisol®
AM:	Advanced manufacturing (technical term for "3-D printing")
AQSC:	air quality control system

ASU:	Air separation unit
ATR:	Autothermal reforming (or reformer)
BBC:	Brown-Boveri Company (forerunner of ABB and Alstom)
BC:	Bottoming cycle
BESS:	Battery energy storage system
BOP:	Balance of plant
CAC:	Cooling air cooling (or cooler)
CAD:	Computer-aided design
CAES:	Compressed-air energy storage
CC:	Combined cycle or carbon capture (depending on the context)
CCS:	Carbon capture and sequestration
CCUS:	Carbon capture, usage, and sequestration
CDP:	Compressor discharge pressure
CDT:	Compressor discharge temperature
CEF:	Compression energy factor (in CAES)
CEMS:	Continuous emissions monitoring system
CES:	Cryogenic energy storage system
CFD:	Computational fluid dynamics
CHP:	Combined heat and power (European terminology for "cogeneration")
CIT:	Compressor inlet temperature
CLE:	Cyclic life expenditure
CMC:	Ceramic matrix composite
COD:	Commercial operation date
COE:	Cost of electricity
COT:	Combustor outlet temperature – same as TIT (not used in this book)
CPC:	Constant-pressure combustion
CPHA:	Constant-pressure heat addition
CRH:	Cold reheat (HP turbine exhaust steam)
CRS:	Central receiver system (solar tower)
CSA:	Contractual service agreement (a GE term, same as LTSA)
CSP:	Concentrated solar power
CT:	Cooling tower[1]
CTQ:	Critical to quality (a GE acronym but used by others, too)
CV:	Control volume
CVC:	Constant-volume combustion (the ideal case of PGC)
CVHA:	Constant-volume heat addition (thermodynamic equivalent of CVC)
DC or dc:	Direct current
DCS:	Distributed control system
DLE:	Dry-low-emissions (same as DLN, used for aeroderivative gas turbine combustors)
DLN:	Dry-low-NOx

---

[1] Not in this book, but in some, especially older, references, CT refers to "Combustion Turbine", i.e., the gas turbine.

DOE:	US Department of Energy
DSSR:	Direct steam generation solar receiver
EGR:	Exhaust gas recirculation
EIA:	Energy Information Administration
ELEP:	Expansion line end point
EOR:	Enhanced oil recovery (e.g., utilizing captured CO_2)
EOS:	Equation of state
EPC:	Engineering, procurement, and construction
EPRI:	Electric Power Research Institute
FEED:	Front-end engineering design
FEM:	Finite elements method
FERC:	Federal Energy Regulatory Commission
FOAK:	First of a kind
FOD:	Foreign object damage
FSFL:	Full speed, full load
FSNL:	Full speed, no load
FUA:	The Fuel Use Act
GE:	General Electric (or gas engine)
GG:	Gas generator (balanced shaft of a multi-shaft gas turbine)
GTCC:	Gas turbine combined cycle (same as CC)
GT(G):	Gas turbine (generator)
GTW:	Gas Turbine World (trade publication) – see https://gasturbineworld.com/
HC:	Heat (i.e., fuel) consumption
HCF:	High cycle fatigue
HGP:	Hot gas path
HHV:	Higher (gross) heating value
HPC:	High-pressure compressor (in aero-derivatives)
HPT:	High-pressure turbine (in aero-derivatives and steam turbines)
HRB:	Heat recovery boiler (same as HRSG)
HRH:	Hot reheat
HSGT:	Hybrid solar gas turbine (same as SHCC)
HRSG:	Heat recovery steam generator
HSS:	Heat-stable salts
IBH:	Inlet bleed heating
ICAD:	Intercooled aeroderivative gas turbine
ICE:	Internal combustion engine
ICV:	Intercept control valve (steam turbine)
IGCC:	Integrated gasification gas turbine
IGV:	Inlet guide vane
IOU:	Investor-owned utility
IPP:	Independent power producer
ISCC:	Integrated solar combined cycle
ISO:	International Standards Organization

ISO:	Independent system operator (similar to RTO)
KWU:	*Kraftwerkunion* (forerunner of *Siemens*)
LAES:	Liquid air energy storage system
LCI:	Load-commutating inverter
LCF:	Low cycle fatigue
LCM:	Lumped capacitance method (in heat transfer)
LCOE:	Levelized cost of electricity
LFR:	Linear Fresnel
LHV:	Lower (net) heating value
LPC:	Low-pressure compressor (in aero-derivatives)
LPT:	Low-pressure turbine (in aero-derivatives and steam turbines)
LSB:	Last stage (gas or steam turbine) bucket
LTSA:	Long-term service agreement (see CSA)
MBC:	Model-based control
MCV:	Main (steam) control valve (steam turbine)
MEA:	Monoethanolamine (chemical solvent)
MECL:	Minimum emissions-compliant load
MHI:	Mitsubishi Heavy Industries (now Mitsubishi Power)
MHPS:	Mitsubishi Hitachi Power Systems (MHI's corporate name prior to MP)
MP:	Mitsubishi Power (MHI's new corporate name)
NDT:	Non-destructive testing
NERC:	North American Energy Reliability Council
NETL:	US DOE's National Energy Technology Laboratory
NFPA:	National Fire Protection Association
NGC:	National Grid Code (United Kingdom)
NGCC:	Natural gas–fired gas turbine combined cycle (same as CC or GTCC)
NPV:	Net present value
OEM:	Original equipment manufacturer
OGV:	Outlet guide vane
O&M:	Operations & maintenance (OPEX)
OT-CL:	Once-through, closed-loop (water-cooled steam turbine condenser with a cooling tower)
OT-OL:	Once-through, open-loop (water-cooled steam turbine condenser without a cooling tower)
OV:	Open ventilated (gas turbine generator)
PCC:	Post-combustion (carbon) capture
PCHE:	Printed circuit heat exchangers
PDC:	Pulse(d) detonation combustor (a special case of PGC)
PDE:	Pulse(d) detonation engine (gas turbine jet engine with PDC)
PED:	Pressure Equipment Directive (European equivalent of the ASME Boiler and Pressure Vessel Code)
PEE:	Primary energy efficiency (in CAES – also known as *roundtrip* efficiency)

PGC:	Pressure-gain combustion
PHES:	Pumped hydro energy storage
POX:	Partial oxidation (reaction or reactor)
PR:	Pressure ratio
PT:	Power turbine (or "free" turbine) in aero-derivatives
PTR:	Parabolic trough
PURPA:	Public Utility Regulatory Policies Act
RAM:	Reliability, availability, and maintainability
RCA:	Root cause analysis
R&D:	Research and development
RH:	Relative humidity or reheat (in the steam cycle)
RICE:	Reciprocating internal combustion engine (i.e., piston-cylinder engine)
RIT:	Rotor inlet temperature[2]
ROI:	Return on investment
RPM (or rpm):	Revolutions per minute (units for rotational speed)
RTE:	Roundtrip efficiency (in energy storage)
RTO:	Regional Transmission Organization (similar to ISO)
SAC:	Single annular combustor
SCC:	Stress corrosion cracking
SCR:	Selective catalytic reduction (emissions control technology)
SCR:	Short circuit ratio (in synchronous ac generators)
SCS:	Solar collector system
SCV:	Stop-control valve (steam turbine)
SHCC:	Solar hybrid combined cycle
SJAE:	Steam jet air ejector
SMR:	Steam methane reforming or small modular reactor (nuclear)
SNG:	Synthetic or substitute natural gas
SPRINT:	Spray-intercooled gas turbine (GE trademark, the technology used in LM6000 aeroderivative gas turbines)
SSG:	Solar receiver steam generator
SSS:	Synchronous, self-shifting (overrunning clutch) – see https://www.sssclutch.com/en/about/history/
SSSF:	Steady-state, steady-flow
STIG:	Steam-injected gas turbine (GE trademark)
ST(G):	Steam turbine (generator)
TAMB:	Ambient temperature
TBC:	Thermal barrier coating
TES:	Thermal energy storage
TET:	Turbine exit temperature – same as TEXH (not used in this book)
TEWAC:	Totally-enclosed water air-cooled (gas turbine generator)

---

[2] Applies only to the first stage rotor of the turbine and is the same as the "firing temperature."

TEXH:	Turbine exhaust (exit) temperature
TIT:	Turbine inlet temperature
TMI:	Turbomachinery International (trade publication)
TRL:	Technology readiness level
UEEP:	Used energy end point
UHC:	Unburned hydrocarbons
UPS:	Uninterruptible power supply
USUF:	Uniform-state, uniform-flow
VB(A):	Visual Basic (for Applications)
VFD:	Variable frequency drive
VIGV:	Variable inlet guide vane (same as IGV)
VS(G)V:	Variable stator (guide) vane
VWO:	Valves wide open (for steam turbine)
WHR:	Waste heat recovery[3]
ZLD:	Zero liquid discharge

16.2   Heat-Mass Balance Simulation

Most of the sample heat and mass balance calculations presented in this book are done using Thermoflow, Inc.'s THERMOFLEX® software (Version 29.0.0.90, Revision: February 17, 2021), which is a *flowsheet simulator* that allows the user to represent a particular process by selecting components from a built-in library and connecting them on the program's graphical user interface to mimic the process flowsheet (e.g., see Figure 16.1).

Company information is provided below:

Thermoflow, Inc.
4601 Touchton Road East
Building 300, Suite 3280
Jacksonville, FL 32246-4485
USA
website: www.thermoflow.com
email: info@thermoflow.com

For simulations that require properties of real fluids such as supercritical CO_2, the program has the option to use the REFPROP (an acronym for *reference fluid thermodynamic and transport properties*) package.[4] REFPROP is a computer program, distributed through *the Standard Reference Data* program of the *National Institute of Standards and Technology*, which provides thermophysical properties of pure

---

[3] Rarely used for electric power generation applications; typically reserved for small-scale cogeneration cases.

[4] For using REFPROP or other property packages see *Evaluation of Property Methods for Modeling Direct-Supercritical CO2 Power Cycles*, Charles W. White, Nathan T. Weiland. 2018. *Journal of Engineering Gas Turbines and Power*, 140(1), 011701.

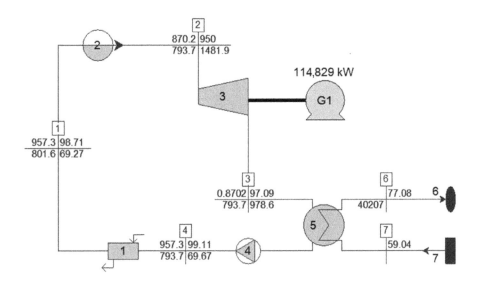

Figure 16.1 Simple Rankine steam cycle model in THERMOFLEX.

fluids and mixtures over a wide range of fluid conditions including liquid, gas, and supercritical phases. Detailed information on REFPROP can be found in https://www .nist.gov/programs-projects/reference-fluid-thermodynamic-and-transport-proper ties-database-refprop (last accessed on December 19, 2021).

16.3 Power System Reserves

Wind and solar power generation resources are commonly labeled as "not dispatchable" or "variable" (alternatively as "intermittent"). While not inaccurate, these descriptions need some qualification. Dispatchability refers to the ability of a power generation resource to be turned on or off, i.e., to adjust its power output supplied to the electrical grid on demand or per the order of the grid operator. To the extent of being available to comply with the dispatch order, there is some variability in *all* power generation technologies, including fossil-fired ones, i.e., not being able to generate power due to expected/planned (e.g., for regular maintenance) or unexpected/unplanned outages. In that sense, wind and solar resources can also be deemed dispatchable to the extent that they can be forecast, albeit not perfectly and with higher uncertainty than their fossil-fired brethren with *storable* fuels, e.g., coal, natural gas, or even distillate.

Consequently, nowadays, grid operators are quite able to incorporate wind and solar generation into their generation (dispatch) plans by means of forecasts in addition to conventional generators. The uncertainty aspect is covered by improved forecasting, energy storage, e.g., BESS, and fully dispatchable reserves to respond to fast load changes created by sudden, disruptive meteorological events. In other words, the problem is one of "controllability" rather than "dispatchability."

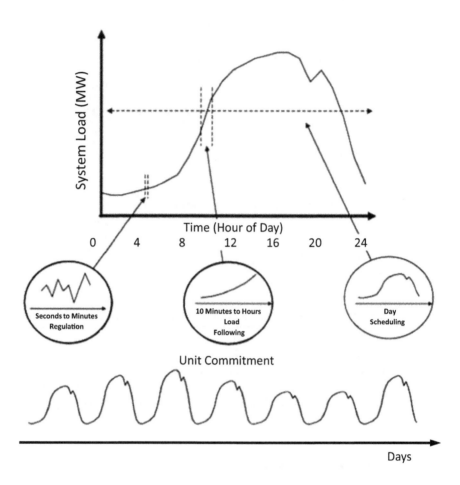

Figure 16.2 Power system operation time frames (from the US DOE NREL report *Operating Reserves and Variable Generation*, Ela, E., Milligan, M., Kirby, B., Technical Report NREL/TP-5500-51978, August 2011).

In terms of power system controllability, grid operations are classified in terms of time frames as follows: regulation, load following, and unit commitment (see Figure 16.2). They are described briefly below.[5]

Regulation covers a time scale ranging from a few seconds to ten minutes and meeting the load demand changes typically not predicted or scheduled in advance, by generation that is online, and synchronized to the grid (*primary reserves*). Primary reserves are typically generating units equipped with *automatic generation control* (AGC), e.g., gas or steam turbine power plants.

Load following covers time periods ranging from ten minutes to a few hours to address load changes that cannot be met by the primary reserves by the *secondary reserves*, i.e., generating units that are either already committed (*spinning* reserves) or

---

[5] From *Operating and Planning Electricity Grids with Variable Renewable Generation*, Madrigal, M., Porter, K., A World Bank Study, 2013, Washington, DC, USA.

can be started quickly (*non-spinning* reserves). These days, spinning reserves are primarily the GTCC power plants. Non-spinning reserves are usually aeroderivative gas turbines or gas-fired recips (see Chapter 7).

Unit commitment covers time periods spanning anywhere from several hours to several days. It refers to the advance scheduling and commitment of all types of generating resources to meet the expected power demand. It is also referred to as "day-ahead dispatch," "operations planning," or "generation programming." Typically, for fossil-fired (thermal) power plants, unit commitment is based on day-ahead or multiday-ahead dispatch, whereas for the hydro-power plants, based on the predicted availability of water, it may be weekly or several weeks ahead. Wind power is, of course, subject to higher uncertainty in longer time frames. Overcommitment due to underprediction can lead to *curtailment*. Undercommitment due to overprediction would result in starting up conventional generators quickly or (expensive) short-term purchases.

16.4   Technology Readiness Level

The *technology readiness level* (TRL) is a method first introduced by NASA in 1974 to assess the technology readiness of a spacecraft design by the *Jet Propulsion Laboratory*. It was formalized in 1989 with *seven* levels identifying different phases of a technology in its lifetime from the basic principles to validation in the field (or in space as in the case of NASA). The current scale with *nine* levels was adopted in the 1990s. Subsequently, different US government organizations and the European Union adopted the TRL methodology for their own use.

For the subject matter of the present treatise, the TRL definitions of the US Department of Energy (DOE) outlined in the *Technology Readiness Assessment Guide* (DOE G 413.3-4A, 9-15-2011, available online https://www2.lbl.gov/dir/assets/docs/TRL%20guide.pdf, last accessed on April 23, 2021) are the most applicable to evaluation of a given new technology. The cited document uses the same TRL scale as that of the US Department of Defense and NASA, and was also adopted by the US DOE Environmental Management in their pilot demonstration program, for technology assessment. Brief definitions of the nine TRLs are summarized in Table 16.1. The reader can consult the document cited above for the details.

There are four distinct system sizes covered by the TRL classification in Table 16.1:

1. Bench or lab scale (TRL 4)
2. Pilot scale (TRL 5–6)
3. Demo scale (TRL 7)
4. Commercial scale (TRL 9)

The exact relationship between the four system scales is highly specific to the technology under investigation. For typical gas-liquid process plants in the chemical process industry, if the pilot plant scale is taken as one, the bench/lab scale is 0.001–0.1,

Table 16.1 Technology readiness levels (TRL) used by the US DOE

TRL	Description
1	Basic principles observed
2	Technology concept formulated
3	Equipment proof of concept
4	Technology validated in the laboratory
5	Technology validated in relevant environment[1]
6	Technology demonstrated in relevant environment*
7	System prototype demonstration in operational environment
8	System complete and qualified
9	Actual system proven in operational environment (competitive manufacturing in the case of key enabling technologies); or in space

[1] Industrially relevant environment in the case of key enabling technologies

the demo plant scale is 100–1,000, and the commercial plant scale is 10,000–30,000. In moving from pilot to demo and, finally, to commercial scale, there are several factors to be reckoned with, e.g.,

- In almost all cases, scale-up from one size to a higher one is a nonlinear process driven by reaction kinetics and relaxation times (where applicable)
- Fluid dynamics change at a nonlinear rate as systems increase in size (e.g., laminar or turbulent), which strongly impacts the heat transfer rates
- Thermodynamic, fluid dynamic, and heat transfer changes can and will lead to re-evaluation of material selection and equipment sizing

For the power plants, a good example of this sequence of scale-up is the closed cycle sCO2 turbine technology development described in Chapter 10. A risky example is the oxy-combustion sCO2 technology, covered in the same chapter, which, it has been announced, will be moving from a 10 MW (guessed) pilot facility with a scantily published (and verified by independent observers) record of success to a 300 MW commercial offering. Caveat emptor.

For pilot tests specifically to de-risk full scale deployment of post-combustion capture plants based on chemical absorption technology with amine-based solvents, the reader is pointed to a recent paper by Elliott et al.[6] The paper discusses possible test unit concepts at ~10 tpd of CO_2 and ~100 kg/day of CO_2 scale, together with proposed test programme objectives and requirements. The smaller unit is a compromise between cost and representativeness, driven by the need for multiple parallel tests requiring multiple years. However, it is critical to verify smaller-scale test plant performance against larger units. The larger unit is intended to reduce the uncertainties stemming from the lack of installed base experience.

[6] *A post-combustion capture deployment derisking pilot plant*, Bill Elliott, Gus Benz, Jon Gibbins, Abby Samson, Mathieu Lucquiaud, Stavros Michaelos, 16th International Conference on Greenhouse Gas Control Technologies, GHGT-16 23rd–27th October 2022, Lyon, France.

16.5   Commercial Readiness Index

The *commercial readiness index* (CRI) is a framework created to complement the TRL methodology by assessing the commercial maturity of technologies across six indicators. It was specifically designed to address the major challenges faced by renewable generation technologies, e.g., high CAPEX (and OPEX), long payback periods, regulatory uncertainty, and FOAK-specific risks. The six indicators, in order of decreasing overall market maturity, are as follows:

- **CRI 6:** *Bankable* asset class driven by the same criteria as other mature energy technologies. The technology is considered a bankable grade asset class with known standards and performance expectations. Market and technology risks are not driving investment decisions. Proponent capability, pricing, and other typical market forces drive the uptake.
- **CRI 5:** *Market competition* is driving widespread deployment in context of long-term policy settings. Competition is emerging across all areas of the supply chain with the commoditization of key components and financial products occurring.
- **CRI 4:** *Multiple commercial applications* are becoming evident locally although are still subsidized. Verifiable data on technical and financial performance in the public domain is driving interest from a variety of debt and equity sources but, however, it is still requiring government support. Regulatory challenges are being addressed in multiple jurisdictions.
- **CRI 3:** *Commercial scale-up* is occurring and is driven by specific policy and emerging debt finance. Commercial proposition is being driven by technology proponents and market segment participants – publicly discoverable data driving emerging interest from the finance and regulatory sectors.
- **CRI 2** (*commercial trial*): A small scale, FOAK project based on the technology is funded by equity and government project support. Commercial proposition backed by evidence of verifiable data typically not in the public domain. (CRI 2 and above corresponds to TRL of 9.)
- **CRI 1** (*hypothetical commercial proposition*): While technically ready, the technology is commercially untested and unproven. Commercial proposition is driven by technology advocates with little or no evidence of verifiable technical or financial data to substantiate claims. (CRI 1 corresponds to TRL of 1–8.)

As an example, consider the progress of CRI for solar PV technology in Germany as summarized in Figure 16.3. Germany created the first mass market for solar PV through the use of a *pull marketing strategy*. As part of this strategy, end users were offered loans with very favorable and simple to understand terms to increase the demand (in 1998). The next step was introduction of *feed-in tariffs* (FIT) in 2000.[7]

---

[7] An FIT is a policy designed to support the development of renewable energy sources by providing a guaranteed, above-market price for producers. FITs usually involve long-term contracts, from 15 to 20 years.

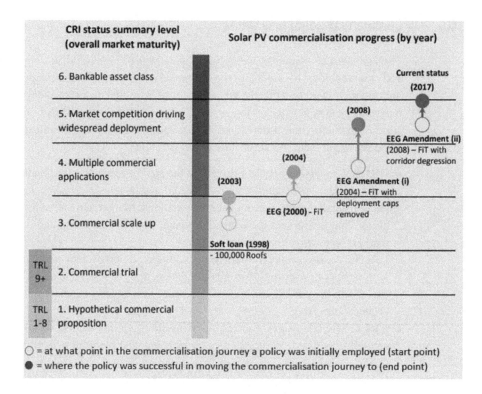

Figure 16.3 Solar PV commercialization timeline in Germany.

A revised FIT structure that reflected the true cost of solar PV units, without limiting the system size or the installed capacity was agreed in 2004. Subsequent FIT reforms followed over a period of almost 15 years and effectively increased support for the commercialization of solar PV.

It should be emphasized that, in comparison to pretty much all emerging technologies covered in this book, solar PV is rather simple. There are no moving parts, there is no combustion, there are no high pressures or temperatures. In short, nothing is *burning*, and nothing is *turning*. The photovoltaic effect was discovered by the French scientist Edmond Becquerel in 1839. Robert Millikan came up with experimental proof of the photoelectric effect in 1916. Finally, in 1954, Chapin, Fuller, and Pearson developed the silicon photovoltaic cell at the Bell Labs, which is generally hailed as the birth of solar PV technology. Thus, by 1998, when the technology was assumed to have reached CRI 3, the technology was nearly a half century old. In spite of this state of things, it took nearly two more decades for solar PV to reach CRI 6.

The author does not have enough information (only the developers have that knowledge) to carry out an accurate TRL assessment for the emerging technologies covered herein. Only educated guesses can and have been made based on publicly available information. Suffice to say that none of them are even close to TRL 8, let alone

TRL 9. Yet, some of them are touted to be in commercial readiness in just a few years. The reader is thus cautioned to take such claims with a pinch of salt. The example provided above is a quite eye-opening cautionary tale.

Let us close this section with a final word on the most hyped feature of the *energy transition*, the *hydrogen economy*, which is projected to replace the prevailing *hydrocarbon economy* and save the earth from the catastrophic events caused by anthropogenic GHG emissions. Its origins, including the coinage of the term itself, go back to the early 1970s.[8] All the possibilities, advantages, and challenges of producing, storing, transporting, and using hydrogen were known at the time. Yet the concept did not catch on and lay dormant until its reemergence (with a vengeance so to speak) in the 2010s. The problems of half a century ago are the same. It is extremely likely that a *centralized* hydrogen economy with large production facilities, a wide-ranging transport network, and end uses covering all aspects of transportation, electricity, and space heating, currently served by a combination of hydrocarbons and some renewables, is a physical impossibility (see Section 8.2.6 for numerical examples). A paradigm shift to a *distributed generation* system combining solar, wind, and hydrogen (for energy storage and utilization in, say, fuel cells) can indeed be a solution – on paper that is. However, such a wholesale overhaul of the existing infrastructure is an undertaking that can create unforeseen problems. Limited deployment will not make a meaningful impact on GHG emissions.

The road to technological and commercial maturity is a very long one and cannot happen by decree. It must happen on its own in an evolutionary manner, at each step building upon the learnings of the previous step and enlarging the application base toward achieving economies of scale culminating in widespread acceptance and affordability. The road is not only long but it can be broken in some places with long periods of inactivity. (Another good example is the MHD technology discussed in Section 13.5.) This book provides the reader with basic principles and tools derived thereof to assess the technical aspects of existing and new technologies in the field of electric power generation toward a more sustainable future. Today, in the third decade of the twenty-first century, we have the technologies to start a meaningful and global fight to curb GHG emissions from electric power resources. Their full deployment, i.e., reaching the stage of full commercial viability, however, is altogether a different matter and depends on factors beyond technical soundness.

16.6   Advancement Degree of Difficulty

Advancement degree of difficulty (AD^2) is a method of systematically dealing with aspects beyond TRL. AD^2 is a *predictive* description of what is required to move a

---

[8] For example, see Chapter 18 in Hammond, A.L., Metz and W.D., Maugh II, T.H. 1973. *Energy and the Future*, Washington, DC: American Association for the Advancement of Science.

system, subsystem, or component from one TRL to another. It identifies risks (defined as the likelihood of occurrence of an adverse event) and their impact, i.e., cost to ensure that such an event does not occur, and the time required to implement the necessary action. AD^2 focuses on five areas: design and analysis, manufacturing, software development, testing, and operation. The levels of risk associated with AD^2 are described in terms of the experience base of the developers, as summarized in Table 16.2.

AD^2 methodology is based on asking questions to assess the readiness level of the technology in question not only in the areas enumerated above but also in the skill set and capabilities of the people developing it and the tools they are using. Another key area is the ability of the organization behind the technology to reproduce or advance existing technology and move its own technology to the next TRL. For instance, in many cases technology developers claim that they are only using "off-the-shelf"

Table 16.2 Levels of risk associated with AD^2

AD^2 Level	Description	Risk, %
9	Requires new development outside of any existing experience base. No viable approaches exist that can be pursued with any degree of confidence. Basic research in key areas needed before feasible approaches can be defined.	90
8	Requires new development where similarity to existing experience base can be defined only in the broadest sense. Multiple development routes must be pursued.	80
7	Requires new development but similarity to existing experience is sufficient to warrant comparison in only a subset of critical areas. Multiple development routes must be pursued.	60
6	Requires new development but similarity to existing experience is sufficient to warrant comparison in only a subset of critical areas. Dual development approaches should be pursued to provide a moderate degree of confidence for success. (Desired performance can be achieved in subsequent block upgrades with high degree of confidence.)	50
5	Requires new development but similarity to existing experience is sufficient to warrant comparison in all critical areas. Dual development approaches should be pursued to provide a high degree of confidence for success.	40
4	Requires new development but similarity to existing experience is sufficient to warrant comparison across the board. A single development approach can be taken with a high degree of confidence for success.	30
3	Requires new development well within the experience base. A single development approach is adequate.	20
2	Exists but requires major modifications. A single development approach is adequate.	10
1	Exists with no or only minor modifications being required. A single development approach is adequate.	0

equipment with minor or no modifications. In other words, what they are implying is that they are proposing something not really FOAK. This has not been (and most likely will never be) the case at all. One obvious example is the technical difficulties experienced in the field (publicized and *not* publicized) with carbon capture pilot/demo plants (e.g., see Section 11.2). The chemical solvent–based absorption and stripping process is nearly a century old (e.g., see the 1930 US patent by Bottoms cited in Chapter 11, footnote 6). It has been used extensively in the *chemical process industry* (CPI) for acid gas removal (sour gas sweetening) for decades. Yet adopting that mature process technology to a new application has not been that simple. This is further complicated by the eagerness of technology developers to use novel (advanced) solvents that introduce uncertainties and unforeseen difficulties into a proven system. All this is done to squeeze the last decimal point of efficiency from the process at the expense of RAM as well as CAPEX and OPEX and makes the adoption of post-combustion technology unnecessarily difficult. The simple fact that goes unrecognized is that an 85% capture plant using basic technology with generic amines costs less to build and operate, has high reliability and availability, and is superior to the 95% capture plant with proprietary, advanced solvents, and complex equipment that costs an arm and leg to build and operate and runs only 50% of the time.

A similar state of mind practically killed the IGCC. Once again, the basic technology and its components have been successfully used in CPI for decades. Yet, extremely complex designs with maximum integration between the power block, ASU, and gasification block, to squeeze that accursed last decimal point of efficiency, competing gasifier technologies with widely differing equipment design and operating conditions prevented the technology from reaching a maturity level with commensurate CAPEX, OPEX, and RAM. Lessons from one project were not carried forward to subsequent ones. As a consequence of that, each spectacular failure (e.g., the Kemper IGCC) fortified the perception that the technology is too complex and expensive to be viable.[9]

The same mistakes are being repeated in the rush to claim the honor of being the developer of the "silver bullet" technology (!) that will save the earth from a climate disaster. In a market-based system, where private organizations and academic institutions compete to grab a slice from the pie of public funds and make a bundle in the process, to expect otherwise would be an exercise in futility.

[9] A significant portion of project cost over-run in Kemper was due to project management issues that are not technology specific (the final bill was more than $7 billion vis-à-vis an estimated $2.5 billion at the onset). Nevertheless, the proverbial "stake in the heart" was the failure of the refractory lining of the gasifier when the plant finally commenced operation. The gasifier technology had been extensively tested at a pilot scale of 50 tpd of coal but was then scaled up by a factor of over two orders of magnitude for the Kemper IGCC. It should be pointed out that the gasification technology in Kemper was the key enabler to allow a proven CO_2 separation technology to be used and thus an integral part of the whole CCS package. (Personal communication from Jon Gibbins, Professor of Carbon Capture and Storage at University of Sheffield.)

The reader is advised to look at each technology with a critical eye and use the tools discussed above, i.e., TRL, AD^2, and CRI to come up with a realistic assessment of what can be accomplished and when. Even with a working pilot plant, TRL 9 can be many years ahead. This is where AD^2 should be used. Furthermore, even with TRL 9, as the IGCC example dramatically showed, CRI 6 may never materialize. *Caveat emptor.*

Index